PLANT BIOTECHNOLOGY

IN VITRO PRINCIPLES, TECHNIQUES AND APPLICATIONS

PLANT BIOTECHNOLOGY

IN VITRO PRINCIPLES, TECHNIQUES AND APPLICATIONS

MAHIPAL SINGH SHEKHAWAT

Assistant Professor of Botany
Department of Plant Biology and Plant Biotechnology
K.M. Centre for Post Graduate Studies
Pondicherry

VIKRANT

Assistant Professor of Botany
Postgraduate Department of Plant Science
Avvaiyar Government College for Women
Karaikal, Pondicherry (UT)

MJP Publishers

ISBN 978-81-8094-084-2

Printed and bound in India

MJP Publishers
New No. 5 Muthu Kalathy Street,
Triplicane,
Chennai 600 005

MJP 222

Publisher : C. Janarthanan

To

Guruji Prof. N.S. Shekhawat
&
Our Parents

PREFACE

Biotechnology is the youngest of sciences and is the fastest growing technical discipline that has advanced at an unprecedented rate. Advances in biotechnology even outpace new developments in computer science, because of which it is called a revolutionary science.

Plant biotechnology has been projected by many to become as dominant in the present century. Now it is playing a very important role in employment, production and productivity, trade, economics and economy, human health and quality of human life throughout the world. Transgenic crops are rapidly growing in popularity and the total area under such crops, the number of countries growing them and the types of cultivars being grown are all on the rise. These crops are likely to become more and more acceptable to consumers as suitable technologies become available to address the concern about their safety and more particularly as transgenic cultivars become increasingly familiar to us. Biotechnology products are becoming increasingly important in terms of human health. We already have many recombinant DNA-based diagnostic and even therapeutic foods.

In writing the first edition of this book, our aim is to explain new and rapidly growing concepts and technologies connected with *in vitro* regeneration and transgenesis, an important area of plant biotechnology. Our basic philosophy is to present the principles of plant growth and development under *in vitro* condition at the molecular level and its associated techniques in sufficient detail to enable the non-specialist reader to understand them. We also intend that the scope of this technology and its potential impact on virtually all areas of plant biology would be evident. It is assumed that the reader would have a reasonable working knowledge of plant *in vitro* biotechnology.

In order to fully appreciate the different and diverse areas of plant biotechnology, the student needs to have good background of not only the new technological aspects but also some basic aspects of plant *in vitro* biotechnology. The University Grants Commission and Department of Biotechnology, Government of India, have independently constituted the expert committee for teaching and research in plant biotechnology and have suggested curricula for M.Sc. teaching in plant biotechnology. From our experiences in teaching and research in plant biotechnology, we have felt that certain very important and fundamental areas are also needed to be given emphasis in order to lay a stronger foundation and to provide a more basic and sound training to the students.

This book has 30 chapters that discuss at length various aspects of the subject starting from history and scope of plant cell and tissue culture followed by the basic concepts of cell totipotency and somatic embryogenesis at the molecular level. As for applied plant biotechnology, synthetic seed technology, protoplast and somatic hybridization technology and production of haploid and triploid plants have also been discussed in detail—both concepts and techniques. Separate chapters have been added at length to deal with transgenic technology and its impacts. Efforts have been taken to introduce some additional chapters connected with the application of *in vitro* plant technology in industry, agriculture and forestry. Moreover an extensive description of a variety of techniques has been added in the concerned chapters, wherever required.

The book is intended to provide wide knowledge on the subject since the field has grown enormously. It is expected that the book would be a useful guide for college and university students, researchers and to those aspiring for various competitive examinations. We hope that the book will provide a useful overview of the state-of-the-art for the readers altogether. We do, in any case, welcome any suggestions/criticisms for further improvement, since there surely remains limitless scope for the same.

We would like to acknowledge the assistance of our colleagues at our respective institutes and elsewhere in the possible task of keeping us up-to-date in this exciting and emerging field of plant biotechnology. We are grateful to Prof. N.S. Shekhawat, Biotechnology Centre, J.N.V. University, Jodhpur, for his generous support and for providing photographs. We are also thankful to Mrs. Saroj Shekhawat for her help throughout the progress of the book, particularly in the preparation of figures. We express our sincere gratitude to Dr. A.K. Dixit for his valuable suggestions.

We would like to express our appreciation for the generous support provided by Mr. C. Sajeesh Kumar, Managing Editor, and the editorial team of MJP Publishers for their cooperation in completion of this work with perfection.

And finally we gratefully acknowledge our families and friends who, throughout this period, provided such strong support, despite having to put up with our frequent absence and distraction.

M S Shekhawat

Vikrant

CONTENTS

1

IN VITRO PLANT BIOTECHNOLOGY: STATUS AND SCOPE

INTRODUCTION

The continuous increase in human population is causing serious concerns. There is a need to achieve high food production. In response to the growing demand of food, to support an increasing population, biotechnology has allowed us to increase agricultural products at global level to meet the challenges in the years to come. This technology allows for a greater production of staple foods and fodder along with many additional products. Most major crops are being genetically modified, for example, corn, wheat, rice, potato, soybean, sunflower, oilseed rape, cotton, tomato etc. Using *in vitro* cell culture technology, we can now produce a whole plant artificially from a clump of cells or even from a single cell.

The cells or tissues obtained from any part of the plant like root, stem, leaf, flower etc. are encouraged to produce more cells in culture. In comparison to the animal system, almost every cell of the plant body has the immense potential to reinitiate cell division even after attaining maturation. Moreover, it has the ability to regenerate any tissue/organ or can undergo different developmental pathways in response to a particular stimulus. When plant cells and tissues are cultured *in vitro* (under controlled conditions), they generally exhibit a very high degree of plasticity. It allows one type of cell or tissue to be initiated from another type, which subsequently leads to the regeneration of whole plants.

Some of the important aspects that are directly concerned with plant *in vitro* culture technology are:

1. The cell or tissue to be cultured must be isolated from an intact healthy plant.
2. The appropriate environment to promote optimal growth must be explored and applied. This may vary depending on the kind of cells or tissues to be cultured.

3. These procedures must be carried out in a sterile environment to prevent growth of microorganisms.

During *in vitro* culture, all the essential requirements of the plant cells, both chemical and physical, are to be provided by the culture vessel supplemented with nutrient growth medium along with the growth conditions such as light, temperature, etc. The nutrient medium has to support all the essential mineral ions required for growth and development. Many plant cell cultures, as they are not photosynthetic, require a fixed carbon source in the form of sugar (most often sucrose) in the nutrient medium. Another vital component that also must be supplied is water, the principal biological solvent. Physical factors such as temperature, pH of the growth medium, the gaseous environment, light (quality, intensity and duration) and osmotic pressure, also have to be maintained within the required limitations.

IN VITRO PLANT CELL REGENERATION

Plant *in vitro* technology is not a separate branch of plant science; it is a collection of experimental methods of growing large number of isolated cells or tissues under controlled and sterile conditions. However, it has been variously defined as follows:

1. Plant tissue culture is the technique of growing plant cells, tissues and organs in an artificially prepared nutrient medium under controlled and aseptic conditions. It is also called *in vitro* culture. The culture medium contains all minerals and growth hormones necessary for the growing cells. The controlled conditions and the medium provide a suitable microenvironment required for the growth of the cells or tissue and for their survival.
2. Plant tissue culture is a collective term for protoplast, cell, tissue and organ cultures, raised under controlled environment on artificial nutrient medium.
3. Plant tissue culture involves the *in vitro* growing of plant tissue from plant cells or tissues taken from a source or donor plant. It has been found that whole plants can be regenerated from fragments of plant tissues or even a single cell, when these are kept in a nutrient medium capable of supporting growth and appropriate hormone treatment.

Cell regeneration relies on the following fundamental inherent abilities of plant cells:

i. ***Cell totipotency*** Totipotency is the inherent potential or capacity of a plant cell or tissue to develop into an entire plant if suitably stimulated. It implies that all the information necessary for growth and development of the organism is contained in every cell.
ii. ***Cell dedifferentiation/redifferentiation*** Dedifferentiation is the capacity of differentiated cells to return to meristematic condition and formation of an unorganized mass of undifferentiated cells. Further, regeneration of a new young plant is possible by redifferentiation of these dedifferentiated mass of cells.

iii. ***Cell competency*** It is the endogenous potential of a given cell or tissue to follow a particular mode of regeneration via organogenesis or somatic embryogenesis. Some somatic cells are highly competent and are capable of developing into fully functional embryos on the addition of a triggering factor. Hence, these cells are termed as pre-determined embryogenic cells (PDEC), whereas, the non-competent or morphogenetically incapable cells can be induced to be embryogenic if these are treated with certain specific inducing factors. Hence, these cells are said to be induced determined embryogenic cells (IDEC). However, in contrast, a group of somatic cells may also follow organogenic pathway of regeneration by differentiating directly into shoot-buds.

In vitro Cell Regeneration Strategies

In order to achieve successful application of this technology, there are some basic and essential approaches, which must be pursued prior to culture initiation. These are:

1. ***Selection of explants*** Culture must be initiated from the most suitable tissue (or explant) for the specific plant species.
2. ***Selection of nutrient formulation*** A nutrient medium should contain organic and inorganic compounds to sustain the optimal growth of cells or tissues.
3. ***Optimization of growth hormones*** The medium should contain an optimal concentration of auxin and cytokinin growth regulators.
4. ***Sterilization or aseptic culture conditions*** Aseptic conditions must be provided during culture to exclude competition from microorganisms.

Plant cells or tissues generally grow on a solid agar medium as friable, pale-brown lumps (callus), or as individual or small clusters of cells in a liquid medium, called as suspension culture. These cells can be maintained indefinitely, provided they are subcultured regularly into fresh growth medium. Culture-grown cells generally lack the distinctive features of most plant cells. They have a small vacuole, lack chloroplasts and photosynthetic pathways and moreover, the structural or chemical features that distinguish various cell types within the intact plant are absent. Furthermore, these cultured cells can also be induced to redifferentiate and develop into whole plants by altering the growth media. Cell culture basically directs and assists the natural potential within the plant to put forth new growth and to multiply in a highly efficient and predictable way. Meristematic tissue is the basis of plant growth and development. Meristematic tissues on culture can differentiate into leaves, stems, roots or any other organs.

In vitro regeneration can be initiated from almost any part of a plant but the physiological state of the plant does have an influence on its response in attempting to initiate cell or tissue culture. The parent plant must be healthy and free from obvious signs of disease or decay. Generally, younger tissues contain a higher proportion of actively dividing cells which are more responsive to a callus initiation procedure than mature tissues. Moreover, the plants which are source of explants must be actively growing and should not be at the stage of dormancy. The exact conditions required to initiate and sustain plant cells in culture or to regenerate whole plants from cultured cells are variable for each plant species. Each variety of a species will often have a particular set of culture requirements. Despite all the knowledge that has been obtained about plant cell and tissue culture, during the twentieth century, these conditions have to be identified for each variety through experimentation. Most plant species are better suited to *in vitro* regeneration and these can be cultivated with time and practice. However, some plants species have been proved relatively recalcitrant and these are difficult to regenerate under *in vitro* conditions.

Different techniques in plant cell culture offer certain advantages over traditional methods of propagation. They are:

i. The production of exact copies of plants that demonstrate particularly good flowers, fruits or other desirable traits.
ii. The growth and development of mature plants very quickly.
iii. The production of multiples of plants in the absence of seeds or necessary pollinators that are involved in seed production.
iv. The regeneration of whole plants from plant cells that have been genetically modified.
v. The production of plants in sterile containers that allows them to be moved with greatly reduced chances of transmitting diseases, pests and pathogens.
vi. The production of plants from seeds that otherwise have a very low chance of germination and growth, for example, orchids and nepenthes.
vii. The achievement of a particular type of plant, which is free from viral and other infections.

In vitro plant regeneration technique generally involves five basic steps:

1. Initiation of cultures or establishment of explants
2. Multiplication stage
3. Root induction
4. Hardening and acclimatization of regenerants
5. Analysis of morphological and physiological variations amongst regenerants

These stages can overlap in certain cases and the requirements of each stage generally vary from species to species.

STRATEGIES OF TRANSGENIC PRODUCTION

Transgenic plants possess a gene or genes that have been transferred from a different species. Although DNA of another species can be integrated in a plant genome by natural processes, the term "transgenic plants" refers to plants regenerated in a laboratory using recombinant DNA technology. The aim is to design plants with specific characteristics by artificial insertion of genes from other species or sometimes entirely different kingdoms.

Agrobacterium-mediated transformation is the method of choice for the transfer of genes. However, alternative methods such as electroporation and particle bombardment have been developed in order to circumvent the poorly understood biological barriers, which prevent its application to certain plants. The prerequisites for the transfer of genes of desirable traits are:

i. Establishment of an efficient plant regeneration system *in vitro* from an explant amenable for transformation and
ii. Standardization of an efficient genetic transformation protocol.

The whole mechanism of transgenic production involves the following major steps:

1. Selection and isolation of the gene of interest.
2. Selection of competent tissues (in terms of regeneration and gene transformation) from the desired target plant.
3. Selection of mode of transgenesis.
4. Identification and selection of putative transgenic cells.
5. Regeneration of putative transformants.
6. Molecular analysis of putative transgenic plants.
7. Analysis of progeny of transgenic parents in order to confirm desired gene transmission to the progeny.
8. Field trial of transgenic plants and evaluation of its efficacy.

HISTORY OF *IN VITRO* PLANT BIOTECHNOLOGY

The important developments in the field of *in vitro* plant biotechnology are:

- Vochtung in 1878, showed that the cells of a stem when cultured in proper position show the formation of roots and shoots.
- Gottlieb Haberlandt (1902), a German botanist, was the first to generate tissue from fully differentiated tissue. This gave an insight into the properties and potentialities possessed by the cell.
- Hanning, E. in 1904, grew nearly mature embryos of crucifers to maturity on mineral salts and sugar solution. Moreover, he also demonstrated that the embryos would not grow to form plants without the addition of growth compounds.
- Simon (1908), cultured callus from stem segments.
- In 1922, W.J. Robbins and W. Kotte prepared cultures of isolated root tips.

- Laibach in 1925, successfully reared embryos that were otherwise unviable, using tissue culture. Subsequently, several new hybrids evolved using cell or tissue culture technology.
- White, P.R. in 1934, successfully propagated tomato root cultures using a medium containing three B-vitamins—pyridoxine, thiamine and nicotinic acid—along with inorganic salts and sucrose.
- In the 1930s, the identification of auxin as a natural growth regulator and the recognition of the importance of the B-vitamins led to improvements in tissue culture practices.
- In the late 1930s with the introduction of colchicine, perennial grasses were hybridized with wheat with the aim of transferring disease resistance and perenniality into annual crops.
- The foundation of plant tissue culture was laid down by three scientists (Gautheret, White and Nobecourt) during 1934–39. They showed that only small pieces of tissue (not individual differentiated cells) could be grown in cultures.
- Gautheret and Nobecourt (1939) cultured first callus from cambial tissue of carrot.
- White (1939) cultured callus of tobacco tumor tissue from interspecific hybrid.
- Snow, in 1935, demonstrated the role of indole acetic acid (IAA) in stimulating the cambial activity for rapid cell proliferation.
- In 1941, Van Overbeek discovered the nutritional value of coconut milk in embryo culture.
- In 1942, White and Braun developed callus culture from bacteria-free crown gall tissue.
- Shoot-bud differentiation from tobacco pith tissues cultured *in vitro* was reported by Skoog in 1944.
- In 1945, Loo introduced the method of shoot meristem culture.
- Caplan and Steward (1948), used coconut milk and 2,4-D for the culture of carrot tissue.
- Morel (1950), first cultured monocot tissue in coconut endosperm.
- Muir (1953), developed a suspension culture technique called the nurse culture technique.
- Tulecke in 1953, produced haploids from the pollen grains of Ginkgo.
- Cell division in tissue had to await the discovery of kinetin. The discovery of cytokinins gave much impetus to tissue culture. The discovery of kinetin by Miller and co-workers in 1955 enabled the initiation of callus cultures from differentiated tissues.
- Culture of gymnosperm (*Sequoia* sp.) tissue was done by E. Ball in 1955.
- In 1957, Skoog and Miller proposed that root–shoot differentiation from callus tissue was regulated by auxin–cytokinin ratio in the medium.
- The first plant from a mature plant cell was regenerated by Braun in 1959.

- Development of somatic embryos was first reported in 1958–59 from carrot tissues independently by Reinert and Steward.
- Plant protoplasts are naked cells from which cell wall has been removed. In 1960, Cocking produced large quantities of protoplasts by using cell wall degrading enzymes.
- Morel (1960 and 1964) developed shoot apex culture technique for micropropagation and also developed orchid propagation techniques by modified shoot apex.
- Haploid plants from pollen grains were first produced by Maheshwari and Guha in 1964 by culture of anthers of *Datura innoxia*. This marked the beginning of anther culture or pollen culture for the production of haploid plants.
- Protoplast culture technique allows the crossing of plants by protoplast fusion. Carlson and co-workers produced the first somatic hybrid plant by fusing the protoplasts of *Nicotiana glauca* and *N. langsdorfii* in 1972.
- Nitsch (1974) developed culture of microspores of *Datura* and *Nicotiana* and production of haploid/diploid plants by chromosome doubling.
- Production of transgenic plants in wide-crosses by plant breeders has been a vital aspect of conventional plant breeding for about a century. The first historically recorded interspecies transgenic cereal hybrid was between wheat and rye and it was carried out by Wilson in 1976.
- Regeneration of plants from cryopreserved tissues was shown by Bajaj *et al.* in 1976.
- Somatic hybrid production by fusion of protoplasts of tomato and potato was conducted by G. Melchers in 1978.
- Mukherjee and Bhaskaran isolated auxotrophic mutants for vitamin B components in *Nicotiana* in the year 1980.
- Electrofusion of protoplasts was conducted by Zimmermann (1982).
- Foreign genes attached to a plasmid vector were delivered into protoplasts. This procedure was performed by Barton *et al.* in 1983.
- Chilton (1983), produced transformed tobacco plants by single cell transformation method.
- Lazar *et al.* (1983), developed cybrids by protoplast fusion method.
- Kohn and coworkers in 1985, produced somatic hybrids in tobacco, mediated by electrofusion.
- In the 20th century, the introduction of alien germplasm into common foods was repeatedly achieved by traditional crop breeders by artificially overcoming fertility barrier. Novel genetic rearrangements of plant chromosomes, such as insertion of large blocks of rye (*Secale*) genes into wheat chromosomes ("translocations"), have been exploited widely for many decades.

- In 1985, Plant Genetic System (Ghent, Belgium), founded by Marc Van Montagu and Jeff Schell, was the first company to develop genetically engineered (tobacco) plants with insect tolerance by expressing genes encoding for insecticidal proteins from *Bacillus thuringiensis* (Bt).
- Sundberg and Glemelius (1986) created somatic hybrids in *Brassicaceae*.
- Isolation and plantation of sugarcane variants, resistance to mosaic virus was carried out by Nadgauda and Mascarenhas in 1986.
- Barton *et al.* (1987), Isolated Bt gene from *Bacillus thuringiensis*.
- The first modern recombinant crop approved for sale in the US in 1994, was the *Flavr Savr* tomato, which had a longer shelf life. The first transgenic cereal is wheat, which itself is a natural transgenic plant derived from at least three parental species.
- *E. coli* genome sequencing was done by Blattner *et al.* in 1997.
- In 1998, the genetic engineering of plants triggered a major food-safety controversy in Europe, where Monsanto's transgenic crops were destroyed by activists.
- In 2000, monsanto Company developed and commercialized the use of many transgenic plant systems and genetically modified corn seed.
- In 2001, Potrykus developed "Golden Rice".
- In 2004, Brazil and South Africa joined the US, Argentina, Canada and China on the list of leading growers. Seven million farmers in 18 countries planted transgenic crops in this year, up from six million in 16 countries.
- In 2006, *B. thuringiensis* was used as insecticide.
- Till 2006 there were around 250 million acres of genetically engineered crops being grown commercially in 22 countries. The US had adopted the technology most widely, whereas Europe had almost no genetically engineered crops. The EU had a formal ban on GM crops, until it was overturned in 2006; in a controversial move, GM crops are now regulated by the EU.
- In 2009, the recommendations of developing transgenic plants other than cotton became the biggest research grant in the history of Chinese molecular biology.
- Sharon (2010), has recently discovered a gene from legume Barrel medic that could be the key to Nitrogen fixation (crop Biotech update, 2010).

HISTORY OF PLANT *IN VITRO* TECHNOLOGY IN INDIA

Research on plant *in vitro* regeneration was first initiated in India in the early 1960s at the Department of Botany, University of Delhi, Delhi, by the late Professor P. Maheshwari. *In vitro* culture of various tissues such as anther, pollen, ovule, ovary and control of fertilization was initiated. This work laid the foundation for several later achievements in the field of plant cell, tissue and organ culture in our country. Attempts were also taken to culture the pollinated ovaries of *Allium cepa, Laminaria maroccana, Althaea rosea, Hyoscyamus niger* etc.

The Indian researchers who are the pioneers in the field of plant tissue culture are:

Joshi (1962)	Grew *Gossypium hirsutum* ovules with 12-celled pro-embryo. He also grew the hybrid ovule of the cross between *G. arboreum* and *G. hirsutum* to maturity and got viable seeds.
Johari and Bhojwani (1965)	Worked on *Exocarpos cupressiformis.* They grew endosperm of this plant and produced plantlets.
Rangaswamy (1967)	Developed plantlets from the excised nucellus of *Citrus microcarpa.* It is a polyembryonic species. He demonstrated callus in pseudo-bulbils which finally developed into complete plantlets.
Guha and Maheshwari (1967)	Worked on anther culture of *Datura innoxia* and produced numerous embryoids and finally plantlets on Nitsch medium.
George and Rao (1967)	Induced triploids through anther cultures of *Physalis* sp.
Mitra and Chaturvedi (1972)	Reported embryogenesis from nucelli of unfertilized ovules of *Citrus sinensis* and *C. aurantifolia.*
Raina and Iyer (1973)	Worked on anther culture of *Solanum melongena.*
Doreswamy and Chacko (1973)	Cultured anther leading to the formation of pollen embryoids of *Petunia axillaris.*
Khuspe (1980)	Demonstrated *in vitro* growth of immature hybrid embryos of *Carica papaya* and *C. cauliflora.*
Bajaj *et al.* (1981)	Worked on anther culture of *Primula obconica.* They have contributed much to cryopreservation of plant tissues and plant regeneration from cryopreserved tissues and also established germplasm banks.
Sharma and Bhojwani (1985)	Demonstrated direct microspore derived embryos in anther culture of two cultivars of *Brassica juncea.*
Kumar *et al.* (1985)	Cultured callus from cellular endosperm of coconut.
Maliga *et al.* (1985)	Transferred chloroplasts to tobacco through cytoplast–protoplast fusion.
Razdan *et al.* (1984–1999)	Performed somatic hybridization in some higher plants to produce somatic hybrids.

Saxena *et al.* (1986–1987)	Performed nuclear transformation in protoplasts of higher plants.
Gupta *et al.* (1987–88)	Worked on protoplast culture and somatic embryogenesis in Pine trees.

STATUS OF PLANT *IN VITRO* BIOTECHNOLOGY IN INDIA

The Government of India has constituted a Scientific Advisory Committee for the establishment of a National Biotechnology Board (NBTB) under the Department of Science and Technology (DST) after the Indian Science Congress held at Mysore in 1982. A separate Department of Biotechnology (DBT) in the Ministry of Science and Technology was created by Government of India in the year 1986. The International Centre for Genetic Engineering and Biotechnology (ICGEB) for developing countries has been established under the auspices of the UN with two centres in New Delhi and Trieste, Italy.

Research centres for biotechnology and plant cell culture in India have been established at the Indian Agricultural Research Institute (IARI), New Delhi. In 1993, the IARI started the National Research Centre on Plant Biotechnology for plant *in vitro* biotechnology research. The Department of Biotechnology, Indian Council of Agricultural Research, Council of Scientific and Industrial Research and University Grants Commission (UGC), New Delhi, are the funding agencies promoting research activities in plant cell and tissue culture. DBT has created seven centres for plant biotechnology along with the following institutions:

1. Jawaharlal Nehru University, New Delhi
2. Madurai Kamaraj University, Madurai
3. Tamil Nadu Agricultural University, Coimbatore
4. Osmania University, Hyderabad
5. Bose Institute, Kolkata
6. National Botanical Research Institute, Lucknow
7. University of Delhi, Delhi

A national facility for plant tissue culture repository has been established at the National Bureau of Plant Genetic Resources (NBPGR), New Delhi, with the objective of germplasm conservation in liquid nitrogen at –196° C. Three national gene banks have also been constructed for medicinal and aromatic plants at NBPGR, New Delhi, Central Institute of Medicinal and Aromatic Plants (CIMAP), Lucknow, and Tropical Botanical Garden Research Institute (TBRI), Trivandrum.

Micropropagation of elite trees has been possible at a commercial scale for many ornamental and medicinal plants. For this, DBT has funded the establishment of some pilot plants at:

1. Tata Energy Research Institute (TERI), New Delhi
2. National Chemical Laboratory (NCL), Pune
3. J.N.Vyas University, Jodhpur

They have successfully undertaken large-scale micropropagation of teak, eucalyptus, poplar, bamboo, jatropha, etc. For manpower development in plant *in vitro* biotechnology, the Department of Biotechnology has initiated an educational and training programme which consists of three components:

1. Masters level training at 37 centres located at different Universities
2. Postdoctoral training
3. Certificate and diploma courses in biotechnology

DBT has established the "biotechnology information network systems" with its apex centre at New Delhi and seven information centres and 38 sub-centres distributed all over India.

Some of the main centres in India for plant *in vitro* research are

1. Bhabha Atomic Research Centre (BARC), Mumbai
2. Central Drug Research Institute (CDRI), Lucknow
3. Central Food and Technology Research Institute (CFTRI), Mysore
4. Indian Institute of Science (IIS), Bangalore
5. IIT, New Delhi, Mumbai, Kanpur, Chennai
6. Banaras Hindu University (BHU), Varanasi
7. Central Potato Research Institute, Shimla
8. Forest Research Institute, Dehradun
9. National Botanical Research Institute, Lucknow
10. Arid Forest Research Institute, Jodhpur
11. Indian Institute of Science Education and Research

SCOPE OF PLANT *IN VITRO* BIOTECHNOLOGY

As an applied branch of plant science, *in vitro* biotechnology has gained profound importance in many ways (Figure 1.1). Plant cell culture, in which cells, tissues and organs are grown on a synthetic solid or in liquid nutrient media, has opened up new avenues in crop improvement in the recent years. Micropropagation, somatic embryogenesis, meristem culture, embryo culture, anther culture, callus culture, single cell culture and protoplast culture are some of the important techniques of plant tissue culture. Such techniques offer quick crop breeding methods than conventional methods, which are time-consuming and labour-intensive. Tissue culture is proving to be useful in a variety of ways including plant propagation, raising and maintenance of high health status plants, germplasm storage, and a valuable technique for plant improvement. A significant advantage of plant regeneration through tissue culture is that million plants can be regenerated from a small piece of plant tissue, throughout the year, irrespective of the season or location.

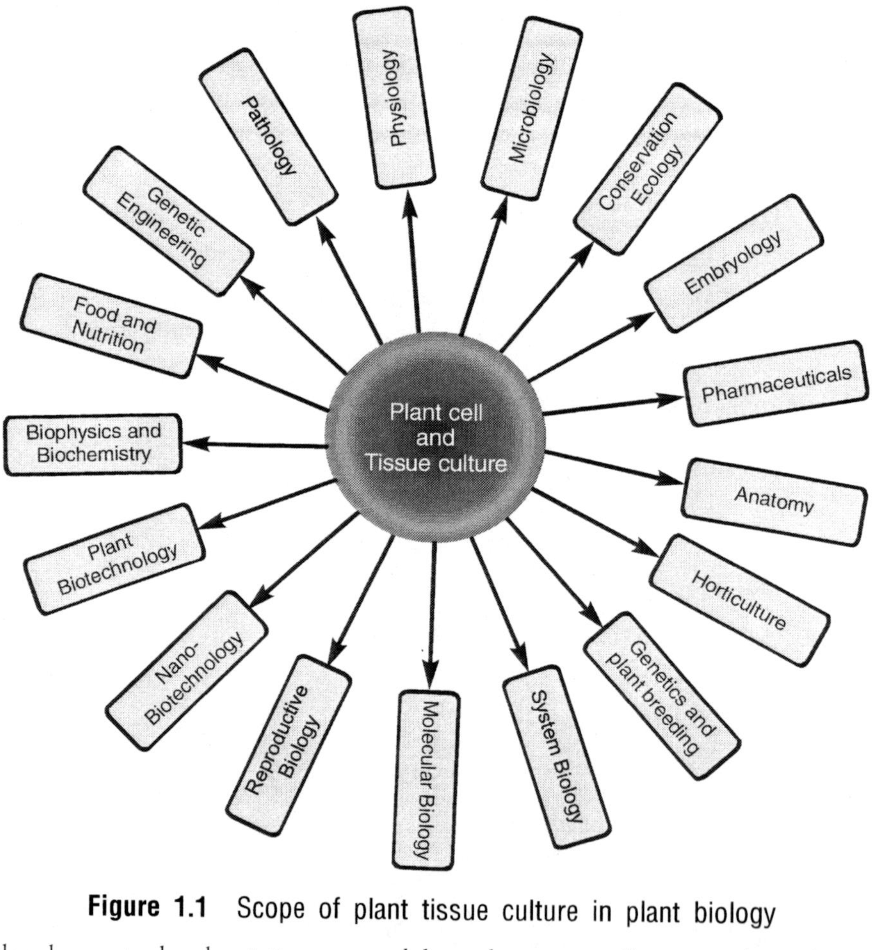

Figure 1.1 Scope of plant tissue culture in plant biology

Somaclonal or protoclonal variations, created through somatic cell or protoplast culture, have often resulted in improved germplasm of field, forest and horticultural crops with respect to biotic (insects, pests and diseases) and abiotic (salts, drought, frost, etc.) stresses. The production of haploids, through the culture of anthers or isolated microspores and protoplasts from higher plant cells, has served as the basic tool for genetic engineering and crop improvement.

Plant cell and tissue culture technique is useful in the study of biosynthesis of secondary metabolites, and provides efficient means of producing economically important plant products. Cultures for the growth and maintenance of plant tissues, for industrial level production of drugs, breweries and other valuable industrial products are produced. Apart from these applications, efforts are being made to increase the nutritional value of many food and fodder plants. These include seed storage proteins and other proteins rich in albumins, prolamins, glutelins, etc.

In vitro plant biotechnology also offers an excellent alternative technique in the production of transgenic crops against abiotic/biotic stresses along with qualitative and quantitative traits improvement in agronomically important crops. The worldwide efforts to transform an ever-increasing number of plant species with high efficiency has resulted in a number of promising gene transfer systems that continue to become more refined. Thus, it is evident that tissue culture tools along with recombinant DNA technology can prove a boon for the betterment of humankind.

Plant Cell Culture in Plant Biotechnology

Cell suspension culture is a relatively juvenile field of wet biotechnology. This technology involves the large-scale culture of isolated plant cells under conditions that induce them to synthesize the natural secondary metabolites characteristic of the parent plants from which they were obtained.

The techniques of callus and cell suspension culture have been reviewed by researchers in recent years, particularly from the viewpoint of studying biosynthesis and metabolism of steroids and cardiac glycosides. Biotechnologists are also trying to increase the synthesis of natural compounds and their derivatives by cell culture of higher plants as a result of mixing or feeding transformable precursors in the culture medium. The metabolic process of transformation or conversion of such added precursors into a natural or new compound within the cell is known as biotransformation. Biotechnologists are also trying to augment the synthesis of medically important alkaloids in culture by means of a fungal elicitor. This means that cells are cultured in a liquid medium by adding the required quantity of a bacterial filter-sterilized extract of certain fungi. Studies have shown that the fungal extract in certain cases helps to increase the synthesis of a desirable compound by the higher plant cell.

Biotechnologists are also trying to modify the genetics of cultured cells by applying the following approaches:

1. Mutagenesis and selection of cell lines in cell suspension culture.
2. Transplantation of foreign genetic material in protoplasts by means of genetic engineering.
3. Somatic hybridization by the fusion of distantly related plant protoplasts to widen the genetic diversity of hybrids.

Biotechnologists are trying to improve plants, for example, by producing crop species that are more resistant to drought, disease, poor soil conditions, chemical pesticides and herbicides by selecting cell lines *in vitro* and by exploiting totipotency.

Somatic hybridization by the fusion of protoplasts brings together in a single plant, genes of different species too unrelated, thereby allowing creation of hybrid or cybrid plants for improvement of the crop species. Another goal of biotechnology using the plant tissue culture technique is to produce plants that would provide their own usable nitrogen.

Plant Cell Culture in Plant Pathology

Many valuable contributions have been made by plant tissue culture to problems concerned with plant pathology. One outstanding success is virus eradication by apical meristem culture and its application to problems of plant tumours, especially crown gall.

It is well known that apical meristems are generally either free from or carry a very low density of viruses. Apical meristem culture is the only way to obtain a clone of virus-free plant, which can be multiplied vegetatively under controlled conditions that protects it from the chance of re-infection. The elimination procedure of the virus can usually be improved by combining it with heat therapy of the host plant or the culture. Virus eradication by apical meristem culture has enormous horticultural and agricultural value, for example, in the production of plants for the cut-flower industry, where stock plants of registered line must be maintained in as near-perfect condition as possible.

In agriculture, the production or yield of a crop can fall drastically as a result of viral infection and render that particular variety no longer saleable or commercially viable. Tissue culture techniques could be of value in restoring the original properties of the variety, by removing the infection and bringing it back into the commercial market. These virus-tested stocks could provide ideal material for the national and international distribution of plants, either for further propagation or for use as breeding material.

MARKET POTENTIAL OF PLANT *IN VITRO* BIOTECHNOLOGY IN INDIA

The first commercial micropropagation laboratory in India was set up in Kerala in 1984 (A.V. Thomas Group of Companies at Kochi) with the basic objective of export. The success story of this group inspired the leading industrial houses in India, and therefore many more tissue culture laboratories were established by Indian Tobacco Company, Indo American Hybrid Seeds, Nath Seeds, Hindustan Uniliver, Hindustan Agrigenetics Ltd., Tata, RPG Enterprises, etc. Currently there are 450–500 units operating in the country churning out more than 850 million plants per year, out of which about 150 units have a production capacity of 20–25 million per year. The production for the domestic market alone constitutes 40% of the total volume.

Incentives from the Government of India

The Indian Government has identified micropropagation of plants through tissue culture as an industrial activity under (D&R) Act, 1951, made effective in 1991, and has provided the following incentives:

1. Import duty reduction
2. Custom duty's exemption
3. Duty-free import for export-oriented units
4. Subsidy on air-freight and simplified clearance procedures

Various Central and State Government departments also have framed financial schemes and announced incentives for assistance tissue culture industry. These are:

1. Ministry of Agriculture.
2. Agricultural and Processed Food Products Export Development Authority (APEDA).
3. Small Farmers Agri-business Consortium (SFAC)
4. State level incentives
5. Financial Assistance by Banks (e.g., NABARD)

The units engaged in floriculture can avail duty-free imports under Export Oriented Units (EOU), Export Processing Zone (EPZ) schemes even if they sell 50% of their products in the domestic market and export the remaining 50%. The National Horticultural Board also gives subsidy of 20% on the total project cost. The micropropagation industry is particularly well-suited for developing countries like India, as it is environment-friendly and labour-intensive.

REVIEW QUESTIONS

1. What is plant tissue culture and why is it more favoured than conventional reproduction methods?
2. Give an outline of the important aspects of the tissue culture technique and its scope in biological studies.
3. Give a brief historical account of the tissue culture technique.
4. Write a brief note on the status of plant tissue culture technology in India.

REFERENCES

Ahuja, M.R. (1992). Biotechnology and clonal forestry. In: Ahuja, M.R. and Libby, W.J. (eds.) *Clonal Forestry I. Genetics and Biotechnology*. Springer-Verlag, Tokyo. pp. 535–544.

Altman, A. (1998). *Agricultural Biotechnology*. Marcel Dekker, Inc., New York.

Bhojwani, S.S. and Razdan, M.K. (1983). *Plant Tissue Culture: Theory and Practice: Developments in Crop Science*. Elsevier, Amsterdam.

Cheliak, W.M. and Rogers, D.L. (1990). Integrating biotechnology into tree improvement programmes. *Can. J. Forest Research*. 20: 452–463.

Gautheret, R.J. (1983). Plant Tissue Culture: A History. *Bot. Mag*. 96: 393–410.

James, A.T. (1984). Plant tissue culture: Achievement and Prospects. *Proc. R. Soc*. (Land.). B. 222: 135–145.

Shekhawat, M.S. and Dixit, A.K. (2007). Micropropagation of *Phyllanthus amarus* Schum and Thom. through nodal shoot segment culture. *Int. J. of Plant Sci*. 2(2): 79–84.

Murashige, T. (1978). Plant tissue culture: History, current status and prospects. *TCA Reports*. 12: 41–47.

Shekhawat, N.S., Mughal, M.H., Johri, B.M. and Srivastava, P.S. (1998). Indian contribution of plant tissue culture and organ culture. In: Srivastava, P.S. (ed.). *Plant Tissue Culture and Molecular Biology: Application and Prospects*. Narosa Publishing House, New Delhi. pp. 751–811.

Shekhawat, N.S., Singh, R.P., Choudhary, N., Yadav, J., Shekhawat, M.S., Rathore, J. S. and Arya, V. (1999). Micropropagation: Achievements and limitations with special reference to plants of arid areas. In: *Proc. National Symp. "Role of Plant Tissue Culture in Biodiversity Conservation and Economic Development*. GBPIHE&D, Kosi-Katarmal, p. 3.

Smith, R.H. (2000). *Plant tissue culture: Techniques and Experiments*. Academic Press, Tokyo.

Thorpe, T.A. (1990). The current status of plant tissue culture. In: Bhojwani, S. S. (ed.). *Plant Tissue Culture: Applications and Limitations*. Elsevier, Amsterdam. pp.1–33.

Thorpe, T.A. (2000). History of plant tissue culture. In: Smith, R.H. (ed.). *Plant tissue culture: Techniques and Experiments*. Academic Press, Tokyo. pp. 1–32.

Zimmerman, R.H. (1996). Commercial application of tissue culture to horticultural crops in the United States. *Jour. Korean Soc. Hortic. Science*. 37: 486–490.

2

IN VITRO PLANT REGENERATION—AN OVERVIEW

INTRODUCTION

Tissue culture is used as a blanket term for cell, tissue, protoplast, or whole plant culture under aseptic conditions. The methodology of tissue culture involves the separation of the cells, tissues and organs of a plant, called "explants", and growing them aseptically in a nutrient medium under controlled conditions of temperature, humidity and light. Each explant generally gives rise to an unorganized, proliferative mass of cells, called callus, which later produces young shoots followed by complete regeneration of the plantlets. Plant *in vitro* morphogenesis techniques are now being used as powerful tools for the study of various kinds of basic problems in the areas of plant physiology, cell biology and genetics.

This chapter provides a basic overview of various techniques of *in vitro* regeneration and some other important aspects of *in vitro* plant biotechnology (Figure 2.1).

IN VITRO MORPHOGENESIS

Changes in the shape, structure and organization during the development of an organism is called morphogenesis. Cells and tissue cultures are viewed as first acquiring competence, which is associated with altered differential gene regulation and expression. Morphogenesis is triggered when competent cells are subcultured into a less complex medium allowing the expression of new developmental potential. The factors affecting *in vitro* morphogenesis are as follows:

i. Explant and its source
ii. Culture environment
iii. Culture medium

The morphogenic potential of explants depends on (i) cell and tissue type, (ii) age of explants/ source and (iii) genotype of explant—genes that regulate morphogenesis. However, explants may have the following inherent abilities to demonstrate morphogenesis when these are encountered with triggering factors during *in vitro* treatment.

i. *Totipotency* Genetic potential of cells to undergo embryo- and organogenesis.
ii. *Ground state* It refers to the normal cell state. Cells may already be competent (meristematic) or incompetent (determined and must re-acquire competence or dedifferentiate before morphogenesis can occur).
iii. *Competency* Cells retain the ability for differentiation and morphogenesis.
iv. *Determinism* Ability of a cell to respond to the stimulus that initiates a developmental process leading to morphogenesis.

Morphogenesis *in vitro* occurs in two different patterns. They are organogenesis and somatic embryogenesis. Organogenesis is the process by which cells and tissues are manipulated to undergo changes, which lead to the production of unipolar structures, namely a shoot or root primordium, whose vascular system is often connected to the parent tissues. On the contrary, somatic embryogenesis leads to the production of bipolar structures containing a root or shoot axis with a closed independent vascular system. Both these patterns can occur directly on explants or indirectly via callus. A basic outline of *in vitro* regeneration or morphogenesis can be well-understood from Figure 2.2.

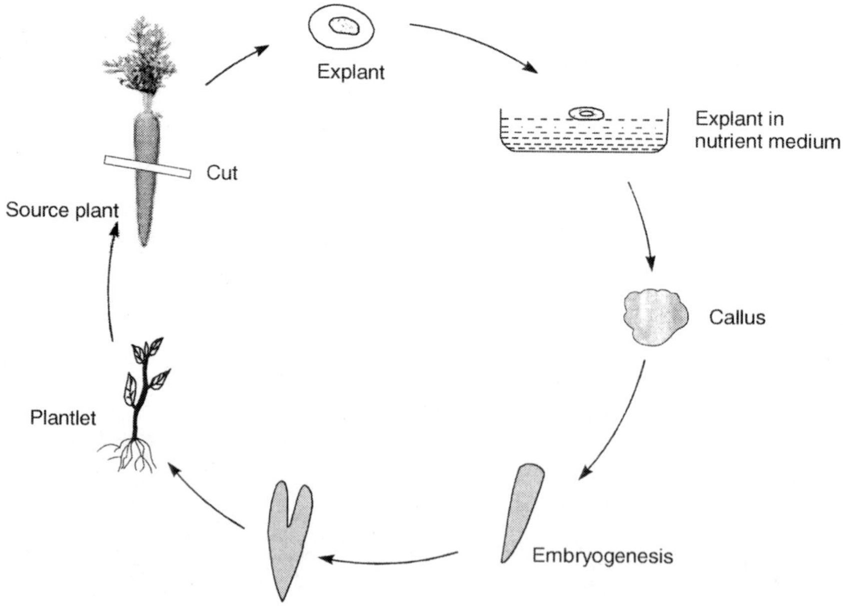

Figure 2.2 Basic outline of *in vitro* regeneration or morphogenesis

Organogenesis

Organogenesis is the development of adventitious organs (root or shoot) or primordia from undifferentiated cell mass (callus) during tissue culture by the process of differentiation. It relies on the production of organs, either directly from an explant or from a callus tissue. The expanded expression of totipotency of the callus tissue offers considerable potential for tissue culture technique as it is possible to grow the root or shoot or both.

There are three methods of plant regeneration via organogenesis. The first two methods depend on adventitious organs arising either from a callus or directly from an explant. Alternatively, axillary bud formation and shoot growth can also be used to regenerate whole plants from some types of explant tissue. Organogenesis relies on the inherent plasticity of plant tissues, and is regulated by altering the components of the medium. When callus tissue is allowed to grow on a medium containing both auxin and cytokinin, it will gradually proliferate by maintaining continuous cell division. If the auxin-to-cytokinin ratio is increased, adventitious roots will differentiate from the callus and this process is called rhizogenesis. Similarly, if the auxin-to-cytokinin ratio is decreased, adventitious shoots will be formed and the process is termed as caulogenesis. Moreover, if the explants are cultured on a medium containing only cytokinin, shoots can be produced directly from the explants without undergoing callus formation. Organogenesis from tobacco pith callus is a classical example which shows how varying plant growth regulator regimes can be used to manipulate the pattern of regeneration from plant tissue cultures.

Somatic Embryogenesis

A plant reproduces naturally through the development of zygotic embryos. Formation of the embryo begins with the division of the fertilized eggs or zygote within the embryo sac of the ovule. Through an orderly progression of divisions, the embryo eventually differentiates, matures, and develops into the new plantlet. Alternatively, an embryo can be originated from a single somatic cell or a group of somatic cells. This regeneration process, which differs from the natural pathway, is called somatic embryogenesis. In somatic (asexual) embryogenesis, embryolike structures, which can germinate and subsequently develop into whole plants, analogous to zygotic embryos, are formed from somatic tissues.

The somatic embryos can be produced either directly or indirectly. In direct somatic embryogenesis, a relatively rare process when compared to indirect embryogenesis, the embryo is formed directly from a cell or small group of cells without the production of an intervening callus. Usually reproductive tissues such as the nucellus, styles or pollen produce somatic embryos by this method. In indirect somatic embryogenesis, callus is first produced from the explant. Somatic embryos can then be differentiated from the callus tissue. Somatic embryogenesis from carrot is a classical example of indirect somatic embryogenesis. Such type of differentiation of somatic embryos either directly on explant tissues or via callus is called primary somatic embryogenesis. Furthermore, if these somatic embryos, on maturation or after germination, again show the next cycle of differentiation of somatic embryo, it is called secondary somatic embryogenesis. It is very common in carrot and *Ranunculus* spp.

Somatic embryogenesis usually proceeds in two distinct stages. In the initial stage (embryo induction), a high concentration of 2,4-dichlorophenoxyacetic acid (2,4-D) is required. In the second stage (embryo development), somatic embryos are produced and developed in a medium supplemented with either very low level of 2,4-D or no 2,4-D at all. Somatic embryogenesis is used to quickly produce many genetically identical plantlets. These plantlets can be genetically modified to enhance certain plant characteristics, such as increased growth rate, disease resistance and improved fibre quality.

METHODS OF *IN VITRO* REGENERATION

Plant cell culture technology can be grouped into the following classes based on the type of explants used for culture initiation.

1. *Callus culture* Induction and development of unorganized mass of cells (callus) on agar or liquid medium, produced from cultured explants.
2. *Cell culture* Culture of cell in liquid medium, usually aerated by agitation.
3. *Organ culture* Aseptic culture of embryos, anthers, roots, shoots, leaves, etc.
4. *Meristem culture* Aseptic culture of shoot meristem tissue in nutrient media.
5. *Protoplast culture* The aseptic isolation and culture of plant protoplast from cultured cell or plant tissue directly.

Formulation of Culture Nutrient Media

Tissue culture media are liquid or semi-solid media used for maintaining the growth of plant cells. MS medium, White medium, B5 medium, N6 medium, LS medium, Nitsch medium and their modifications are commonly used in plant tissue culture experiments. Culture media used for the *in vitro* cultivation of plant cells are composed of the following basic components:

1. Essential elements or mineral ions (macroelements and microelements), supplied as complex salts.
2. An organic supplement such as vitamins and/or amino acids.
3. A source of fixed carbon usually supplied as sugar/sucrose.
4. Plant growth regulators.

Macroelements As the name implies, these elements are required in large amounts for plant growth and development. Nitrogen, phosphorus, potassium, magnesium, calcium and sulphur (also carbon, which is added separately) are usually regarded as macroelements. These elements usually comprise at least 0.1% of the dry weight of plants.

Microelements These elements have specific roles and are required in trace amounts for plant growth and development. Manganese, iodine, copper, cobalt, boron, molybdenum, iron and zinc are the microelements. Nickel and aluminium are also frequently found in some formulations.

Organic supplements Only two vitamins, thiamine (Vitamin B_1) and myo-inositol are considered to be essential for the culture of plant cells *in vitro*. However, other vitamins are also added. Amino acids are also commonly included in the organic supplement. The most frequently used is glycine but arginine, asparagine, aspartic acid, alanine, glutamic acid, glutamine and proline are also sometimes used as adjuvants. Amino acids provide a source of reduced nitrogen and their uptake causes acidification of the medium. Casein hydrolysate can be used as a relatively cheaper source that contains a mixture of amino acids.

Carbon source Sucrose is the cheapest, easily available, readily assimilated and relatively stable sugar. Therefore, it is most commonly used as a carbon source. However, the other carbohydrates like glucose, maltose, galactose and sorbitol can also be used and in some cases they may prove superior to sucrose.

Gelling agents Media for plant cell culture *in vitro* can be used in either liquid or solid forms, depending on the type of culture being grown. For any culture type that requires the plant cells or tissues to be grown on the surface of the medium, the medium must be solidified. Agar, produced from seaweeds, is the most common type of gelling agent and is ideal for routine applications. For more demanding applications, a range of purer (but expensive) gelling agents are also available. Isabgol powder has also been extensively used as an alternative gelling agent for *in vitro* studies.

Plant growth regulators (PGRs) Plant growth regulators are plant hormones or their synthetic analogues, which regulate the growth of plant cells and tissues. PGRs are the critical media components in determining the developmental pathway of the plant cells. The major classes of plant growth regulators, used extensively in plant cell culture studies are:

1. Auxins
2. Cytokinins
3. Gibberellins
4. Abscisic acid
5. Ethylene
6. Brassinosteroids

Auxins Auxins promote both cell division and cell growth. The most important naturally occurring auxin is IAA (indole 3-acetic acid). But its use in plant cell culture media is limited because it is unstable in both heat and light. Occasionally, amino acid conjugates of IAA (such as indole-acetyl-L-alanine and indole-acetyl-L-glycine), which are more stable, are used to partially alleviate the problems associated with the use of IAA. 2,4-D is the most commonly used auxin and is extremely effective in most circumstances. IBA (indole 3-butyric acid) and NAA (1-naphthalene acetic acid) are also used.

Cytokinins Cytokinins promote cell division. They are purine derivative compounds. Zeatin and 2-isopentyl adenine are naturally occurring cytokinins, generally used in nutrient medium. But these are relatively unstable. Therefore, BAP (6-benzylaminopurine), kinetin (6-furfurylaminopurine), thidiazuron [1-phenyl-3-(1,2,3-thiadiazol-5-yl)urea], etc. are used more frequently.

Gibberellins There are numerous, naturally occurring, structurally related compounds termed "gibberellins". Gibberellins are involved in regulating cell elongation, and are agronomically important in determining plant height and fruit set. Only a few of the gibberellins are used in plant tissue culture media, GA_3 being the most common.

Abscisic acid Abscisic acid (ABA) inhibits cell division. It is most commonly used in plant tissue culture to promote distinct developmental pathways such as induction of somatic embryos.

Ethylene Ethylene is a gaseous, naturally occurring, plant growth regulator. It is commonly associated with controlling fruit ripening in climacteric fruits. It is not useful in tissue culture medium directly. But some plant tissues produce ethylene, which can inhibit the growth and development of the culture. Such ethylene-producing tissues are subjected to ethylene inhibitors supplemented in the nutrient media. These ethylene inhibitors such as $AgNO_3$, $CoCl_2$ or $NiCl_2$ are effective in the stimulation of tissue potential for somatic embryogenesis.

Brassinosteroids These are a new group of plant growth regulators of steroidal nature with significant growth-promoting activity. These were first isolated and characterized from the pollen of *Brassica napus* L. These are considered as hormones with pleiotropic effects, as these influences varied developmental processes (e.g., seed germination, growth, rhizogenesis, flowering, senescence, etc.).

Today over 40 naturally occurring brassinosteroids are known. The most abundant one is named as brassinalide.

One of their most dramatic effects is on cell elongation. These are far more potent than the other PGRs.

Selection of Explants

Explant is the excised piece of tissue or organ used for culture. Explants range in size and may vary from microscopic single cell or protoplast to stem piece of several centimetres in length. Some of the explants which can be used commonly for *in vitro* culture are:

- Embryonal tissues
- Fleshy storage organs
- Aerial tubers
- Shoots
- Buds and base of apical meristems
- Pith
- Segments of root and stem

- Floral parts
- Young inflorescences
- Nucellus
- Vascular cambia of trees
- Petiole and leaf mesophyll
- Endosperm and mesocotyl
- Ovular tissues
- Seedling hypocotyl and immature embryos
- Leaves and cotyledons
- Pollen
- Epidermis, etc.

Selection of explants is the first critical step in the practice of initiation and establishment of a culture. For some species and certain explants, this is extremely an easy job, whereas for others, it can be a desperately frustrating experience. Only disease-free plants must be used to collect explants for culture initiation. With suboptimal material, problems can be encountered in obtaining sterile cultures. This may lead to excessive variability in culture response and at worst, a complete failure of the experiment may occur. In general, explants excised from very young plants prove to be the best response. Germplasm repositories are good sources of explants that are free from pathogens. Otherwise, pathogen detection kit has to be used, which is commercially available to identify viral, bacterial and fungal infections.

It is considered the best to bring the source plant to the green house, screen house and grow there in dust free and healthy condition for six months. The new shoots that appear will provide relatively clean explants and in addition, may provide some increased juvenility. They should be cut longer than the final size of explant to be used. Smaller explants contain relatively less contaminants. Explants should be taken in the morning time rather than later in the day. They must be cut with a sterilized cutter. After excision of explant material from the source plant, the explants are placed in a plastic bag containing a moist paper towel and then brought to the laboratory as such. If it is not possible to use explant immediately then it stored is in refrigerator for further use.

For specific applications, precise growth conditions may be essential, particularly with regard to the period before the plants are to be used. Similarly, even when plants are healthy and at the desired stage for use, often only a specific part of these plants will give the best explants, for example, a particular internode, the youngest fully expanded leaf, flower buds within a certain size range and so forth. Again the choice of tissue depends upon an ultimate goal of the experiment. There are various factors which are directly concerned with explant source and can influence the culture response. These are as follows:

1. Physiological age of tissue
2. Season in which explant tissue is collected

3. Quality of source plant
4. Size of the explants
5. Aim of the culture

Types of Explants for Culture Initiation

The tissue or cell that is obtained from a plant to initiate culture is called an explant. It has often been claimed that totipotent explants can be grown from any part of the plant (Figure 2.3). The various explant types commonly employed in order to initiate the different kinds of cultures are:

1. Meristem culture: Apical meristems
2. Micropropagation: Shoot apex or lateral bud
3. Bud culture: Apical and axillary bud
4. Root culture: Lateral roots
5. Callus culture and somatic embryogenesis: Cotyledons, hypocotyl, stem, leaf, root, etc.
6. Haploid culture: Anther, pollen or ovary

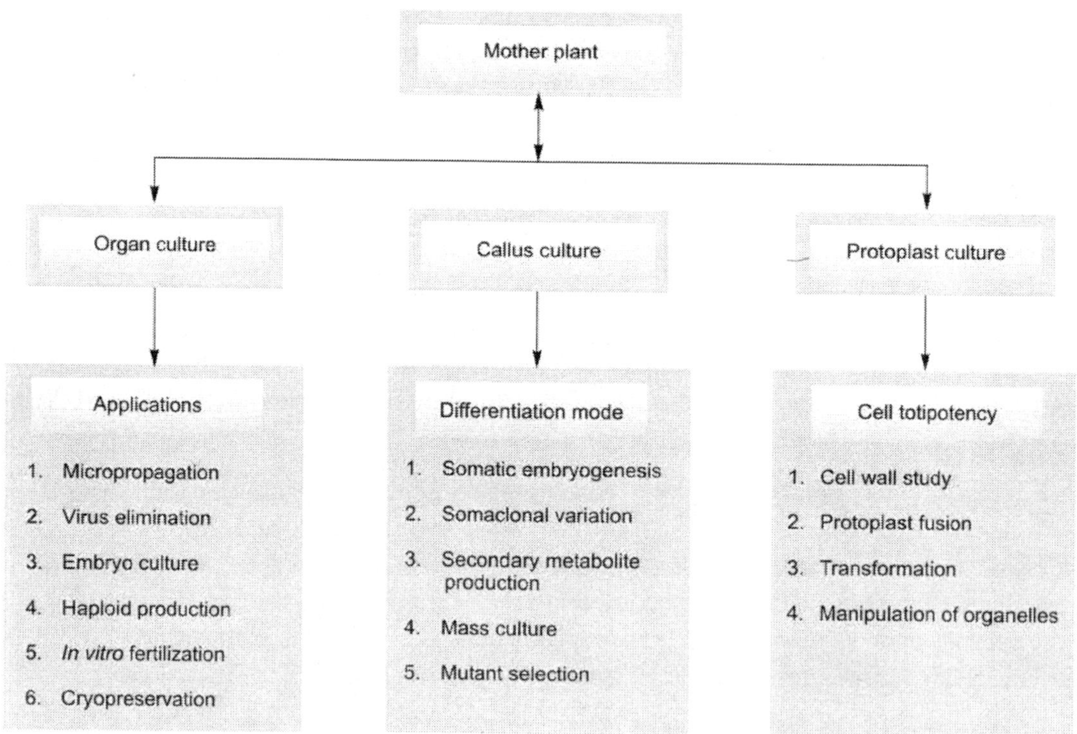

Figure 2.3 Levels of tissue organization in culture and their applications

Callus Culture

Explants, when cultured on the appropriate medium, usually with both auxin and cytokinin, can give rise to an unorganized, growing and dividing mass of cells. This type of culture is termed as callus culture. Callus is usually composed of unspecialized parenchyma cells. It is thought that any plant tissue can be used as explant, if the favourable conditions are given. In culture, tissue proliferation can be maintained more or less indefinitely, provided that the callus is subcultured on to fresh medium periodically.

Callus cells are cytologically, morphologically, biochemically and physiologically different from each other. During callus formation there is some degree of dedifferentiation in both morphology and metabolism. Callus culture is often performed in the dark as light can encourage differentiation of the callus. During long-term culture, the culture may lose the requirement for auxin and/or cytokinin. This process, known as "habituation", is common in callus cultures of some plant species, such as sugar beet.

Callus cultures are extremely important in plant biotechnology. Manipulation of the auxin-to-cytokinin ratio in the medium can lead to the development of shoots, roots or somatic embryos from which whole plants can subsequently be produced. Callus cultures can also be used to initiate cell suspensions, which are used in a variety of ways in protoplast technology and plant genetic transformation studies. This indeed provides a rich source of variation created by somatic cell culture and shoot regeneration, leading to an incremental improvement in the existing commercial varieties with respect to various traits (somaclonal variation). Superior clones possessing herbicidal and disease-resistant qualities have been developed in sugar cane and potato through this method.

Cell Suspension Culture

Tissues and cells cultured in a liquid medium produce a suspension of single cells and cell clumps of a few to many cells; these are called suspension cultures. Callus cultures fall under two categories, (i) compact and (ii) friable. In compact callus, the cells are densely aggregated. But in friable callus, the cells are loosely associated with each other and the callus becomes soft and can break apart easily. Friable callus provides the inoculum to form cell suspension cultures. Explants from some plant species or particular cell types do not form friable callus, making cell suspension initiation a difficult task.

The friability of callus can sometimes be improved by manipulating the medium components, or by repeated subculture. The friability can also be improved by culturing it in a "semi-solid" medium (medium with a low concentration of gelling agent). When friable callus is transferred to a liquid medium and then agitated, single cells and/or small clumps of cells are released into the medium. Under suitable conditions, these released cells continue to grow and divide, eventually producing a cell suspension culture.

Suspension cultures are of three types:

1. *Batch culture* In a batch culture, the same medium and all the cells produced are retained in the culture vessel. The cell number or biomass of a batch culture exhibits a typical sigmoid curve. Batch cultures are maintained on frequent subculture. They are used for initiation of cell suspensions, cloning, cell selection or as seed cultures for scaling up or for continuous culture.
2. *Continuous culture* In a continuous culture, the cell population is maintained in a steady state by regularly replacing a portion of the used or spent medium by fresh medium. Such suspension systems are of either closed or open type.
3. *Immobilized cell culture* Plant cells and cell groups may be encapsulated in a suitable material (like agarose, calcium alginate, etc.) or trapped in membranes or stainless steel screens. The gel beads containing cells may be packed in a suitable column or, alternatively, cells may be packed in a column of a membrane or wire cloth. Liquid medium is continuously run through the column to provide nutrients and aeration to cells. Such types of cultures are used mainly for biochemical production.

Single Cell Culture

Single cell culture is a method of growing an isolated single cell aseptically on a nutrient medium, under controlled conditions. The single cells are traditionally isolated from the established friable callus tissue and cell suspension culture. Mechanically, single cells are carefully isolated from cell suspension or friable callus with a needle or fine glass capillary tube. Alternatively, the friable tissue is transferred to liquid medium and the medium is continuously agitated by a shaker. Agitation of liquid medium breaks and dispenses the single cells and cell clumps in the medium. As a result, it makes a cell suspension. The cell suspension is first filtered aseptically to remove cell clumps and the filtrate is then centrifuged to collect the single cells from the pellet. Some of the basic methods that can be used in single cell culture technology are

1. Filter paper raft-nurse tissue technique
2. Microchamber method
3. Microdrop method
4. The Bergmann's plating technique
5. Thin layer liquid medium culture technique

The advantage of a single cell culture over callus culture or cell suspension culture or intact organ culture is that single cell culture system is an ideal system for studying cell metabolism and the effects of various substances on cellular responses, and for obtaining single cell clones.

Embryo Culture

The culture of mature or immature embryos and their regeneration into plantlets in a nutrient medium is known as embryo culture. It is usually practised in plants in which fertilization takes place as usual but the fertilized egg fails to develop into mature embryo. Since this method helps to overcome the non-viability of seeds, it is also called embryo rescue technique. The technique has been employed basically for cereals and pulses. It is very useful in cases where embryo abortion takes place due to some sexual barrier or incompetency after pollination and fertilization.

In wheat, intergeneric hybrids among *Triticum* and *Hordeum, Triticum* and *Elymus, Triticum* and *Secale* and *Hordeum* and *Secale* have been produced using embryo culture. In pulses, interspecific hybrids among *Vigna* species have also been obtained using this technique. Embryo culture has several applications like overcoming embryo inviability and seed dormancy, seed germination of obligatory parasites without the presence of host, monoploid productions and vegetative propagation. It acts as a tool for research in experimental embryogenesis, etc.

Root Culture

Root cultures can be established *in vitro* from explants of the root tip of either primary or lateral roots. They can be cultured on simple media. The growth of roots *in vitro* is potentially unlimited, as roots are indeterminate organs. Although the establishment of root cultures was one of the first achievements of modern plant tissue culture, it is not widely used in plant transformation studies.

Synthetic Seed Production

Synthetic seeds are produced by encapsulation of the somatic embryos obtained by either direct embryogenesis or indirect embryogenesis, with a gel rich in sugars. This technique has opened up the opportunity for mass cloning of field, forest and horticultural crops. Synthetic seeds have been developed in celery, *Brassica*, carrot, lettuce, alfalfa, sandalwood, maize, etc., using this technique. Such seeds could be germinated both *in vitro* and *in vivo* (in natural conditions). However, the percentage of germination varies from species to species.

Haploid Production

Haploid tissue can be cultured *in vitro* by using pollen or anthers as explant. Pollen contains the male gametophyte, which is termed as the "microspore". Both callus and embryos can be produced from pollen. There are two main methods to produce *in vitro* cultures from haploid tissue. These methods are as follows.

Anther culture The culture of immature anthers on nutrient media to achieve embryos and haploid plantlets is called anther culture. Anthers (somatic tissue that surrounds and contains the pollen) can be cultured on solid medium. Pollen-derived embryos are subsequently produced via dehiscence of the mature anthers. Haploid plants from pollen grains were first produced by Maheshwari and Guha in 1964 by anther culture of *Datura innoxia*. In young anthers, the cells are diploid. Later it was observed that embryoids originated from pollen grains through a series of divisions.

The process of growing plantlets from male gametophytes is known as androgenesis. The cut flowers are surface-sterilized and opened in sterile conditions under a binocular microscope. The anthers are dissected and transferred to a solid nutrient medium. Large numbers are placed on each Petri dish. Furthermore, developed callus is transferred to a different medium to differentiate embryos and regenerate haploid plants.

Haploids and doubled haploids enable breeders to develop completely homozygous genotypes in the shortest possible time as compared to conventional breeding. This technique has been successfully used to develop new varieties of *Triticum aestivum, T. durum, Oryza sativa, Zea mays, Nicotiana* spp., etc.

Microspore/pollen culture Immature pollen can also be obtained from developing anthers, and by direct culturing into a suitable nutrient media, these can produce haploid tissue. This process is called microspore or pollen culture. It is a very time-consuming process. In microspore culture, the condition of the donor plant is of critical importance, as is the flower stage for pollen isolation. Regeneration from microspore explants can be obtained by direct embryogenesis, or via a callus stage and subsequent embryogenesis. There is, however, the danger that some of the embryos produced from anther culture will originate from the somatic anther tissue rather than the haploid microspore cells. If isolated pollen is used, there is no danger of mixed embryo formation, but the efficiency is low. Pretreatments, such as a cold treatment, are often found to increase the efficiency.

Haploid tissue cultures can also be initiated from the female gametophyte (the ovule). In some cases, this is a more efficient method than using pollen or anthers.

Ovary culture Culture of unfertilized ovaries to obtain haploid plants from egg cell or other haploid cells of the embryo sac is called ovary culture, and the process is termed as gynogenesis. Ovary is an ovule-bearing region of a pistil. Excised ovaries can be cultured *in vitro*. For many species (e.g., tomato, *Cucumis anguria*, etc.,) the fertilized ovaries are excised and then grown in culture to form fruits that ripen and produce viable seeds. This development takes place on a simple nutrient medium containing only mineral salts and sucrose, provided the flowers have been fertilized two or more days before excision. But remarkably, the ovaries of unpollinated flowers do not grow on simple nutrient medium. It requires some hormonal combination for development.

Often, in culture, the ovaries fail to grow into full size fruits because of the restricted space of the culture vessel. To overcome this problem, a partial sterile culture technique is devised in which only a long flower stalk is inserted into the aseptic medium through an opening in the stopper, thus leaving the ovary free to grow outside the culture vessel. Ovary culture is useful in the study of the earlier stages of embryo growth and fruit development. The effect of phytohormones on parthenocarpic fruit development can be studied from the culture of unpollinated pistil.

Protoplast Culture

Protoplast culture and manipulation are the recent advances in plant tissue culture. Protoplasts are the living cells of the plant enclosed in plasma membrane and devoid of cell walls. Protoplasts are most commonly isolated either from leaf mesophyll cells or cell suspensions, although other sources can also be used. There are two general approaches in removing the cell wall —mechanical or enzymatic isolation.

1. *Mechanical isolation* This type of isolation often results in low yields, poor quality and poor performance in culture due to substances released from damaged cells.
2. *Enzymatic isolation* It is usually carried out in a simple salt solution with a high osmoticum, and cell wall degrading enzymes. This technique involves the use of a mix of both cellulase and pectinase enzymes of high quality and purity.

Protoplasts are fragile and easily damaged and therefore, must be cultured carefully. The liquid medium is not agitated and a high osmotic potential is maintained, at least during the initial stages. The liquid medium must be shallow enough to allow aeration in the absence of agitation. Protoplasts can be plated out on to a solid medium to produce callus. Further, whole plants can be regenerated even by organogenesis or somatic embryogenesis from this callus.

A protoplast can be fused with protoplasts of same or other species, leading to the development of somatic hybrid (cytoplasm and nucleus of both species) and cybrid (cytoplasm of both species and nucleus of one species) plants. Protoplast fusion is achieved through chemical (PEG) or electrical methods. It is possible even among different genera or species which are sexually incompatible.

Somatic hybrids using protoplast culture technique have been developed in potato, rice, citrus, tomato, petunia, carrot, *Brassica,* etc. Further, protoplast culture is very useful for non-sexual gene transfer to get transgenic plants. Transgenic plants have been developed in more than 350 plant species, including peach, apple, carrot, rice, corn, potato, sunflower, cotton, soybean, cucumber, citrus, oilseed rape, etc.

MICROPROPAGATION

Micropropagation means growing a plant propagule on a nutrient medium, resulting in rapid shoot-buds/shoot proliferation, thereby enabling faster multiplication of true-to-type disease-free plantlets of elite germplasm. The differentiation of the meristem leads to the development of multiple shoot/shoot-bud formation which can be rooted *in vitro* for the development of complete plantlets. This method can be used for clonal propagation. Shoot meristem cultures are potential alternatives to the more commonly used methods for cereal regeneration as they are less genotype-dependent and more efficient (seedlings can be used as donor material).

Further, meristem culture is being exploited commercially for the production of pathogen-free plants of potato, dahlia, strawberry, carnations, chrysanthemums, orchids, etc. This technique, when combined with chemotherapy and thermotherapy, has resulted in the production of plants free from even those pathogens, which otherwise are difficult to eliminate.

SOMACLONAL VARIATION

Plants derived from tissue culture have been variously referred to as somaclones or calliclones and the variations displayed by such plants are simply called somaclonal variation. Generally the term somaclonal variation is used for genetic variability present among all kinds of cells/plants obtained from cells cultured *in vitro.* Plants regenerated from tissue and cell cultures show heritable variation of both qualitative and quantitative traits in many plants. There are several possible mechanisms causing somaclonal variation. These are chromosomal aberrations, DNA amplification and transposable elements.

Somaclonal variation has been suggested as a useful source of potentially valuable germplasm for plant breeding and plant improvement. The major benefit of somaclonal variation is to create variations in adapted genotype without treatment with mutagenic substances.

IN VITRO POLLINATION AND FERTILIZATION

When pollen is applied to the stigma of ovaries cultured *in vitro,* or directly onto ovules cultured with or without placental tissue, it is called *in vitro* pollination. Ovaries are collected from emasculated flowers usually 1–2 days after anthesis. These can be cultured intact or with the ovarian wall removed to expose the placenta. Alternatively, the entire placenta or pieces of placenta bearing ovules may be cultured. In species like maize, small pieces of cob (bearing 10 ovaries) may be suitable for culture after 2–6 days of silk emergence. Generally, ovaries/ovules are cultured along with the pedicel. In some dicot species, it may be desirable to retain the calyx, while in monocots like barley, the glumes need to be retained since their removal is deleterious to embryo development. Wetting of ovules and stigma should be avoided. It may reduce pollen germination and interfere with normal pollen tube growth.

This technique is useful in breeding and hybridization programmes when the zone of incompatibility lies in the stigma, style or ovary. In self-incompatible species, pollen tube does not enter the ovary, though pollen germination is good during self-pollination. In this case self-placental pollination brings about fertilization and seed formation. Self-incompatibility may be overcome by this process.

Success of *in vitro* pollination depends upon two basic considerations, (i) correct stage of development of pollen grain and ovule and (ii) nutrient medium composition.

CRYOPRESERVATION

Some techniques involve storage of living tissues at the growing stage whereas others relate to the suspension of growth of the tissue. Growth inhibition may be carried out by temperature reduction, use of growth-retardant chemicals or hormones, reduction in oxygen concentration, etc. These methods require periodic renewal at regular intervals. Another method is cryopreservation which can often store plant materials for virtually indefinite periods at low temperature. This technique preserves original genetic stocks in a limited area and ensures genetic stability of cryopreserved material.

The preservation of cells, tissues and organs by cooling to sub-zero temperatures, such as (typically) 77K or –196°C (the boiling point of liquid nitrogen) in liquid nitrogen is called cryopreservation and the science pertaining to this activity is known as cryobiology. Cell suspensions, hybrid protoplasts, pollen grains, seeds and meristems of desired plant species are stored by this method for the establishment of germplasm bank. Liquid nitrogen maintains low temperature for long-term storage. The low temperature reduces the growth rate of the stored cells and delays ageing. Therefore, the plant materials do not lose their viability for a long time. After a long period of storage, new plants can be regenerated from the cryopreserved plant materials.

TRANSGENESIS

Transgenic technology enables plant breeders to bring together in one plant useful genes from a wide range of living sources, not just from within the crop species or from closely related plants. This technology provides the means for identifying and isolating genes controlling specific characteristics in one kind of organism, and for moving copies of those genes into a different organism, which will then acquire those characteristics. This powerful tool enables plant breeders to generate more useful and productive crop varieties containing new combinations of genes but it expands the possibilities beyond the limitations imposed by traditional cross-pollination and selection techniques.

There are two main methods of transforming plant cells and tissues:

1. *The "gene gun" method* It is also known as microprojectile bombardment or biolistics. This technique involves the bombardment of microparticle (gold/tungsten), coated with desired foreign DNA to the target tissues and furthermore selection and identification of putative transformants. It has been especially useful in transforming monocot species like corn and rice.
2. *The Agrobacterium method* Transformation via *Agrobacterium* has been successfully practiced in dicots but only recently has it been effective in monocots (such as grasses). In general, the *Agrobacterium* method is considered preferable to the gene gun, because of the greater frequency of single-site insertions of the foreign DNA, making it easier to monitor.

These putative transgenic cells or tissues are further subjected to different nutrient media for their regeneration and development into complete transgenic plants. Simultaneously, molecular analysis should also be conducted with transgenic tissues in order to confirm the foreign gene integration into the host genome, desired gene expression in target tissue and finally transmission of transferred gene from parents to the progeny crops.

PROBLEMS IN PLANT TISSUE CULTURE PRACTICES

Plant tissue culture is considered to be a simple technique but it has some limitations, which include the following.

1. Contamination is one of the biggest problems when dealing with cell and tissue cultures. Microorganisms can grow much faster than plant cells and take up all the nutrients preventing the plants from growing. Steps should be taken to minimize the possibility of contamination.
2. There is at present no way to predict the exact growth medium composition and growth conditions, to generate a particular type of callus. These characteristics have to be determined through a systematically designed and observed experimental data for each new plant species.
3. In many species, phenolics move into the medium from the cut surfaces of explants. These phenolics turn brown on oxidation and are harmful to the cultures. This problem is very common in case of woody species, where explants are generally taken from mature trees.
4. Some shoots developed *in vitro* appear brittle, glassy and water-soaked. This phenomenon is called vitrification or hyper-hydration. In many plant species, vitrification cannot be seen through naked eyes and the symptoms are internal or very small (poorly developed vascular bundles, abnormal functioning of stomata, etc.). This problem is the consequence of culture conditions, which leads to loss of plantlets.

REVIEW QUESTIONS

1. What is culture medium? State the basic composition of a general plant tissue culture medium.
2. Define callus and suspension cultures. Briefly describe the different types of suspension cultures.
3. What is haploid culture? Describe haploid culture under the following heads:
   i. Anther culture
   ii. Microspore culture
4. Write a brief note on organ culture.
5. Give a brief outline of plant tissue culture techniques.

REFERENCES

Ballester, A., Sanchez, M.C., San-Jose, M.C., Vieitez, F.J. and Vieitez, A.M. (1990). Development of rejuvenation methods for *in vitro* establishment, multiplication, and rooting of mature trees. In: Rodriguez *et al.* (eds.). *Plant Aging: Basic and Applied Approaches*. Plenum Press, New York. pp. 43–49.

Bhojwani, S.S. and Razdan, M.K. (1996). Studies in plant Science, 5. *Plant tissue culture: Theory and Practice*, (revised edition) Elsevier Science B.V., The Netherlands. pp. 537–562.

Cocking, E.C. (1989). Plant Cell and Tissue Culture. In: Max, J. L. (ed.). *A Revolution in Biotechnology*. Cambridge University Press, Cambridge. pp. 119–129.

Griesback, R.J. (1984). Introduction to somatic cell genetics. *Hort Science*. 19: 367–371.

Hartel, O. (1999). Gottlieb Haberlandt—A portrait. *In vitro Dev. Biol. Plant*. 35: 157–158.

Hammatt, N. (1992). Progress in the biotechnology of trees. *World J. Microbiology and Biotechnology*. 8: 369–377.

Huetteman, C.A. and John E. Preece. (1993). Thidiazuron: a potent cytokinin for woody plant tissue culture. *Plant Cell, Tissue and Organ Culture*. 33: 105–119.

Lila-Smith, M.A. (1994). Plant Tissue culture. *Encyclopedia Agric.* Science. 3: 369–384.

Murashige, T. (1978). The impact of plant tissue culture on agriculture. In: Thorpe, T.A. (ed.). *Frontiers of plant tissue culture*. University of Calgary, Canada. pp. 15–26.

Pierik, R.L.M. (1988). *In vitro* culture of higher plants as a tool in the propagation of horticultural crops. *Acta Horticulturae*. 226: 25–40.

Thorpe, T.A. (1985). Applications of tissue culture technology to forest tree improvement. *Forestry chronicle*. 61: 436–438.

Vasil, I.K. and Thorpe, T.A. (1994). *Plant Cell and Tissue Culture*. Kluwer Academic Publishers, Dordrecht, The Netherlands.

3

IN VITRO CULTURE LABORATORY–ORGANIZATION AND MANAGEMENT

INTRODUCTION

A laboratory devoted to *in vitro* experiments with plant cells or tissues must have adequate space for the performance of several activities such as nutrient media preparation, sterilization, cleaning and storage of supplies, aseptic manipulation of plant material, growth of the cultures under controlled environmental conditions, observation and evaluation of the cultures, etc. Simultaneously, a proper facility must be provided to run the experiments connected with production of transgenic crops and their molecular analysis.

This chapter deals with the basic laboratory set-up that is required to initiate plant *in vitro* regeneration and its associated experiments. In recent years, there has been a large increase in the number of research laboratories using these techniques to investigate many fundamental and applied aspects of higher plants. The *in vitro* cultivation of plant cells and tissues is primarily devoted to keeping the plant cells and tissues free from microbes. It also ensures the desired development in the cells and tissues by providing suitable media and other supporting conditions. However, the use of these techniques is not confined to research alone but is also applicable to industrial avenues.

LABORATORY STRUCTURE

The laboratory set-up for *in vitro* culture depends on the nature of the research undertaken and the availability of funds. For a standard cell and tissue culture laboratory (Figure 3.1), the following minimum facilities are required:

1. Washing and storage facilities
2. Media preparation, sterilization and storage room

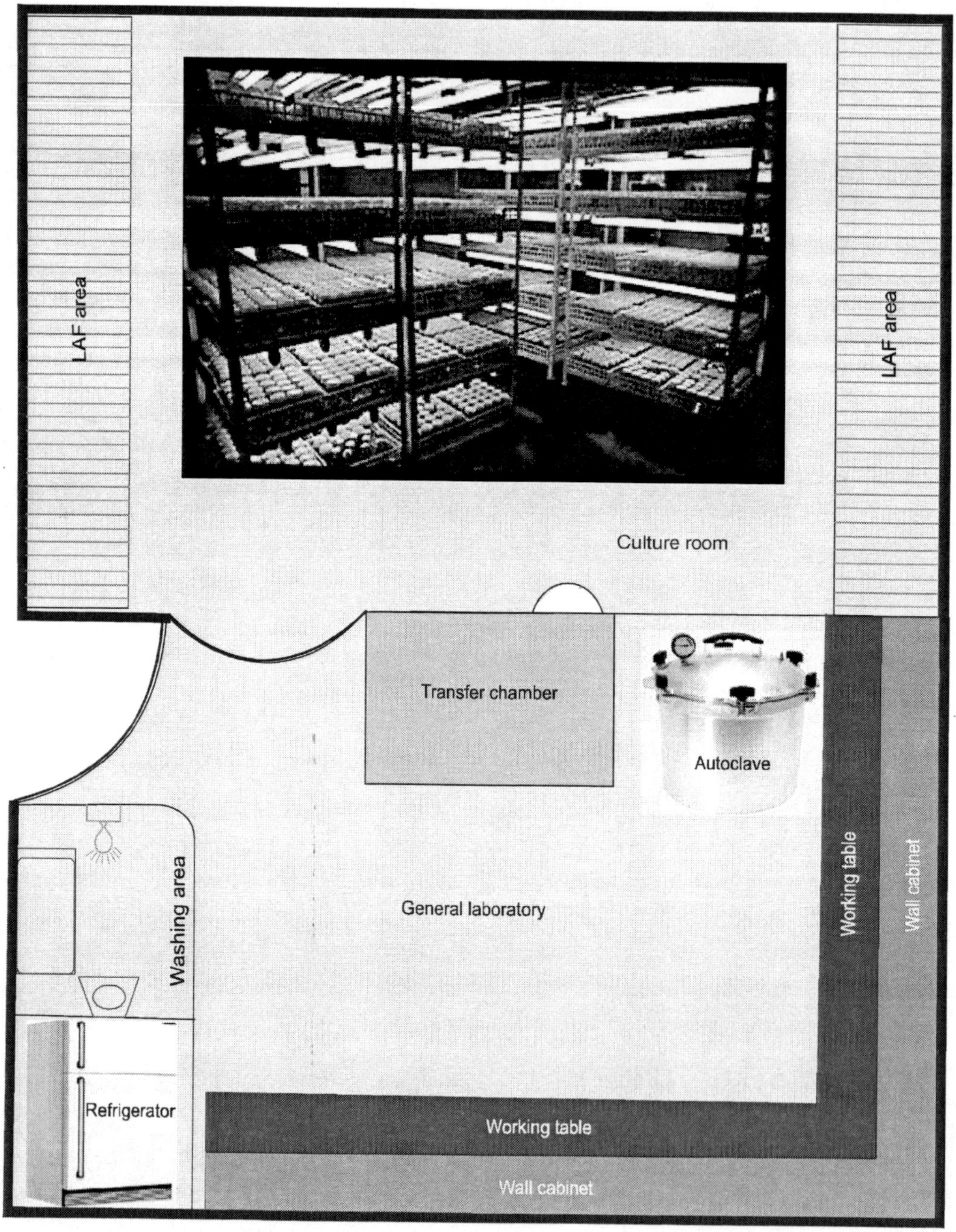

Figure 3.1 A general view of cell/tissue culture lab organization

3. Transfer area for aseptic manipulations
4. Culture rooms or incubators for maintenance of cultures under controlled conditions of temperature, light and humidity
5. Observation or data collection area
6. Acclimatization room (greenhouse)

In a modest laboratory, provisions for activities 1 and 2 (in above list) can be made in a single room (media room), while the remaining work can be done in another room for example, in culture room or inoculation room. During lab set-up, the following precautions must be followed:

i. The washing area must be physically separated, at least by a temporary partition, from the medium preparation room.
ii. The weighing balances should be kept in a separate enclosed space.
iii. Autoclave, refrigerator, deep freezer and incubator may be located in the corridor.
iv. For aseptic manipulation, a laminar airflow (LAF) cabinet is commonly used, which can be housed in the culture room.
v. A small table having a stereoscopic microscope may be adequate for culture observation.

Washing and Storage Facilities

This facility is very important for a cell culture laboratory. It should be in a separate room with a large sink supplied with running hot and cold tap water facility, brushes of various sizes, detergents and buckets and distilled water for a final rinse of the washed glassware. Used media and media vessels also are cleaned in this area. New glassware must be washed prior to use, using detergents especially designed for the purpose to remove all traces of acids, etc. Finally, the glassware is rinsed in tap water followed by a final rinse in distilled water. For this, sufficient area is required to accommodate large sinks (acid-lined), draining boards/racks and sufficient distilled/demineralized water. There should also be enough space to set up drying ovens. The washing area should also have dustproof cupboards for storage of labware.

In cell culture study, glassware should be resistant to heat. Cleaning of glassware is done by soaking it in sodium dichromate/conc. sulphuric acid for 4 hours and then washing it properly under tap water. The glassware is then soaked in a detergent solution for 16 hours, then rinsed, first in tap water followed by a second rinse in distilled water. Glassware is then dried in an oven at high temperature. Instead of glassware, a wide range of pre-sterilized polystyrene culture containers (plastic labware) are now available which can be disposed off after use. Reusable plastic labware is also available now. These can be washed with a mild detergent and rinsed with tap and distilled water. Washed and dried labware is finally stored in a closed and dust-free cupboard.

Washing and handling of used glassware

1. Reusable glassware for culture work should be emptied immediately after use and soaked. The media or agar should never be allowed to get dried on/in the glassware.
2. All glassware containing corrosive chemicals or fixatives should be segregated from the rest of the glassware used for culture work.
3. All contaminated glassware or labware, coming into contact with microorganisms, should be decontaminated before washing.
4. The labels and markings must be removed before washing. Marking ink may be wiped out by abrasive cleanser or acetone and then rinsing immediately.
5. Melted agar in the culture vessels should be poured into a collecting sieve and discarded.
6. Rubber-lined screw caps should be soaked only in distilled water. They should never be soaked in detergent or soap solution.
7. Glassware for growing cells should be acid-cleaned if proteinaceous deposits are not removed by conventional washing.
8. Silicone-treated glassware should be permanently labelled and placed separately from the rest of the glassware.

Glass-washing and acid cleaning

1. Culture glassware should be soaked in 5% detergent solution for a minimum of one hour.
2. Automatic washer is used only as a rinsing unit and no detergent is added to the washer. Glassware should be treated with detergent prior to its transfer into the washing unit.
3. The automatic washer is programmed for not less than six tap water rinses and a minimum of one minute distilled water rinse.
4. The glassware that does not fit into an automatic unit is rinsed by hand. Hand rinsing requires a minimum of six thorough tap water rinses and four rinses in quality water.
5. Personnel concerned should wear a full-face mask and acid-resistant apron and gloves.
6. Handling of acid and processing of glassware in the acid should be done in a fume hood.
7. Water should never be added to acid. To dilute acid, acid must be slowly added to water, while stirring.
8. After acid cleaning, if the solution in the acid bath becomes dark coloured, it should be discarded.

Media Preparation Room

This room should have sufficient space for storage of chemicals, labware, culture vessels, pH meters, balances, water baths, burners, etc. (Figure 3.2). A microwave oven, autoclave (for sterilizing media), culture vessels and instruments are appliances that are essentially needed for media preparation. A detailed list of equipment and accessories needed for media preparation is as follows.

i. Gas and water supplies
ii. Compressed air and vacuum line supplies
iii. Water heater and gas stove
iv. Hot plate with magnetic stirrer
v. Glass or stainless steel containers for heating and dissolving media.
vi. Autoclave and pH meter
vii. Balances of different capacities
viii. Measuring cylinders, flasks, beakers, Petri dishes and pipettes
ix. Culture tubes, bottles and other glassware
x. Small instruments such as spatulas, scalpels, forceps and dissecting needles
xi. Hot-air oven for rapid heating of media
xii. Distilled and double-distilled water units
xiii. Pipette washers
xiv. Drying and draining racks

Figure 3.2 Media pouring in media preparation room (Plate 1.2)

A culture laboratory requires analytical loading single pan balance with precision of ± 0.001 g (weighing range from 0.1 mg to 200 g) with digital read out, hot plate and magnetic stirrers, top pan loading balance for quick weighing (capacity ranging from 100 mg to 500 g), refrigerator and freezer and distilled water unit. Glassware washing facility with proper drainage, gas outlet, electric hot air oven (range up to 250 ± 2°C and microwave oven, digital pH meter (range 0–14 pH) with accuracy 0.1 pH, an autoclave with temperature compensation 0°C to 120°C and water purification unit with storage system, are also required.

The other appliances that are also necessary include culture tubes/conical flask/Petri dishes of various capacities, measuring cylinders (25 ml, 100 ml, 500 ml, 1000 ml), cotton for plugs/plastic caps (autoclavable), general glassware/plasticware of various capacities such as volumetric flasks, beakers, reagent bottles, pipettes, vacuum filtration system and glass rods. A small spirit lamp made up of glass will be required for the flame sterilization of instruments during inoculation in LAF cabinet.

REQUIREMENTS FOR MEDIA PREPARATION

Although media preparation requires a balance that is sensitive to milligram quantities for weighing hormones and vitamins, a less sensitive scale for weighing agar and carbohydrates is also essential. The media reagents should be shelved near the balance for convenience. A refrigerator and a deep-freezer in the media room are necessary for storing stock solutions and chemicals, otherwise they may degrade at room temperature. A combined hot plate and magnetic stirrer is a time-saver in dissolving inorganic reagents. Either a pH meter or pH indicator paper is required for adjusting the final pH of the medium. Relatively large quantities of single- and double-distilled water must be available in the media room.

Sterilization of equipment is an integral part of media preparation. A commercial electric oven is the most economical type of dry sterilization. Wet heat sterilization involves either an autoclave or a pressure cooker. Some hormones and vitamins are sterilized by ultrafiltration at room temperature. After sterilization of the culture vessels by dry heat and autoclaving the medium, the culture tubes or Petri dishes are kept in the transfer chamber. A microwave oven is also useful to dissolve agar and melt medium.

Aseptic Transfer Chamber/Transfer Area/Inoculation Chamber

The simplest type of transfer area is an enclosed plastic box which can be sterilized with a UV light and by cleaning the floor surface with 70% ethyl alcohol. A small wooden hood may also be used for culture work. When a large number of cultures or transfers are being performed, larger equipment are required. The most desirable arrangement is a dust-free room equipped with an overhead UV and positive pressure ventilation unit. The ventilation unit should possess a high efficiency particulate air (HEPA) filter. It has been observed that a 0.3 μm HEPA filter shows almost 100% efficiency. All the surfaces in the room should be thoroughly cleansed and disinfected regularly. The room should be air-conditioned.

Laminar air flow cabinet Another type of transfer area used in most laboratories is a laminar air flow cabinet (LAF cabinet). A small motor blows air into the unit first through a coarse filter, where large dust particles are separated, and then, passes through a 0.3 µm HEPA filter, (the horizontal/vertical flow unit) over the working surface. The air coming out of the fine filter is ultra clean (free from fungal or bacterial contaminants). Its velocity (27 ± 3 m/min) adequately prevents the microcontamination of the working area in front of the cabinet. Various other contaminants, coming from the researchers are also blown away by the ultra clean airflow. In this way, an aseptic environment is maintained when the cabinet is switched on (Figure 3.3).

Figure 3.3 A picture of a Laminar Air Flow cabinet for culture work (Plate 1.1)

LAF cabinets are commercially available in various sizes and shapes. The advantage of using such cabinets is that the flow of air does not hamper the use of a spirit lamp or Bunsen burner. Each cabinet occupies a relatively small space within an ordinary lab. If atmospheric dust is unfortunately very high, it is advisable to keep the LAF cabinet in a culture room fitted with double doors to prolong the life of the filter. An important precaution to be taken is that a LAF cabinet should never face a window or door that is frequently used. Instruments like lens magnifier, micro-dissecting scissors, scalpel, handles with blades, forceps, catheter tray, needles and inoculating loops, gas outlet, vacuum facility, tissue paper/filter paper sterilizer (dry heat with glass beads) and pipette dispenser should be in or near the LAF cabinet.

The essential requirements for culture inoculation are as follows:

1. Laminar air flow cabinet
2. Spirit lamp/Bunsen burner in the LAF cabinet

3. Various types of microscopes
4. Ethyl alcohol for sterilization and flaming of small metal instruments
5. Mercuric chloride solution/hypochlorite solution for sterilization of plant material

Operation of laminar air flow cabinet

1. Switch on the LAF cabinet 30–45 minutes before making inoculation activities.
2. Make sure that no object blocks the filter of air intake at the bottom of the unit and that the sterile air coming out of the filter flows uniformly in the hood.
3. Use 70% ethanol to wipe the working surface of the hood with the help of cotton.
4. Place only sterile media, instruments, appliances and culture vessels inside the cabinet.
5. Switch on the UV lamp of LAF cabinet for a minimum of 30 minutes before starting the inoculation.
6. Prior to inoculation, switch off the UV lamp and switch on the working lamp.
7. During flame sterilization of the small instruments, keep the spirit lamp at a distance from the alcohol bottle to avoid fire.
8. Rewipe LAF table after use and then switch it off properly.

Culture Room/Growth Chamber

Cultures are generally incubated at 25 ± 2°C under 14:10 light/dark photoperiod. The source of light for the cultures in racks should be cool daylight which is provided by a florescent tube light. The culture room should have the following facilities:

1. Controlled temperature (at 25 ± 2°C) with the help of air conditioners and room heaters.
2. Culture racks fitted with light (generally 2000–4000 lux or lower light generated by fluorescent tubes; automatic timer can be used to regulate photoperiod).
3. A shaker for agitation of suspension cultures. It is desirable to have a good quality generator set for providing power to the culture room in the event of power failures.

All types of cultures are incubated under conditions of well-controlled temperature, humidity, illumination and air circulation. Usually, air conditioners and heaters are used to maintain the temperature around 25 ± 2°C. Cultures are generally grown in diffused light. Other requirements are a humidity range of 60–90% (controllable to ± 3%) and uniform forced air ventilation. Culture room should have enough shelves illuminated by a set of fluorescent tubes for storing cultures (Figure 3.4). All the cultures should be labelled, giving details of the experiment (name of the plant, explant, medium, date of culture and other information) to ensure identity and for recording monitored results. Shakers with controlled temperature and light are also installed in the culture room.

Figure 3.4 A picture showing arrangement of cultures in a growth/incubation chamber (culture room) (Plate 1.6)

Several engineering aspects should be considered in designing a culture room. They are:

i. Safety and convenience of the electrical system
ii. Air flow for uniform temperature regulation
iii. Arrangement of the shelving
iv. Elimination of airborne contaminants

Optimal environmental conditions will vary depending on the species and the purpose of the experiment. Special attention should be given to diurnal temperature variations, light intensity, light quality and photoperiod. Fluorescent lamps have certain advantages over incandescent sources (bulbs) in having better spectral quality, a more convenient shape and a lower heat output.

Incubators, which are large plant growth chambers, are readily available in the market. They occupy less space and have the range of control and flexibility desirable for growth of tissues under *in vitro* conditions.

When constructing a growth room, it is essential to keep the light bulbs outside, as they generate a lot of heat. Some cultures, however, appear to show the best growth in the presence of a mixture of both types of illumination. It is advisable to equip the culture room with a clock-operated timing switch for the regulation of photoperiods. Although, some investigations may need to expose the cultures to thermo-periodic cycles, most of the experiments are conducted at constant temperatures.

The general characteristics of incubators or growth chambers are as follows:

1. Temperature range, 2–40°C
2. Temperature control, ± 0.5°C
3. Continuous temperature recorder
4. Twenty-four-hour temperature and light programming
5. Adjustable fluorescent lighting up to 10,000 lux
6. Relative humidity range, 20–95%
7. Relative humidity control, ± 3%
8. Uniform forced air distribution
9. Capacity up to 0.7 m^3 of 0.5 m^2 shelf space

The general equipment required for *in vitro* culture study are:

1. Refrigerator, deep-freezer, automatic dish-washer, glass atomizer
2. Dispensing devices like wire-mesh baskets, trolleys with trays and metal racks for holding test-tubes or culture vials in the autoclave
3. Microscopes (for example, compound, inverted) with micro-photographic facilities
4. Markers, labels and aluminium foil (for wrapping culture vessels and glassware)

GENERAL TECHNIQUE FOR PLANT CELL AND TISSUE CULTURE

Several techniques are adopted for plant cell and tissue culture and they vary depending upon the objective of the experiment. Some of them are general techniques that are essentially followed in almost all of the experiments such as preparation of nutrient medium, sterilization, aseptic manipulation, maintenance of cultures, etc. MS (Murashige and Skoog, 1962) medium is the most commonly used nutrient medium in plant cell and tissue culture technology.

Protocol of MS Nutrient Medium (1 litre) Preparation

1. Dissolve 30 g of sucrose or sugar in 200 ml double-distilled water.
2. Take some double-distilled water in another flask and add the appropriate amount of stock solutions (Refer Chapter 4 for details on stock solutions).
3. Add the desired concentration and amount of auxin and/or cytokinin to the above solution.
4. Mix the solutions in step 1 and 2 in a one-litre measuring cylinder.
5. Make the final volume to one litre with double-distilled water and shake well for uniform mixing.
6. Adjust the pH of this medium in the range 5.6–5.8 with 0.1N HCl or 0.1N NaOH with the help of a pH meter.

7. Add 6–8% agar to this to make a semi-solid medium and dissolve this agar by heating to 60°C.
8. Liquid medium can be used without the addition of agar.
9. Dispense the culture medium into culture vessels and close with non-absorbent cotton plug wrapped in muslin cloth.
10. This medium is finally sterilized in an autoclave at 121°C with 15 lb/in^2 pressure for 15 to 20 minutes.

Sterilization of Explants

The explant which is to be cultured is surface-sterilized to remove the surface-borne microorganisms. This process involves the following steps:

1. Thoroughly wash the explants in tap water and then immerse in 5–6% solution of liquid detergent such as "Teepol" for 10–15 minutes.
2. Wash the explants thoroughly in tap water followed by distilled water washing. Perform the further sterilization treatments of these explants inside the LAF cabinet.
3. Dip these pre-sterilized explants in 70% ethyl alcohol for 30 seconds.
4. Transfer these explants into a sterilized beaker containing 0.1% mercuric chloride ($HgCl_2$) and keep for 3–6 minutes or that containing 5–10% sodium hypochlorite solution for 10–15 minutes, and shake well.
5. Decant the sterilant and wash the explants thoroughly 5–7 times with sterile distilled water to remove all traces of sterilant.

Inoculation of Explants

The technique which involves the transfer of surface-sterilized explants onto the nutrient medium, using sterilized instruments is called inoculation. Starting from surface-sterilization to inoculation, all operations must be done aseptically in an LAF cabinet as follows:

1. Keep all the sterilized articles (instruments, glass goods, media, etc.) on the table of the LAF cabinet and switch on the UV lamp and LAF. Sterilized air should blow over the articles for 15 minutes before starting the work.
2. Put off the UV lamp of the LAF cabinet when work is about to start but do not put off LAF. Wipe the table of the LAF cabinet and hands with alcohol before starting the inoculation activity.
3. Pour alcohol in a coupling jar and put all instruments into it. Light the spirit lamp and take the sterilized explants in a sterile Petri dish.
4. The instruments must be passed through the flame of the spirit lamp prior to inoculation.
5. Flame the neck of the culture vial and quickly remove the plug of the glass vial. Transfer the tissue onto the medium and replace the plug.

Thus the sterilized explants are transferred to the autoclaved medium.

Precautions during inoculation

1. Keep the alcohol-moistened hand away from the spirit lamp.
2. Switch off the UV lamp 20–30 minutes before starting work in the inoculation chamber because UV light converts oxygen into toxic ozone gas.
3. Do not use hot instruments to cut or hold the plant materials.
4. Make sure that the plant material is in direct contact with the nutrient medium.

Incubation of Cultures

The freshly prepared cultures are grown under carefully regulated environmental conditions, that is, temperature, light and humidity. This is accomplished with an incubator, plant growth chamber or controlled environment room. If cell suspensions are to be grown, some type of shaker or aeration equipment will be necessary.

The cultures are incubated on culture rack at 25 ± 2°C temperature. Light is provided by cool white fluorescent tube lights placed about 20 inches above the cultures to give a light intensity of 2000 to 4000 lux for 14 hours. Humidity should be maintained properly (70–85%) because low humidity causes quick desiccation of culture medium and high humidity favours the contamination. In some cultures (callus and cell cultures) where light is not necessary, put off the light and cover the whole rack with a black cloth.

Data Collection or Observation of Cultures

The growth and development of cultures is generally monitored by observing cultures at regular intervals. The observation or data collection area or analytical room is the place where cultures can be analysed for results and further experimentation. Instruments like inverted microscope for bright field, dark field with compensating wide angle eyepieces, preferably with photo-micrographic attachment, low and high speed centrifuge with continuous variable electronic speed control, colorimeter, viscosity meter, etc. are required.

Greenhouse or Acclimatization Room

This is also known as hardening chamber, and it needs high illumination (4000 to 10,000 lux light intensity) and high humidity (90–100%, through mist and fog system). Humidity is required for conditioning of culture plants after taking them out from the rooting media and transferring to pots in the greenhouse.

A greenhouse is a structure with a glass or plastic roof and frequently also with glass or plastic walls. It heats up because of incoming solar radiation from the sun, which warms plants, soil, and other things inside the building faster. Air warmed by the heat from hot interior surfaces is retained in the building by the roof and wall. These structures range in size from small sheds to very large buildings. Greenhouses can be divided into glass greenhouses and plastic greenhouses (or constructed greenhouse with polycarbonate sheets on sides and roof) (Figures 3.5 and 3.6). Plastics mostly used are PE film and multiwall sheet in PC. The glass greenhouses are fitted with equipment like screening installations and heating, cooling and lighting and may be automatically controlled by a computer.

Figure 3.5 Outer view of a constructed greenhouse (Plate 1.3)

Figure 3.6 Inner view of a greenhouse (fan section) (Plate 1.5)

Figure 3.7 A general view of a net-house for acclimatization of regenerated plants (Plate 1.4)

Because the temperature and humidity of greenhouses must be constantly monitored to ensure optimal conditions, a wireless sensor network can be used to gather data remotely. The data is transmitted to a control location and used to control heating, cooling, and irrigation systems. Net (agri-net) houses are also useful for acclimatization of regenerated plants before transferring them to the field (Figure 3.7). It may increase the survival of *in vitro*-generated plants in field conditions.

REVIEW QUESTIONS

1. Write an essay on the general laboratory organization for plant cell and tissue culture.
2. Write a detailed note on laminar air flow cabinet.
3. What are the essential conditions of a growth chamber/incubation room?
4. Write in detail about the general plant tissue culture technique.

REFERENCES

Bhojwani, S.S and Razdan, M.K. (1983). *Plant Tissue Culture: Theory and Practice.* Elsevier, Amsterdam.

Collins, C.R. and Lyne, P.M. (1976). *Microbiological Methods,* 4th edn. Butterworths, London.

Davis, L., Dibner, M.D. and Battey, J.F. (1986). *Basic Methods in Molecular Biology.* Elsevier, Amsterdam.

Evans, D.A. *et al.* (1983). *Hand Book of Plant Cell Culture*-1. Macmillan, New York.

Levin, R. (1988). Automated Plant Tissue Culture for Mass Propagation. *Biotechnology.* 6:1035–1039.

Reinert, J. and Yeoman, M.M. (1982). *Plant Cell and Tissue Culture: A Laboratory Manual.* Springer-Verlag, Berlin.

Shekhawat, M.S. and Dixit, A.K. (2008). Micropropagation of Eucalyptus camaldulensis Dehn. Clones selected for the Tsunami affected areas. *Int. J. of Plant Sci.* 2 (3)31–35.

Street, H.E. (1977). *Plant Tissue and Cell Culture.* Blackwell, Oxford.

4

STERILIZATION TECHNIQUES

INTRODUCTION

Sterilization is the opposite of contamination. It is an aseptic manipulation technique that frees an article from all living organisms, including viruses, bacteria, fungi and their spores. Culture media, reagents and equipment used in tissue culture are made free of contamination by this technique. The microorganisms may be present in glass vials, instruments, nutrient medium used for culture and even in plant material. Therefore, the surface of plant tissue and all non-living articles including nutrient medium should be sterilized.

Sterilization is absolutely necessary for the successful establishment and maintenance of plant cell cultures. The *in vitro* environment in which the plant material is grown is also ideal for the proliferation of microorganisms. In most cases, the microorganisms outgrow the plant tissues, resulting in their death. Contamination can also spread from culture to culture. The purpose of aseptic technique is to minimize the possibility of propagation of microorganisms in or around the cultures. The basic sterilization techniques are represented in Figure 4.1.

Aseptic manipulations can be categorized into the following major groups:

1. *Sanitation* It is the process of substantial reduction and then maintenance of the microbial population in the air and on the objects in the laboratory to a minimal level, for example, cleaning by water, soap or detergent, etc.
2. *Disinfection* It implies the use of chemical agents to kill pathogenic microbes, without damaging the plant material to which the chemical is applied.

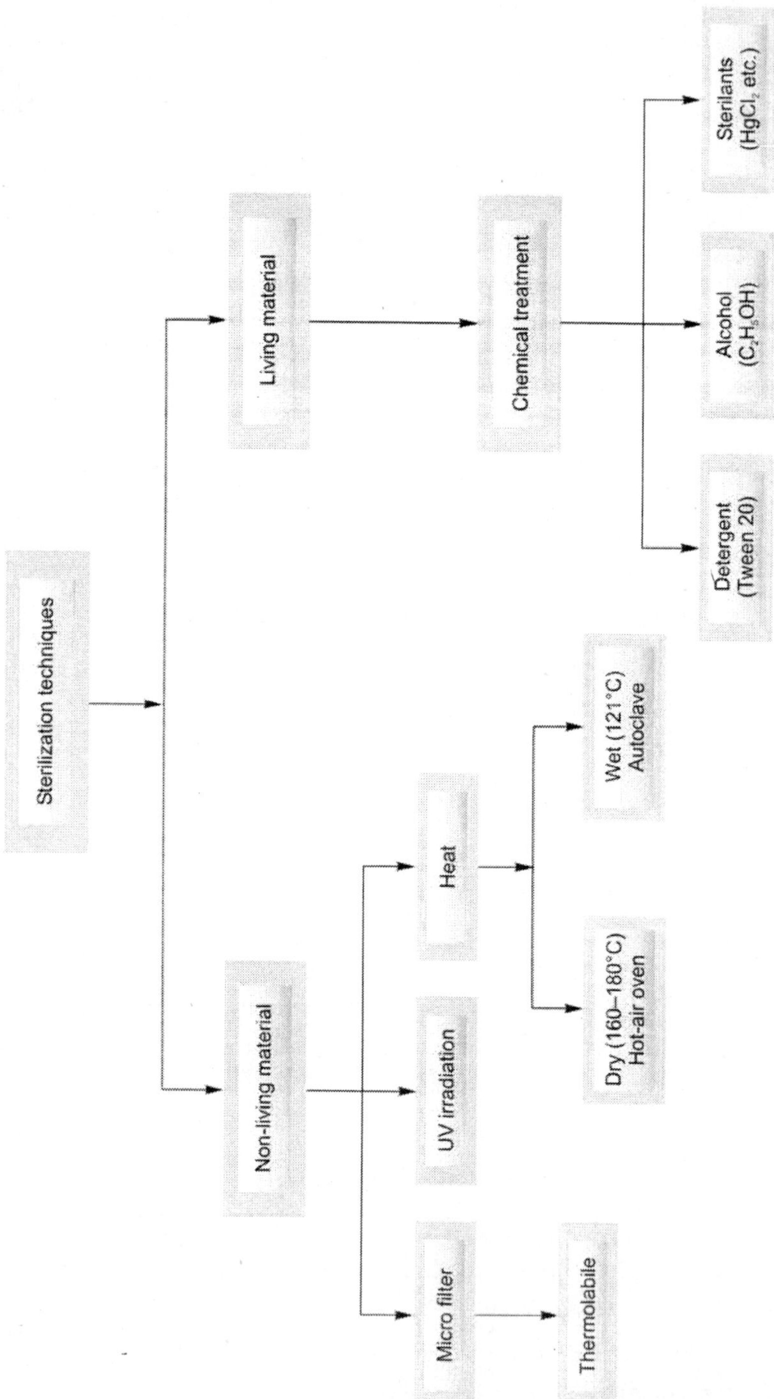

Figure 4.1 Basic methods of sterilization

3. *Sterilization* It is the destruction of all living matter. Sinks are cleaned or disinfected, explants are disinfected, but media and transfer tools are sterilized.

The air in the laboratory is also of concern because it may be highly contaminated. For example, sneezing produces 100,000–200,000 aerosol droplets that can attach themselves to dust particles and these contaminated particles may be present in the air for weeks. The air may also contain bacterial and fungal spores.

COMMON CONTAMINANTS

The common contaminants include viruses/mycoplasma, bacteria, fungi and insects.

Viruses/mycoplasma Viruses, mycoplasma-like organisms, spiroplasmas and rickettsias are extremely small organisms that are not easily detected. Thus, plant culture is not necessarily pathogen-free even if microorganisms are not detected. Special measures such as meristem cultures are often necessary to eradicate such contaminants.

Bacteria Bacteria are the most frequent contaminants, usually introduced with the explants. They may survive even after surface-sterilization of the explants because they might be present in the interior tissues. So, bacterial contamination can first become apparent long after a culture has been initiated. Some bacterial spores can also survive even after proper sterilization treatment. Many kinds of bacteria have been found in plant cell cultures including *Agrobacterium, Bacillus, Corynebacterium, Enterobacter, Lactobacillus, Pseudomonas, Staphylococcus, Xanthomonas,* etc. Bacterial contamination can be recognized by characteristic "ooze". The ooze can be of many colours including white, cream, pink and yellow. There is also often a distinctive foul odour.

Fungi Fungal hyphae present on explants may enter cultures and spores may be airborne. Fungi are frequently present as plant pathogens and in soil. They may be recognized by their "fuzzy" appearance and occur in a multitude of colours. Yeast is a common contaminant of plant cultures. Yeasts live on the external surfaces of plants and are often present in the air.

Insects The insects that are the most troublesome in plant cultures include ants, thrips and mites. Thrips often enter cultures as eggs present on the explants. Ants and mites, however, usually infest already established cultures. Mites feed on fungus and mite infestations are often first detected by observing lines of fungal infection that lead from the edge of the culture vessel to the plant tissue, having been introduced by the insect. It is very difficult to eradicate insect infestations. Careful and effective lab practices and cleanliness should be followed to prevent the infestations.

RANGE OF CONTAMINANTS

Initial Contaminants

Most contaminations are introduced with the explants themselves because of their inadequate sterilization. These can be of fungal or bacterial origin. Such contaminations are very common when the explant material is harvested from the field or greenhouse. Initial contamination is apparent within a few days after cultures are initiated. Bacteria produce "ooze" on solid medium and turbidity in liquid cultures. Fungi look "fuzzy" on solid medium and often accumulate as little balls in liquid medium.

Latent Contamination

This kind of contamination is usually of bacterial origin and is often observed long after cultures are initiated. Apparently, the bacteria are present endogenously in the initial plant material and are not obviously pathogenic *in situ.* Once they are brought for *in vitro* culture, they increase in number and overrun the cultures. Latent contamination is particularly dangerous because it can easily be dispersed among cultures.

Introduced Contamination

Contamination can also occur as a result of poor sterilization protocol or dirty lab conditions. This kind of contamination is largely preventable with proper care.

DETECTION OF CONTAMINANTS

Contamination is usually detected by the "eyeball" method in research labs. However, indexing is frequently done in commercial settings. This involves taking a part of the plant tissue and culturing it in media that are specific for bacteria and fungi. Media that have been used for this purpose include PDA (potato dextrose agar) and NB broth (with salts, yeast extract and glucose). This is the most reliable method for detecting bacteria and fungi, but as indicated above, there may be infecting organisms that cannot be detected.

METHODS OF STERILIZATION

Several sterilization methods are used by researchers to avoid contamination. Broadly these can be categorized into two groups:

1. Physical methods—Heat, filtration, radiations, etc.
2. Chemical methods—Disinfectants, for example, formaldehyde, ethylene oxide, etc.

Heat Sterilization

Heat sterilization can be again divided into two categories:

1. Moist heat sterilization
2. Dry heat sterilization

Moist heat is the most efficient sterilization technique. Spores are killed by exposure to moist heat at 121°C for 10–20 minutes. Dry heat kills pathogens by oxidative destruction of the cell constituents. Spores can be killed by exposure to dry heat at 160°C for one hour.

In dry heat sterilization, oven is used for glassware and instruments, wrapped in aluminium foil, and further treated for 2–3 hours at 150°C as a routine procedure. After sterilization, scalpels and forceps are brought to the LAF cabinet and their distal ends are kept immersed in a beaker containing 70% ethyl alcohol. They are flamed using a spirit lamp and kept under sterilized sheet of paper towel, and this is an effective barrier against airborne contamination.

Sterilization of Non-living Items by UV Radiation

It is possible to use germicidal lamps to sterilize items in an LAF cabinet. UV lamps should not be used when people are present because the UV light is harmful to eyes and skin. Plant tissues left under UV lamps will die.

STERILIZATION OF EQUIPMENT

The Laminar Air Flow Cabinet/Hood Sterilization

Laminar air flow cabinets (Figures 4.2a and 4.2b) are sterile air cabinets. They are used for commercial and research purposes in tissue culture laboratories. A laminar flow unit is designed to remove particles from the air. Room air is pulled into the unit and pushed through a HEPA (High Efficiency Particle Air) filter with a uniform velocity of 90 feet/minute across the work surface. The air is filtered by the HEPA filter. This renders the air sterile. The positive pressure of the air flow from the unit also discourages any fungal spores or bacteria from entering. Depending on the design of the cabinet, the filters are located at the back or at the top of the cabinet.

Following steps are taken while working in LAF cabinet:

i. Run the LAF cabinets in the lab continuously. If for some reason one has turned off, turn it on again and let it run for at least 15 minutes before use.

ii. Make sure that everything needed for the work is in the cabinet and all unnecessary things are removed. Only necessary things should be kept inside the cabinet.

iii. Check the bottom of the hood to make sure, there is no paper or other debris blocking air intake.

iv. Remove watches, etc. roll up long sleeves and wash hands thoroughly with soap (preferably bactericidal) and water before starting the inoculation work.

v. Spray or wipe the inside of the LAF cabinet (bottom, not directly on the filters) with 70% ethyl alcohol. Wipe the work area and let it dry.

vi. Wipe hands and lower arms with 70% ethyl alcohol.

vii. Spray everything that goes into the sterile area with 70% ethanol. For example, spray bags of Petri dishes with 70% alcohol before you open them and place the desired number of unopened dishes in the sterile area.

viii. Movements in the hood should be kept limited to small areas.

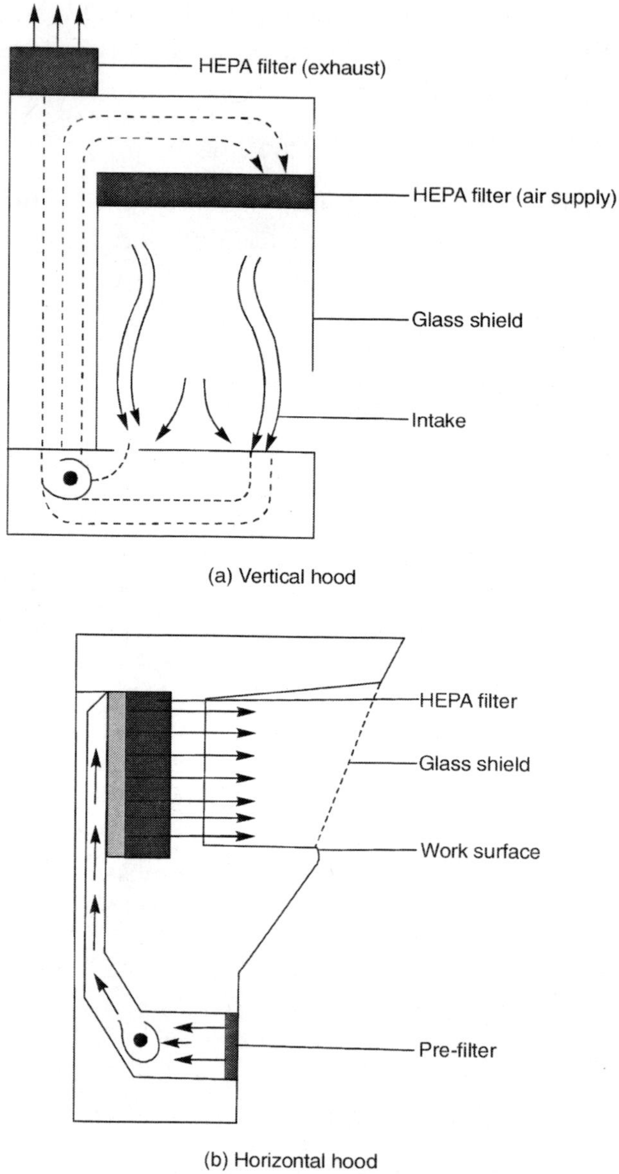

Figure 4.2 Laminar air flow cabinets

ix. Make sure that materials in use are to the side of your work area, so that airflow from the cabinet is not blocked.

x. Don't touch any surface that is supposed to remain sterile with your hands. Use forceps.

xi. Instruments (scalpels, forceps) can be sterilized by flaming, or dipping them in 70% ethyl alcohol and then immediately placing them in the flame of an alcohol lamp or gas burner. A fairly deep container should be used to contain the ethanol. Use enough ethanol to submerge the working ends of the instruments.

xii. Some people wear gloves in the hood for certain procedures. Be very careful not to get them near the flame. Another method of sterilization that does not require alcohol is the glass bead sterilizer. There is not much risk from fire with these but the instruments can still get extremely hot, causing burns.

xiii. Arrange tools and other items in the cabinet so that researchers' hands do not have to cross over each other while working. For a right-handed person, it is best that sterilizing tools be placed to the right. The plant materials should be placed to the left.

xiv. Plant material should be placed on a sterile surface when manipulating it in the LAF cabinet.

xv. Sterilize instruments often. The tools should be placed on a holder in the cabinet to cool or should be cooled by dipping in sterile water or medium before handling plant tissues.

xvi. Wipe up any spills quickly. Use 70% ethyl alcohol for cleaning. Clean table surface periodically while working.

xvii. Sterilize culture tubes with lids or caps on. When opening a sterile tube, touch only the outside of the cap and do not set the cap on any surface. Instead, hold the cap with your fingers during the complete operation and then replace it on the tube. This technique usually requires some practice, especially if technician is simultaneously opening tubes and operating a sterile pipette.

xviii. After removing the cap/plug from the test tube, pass the mouth of the tube through the flame. If possible, hold the open tube at an angle. Put only sterile objects into the tube. Complete the operation as quickly as possible, and then flame the mouth of the tube again. Replace the lid.

xix. Remove items from the hood as soon as their need is over. All cultures must be sealed before leaving the LAF cabinet.

xx. During subcultures, transfer the contaminated cultures at the last and arrange the cultures in such a way so that the contaminated part is closest to the front of the cabinet (so that contaminants blow outward away from the interior of the LAF cabinet).

xxi. When finished, remove all unnecessary materials and wipe the hood down with 70% ethyl alcohol.

Sterilization of Tools/Media by Autoclaving

Two methods—autoclaving and membrane filtration under positive pressure—are commonly used to sterilize culture media. Autoclaving is the method most often used for sterilizing heat resistant items. This is the easiest and the most widely used method to sterilize culture media. It is autoclaved for 15–20 minutes at 121°C and 15 psi pressure. However, this is only possible if all the components in the medium are heat-stable. Longer times are to be avoided to prevent the risk of chemical modification/ decomposition.

It is important that items reach 121°C temperature before the timing of 15 minutes begins. Therefore, time in the autoclave will vary, depending on volume in individual vessels and number of vessels in the autoclave. Most autoclaves automatically adjust time when temperature and pressure (psi) are set and include time in the cycle for a slow decrease in pressure. There are tape indicators that can be affixed to vessels, but they may not reflect the temperature of liquid within them.

Empty vessels, beakers, graduated cylinders, etc. should be closed with a cap or aluminium foil. Tools should also be wrapped in foil or paper or put in a covered sterilization tray. It is critical that the steam penetrates the items, for sterilization to be successful. For certain components, for example, when glucose is used instead of sucrose, a lower temperature (110°C) is often recommended. The autoclaving time should be measured from the moment the desired pressure is reached and not from the time, the autoclave is switched on. To avoid excessively long periods before maximum pressure is reached, it is advisable not to overload the autoclave or fill large volumes in single flasks. Divide the medium into a number of smaller flasks (preferably 500 ml flasks and only if absolutely necessary, one litre flasks). This increases the surface area/volume ratio and therefore, allows the medium to heat through quickly. This also reduces the time needed to reach the desired steam pressure.

Thus larger volumes need longer autoclaving times (30 and 40 minutes are recommended for volumes of one litre and two litres respectively) as shown in Table 4.1. After autoclaving, the pressure should be allowed to fall slowly to avoid the media from boiling over in the flasks. In this regard, flasks should never be filled to more than 60% of their total volume.

Table 4.1 Minimum autoclaving time for plant tissue culture media

Volume of medium per vessel (ml)	Minimum autoclaving time (min)
25	20
50	25
100	28
250	31
500	35
1000	40
2000	48
4000	63

Precautions

- Sterile things which have been autoclaved should be wrapped with some kind of protective covering, for example, aluminium foil, for transport from the autoclave to the LAF cabinet.
- Usual autoclave time of 20 minutes is intended for relatively small volumes. Large flasks of media, water, etc. may require longer autoclaving periods. It is preferable to put not more than one litre of liquid in a container to be autoclaved. It should also be ensured that there is enough space in the container so that the liquid does not boil over.
- Sterilized items should be used within a short time (a few days at most).
- Items that are packaged sterile, for example, plastic Petri plates should be examined carefully for damage before use. If part of a package is used, the remainder should be sealed with date and label.

Filter Sterilization of Media and its Components

In this method, a filter simply removes the contaminating microorganisms rather than destroying them directly. Disposable filter units available from a wide range of sources are the ideal choice, but a cheaper alternative is to buy autoclavable filter units, which can be used with a wide range of membrane inserts. The solution to be sterilized is placed in a syringe that is connected to the sterile filter unit in the sterile transfer area. The solution is forced through the filter with a vacuum pump or with pressure from a syringe, peristaltic pump, or nitrogen gas bottle and collected in previously sterilized containers.

For filtration, various filters are now commercially available to filter different types and volumes of media. They include Asbestos and asbestos paper disks, sintered glass filters, micro filters, cellulose membrane filters, etc.

The usual pore size of the filter is 0.22 μm, which is appropriate for excluding microbial contaminants while allowing the medium to flow through easily. Care should be taken to choose the suitable membrane type to use for certain solutions (for example, those containing proteins, DNA or alcohol need special requirements). Solutions that contain heat-labile components must be filter sterilized. Organic compounds such as some growth regulators, amino acids and vitamins may be degraded during autoclaving. These compounds require filter sterilization through a 0.22 μm membrane. Several manufacturers make nitrocellulose membranes that can be sterilized by autoclaving. They are placed between sections of a filter unit and sterilized as one piece.

Other filters which come pre-sterilized are also available in the market. Larger ones are set over a sterile flask and a vacuum is applied to pull the compound dissolved in liquid through the membrane and into the sterile flask. Smaller membranes fit at the end of a sterile syringe and liquid is pushed through by depressing the top of the syringe. The size of the filter selected, depends on the volume of the solution to be sterilized and the components of the solution. Membrane filters are of two types:

1. Gradocal membrane (made of cellulose nitrate)
2. Cellulose acetate (membrane is about 120 μm thick and placed in two layers)

Media containing heat labile components should either be filter sterilized in their entirety or the heat labile components should be dissolved separately and added after autoclaving the other components. In the latter case, care must be taken to ensure the following:

i. The pH of the solution to be filter sterilized is the same as that of the desired final pH of the medium.
ii. All components are fully dissolved prior to filtration.
iii. The temperature of the autoclaved fraction should be as low as possible before adding the filter sterilized components that is, room temperature for liquid media, 50°C for agar-based and 40°C for agarose-based media.
iv. If one or more of the components are poorly soluble, they require a significant volume to fully dissolve. The volume of the autoclaved components should be reduced accordingly in order to end up with the desired final volume and concentration of all components.

For example, when requiring solidified versions of heat-labile media, it is standard practice to make a double concentration of stock medium for mixing with an equal volume of double concentrated agar/agarose stock in water. After this, the latter is autoclaved and allowed to cool to the required temperature.

Nutrient media that contain thermolabile components are typically prepared in several steps:

1. A solution of the heat stable components is sterilized in the usual way by autoclaving and then cooled to 35°–50°C under sterile conditions.
2. Solutions of the thermo-labile components are filter sterilized.
3. The sterilized solutions are then combined under aseptic conditions to give the complete medium.

In spite of possible degradation, however, some compounds that are thought to be heat-labile are generally autoclaved if results are found to be reliable and reproducible. These compounds include ABA, IAA, IBA, kinetin, pyridoxine, 2-ip and thiamine. Various sterilization techniques used in tissue culture are given in Table 4.2.

Glassware Sterilization

Before starting glassware sterilization, the glasswares are opened, (for example, beakers, Erlenmeyer flasks, etc.) capped with a double layer of aluminium foil to ensure that sterility is maintained even after treatment. Glassware with screw caps should always be kept partially loose before sterilization treatment to prevent high pressures building up, which can lead to vessel explosion. Glassware can routinely be autoclaved at 121°C with a pressure of 15 psi for 15 minutes. Alternatively, dry heat can be used at 160°C for 3 hours. The later should, however, be avoided when plastic caps are used, since these cannot withstand the prolonged high temperatures. Dry heat sterilization is also recommended for glassware destined for use with protoplast culture media. The osmolarity of these media is often very critical and even small amounts of condensation caused by autoclaving, can prove to be detrimental.

Instruments Sterilization

The lower parts of instruments (for example, scalpels, forceps, scissors, etc.) are routinely flamed in LAF cabinet directly before use. These are then allowed to cool before bringing into contact with plant tissue. Between manipulations, the instruments are immersed in 70% ethanol in a glass vessel (for example, a 100 ml measuring cylinder or beaker) kept in the LAF cabinet for this purpose. The alcohol is replaced at least once in a day. Instruments and other metal objects can also be sterilized by using dry heat technique after wrapping them in aluminium foil. Autoclaving is to be avoided, since the combination of elevated temperatures and steam quickly leads to corrosion of heat-labile components. Certain plastics (for example, PVC, polystyrene) and other materials may not tolerate the high temperatures generally required for sterilization.

Plasticware Sterilization

Plastic containers that cannot be heated are sterilized commercially by ethylene oxide gas. These items are sold already sterile and cannot be re-sterilized. Examples of such items are plastic Petri dishes, plastic centrifuge tubes, etc.

Table 4.2 Various sterilization techniques used in tissue culture

Sterilization technique	Materials sterilized
Dry heat (160–180°C for 3 hours)	Empty glassware and certain plasticware; instruments like scalpels, forceps, needles, etc. (glass bead sterilizer used).
Flame sterilization	Instruments like scalpels, forceps, etc., mouth of culture vessels.
Autoclaving (121°C at 15 psi for 15–40 minutes)	Medium, culture vessels and contaminated cultures.
Filter sterilization	
Liquid (membrane filter of 0.45 μm or smaller pore size)	Heat labile compounds like GA_3, ABA, zeatin, urea, etc.
Air (HEPA filter)	Air blown through laminar air flow cabinets.
Wiping with 70% ethanol	Platform of laminar air flow cabinets, hands of the researchers', etc.
Surface sterilization (using one of several sterilizing agents)	All plant materials to be cultured.

WORKING STRATEGY IN THE LAMINAR AIRFLOW CABINET/HOOD

SURFACE STERILIZATION OF PLANT MATERIAL

1. Preparation of Stock Plants

Treatment of stock plants with fungicides and/or bactericides is helpful in reducing contamination in the explant. It is sometimes possible to harvest shoots and buds from them in clean conditions. The shoots may then be free of contaminants when surface sterilized in a normal manner.

Seeds may be sterilized and germinated *in vitro* to provide sterile material. Covering of growing shoots for several days or weeks prior to harvesting the tissue for culture, may supply cleaner material. Plant materials, from which explants will be procured, can be washed in detergent water and then placed under running water for 10 to 20 minutes.

2. Major Explant Sterilants

Ethanol (C_2H_5OH) Ethanol is a powerful sterilizing agent but it is extremely phytotoxic. Therefore, plant material is typically exposed to it only for a few seconds to a minute. The more tender the tissue, the more it will be damaged by alcohol. Tissues such as dormant buds, seeds or unopened flower buds can be treated for a longer time. Generally 70% ethanol is used prior to treatment of tissues with other compounds.

Sodium hypochlorite (NaOCl) Sodium hypochlorite, usually purchased as laundry bleach, is the most frequent choice for surface sterilization. It is readily available and can be diluted to proper concentrations. Commercial laundry bleach is 5.25% sodium hypochlorite. It is usually diluted to 10–20% of the original concentration, resulting in a final concentration of 0.5–1.0% sodium hypochlorite.

Plant material is usually immersed in this solution for 10–20 minutes. A balance between concentration and duration of treatment must be determined empirically for each type of explant, because of phytotoxic nature of sterilizing agents.

Calcium hypochlorite (CaOCl) Calcium hypochlorite is used as an alternative to sodium hypochlorite. It is obtained as a powder that can dissolve in water. The solution must be filtered prior to use. Calcium hypochlorite may be less injurious to plant tissues than sodium hypochlorite. It is commonly known as bleaching powder, with maximum available chlorine content of 37– 44%. This may be used as an aqueous solution. Freshly prepared chlorine solution is equally good. They are used in a range of concentrations varying from 5.0 to 15% depending on the virulence of contamination and hardness of explant. The concentration that is generally used is 3.25%. The explant is immersed in the solution to which a few drops of liquid detergent is added, to enhance penetration of the sterilizing agent. This is carried out for 20 minutes with intermittent shaking. A high concentration is necessary in the case of heavy infection. It will be more effective if a weaker solution is used for a longer period than vice versa. Subsequently, the explants are washed with double distilled water for 4–6 times.

Hydrogen peroxide (H_2O_2) Explants can also be surface-sterilized using H_2O_2 (3.0–10%). It is much easier to remove from the tissue than NaOCl and CaOCl. It is ten times stronger than that obtained in a pharmacy. Some researchers have found that hydrogen peroxide is useful for surface sterilization of materials that grow with source plant in the field.

Mercuric Chloride ($HgCl_2$) Mercuric chloride is used only as a last resort. It is extremely toxic to both plants and human beings, and must be disposed off with care. Since mercury is phytotoxic, it is critical to perform many rinses to remove all traces of the sterilant from the plant material. A solution with 0.1 to 1.0% concentration for 2–8 minutes (depending upon the hardness of tissue and nature of contamination) is used for sterilizations.

Bromine water Bromine water is used for surface sterilization. Generally 1.0–2.0% solution of bromine water for 5–20 minutes is preferred.

Antibiotics These are nontoxic substances. The incorporation in medium may prove effective for internal infections. It is not practical on a commercial scale. Some useful antibiotics are terramycine, oxytetracycline, rifampicin, gentamicin, etc.

Though many sterilants are available, ethanol and sodium hypochlorite are the most commonly used because of their low toxicity and treatment does not show any growth inhibition.

Effectiveness of sterilization procedure can be enhanced by the following methods:

i. Surfactant (for example, Tween 20 or 80) is frequently added to sodium hypochlorite.

ii. A mild vacuum may be used during the procedure.

iii. The solutions that contain explants are often shaken or continuously stirred.

3. Rinsing

After the plant material is sterilized with one of the above compounds, it must be rinsed thoroughly with sterile water. Essentially, three to four separate rinses with double sterilized distilled water are required at the end of final sterilization treatment.

4. Plant Preservative Mixture

Plant preservative mixture (PPM) is a proprietary broad spectrum biocide, which can be used to control contamination in plant cell cultures, either during the sterilization procedure or as a medium component.

PPM has several advantages over antibiotics. They are:

i. It is effective against fungi as well as bacteria. So it can be substituted for a mixture of antibiotics and fungicides.

ii. PPM is less expensive than antibiotics, which makes it affordable for wide and routine use.

iii. The formation of resistant mutants towards PPM is very unlikely because it targets and inhibits multiple enzymes.

iv. PPM does not adversely affect *in vitro* seed germination, callus proliferation or callus regeneration.

v. Seeds and explants with endogenous contamination can be sterilized at doses of 5–20 ml/l of PPM. This is useful when routine surface sterilization is insufficient.

STERILIZATION OF EXPLANTS

Sterilization of explants is directly related to the way the source has been grown. Diseased or pest-infested plants are likely to be contaminated both externally and internally. It may prove impossible to obtain microbe-free material without the continued/prolonged use of antibiotics. However, seed is a favoured choice since it can generally survive even on stronger sterilization treatments than softer tissues, such as leaves or stems. However, weather conditions before and during seed harvest can also influence to the subsequent success of sterilization.

Explants that are endogenous in nature are supposed to be naturally sterile and therefore, in such a case, the usual procedure of sterilization is to apply an exogenous chemical treatment for a specified period. Different sterilants are available for use at a range of different concentrations (Table 4.3), and the choice of protocol is usually determined by the nature of the explant and the extent of external contamination. A wetting agent is usually included (a drop of household detergent or Tween 80) to improve contact with the sterilant, since the ease of wetting can be critical. Very heavy or heavily waxed tissues are relatively difficult to sterilize successfully.

Table 4.3 Most commonly used sterilants for sterilizing plant material

Disinfectant/Sterilizing agents	Concentration (%)	Exposure duration (minutes)
Calcium hypochlorite	9.0–10	5–20
Sodium hypochlorite *	0.5–1.0	5–20
Hydrogen peroxide	3.0–10	5–15
Mercuric chloride	0.1–1.0	2–8
Ethyl alcohol	70–95	0.1–1
Bromine water	1.0–2.0	5–20

*Commercial bleach contains about 5.0% sodium hypochlorite, and thus should be used at a lower concentration of 0.5–1.0% sodium hypochlorite.

Steps in Explant Sterilization

1. Thoroughly wash the explants in tap water and immerse in 5.0% solution of liquid detergent such as Tween 80 or Teepol for 10 to 15 minutes.
2. Dip the explants in 1.0% solution of Bavistin for 5–10 minutes, and then wash the explants thoroughly in tap water followed by distilled water.

   These steps can be done in the general laboratory without the use of LAF cabinet; however, the subsequent steps of sterilization should be carried out in front of an LAF cabinet.
3. Treat the explants with various sterilants based on the nature of explant, contamination nature, availability of sterilant, etc.
4. Swirl the bottle frequently so that all surfaces of plant material come equally in contact with the sterilant.
5. Decant the sterilant and wash the explants thoroughly with several changes of sterile distilled water to remove all traces of sterilant.

Treatment of Leaching (Phenolic Explants)

Many explant tissues, especially woody explants produce phenolic compounds in the culture medium. This process is called phenolic oxidation. It inhibits the growth of the tissue in culture. For example *Phyllanthus* and *Eucalyptus* explants can discolour the medium within two hours. This is inhibitory to the growth of the cultures and in order to overcome this, certain pretreatment procedures are adopted. Washing in running water for one to two hours before sterilization in antioxidant solution such as L-ascorbic acid (50 mg/l) and citric acid (100 mg/l) were found to be good for pretreatment to check or minimize the browning problem.

If excessive browning problems exist, addition of antioxidants (ascorbic acid 50 mg/l, adenine sulphate, citric acid and arginine 25 mg/l each) in culture medium reduces it. Further addition of 50 mg/l activated charcoal (Grade G 40) reduces browning to a significant extent. Again addition of 1.0% sucrose in liquid medium, culture incubation in dark and maintenance of low temperature in the transfer chamber have also been found to be favourable in reducing browning of woody explants.

PRECAUTIONS DURING STERILIZATION

The following precautions and recommendations should be followed to avoid some of the common pitfalls:

1. Prepare the culture medium several days before it is required; if the autoclave has been faulty, contamination will become evident before inoculation.
2. When preparing solid media, allow the temperature to fall to at least 50°C before pouring, in order to prevent excessive condensation forming in the dishes and jars.

3. When initiating new cultures, always begin with healthy plant material.
4. To maintain maximum sterility in a culture laboratory, always autoclave contaminated cultures before discarding.
5. Never try to "rescue" cultures infected with a fungus that is already sporulating.
6. The waste bin is a frequent source of contamination in a laboratory, clean it thoroughly, and sterilize it with ethanol regularly.
7. When autoclaving media in bottles with screw caps, always loosen them, since this will avoid both the risk of the bottles exploding and also the creation of a vacuum after cooling. The vacuum created can be so strong that it becomes impossible for the cap to be removed.
8. When adding filter-sterilized supplements to autoclaved medium, always allow the latter to cool to approximately 10°C above gelling temperature. After mixing, pour the medium immediately into vials to ensure that it cools as rapidly as possible.

REVIEW QUESTIONS

1. What is sterilization? Explain the various methods of sterilization.
2. Write notes on sterilization of the following:
   i. Laminar air flow cabinet
   ii. Glassware
   iii. Instruments
3. What are the methods of sterilization of culture medium?
4. Write an essay on sterilization of explants.
5. What are the precautions that must be taken to successfully avoid contamination in tissue culture practices?

REFERENCES

Ball, E. (1946). Development in sterile culture of stem tips and subjacent region of *Tropaeolum majus* L. and of *Lupinus albus L. Amer. J. Bot.* 33: 301–318.

Ballester, A., Sanchez, M.C., San-Jose, M.C., Vieitez, F.J. and Vieitez, A.M. (1990). Development of rejuvenation methods for *in vitro* establishment, multiplication, and rooting of mature trees. In: Rodriguez *et al.* (eds.). *Plant Aging: Basic and Applied Approaches.* Plenum Press, New York. pp. 43–49.

Cassells, A.C. (1998). *In vitro* production of pathogen-free and contamination-free plants. In: Altman, A. (ed.). *Agricultural Biotechnology.* Marcel Dekker, Inc., New York. pp. 43–56.

Deora, N.S. and Shekhawat, N.S. (1995). Micropropagation of *Capparis decidua* (Forsk.) Edgew - a tree of arid horticulture. *Plant Cell Reports.* 15: 278–281.

First, N.L., Schell, J. and Vasil, I.K. (1998). Prospects and limitations of Agricultural Biotechnologies: An update. In: Altman, A. (ed.). *Agricultural Biotechnology.* Marcel Dekker, Inc., New York.. pp. 743–748.

Krikorian, A.D. (1982). Cloning higher plants from aseptically cultured tissues and cells. *Biol. Review.* 57: 151–218.

Krikorian, A.D. (1988). Plant Tissue Culture: perceptions and realities. *Proc. Indian Acad. Sci. (Plant Science).* 98: 425–464.

Krikorian, A.D. (1996). Strategies for "minimal growth maintenance" of cell cultures: A perspective on management for extended duration experimentation in the microgravity environment of a space station. *The Botanical Review.* 62: 42–108.

Pierik, R. L. M. (1988). *In vitro* culture of higher plants as a tool in the propagation of horticultural crops. *Acta Horticulturae.* 226: 25–40.

Shekhawat, M.S. (2007). New methods for sterilization of explants and hardening of Cucurbitaceous plants. *Int. J. of Plant Sci.* 2(2): 228–230.

5

PLANT CELL *IN VITRO* NUTRITION: CULTURE MEDIUM

INTRODUCTION

Plant cell or tissue *in vitro* culture refers to the technique of growing plant cells, tissues, organs, seeds or other parts in a sterile environment on a nutrient medium. Explants grow on a suitable, artificially prepared nutrient solution, which is known as culture medium.

Plant scientists have proposed various compositions of nutrient medium for the growth of plant tissues. The first attempt to initiate culture of plant cells was made by the German botanist G. Haberlandt at the turn of the 20th century. It is only during the last three decades that rapid developments in plant cell, tissue and organ culture have taken place. The first nutrient medium formulations used for plant culture work were inevitably based on the experiments with microbial cultures. The most commonly used culture media are based on the established formulations defined by Gamborg, Heller, Linsmaier and Skoog, Murashige and Skoog, Schenk and Hildebrandt and White.

Culture media used for *in vitro* cultivation of plant cells are composed of the following basic components.

1. Complex mixture of salts: Essential elements, or mineral ions
2. Organic supplement: Vitamins and/or amino acids
3. Gelling agents
4. Plant growth regulators
5. Adjuvants

As work on plant cell cultures progressed, efforts were made to define media that are more suited to the growth of plant cell cultures. The addition of undefined sources of nutrients and hormones, such as coconut milk, are avoided nowadays. Several media were formulated by varying the basic components. The choice of nitrogen source varies depending upon the kind of tissues or species being cultured and may be in the form of ammonium or nitrate salt, amino acids, casein hydrolysate, or urea. Since few plant cell cultures are autotrophic in nature, the provision of an exogenous energy source is of key importance. The most common carbon source used in culture media is sucrose, although glucose is also sometimes used as an alternative.

A high rate of growth is not always of paramount importance for cultures. Often the aim of initiating a cell culture is to obtain secondary products that are generally accumulated during the stationary phase of growth or in differentiated cultures only. Therefore, high growth rates may not necessarily be desirable in all cases. Growth regulators or plant hormones such as auxins, cytokinins, gibberellins, abscisic acid and ethylene are known to influence various stages of growth in the whole plant. Of these, only auxins and cytokinins are routinely incorporated into the culture medium.

SELECTION OF CULTURE MEDIUM

The number of plant species that have been cultured and the different media used are extensive. Once the plant species to be cultured is decided upon, a suitable nutrient medium formulation must be selected.

The first step would be to follow up appropriate references, subsequent to a literature search and/or consult a handbook dealing with plant cell culture. If a literature search is inconclusive or if access to specialized sources is limited, then it is recommended to attempt initiating cultures on several of the well-known medium formulations. In addition, it is advisable to experiment with different auxin and cytokinin combinations. A single nutrient medium is not enough to maintain optimum growth for all plant tissues. Consequently, the most suitable culture medium for a particular tissue must be determined by trial and error. Proposed composition of a culture medium has often been modified to stimulate the growth of plant material. Of great importance to the growth and productivity of a cell culture are the carbon/nitrogen ratio and the form in which these media components are supplied. It is possible to investigate the effects of the carbon/nitrogen ratio by altering the concentration and/or the type of these important medium constituents.

CHEMICAL COMPOSITION OF CULTURE MEDIUM

On the basis of composition of chemical constituents, culture media are generally divided into two major categories:

i. *Chemically defined medium* In this category, the exact composition and the concentration of all constituents are known and mainly prepared from inorganic and organic chemicals.

ii. *Chemically undefined medium* In this medium, the exact composition and the concentration of all constituents are not known due to the addition of natural products, for example, coconut milk.

A chemically defined nutrient medium generally contains inorganic salts, vitamins, growth regulators, a carbon source and/or gelling agent. Other components added for specific purposes include organic nitrogen compounds, hexitols, amino acids, antibiotics and plant extracts. The decision of using a suitable type of nutrient media for the metabolic needs of the cultured cells and tissue is a major factor of success in the plant regeneration process.

The inorganic salts needed for plant culture belong to two groups:

1. Those inorganic salts needed in higher amounts, called the macro salts (macronutrients/macroelements).
2. The other essential inorganic salts needed in little or trace amounts, called the micro salts (micronutrients/microelements).

Macroelements

As the name indicates, these elements are required in large amounts for plant growth and development. Nitrogen, phosphorus, potassium, magnesium, calcium and sulphur are usually regarded as macroelements or major salts (Table 5.1). These are inorganic nutrients. They are added to the culture media in the form of one or more organic salts. Nitrogen is mostly provided either in the form of nitrates (in the form of KNO_3) or ammonium compounds (in the form of NH_4NO_3). There is an advantage in supplying nitrogen in the form of ammonium ions, as nitrogen must be incorporated in the reduced form into macromolecules. Nitrate ions, therefore, need to be reduced before incorporation. However, at high concentrations, ammonium ions can be toxic to plant cell cultures and uptake of ammonium ions from the medium causes acidification. These ions can also cause problems by increasing the frequency of vitrification (the culture appears "glassy" and is usually not suitable for culture). In order to use ammonium ions as the sole nitrogen source, the medium needs to be buffered. Use of nitrate and ammonium ions as a mixture has the advantage of slightly buffering the medium. This method causes the uptake of nitrate ions whereby OH^- ions are excreted. Phosphorus is usually supplied as phosphate ion of ammonium, sodium or potassium salts. High concentrations of phosphate may lead to the precipitation of medium elements as insoluble phosphates.

Microelements

These elements are required in trace amounts as compared to macroelements for plant growth and development. Manganese (Mn), iodine (I), copper (Cu), cobalt (Co), boron (B), molybdenum (Mo), iron (Fe) and zinc (Zn) usually comprise the microelements frequently used in tissue culture media. They are supplied in the form of inorganic salts. Their concentration is expressed in µmol per litre.

Table 5.1 Important elements for plant nutrition and their physiological function

Element	Functions
Nitrogen	Component of proteins, nucleic acids and some coenzymes, required in large amounts
Potassium	Regulates osmotic potential, principal inorganic cation
Calcium	Cell wall synthesis, membrane function, cell signaling
Magnesium	Enzyme cofactor, component of chlorophyll
Phosphorus	Component of nucleic acids, energy transfer, component of intermediates in respiration and photosynthesis
Sulphur	Component of some amino acids (methionine, cysteine, etc.) and some cofactors
Chlorine	Required for photosynthesis
Iron	Electron transfer as a component of cytochromes
Manganese	Enzyme cofactor
Cobalt	Component of some vitamins
Copper	Enzyme cofactor, electron-transfer reactions
Zinc	Enzyme cofactor, chlorophyll biosynthesis
Molybdenum	Enzyme cofactor, component of nitrate reductase

Iron Source

Iron is usually added as iron sulphate, although iron citrate can also be used. In a majority of the nutrient media, iron is used in chelated form as ferric-sodium ethylene diamine tetra-acetate (Fe-EDTA). In this state, iron is gradually released into the culture medium, as it is utilized by the living cells. Uncomplexed iron can be precipitated out of the medium as ferric oxide.

Organic Supplements

Thiamine (vitamin B_1), nicotinic acid (B_3), pyridoxine (B_6), calcium pantothenate (B_5) and myo-inositol are considered to be essential for the culture of plant cells *in vitro*. Vitamins are usually added at the concentration of 0.1–10.0 mg/l. Simultaneously, amino acids are also included in the organic supplement. The most frequently used amino acid is glycine, but arginine, asparagine, aspartic acid, alanine, glutamic acid, glutamine and proline are also used. Amino acids are sources of reduced nitrogen and this uptake causes acidification of the medium. They are added at the concentration of 0.05–100 m/mol to the culture media. Casein hydrolysate, malt extract, potato extract, ground banana, coconut milk, yeast extract, orange juice and tomato juice can also be used as organic supplements. They may promote the growth and differentiation of plant cell and tissue *in vitro*.

Carbon Source

Sucrose is a cheap, easily available, readily assimilated and a relatively stable compound. Therefore, it is the most commonly used carbon source in culture media. Other carbohydrates like glucose, maltose, galactose and sorbitol can also be used and sometimes they may even prove superior to sucrose.

Solidifying or Gelling Agents

Gelling or solidifying agents are commonly used for preparing semi-solid or solid tissue culture media. In static liquid cultures, the tissue or cells become submerged and die due to lack of oxygen. Gels provide a support to tissues growing in static conditions. Agar, a polysaccharide obtained from seaweed, has several advantages over other gelling agents. Agar gels do not react with media constituents and significantly they are also not digested by plant enzymes and remain stable at all feasible incubation temperatures.

Normally, 1–8% agar is used in the nutrient medium to form a firm gel at the pH typical of plant cell culture media. In quantitative analysis, the use of agar should be avoided because commercially available agar contains impurities in the form of Ca, Mg, K, Na and trace elements. Consequently, a change in the agar concentration also affects the nutrients present in it as well as the overall nutrient concentration in the experiment. Impurities can be removed, however, for such critical experiments by washing agar in double-distilled water for at least 24 hours followed by rinsing in ethanol and drying at 60°C for 24 hours.

Gelatin at higher concentrations (10%) has been tried as a gelling agent but has limited use because it melts at low temperature (25°C). Other compounds that have been successfully tested include, biogel (polyacrylamide pellets) alginate, phytagel and gelrite.

A nutrient media may be solid or semi-solid. When 8% agar agar is dissolved in the liquid nutrient medium, it becomes a solid medium. A partially solidified medium is known a semi-solid medium, where reduced amount of agar is added. A medium without agar remains in liquid form and is known as liquid medium. Generally, solid and semi-solid media are prepared for callus culture and liquid medium is used for cell suspension cultures. A comparison of the chemical composition of various nutrient media is shown in Table 5.2.

Table 5.2 Comparative chemical composition of various nutrient media (all values in mg/l)

Components	Murashige–Skoog (1962)	White (1963)	Gamborg (1968)	Nitsch (1951)	Heller (1953)	Schenk–Hildebrandt (1972)	Nitsch–Nitsch (1967)	Kohlenbach–Schmidt (1975)	Knop (1865)
$(NH_4)_2SO_4$	–	–	134	–	–	–	–	–	–
$MgSO_4 \cdot 7H_2O$	370	720	500	250	250	400	125	185	250
Na_2SO_4	–	200	–	–	–	–	–	–	–
KCl	–	65	–	1500	750	–	–	–	–
$CaCl_2 \cdot 2H_2O$	440	–	150	25	75	200	–	166	–
$NaNO_3$	–	–	–	–	600	–	–	–	–
KNO_3	1900	80	3000	2000	–	2500	125	950	250
$Ca(NO_3)_2 \cdot 4H_2O$	–	300	–	–	–	–	500	–	1000
NH_4NO_3	1650	–	–	–	–	–	–	720	–
$NaH_2PO_4 \cdot H_2O$	–	16.5	150	250	125	–	–	–	–
$NH_4H_2PO_4$	–	–	–	–	–	300	–	–	–
KH_2PO_4	170	–	–	–	–	–	125	68	250
$FeSO_4 \cdot 7H_2$	27.8	–	27.8	–	–	15	27.85	27.85	–
Na_2EDTA	37.3	–	37.3	–	–	20	37.25	37.25	–
$MnSO_4 \cdot 4H_2O$	22.3	7	10 $(1H_2O)$	3	0.1	10	25	25	–
$ZnSO_4 \cdot 7H_2O$	8.6	3	2	0.5	1	0.1	10	10	–
$CuSO_4 \cdot 5H_2O$	0.025	–	0.025	0.025	0.03	0.2	0.025	0.025	–
H_2SO_4	–	–	–	0.5	–	–	–	–	–
$Fe_2(SO_4)_3$	–	2.5	–	–	–	–	–	–	–
$NiCl_2 \cdot 6H_2O$	–	–	–	–	0.03	–	–	–	–
$CoCl_2 \cdot 6H_2O$	0.025	–	0.025	–	–	0.1	0.025	–	–
$AlCl_3$	–	–	–	–	0.03	–	–	–	–
$FeCl_3 \cdot 6H_2O$	–	–	–	–	1	–	–	–	–
$FeC_6O_5H_7 \cdot 5H_2O$	–	–	–	10	–	–	–	–	–
KI	0.83	0.75	0.75	0.5	0.01	1.0	–	–	–
H_3BO_3	6.2	1.5	3	0.5	1	5	10	10	–
$Na_2M_0O_4 \cdot 2H_2O$	0.25	–	0.25	0.25	–	0.1	0.25	0.25	–
Sucrose	30000	20000	20000	50000/ 36000	20000	30000	20000–30000	10000	–
Glucose	–	–	–		–	–	–	–	–
Myo-Inositol	100	–	100	–	–	1000	100	100	–
Nicotinic Acid	0.5	0.5	1.0	–	–	0.5	5	5	–
Pyridoxine HCl	0.5	0.1	1.0	–	–	0.5	0.5	0.5	–
Thiamine HCl	0.1-1	0.1	10	1	1	5	0.5	0.5	–
Ca-pantothenate	–	1	–	–	–	–	–	–	–

(Contd.)

Table 5.2 (Continued)

Components	Murashige–Skoog (1962)	White (1963)	Gamborg (1968)	Nitsch (1951)	Heller (1953)	Schenk–Hildebrandt (1972)	Nitsch–Nitsch (1967)	Kohlenbach–Schmidt (1975)	Knop (1865)
Biotin	–	–	–	–	–	–	0.05	0.05	–
Glycine	2	3	–	–	–	–	2	2	–
Cysteine HCl	–	1	–	10	–	–	–	–	–
Folic acid	–	–	–	–	–	–	0.5	0.5	–
Glutamine	–	–	–	–	–	–	–	14.7	–

Plant Growth Regulators

Plant growth regulators are naturally occurring (or synthetic) substances which control growth and development of cells and tissues at extremely low concentrations (Figure 5.1). They are used to induce cell division, growth, proliferation, differentiation and organogenesis in plant tissue cultures. There are three classes of growth regulators frequently used in *in vitro* plant biotechnology. They are auxins, cytokinins and gibberellins. Auxins and cytokinins are the most widely used and they are usually used together. The ratio of auxin to cytokinin determines the type of culture to be established or regenerated. A high auxin-to-cytokinin ratio generally favours root formation, whereas a high cytokinin to auxin ratio favours shoot formation. However, an intermediate ratio of auxin/cytokinin stimulates callus induction.

Auxins Auxins are the growth substances that typically stimulate cell elongation but they also cause a wide range of *in vitro* growth responses. These are generally dissolved in NaOH or ethyl alcohol and added either before or after the sterilization of media. The commonly used auxins in culture experiments are given in the Table 5.3.

Cytokinins Cytokinins are N_6-substituted adenines which induce cytokinesis, that is, cell division in plant tissue grown in a medium containing an optimum concentration of auxin. They are usually dissolved in dilute HCl or NaOH. The commonly used cytokinins in nutrient culture media are given in Table 5.4.

Gibberellins Gibberellins are a large family of diterpene acids. The systematic nomenclature is based on the gibbane ring system, which is common to all known gibberellins. They are involved in regulating cell elongation and important in determining plant height and fruit-set. Only a few of the gibberellins are used in plant cell culture media, of which GA_3 is the most common one. GA_3 should be filter-sterilized before adding it to the culture nutrient media.

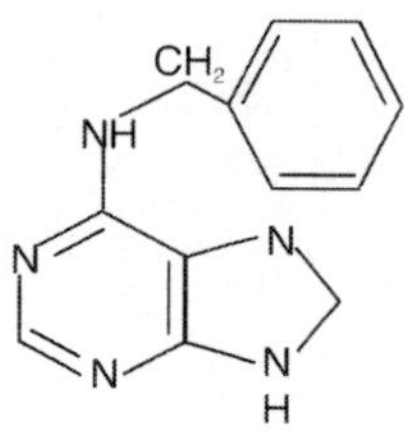

(a) BAP

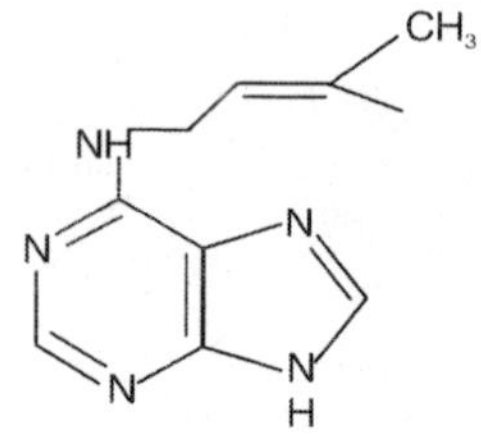

(b) 2-ip

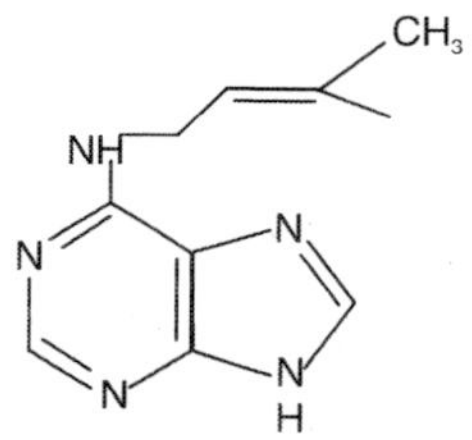

(c) *Trans*-zeatin

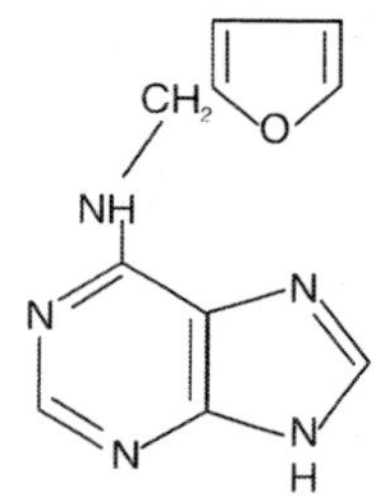

(d) Kinetin

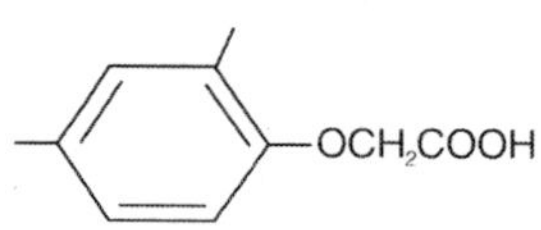

(e) 2,4-D

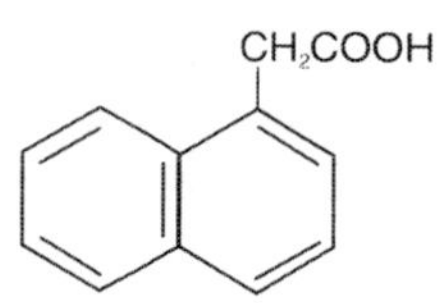

(f) NAA

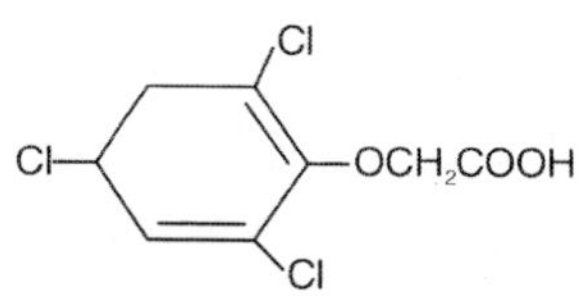

(g) 2,4,5-T

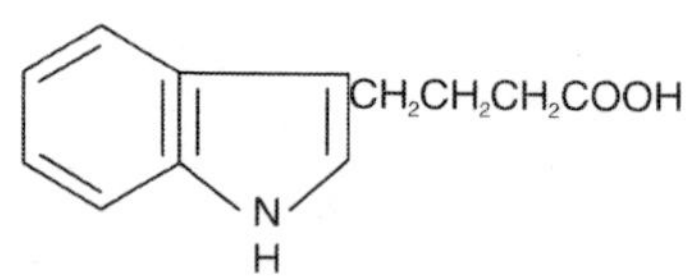

(h) IBA

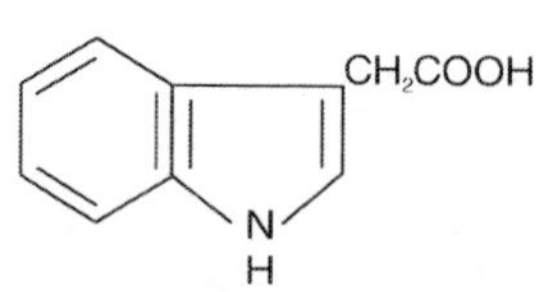

(i) IAA

Figure 5.1 Chemical structures of plant growth regulators

Table 5.3 Commonly used auxins

Abbreviation/name	Chemical name
2,4-D	2,4-dichlorophenoxyacetic acid
2,4,5-T	2,4,5-trichlorophenoxyacetic acid
Dicamba	2-methoxy-3,6-dichlorobenzoic acid
IAA	Indole-3-acetic acid
IBA	Indole-3-butyric acid
MCPA	2-methyl-4-chlorophenoxyacetic acid
NAA	1-naphthaleneacetic acid
NOA	2-naphthyloxyacetic acid
Picloram	4-amino-2,5,6-trichloropicolinic acid

Table 5.4 Commonly used cytokinins

Abbreviation/name	Chemical name
BAP	6-benzylaminopurine
2iP	(IPA) [N_6-(2-isopentyl) adenine]
Kinetin	6-furfurylaminopurine
Thidiazuron	1-phenyl-3-(1,2,3-thiadiazol-5-yl)urea
Zeatin	4-hydroxy-3-methyl-*trans*-2-butenylaminopurine

MEDIUM PREPARATION TECHNIQUES

Nutrient Medium (Stock Solutions)

Initial nutrient formulations required for plant *in vitro* culture were commercially available in the form of mineral salt mixtures and were prepared as stock solutions ranging from 10 to 100 times of the final concentrations used in the medium. Stock solutions can be prepared as two solutions—one containing all of the macronutrients and the other containing all of the micronutrients. These solutions must be kept fairly diluted (10–20×) in order to avoid precipitation of calcium and magnesium phosphates and sulphates.

A more common method is to arrange for mineral salt stock solutions according to the ions they contain. A series of solutions containing the inorganic components of the medium is prepared (precise combinations may vary from lab to lab). Using this method, salt stocks can be prepared in 100× concentrations. Iron–EDTA chelate is prepared by mixing the proper amounts of the two compounds, iron sulphate and Na–EDTA, with water and then autoclaving. This stock of iron–EDTA must then be stored in a dark container. These prepared stock solutions are usually stored in the refrigerator.

Although the initial cost of chemicals may be substantial, the ongoing cost of using stock solutions is probably less than that of using prepared mixes. However, the stocks must be prepared over and over when they are used up. Stock solutions may be useful if many different media are made in the laboratory or frequent changes in concentration are made in individual components.

Prepared Mixes (Powdered Medium)

Several companies sell prepared salt and vitamin mixtures as powders. These are easily handled by adding the proper amount of powder to water. The mixtures can be purchased as complete media or as salts alone. The packs often contain the necessary ingredients for one litre of medium. The contents are extremely hygroscopic. Once a pack is opened, it is best to use the contents all at once. A small amount of precipitate may be seen in media prepared using either of these methods. This results from iron being displaced from the chelate over time as media is stored. However, it does not appear to be deleterious.

Weighing The preparation of nutrient media usually requires careful weighing of at least some components, even if a commercially prepared medium is used. Sucrose, myo-inositol and the gelling agent, which are used almost frequently, must be weighed properly.

Different types of balances are available in different laboratories; however, their details of operation vary. Moreover, unless very specific types of medium and other reagents are required to be made, it is necessary to have access to a balance that can weigh in the milligram range accurately. The balance should be located on a hard, stable, level surface that is free of vibrations and excessive air drafts. A piece of weigh paper or a plastic weigh boat, depending on the amount to be weighed, is placed on the balance pan and the balance is tarred (zeroed). Then the required amount is carefully weighed using an appropriate instrument. Excess material that has to be removed from the balance should be discarded rather than returned to the chemical container. The balance and weighing area around it must be kept clean. This involves cleaning the balance after each use. Often a brush is used for this. Balances are accurate for specific weight ranges and should never be overloaded.

Measuring of liquids Calibrated glassware (for example, graduated cylinders and pipettes) are required for the preparation of culture media. Graduated cylinders of 10 ml, 25 ml, 100 ml, 500 ml and 1000 ml capacities are used for many measuring operations. Pipettes are used for smaller volumes. Volumetric flasks are less commonly used.

When one needs to measure solutions with pipettes or graduated cylinders, the bottom of the curved air–liquid interface should be aligned with the measuring mark. Pipettes should be filled with a hand-operated pipettor, never by mouth.

Water The quality of water used in stock preparation can show a significant difference in the final medium. Type I water is the most desirable. Type I water is produced by distillation, ion exchange, reverse osmosis, or a combination of these methods. The water can be tested for purity.

pH pH is the negative logarithm of hydrogen ions. The pH of a solution is a measure of the concentration of hydrogen ions in the solution. The pH scale extends from very acidic (0) to very alkaline (14) with 7 being the "neutral" point. The pH of most culture media is adjusted to 5.8 ± 0.1 before autoclaving. The pH can influence the following:

- The availability of ions in nutrient media
- The ability of agar to gel
- The subsequent growth of cells

For example, if the pH is lower than 4.5, agar medium will not gel. Therefore, accurate determination and control of the pH of culture nutrient media are necessary. Generally, pH is determined with a pH meter, which may be digital or analog. The pH meter is calibrated before use by adjusting it using standards of known pH. Although pH drops a little during autoclaving, the pH of the medium is not typically readjusted after autoclaving.

Pouring and Storing of Media

Once media is made and pH adjusted, it may be heated to dissolve the gelling agent and dispensed into culture vessels such as tubes, vials, baby food jars or magenta boxes. Then these vessels are autoclaved, if no filter-sterilized components need to be added. Alternatively, the media may be placed in a flask, which is covered with aluminium foil and autoclaved. The autoclaved medium should be stirred or shaken frequently while cooling, prior to addition of filter-sterilized reagents.

Filter-sterilized reagents, if any, are then added and mixed well. The media is then poured into vessels (for example, test tubes, Petri dishes, etc.). The vessels in either case are cooled inside the transfer hood; lack of movement at this point promotes even gelling. All media should be labelled before it is removed from the hood. It should then be sealed up to prevent contamination. It should be stored in the refrigerator or at least in the dark (light causes some reagents to break down). Under good conditions, most media can be stored for at least a month.

Organic Compounds

Most of the organic compounds are added at relatively very low concentrations. Stock solutions ranging from 100 to 1000× concentration of final concentration are generally prepared. Vitamins may be prepared as stocks or purchased as pre-made. Vitamin solutions can be divided into aliquots and stored in the freezer at –20°C. Most growth regulators are stable up to a month in stock solutions. The solvents used for dissolving growth regulators vary depending on the compound under consideration.

Concentration of Media Components

There are three ways to describe concentration of chemicals in a stock:

1. Parts per million (rarely used)
2. Grams or milligrams per litre
3. Molar units (journals usually require concentrations to be expressed this way).

Units used in solutions

1. Mg/l = milligrams of substance in one litre (1000 ml) of solution.
2. ppm = parts per million = 1 unit of substance in one million units of solution
   1 mg/l = 1 ppm
3. Mole = gram molecular weight
   It is the formula weight of the substance in grams
   FW (formula weight) = sum of weights of atoms in the formula
   For example, for H_2O, $1 \times 2 + 16 = 18.1$ mole
   $H_2O = 18$ g
   For preparing molar solutions, it is necessary to know the molecular weight to calculate the amount of a chemical to be weighed, to make a stock solution:
   FW/desired molar concentration × the number of litres desired = amount to be dissolved.

Volume metrics

1 molar (1 M or 1 mol/l) = 1 molecular weight in gram of substance in 1 litre of solution.

1 M is equal to 10^{-3} mMol/l or 10^{-6} µmol/l or 10^{-9} nmol/l.

1 millimolar (1 mM) = 0.001 molecular weight of substance in one litre solution (molecular weight in mg/litre).

1 micromolar (1 µM) = 0.00001 molecular weight of substance in one litre solution (10^{-3} mg molecular weight/litre).

When working in molar units, it is often convenient to prepare a stock as, for example, 10 ml of a 10 µM solution. This is easily remembered as 0.1X the formula (molecular) weight (in mg) per 10 ml solvent. So, for example, to prepare 10 ml of a 10 µM solution of IAA, weigh 17.5 mg of IAA and add it to 10 ml of solvent.

Calculating dilutions of stock solutions:

M_1 = molarity of stock solution

M_2 = molarity of final solution desired

V_1 = volume of stock solution needed

V_2 = volume of final solution desired

Usually, all values are known but V_1 is to be calculated by using the simple formula:

$$M_1 \times V_1 = M_2 \times V_2$$

It can be rearranged as

$$V_1 = (M_2 \times V_2)/M_1$$

For example, to make a 10 mM solution out of 1N stock of NaCl (in other words a 0.01 M solution), 100 ml of NaCl is needed.

So using the formula:

V_1 = (0.01 M × 100 mL)/1M = 1 ml, we understand that we need 1 ml of the stock, mixed with 99 ml H_2O to have 100 ml of solution.

To prepare stock solutions of macroelements and microelements, the required amount of the compound is weighed and put in a clean flask. It is dissolved in a small amount of the proper solvent, usually water. Next, double-distilled water is slowly added to the flask and it is agitated until the proper volume is attained. The flask is labelled with the name of the solution, date of preparation and expiry date and the name of the person who prepared the solution.

It is a time-consuming job to weigh and mix all the constituents just before the preparation of medium. Again if less than one litre medium is to be prepared, it is very difficult to weigh some constituents that are used in very small quantities (vitamins, amino acids, hormones, etc). The number of stock solutions can be varying according to the time, requirement, space and nature of work. Again it depends on the choice of media and type of the experiment.

Stock solution of MS Medium

The most widely used culture medium was formulated by Murashige and Skoog in 1962 (MS medium), so the procedure for the preparation of stock solution of MS medium is given below.

A. *Macro salts stock solution* Macro salts should be made at twenty times of their final concentration in medium.

1. To make one litre of this stock solution, the salts are dissolved one after another (as given in Table 5.5) in 500 ml of double-distilled water and then made up to the final volume.
2. The solution is filtered. This can be stored in the refrigerator (10°C) for a long period.

It is necessary to add some components separately. For example, in this medium, calcium chloride must be thoroughly dissolved before any other salt is added to the nutrient solution. To prevent precipitation of other medium components, magnesium sulphate should be added last.

The solution should remain usable at this temperature for three months. However, at the first sign of microbial growth, it should be replaced with a new solution.

Table 5.5 Stock chemicals of macro salts (Twenty times)

Constituents	Amount (mg/l) present	Amount (g) to be taken for solution	Final volume of stock (ml)
NH_4NO_3	1650	33.8	—
KNO_3	1900	38.0	—
$CaCl_2 \cdot 2H_2O$	440	8.8	1000
KH_2PO_4	170	3.4	—
$MgSO_4 \cdot 7H_2O$	370	7.4	—

B. *Micro salts stock solution* Micro nutrients can be prepared at thousand times of their final strength (Table 5.6). While making the stock solutions, it is advisable to dissolve each constituent completely before adding another, otherwise, precipitation of salts may occur. One suitable way of prolonging the life of inorganic salt stock solution is to filter-sterilize before storage. This is particularly advisable if the solution is to be used intermittently.

Table 5.6 Stock chemicals of micro salts (Thousand times)

Constituents	Amount (mg/l) present in original medium	Amount (X100) to be taken for stock solution (X20) (value expressed in mg)	Final volume of stock (ml)
H_3BO_3	6.2	620	—
$Na_2MoO_4 \cdot 2H_2O$	0.25	25	—
$CoCl_2 \cdot 6H_2O$	0.025	2.5	—
$CuSO_4 \cdot 5H_2O$	0.025	2.5	100
$ZnSO_4 \cdot 7H_2O$	8.6	860	—
$MnSO_4 \cdot 4H_2O$	22.3	2230	—

C. *Vitamins stock solution* MS medium contains only three vitamins. They are dissolved at thousand times of their final strength one by one as given in Table 5.7.

Table 5.7 Stock of vitamins (Thousand times)

Constituents	Amount (mg/l) present in original medium	Amount (×50) to be taken for stock solution (in mg)	Final volume (ml)	Storage temperature (°C)	Duration of storage (in days)
Thiamine–HCl	0.1	5			
Nicotinic acid	0.5	25	50	0	15
Pyridoxine–HCl	0.5	25			

There is no need to autoclave the organic compound stock solution. It can be stored in 500 ml bottle in the refrigerator.

D. *Iron stock solution (two hundred times)*

1. Take 745 mg of Na_2 EDTA (37.25 mg/l in original). Dissolve in 75 ml boiling double-distilled water and then add gradually 557 mg of $FeSO_4 \cdot 7H_2O$ (27.85 mgl^{-1} in original).
2. Dissolve it by using a magnetic stirrer for at least 1 hour in hot condition until the colour of the solution changes to golden yellow.
3. Finally make the volume to 100 ml, filter and store the solution at 5°C for long periods. Keep this solution in amber coloured bottle.

E. *KI stock solution (thousand times)* Dissolve 83 mg of KI (0.83 mg/l present in original) in 100 ml of double-distilled water and store it at 10°C.

F. *Meso-inositol stock solution (five hundred times)* Dissolve 1 g of meso-inositol (100 mg/l in original) in 20 ml of double-distilled water and store it at 0°C.

G. *Glycine stock solution (thousand times)* Dissolve 40 mg of glycine in 20 ml of double-distilled water and store it at 0°C.

H. *Hormones stock solution* Stock solution of hormones is common to all media and can be used for any nutrient medium at various combinations and concentrations. Both auxins and cytokinins are not directly dissolved in water. All auxins are first dissolved in ethyl alcohol (absolute) and then water is added to make final volume. Similarly cytokinins are generally first dissolved in 1N NaOH solvent and then water is added to get the final volume (Table 5.8). The preparation of stock solution of hormones is listed as follows:

1. **BAP (1 mg/ml)** Add 500 μl KOH to 50 mg BAP (6-benzylaminopurine) to dissolve the BAP (prepare in 50 ml centrifuge tube). Bring final volume to 50 ml by adding distilled water and it can be stored in refrigerator.
2. **NAA (0.1 mg/ml)** Dissolve 5 mg of NAA (naphthaleneacetic acid) in 45 ml double-distilled water in a 50 ml centrifuge tube. Bring final volume to 50 ml. Store in refrigerator. Dissolve in NaOH, if necessary.
3. **Gibberellins (GA_3)** Stock solution of gibberellins (GA_3) can be prepared by dissolving in water and adjusting the pH to 5.7. They are not thermostable, therefore, should be filter-sterilized.

I. Gelling agent (agar agar, etc.) Agar agar is extracted from several species of red algae. It is a mixture of polysaccharides. As a gelling agent, agar is strong enough to support the culture, yet liquid enough to allow the nutrients to diffuse through the medium to the cultured tissues. Agar must be completely dissolved before it is dispensed into the culture vessels. Gel strength will vary with the medium formula being used and the source and grade of agar. Media with low pH (around 4.5) will tend to be softer than media with a higher pH (5.8). Hard agar may slow down the nutrient movement through the medium. Some highly refined polysaccharides like gelrite, phytagel, agargel, gellon, etc. are commercially available, which can be utilized as substitutes for agar. These can be used at 3.5 to 5 g/l. Although agar is supposed to be an inert material, it contains greater trace amounts of chloride, sulphate, calcium, magnesium and iron ions.

It has been noted that hormones present in the medium also have some effect on the setting of agar powder.

J. *Stock solution of antibiotics and other reagents*

1. **Kanamycin sulphate (K75)** To prepare kanamycin sulphate, 750 mg kanamycin is dissolved in 10 ml of double-distilled water (= 75 mg/ml). This solution is filter-sterilized and stored in small vials.
2. **Tetracycline (tet10)** To prepare tetracycline, 2.5 ml of 95% ethanol is added to 2.5 ml of double-distilled water. 50 mg tetracycline is then added. This solution is passed through a sterile filter into 1.5 ml microcentrifuge tubes (1.0 ml/amber) and stored.
3. **Gentamicin sulphate (G60)** To prepare gentamicin sulphate, 300 mg of gentamicin is dissolved in 5.0 ml of double-distilled water.
4. **1N HCl** To make 1N HCl, 42.5 ml of concentrated HCl is poured into 457.5 ml of deionized water.
5. **5N NaOH** To make 5N NaOH, 20 g of NaOH pellets is dissolved in 100 ml of deionized water. It is then stored in plastic bottles.

All stock solutions of salts should be stored in the refrigerator so that they remain stable for several months. Always stocks must be prepared with double-distilled water. Reagent-grade chemicals should be used to ensure maximum purity. The nitrate stock usually precipitates out and must be heated until crystals are completely dissolved before using. Any stock solution showing cloudy appearance because of bacterial contamination or has fungal growth should be discarded. Stock solutions of vitamins, amino acids and hormones should not be stored for indefinite period and should be kept in a deep freezer.

Protocol of Culture Medium Preparation

MS medium is the most commonly used medium. One litre of MS medium can be prepared as follows:

1. Dissolve 30 g of sucrose (cane sugar) in 200 ml of double-distilled water in a conical flask.
2. Take double-distilled water in another flask and add in sequence the appropriate amount of stock solution as given in Table 5.9.

Table 5.9 Stock solutions and their amount to be used for the preparation of one litre of MS medium

Stock solution	Amount to be used in ml
Stock solution of macro salts	50.0
Stock solution of micro salts	1.0
Stock solution of KI	1.0
Stock solution of Fe–EDTA	5.0
Stock solution of three vitamins	1.0
Stock solution of glycine	1.0
Stock solution of myo-inositol	2.0

3. Add the desired concentration of hormones from the stock solution.
4. Pour sugar solution and salt along with vitamins, amino acid and hormone solution into a one-litre measuring cylinder.
5. Make the final volume to one litre with double-distilled water. All the ingredients are shaken well to mix uniformly.
6. Adjust the pH of the liquid medium to 5.8 with the aid of 0.1N HCl or 0.1N NaOH.

Liquid medium can be used as such directly but if solid medium is required, 5–8% agar is added to the heated medium to dissolve it completely.

1. When preparing liquid media, dispense 100 ml aliquots into 250 ml of Erlenmeyer flasks for sterilization.
2. Insert non-absorbent cotton plug wrapped with muslin cloth to culture flask. Cover the plug with the help of an aluminium foil.

3. Label each flask with a marker pen, and state the medium type and date of preparation.
4. For solid medium (containing agar), there exist two alternatives for dispensing into culture vessels:
   i. One alternative is to dispense approximately 30 ml of liquid medium containing agar into 100 ml Erlenmeyer flasks.
   ii. On the other hand, the medium is first autoclaved in 500 ml lots in pyrex bottles and then dispensed into commercially purchased sterile culture pots inside the LAF cabinet, when the solution temperature comes down to around 50°C.
5. Autoclave the media, both solid and liquid, at 121°C and 15 psi pressure for 15–20 minutes. If volume of medium is one litre or over and is to be autoclaved in a single container, increase the sterilization time to 40 minutes.
6. If the medium requires addition of any thermolabile (heat-unstable) components (for example, zeatin), it is necessary to add them to the individual flasks containing autoclaved nutrient solution in an LAF cabinet. These additions should be made by passing the solution through a sterile 0.2 μm filter attached to a 5 ml syringe.
7. In the case of solid media, add the thermolabile chemicals to the medium after its temperature gets cooled down sufficiently and before it has begun to set (40–50°C). Additions can be made either before or after dispensing the medium into individual containers.

PRECAUTIONS

1. Care should be taken when supplying amino acids as the nitrogen source. In certain cell cultures, amino acids may be deaminated and used as a source of carbon.
2. Even when a published method for the culture initiation of a given plant species is adopted, it is still advisable to use and analyse other defined media formulations and hormone combinations. Occasionally, an alternative medium may emerge as the most suitable formulation for growth.
3. If for some reasons, agar is not poured into appropriate containers before it sets, then it is possible to melt the agar by placing the bottle containing the solid medium in a water bath at 80°C before pouring.
4. Sterilized liquid medium must be stored in a cupboard sheltered from light, dust and extremes of temperature. Certain medium salts, especially those of Murashige and Skoog's and related media tend to precipitate following extended periods of exposure to light.
5. Liquid medium should be used within a period not exceeding three months from the date of preparation.

6. For most media, autoclaving results in the release of sugar acids that cause a slight drop in the final pH of the medium.
7. To check for visible signs of microbial contamination, several days should be allowed between the preparation and use of solid as well as liquid media, and the media must be stored at room temperature.

REVIEW QUESTIONS

1. Define culture medium and state the basic chemical composition of a general plant tissue culture medium.
2. Write notes on the preparation and sterilization of culture media.
3. What are the precautionary measures that must be taken during media preparation?
4. How many types of plant growth regulators are generally essential in tissue culture practice?

REFERENCES

Caplin, S.M., and Steward, F.C. (1948). Effect of coconut milk on the growth of explants from carrot root. *Science*. 108: 655.

Griesback, R.J. (1984). Introduction to somatic cell genetics. *Hort. Science*. 19, 367–371.

Krikorian, A.D. (1996). Strategies for "minimal growth maintenance" of cell cultures: A perspective on management for extended duration experimentation in the microgravity environment of a space station. *The Botanical Review*. 62, 42–108.

Murashige, T. and Skoog, F. (1962). A revised medium for rapid growth and bioassay with tobacco tissue culture. *Physiol. Plant*. 15: 473–497.

Pierik, R.L.M. (1988). *In vitro* culture of higher plants as a tool in the propagation of horticultural crops. *Acta Horticulturae*. 226: 25–40.

Skoog, F. and Miller, C.O. (1957). Chemical regulation of growth and organ formation in plant tissue cultured *in vitro*. *Symp. Soc. Exp. Biol.*, 11: 118–131.

Smith, M.A.L. and Art Spomer, L. (1995). Vessels, gels, liquid medium and support system. In: Aitken-Christie, J., Kozai, T. and Lita Smith, M. (eds.). *Automation and Environmental Control in Plant Tissue Culture*. pp. 371–404.

Thomas, J.C. and Katterman, F.R. (1986). Cytokinin activity induced by thidiazuron. *Plant Physiology*. 81: 681–683.

6

CELL DIFFERENTIATION AND TOTIPOTENCY

INTRODUCTION

In animals, cellular differentiation is an irreversible phenomenon. In plants, on the other hand, even fully mature, old and well-differentiated cells have the capability of regeneration. Thus, cell differentiation is the basic event of development in higher organisms and is usually referred to as cytodifferentiation.

The development of a whole organism from a single cell (zygote) is the result of the integration of cell division and cellular differentiation. The cells isolated from differentiated tissues are normally non-dividing and quiescent and in order to express cellular totipotency, the differentiated and mature cells first undergo the process of dedifferentiation, followed by redifferentiation. As the word itself indicates, the phenomenon of a mature cell reverting to meristematic state and forming undifferentiated callus tissue is referred to as dedifferentiation, whereas the ability of a dedifferentiated cell to form a whole plant or its organs is known as redifferentiation.

In plants, the zygotic cell retains the ability to regress to a meristematic state and the nuclear particles store the information for producing identical offspring. Exact copies of such information are duplicated during mitosis and the daughter cells also, after attaining the highly differentiated state, contain the same information for the regeneration of a new plant. This phenomenon made botanists to think why somatic cells of mature plants, like zygotic cell, are capable of recapitulating the embryonic stages of development with the eventual regeneration of a mature plant.

Moreover, cell totipotency is the inherent ability of a single cell to divide and produce all the differentiated cells in an organism, including extra-embryonic tissue which leads to the production of entirely new individuals of the same type. Totipotent cells formed during sexual and asexual reproduction include spores and zygotes. Zygotes are the products of fusion of two gametes (fertilization). In some organisms, mature and fully differentiated cells can dedifferentiate and regain totipotency. For example, a plant cutting or a dedifferentiated mass of callus tissue can grow into an entire plant (Figure 6.1). Human stem cells are considered to be totipotent only if they can develop into any cell type in the body or into placental cells that do not become part of the developing foetus. This is an important aspect of the stem cell controversy.

Totipotent cells in general have the total potential to regenerate into any cell type or even the entire organism. They specialize into unipotent cells which can grow into single type of tissues and pluripotent cells that can give rise to most, but not all, of the tissues necessary for foetus development. Pluripotent cells undergo further specialization into multipotent cells that are committed to give rise to cells that have a particular function. For example, multipotent blood stem cells give rise to the red cells, white cells and platelets in the blood.

The regeneration of isolated plant cells during *in vitro* culture has amply demonstrated the fact that mature and well-differentiated somatic cells, under suitable conditions, have inherent potential to regenerate into a whole plant. This also proves the genetic fact that all genes required for the development of a whole plant or organism are located within each cell, but some of them remain inactive in differentiated cells or tissues or even organs and these transiently inactive genes are able to express only if suitable culture conditions are provided to the cells or tissues.

HISTORY

The concept of totipotency is implicit in the statement of cell theory as proposed by Schlieden and Schwann (1838), which expressed the view that each living cell of a multicellular organism is capable of developing independently if provided with the proper external conditions.

- Gottlieb Haberlandt (1902) reported that an isolated cell, if given proper nutrition and environmental conditions has the capacity to regenerate into entire plant. The isolated non-dividing cells undergo certain changes like replacement of non-functional cellular components. Haberlandt's publication on cell culture appeared in 1902 and was apparently "one of its kind". He chose to cultivate single cell under virtually aseptic conditions on culture medium containing Knop's nutrient solution and sucrose; asparagine and peptone served as nutrients. Haberlandt was not successful in demonstrating cellular totipotency or regeneration potential of the cultured cell because of the wrong choice of plant material.
- Kotte (1922) in Germany and Robbins (1922) in the US postulated that true *in vitro* culture could be easily prepared by using meristematic cells such as those in root tips or buds.
- White (1934), working with tomato roots, obtained unlimited cultures with medium containing yeast extract.

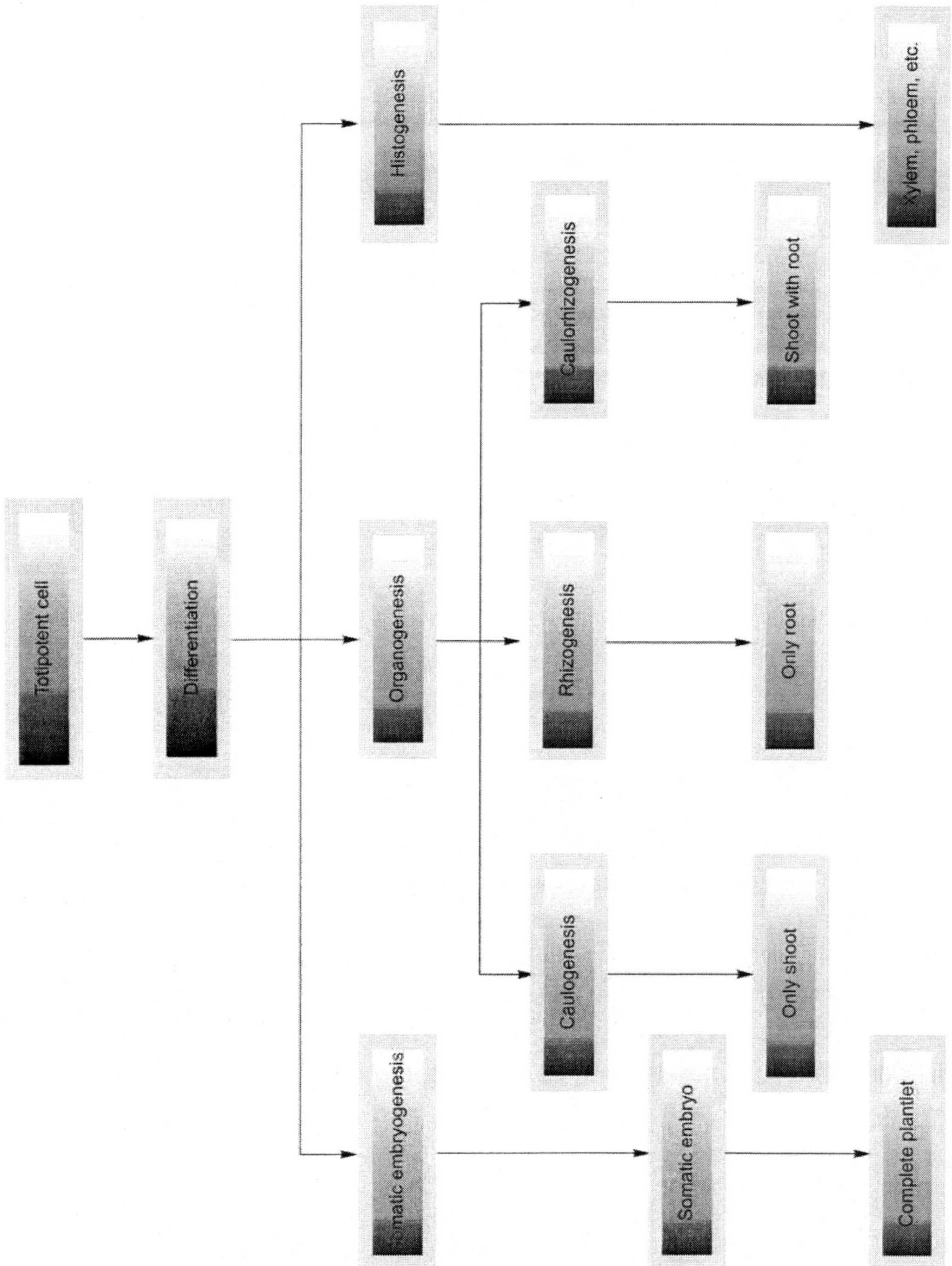

Figure 6.1 Different aspects of cell totipotency

- Ball (1946) showed that a part of shoot meristem was able to give rise to a whole plant of *Tropaeolum majus*.
- The possibility of cultivating plant tissues for an unlimited period was announced almost simultaneously but independently by White (1939), Nobecourt (1939) and Gautheret (1939).
- Overbeek, Conklin and Blakeslee (1941) suggested that a liquid medium such as coconut milk would be a good medium for embryo culture.
- Caplin and Steward (1948) applied coconut milk in the plant cell culture for the first time.
- Muir *et al.* (1954) undertook an exercise and set a cell of tobacco and *Tagetes erecta* on a small fragment of filter paper and placed this on top of a colony. Eight percent of these isolated cells multiplied.
- Miller *et al.* (1955) extracted a derivative of adenine (6-furfurylaminopurines or kinetin) from autoclaved yeast extract, which induced proliferation of which induced proliferation of tobacco cells and bud formation in the presence of indoleacetic acid.
- Skoog and Miller (1957) performed the famous experiment with tobacco tissues. They observed that organogenesis was controlled by a balance between auxins and cytokinins.
- In 1958, Reinert described for the first time, somatic embryogenesis in carrot tissues.
- Several methods enabled the observation of isolated cell multiplication and offered the possibility to demonstrate cellular totipotency (Bergmann, 1959). But they did not prove that the whole plant resulted from somatic embryogenesis affecting a single cell of the colony. Proof for this last stage was achieved by Backs-Husemann and Rienert (1970) when they proved that a single cell of carrot transformed into an embryo.
- Vasil and Hildebrandt (1965) observed that an isolated tobacco cell that could divide and form a colony are able to produce whole plants.
- Kao *et al.* (1970) observed wall formation around protoplasts and subsequently the division of regenerated cells.
- In 1971, Takebe *et al.* regenerated whole plant from protoplasts and Carlson *et al.* (1973) were able to regenerate plants from fusion between protoplasts of *Nicotiana glauca* and *N. langsdorffii*.
- Melchers *et al.* (1978) obtained a hybrid between potato and tomato through fusion of protoplasts of the two species. The products of such fusion of two types of protoplasts gave a plant not attainable from usual cross breeding.

THEORIES OF TOTIPOTENCY

Totipotency is the capacity of a living cell to differentiate into any type of cell, forming a new organism or regenerating a lost part of the body. A cell exhibits totipotency because it contains the entire genetic information in its nucleus. This information is expressed when it becomes necessary.

A German biologist, Haberlandt, suggested the concept of totipotency in 1902 on the assumption that each cell of an organism being derived from the mitotic divisions of a zygote, must be able to produce the entire organism. In the 1950s, Steward and his coworkers succeeded in growing carrot plants from isolated phloem cells. Totipotency is known to be present in plant cells and has today formed the basis for tissue culture (Figure 6.2). But this capacity is comparatively limited in animal cells.

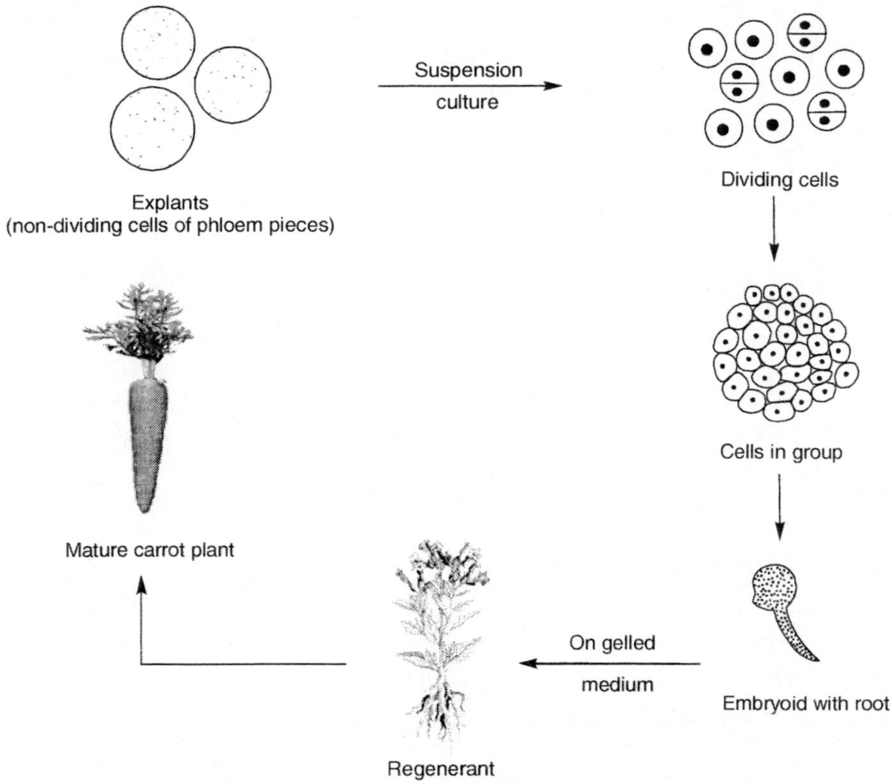

Figure 6.2 A simple experiment showing totipotency of cells

The molecular mechanisms controlling totipotency are not well-understood and are a subject for current research. It is suggested in the model organism *Caenorhabditis elegans* that multiple mechanisms including RNA regulation maintain totipotency at different stages of development. It is possible to raise whole plants from cuttings or even from a single or a clump of few cells. This indicates that it may be possible to raise a whole plant from a root (or shoot) cutting, or a complete plant from some pith cells. Now, there is a serious debate as to why cells lose this inherent ability when they become differentiated or mature.

Researchers believe that all plant cells retain the ability to use all of their genes and thereby can produce any type of tissue and eventually the whole plant. This ability to generate any cell from such starting tissue is the property of "totipotency". If one could only produce a limited range of cells from companion cells it indicates that such cells have lost this ability and thereby, they would have lost their "totipotency". Thus fertilized eggs (zygotes) are totipotent because they produce a population of differentiated cells forming an entire organism, whereas under normal circumstances, human skin cells are unipotent and not totipotent since in culture they divide to produce only skin cells (not nerve cells, muscle cells, etc.). The significance of the term has increased in the recent years with respect to cloning (particularly of animals). However, in plants, it has been used for many years to show that a new plant grows from a cutting or an explant used under *in vitro* culture. Meristematic cells (cells that divide to give rise to cells which differentiate into different cell types) are found in the cambium and at the apices of stems and roots. In many plants, this can be exploited to produce new plants (clones) by taking cuttings from leaves, stems and roots. Such cells are thus totipotent.

Dolly, the sheep, was cloned from a cell isolated from another sheep's udder cell (a mature differentiated cell), which in normal circumstances would have remained as such. Here the cytoplasm of the normal egg cell stimulates the nuclear genes of the diploid adult cell.

One of the hardest things in plant cell culture is to regenerate tissues excised from mature trees to illustrate totipotency, but by forcing a rejuvenation step, now it has also been possible to achieve *in vitro* regeneration with recalcitrant tissues. It can be done by adjusting the levels of hormones in the culture medium.

DEGENERATION OF TOTIPOTENCY

Cells of higher plants often retain their usual potentialities when cultivated *in vitro*. They are able to regenerate organs or whole plantlets. Yet these properties are not stable and are gradually lost some time after isolation of the tissues or in due course of culture. The cause of this phenomenon is not exactly known and it has not yet been possible to influence or control it. However, in some cases, exogenous triggering factors may play a major role and they can influence the gene expression of concerned cells, which leads to regaining of cell totipotency. However, scientists are now successful in regulating and maintaining the totipotency of carrot cells over prolonged periods by adjusting hormonal levels in the medium and changing the culture conditions.

VASCULAR DIFFERENTIATION DURING CELL CULTURE

The best way to study the process of vascular differentiation in culture is xylogenesis (differentiation of xylem or tracheary elements from simple meristematic cells). It is known that differentiation of tracheary elements involves deposition of localized secondary wall thickenings. The study of tracheary element differentiation by tissue culture technique helps us in understanding the mechanism of differentiation of higher plant cells. This is supported by the following reasons:

1. The morphological characteristics of tracheary elements such as annular, spiral, reticulate and pitted secondary wall thickenings enable us to distinguish differentiated cells from undifferentiated cells.
2. The formation of tracheary elements can be induced in tissue and cell cultures of many species of plants.
3. Specific biochemical events leading to deposition of cell wall polysaccharides and lignin make it possible to trace marker proteins associated with the process of cytodifferentiation.
4. Due to the loss of nucleus, the tracheary element loses its capacity for dedifferentiation or redifferentiation. Inflicting mechanical injury causing wounds in the internodes of intact healthy plants is often used to initiate differentiation of tracheary elements under *in vitro* treatments. This is one reason why such tracheary elements are often referred to as wound elements. Explants without pre-existing xylem tissues such as phloem, pith and tubers can also be used for this purpose. However, the explants being used for the purpose must have their cells in a uniform physiological state so that the differentiation would be in synchronous pattern.

Generally, the following *in vitro* culture types can be used to study the molecular mechanism of vascular differentiation during culture:

1. ***Callus culture*** Callus obtained after a series of subcultures is the best to observe cytodifferentiation in calli. Wetmore and Sorokin (1955) observed scattered groups of tracheary elements in the calli raised from cambial region of the stem of *Syringa*. However, many calli could not show vascular differentiation because of continuous subcultures and exogenous hormone supply.
2. ***Suspension culture*** In old suspension cultures, sugar and auxin induce cytodifferentiation of parenchyma cells into tracheary elements. They show synchrony in differentiation of tracheids. They are ideal for studying biochemical pathways associated with vascular differentiation. Enzymes taking part in shikimate and cinnamate pathways, peroxidases and phenylalanine ammonia lyase are associated with vascular differentiation in plant cell cultures.
3. ***Single cell culture*** These are ideal for studying any cell differentiation *in vitro*. Single cell culture raised from mesophyll cells of *Zinnia elegans* shows vascular differentiation in three days when it is grown in a medium supplemented with 1 mg/l 2,4-D and kinetin.
4. ***Protoplast culture*** These are ideal for the investigation of cell wall regeneration, cell fusion and transfer of exogenous DNA into protoplasts. In protoplast cultures, vascular differentiation occurs without cell division in the presence of auxin and cytokinin.

MOLECULAR BASIS OF DIFFERENTIATION

All the cells of an organism receive the complete set of genes (total information) present in the fertilized egg through mitotic divisions. However, cells of an early embryo lose the potential for the expression of all the genes through the processes of induction and repression, perhaps by mechanisms similar to those explained in gene regulation. Only certain genes remain functional in particular cells. Thus, the cells synthesize some proteins and lose the ability to synthesize others. This enables the cells containing similar genes to assume different structure and function, which in turn results in differentiation. Thus, only a small proportion of the total genetic information present in any cell is actually used, and the more specialized the cell, the fewer the genes utilized. This is why only a muscle cell synthesizes myosin, an erythroid cell forms globin, brain cell produces myelin, and a leaf cell manufactures RuBP carboxylase. Thus, altered gene activity or expression is the molecular basis of differentiation.

Development involves precisely a coordinated series of events. Each gene is switched on in its turn, and then as a result of the expression of that gene, or due to some environmental influence, the next gene in the programme is switched on. The exact mechanism of differentiation is not clear, but certainly the embryonic cells themselves cause differentiation of one another by passing chemical substances between them. The fate of a developing cell, thus, depends on the influence of its neighbouring cell, or simply on its position in the embryo. The cells that induce or control the development of other cells in an embryo are called "inducers" or "organizers", and the interactions of embryonic cells are termed "inductions".

Experiments on certain aspects of differentiation indicate that

1. some changes occur in the nuclei of cells during differentiation.
2. differentiation does not result in the loss of, or permanent inactivation of, the genes; in fact it "turns off" the genes in a reversible manner. This means that a differentiated cell can dedifferentiate itself and become a totipotent embryonic cell capable of expressing all of its genes. However, dedifferentiation is often limited in animals to early stages of differentiation whereas in plants it can take place even in mature cells.
3. it is the cytoplasm of the differentiated cell that determines the expression of the genetic information, and prevents the cells from acting as a fertilized egg and forming a new individual. The nucleus of an embryonic cell contains the full genetic information but it expresses only that information which is needed for the embryo while in the cytoplasm of an embryonic cell, though it expresses the entire information to form a new offspring when introduced into the cytoplasm of an egg.

Thus, the change in gene activity is brought about by its interaction with the immediate environment, the cytoplasm. The cytoplasm is modulated by factors such as nutrition, temperature, light, cell interaction, medium, hormones, etc. These factors may exert an influence by modulating gene activity. Hence, the modulation of the gene and its cytoplasm are interdependent.

It is now clearly established that alteration in the expression of specific genes brings about differentiation, and change in the gene expression results from the interaction of three factors namely, the nucleus, cytoplasm and environment.

Differentiation of Cells

Generally, in the body of multicellular plants or animals, three types of cells are recognized. They are:

1. ***Undifferentiated cells*** Undifferentiated cells or stem cells are unspecialized cells, which give rise to new cells by mitotic divisions. These go to form new tissues or are used up in the maintenance of existing tissues. Examples for these are the meristematic tissue in plants, Malpighian layer in the epidermis of skin, the germinal epithelium found in the gonads, the stem cells in the bone marrow, etc.
2. ***Differentiated cells*** Differentiated cells are specialized cells, which carry on specific functions. They have a specific form, structure and function, which normally do not change. Differentiation increases the functional efficiency through division of labour.
3. ***Dedifferentiated cells*** Dedifferentiated cells are specialized cells, which can revert to an embryonic state. Dedifferentiation is seen in dicot plants, particularly at the time of secondary growth. It is also seen in the process of regeneration that involves the ability of an animal to develop the lost parts of its body. Such a phenomenon is seen in coelenterates and echinoderms.

The capacity of cells to undergo dedifferentiation indicates that cells retain their complete genetic information. Organogenesis or organogenic differentiation follows the processes of dedifferentiation and redifferentiation of cells. Cell culture is an excellent technique for differentiation of plant tissues or organs, and uses the meristematic centres of the isolated explant tissues. An alternative mode of differentiation is by the induction of somatic embryos on excised explant directly, or by formation of intervening callus phase. These somatic embryos regenerate into the whole plant, the morphological differentiation being achieved by the adjustment of nutritional and hormonal conditions of the culture. With the application of auxin (higher or lower), the development of roots or shoots has been achieved by many workers. The effect of hormones on root and shoot development from callus is shown in Figure 6.3. The meristematic centres involve two growth phases and tissue differentiation takes place either by root or shoot buds. These are as follows:

1. The dedifferentiation takes place from the excised plant tissue and the phase has the potentiality of formation of embryoid-like structure.
2. The differentiation of specialized structure occurs if exogenous manipulations are made in terms of modification in culture nutrient medium as well as suitable stage of explant tissues. This is an important phase and nutritional and hormonal substances determine the mode of the development process. This demonstrates the persistence of the potencies and of the plasticity of form in plant cells in cultured condition.

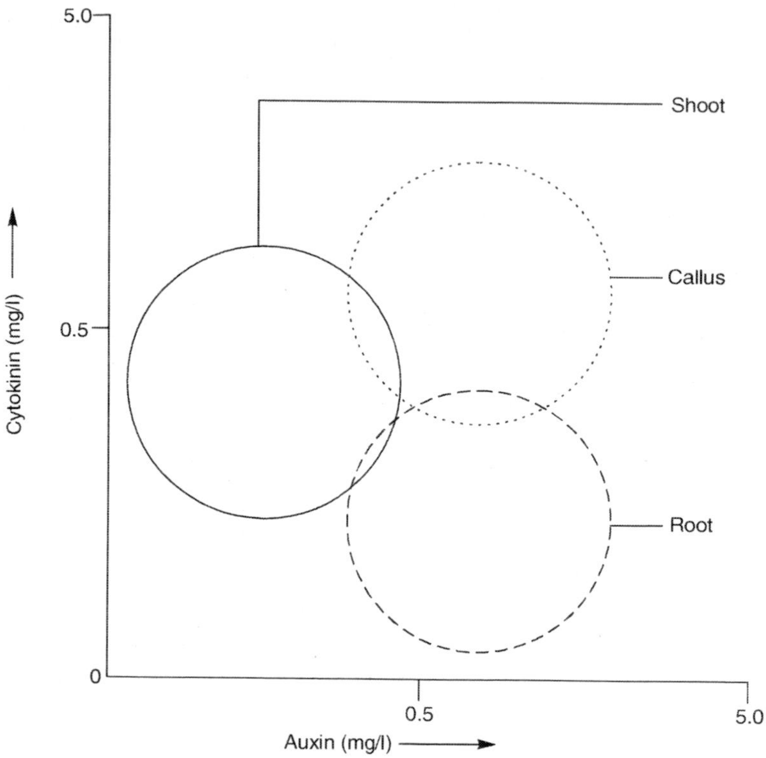

Figure 6.3 Relationship between the concentration of phytohormones and mode of differentiation during culture

Thus, in plant cell culture, the meristematic centres are processed in developmental programming for the initiation and formation of plant organs or organoids that differentiate into dermal, vascular and ground tissues. This process of formation of differentiated tissues from undifferentiated ones is known as organogenesis. The differentiation of organs is possible via various modes. The meristematic centre that develops within or from the cultured tissues has the ability of producing root and shoot and is known as meristemoid. In the process of organogenesis, the initiation of formation of only shoot buds or shoots in cultured condition is known as caulogenesis whereas the formation of only root in the process of differentiation in the cultured tissues is known as rhizogenesis.

TOTIPOTENCY OF EPIDERMAL CELLS

Totipotency of epidermal cells from shoot (particularly hypocotyl) and cotyledons has been demonstrated in many plants, such as *Linum usitatissimum, Daucus carota, Nicotiana tabacum, Torenia fournieri, Ranunculus sceleratus* and *Exocarpus cupressiformis.* The potential of forming shoot buds by epidermal cells of hypocotyls in *Linum usitatissimum* depends on the age of the seedling.

In *Linum usitatissimum,* most of the epidermal cells of the hypocotyl from a seedling less than 15 days old are totipotent and can form shoot buds *in vitro*. The *in vitro* formation of shoots from the superficial cell layer of cotyledons of *Pinus radiata* depends on the developmental layers of younger cotyledons, which possess less developed stomatal complexes, thinner cell walls, absence of epicuticular wax, partially reduced protein reserves and unhydrolysed lipid reserves. On the other hand, epidermal cells from older cotyledons which are not capable of forming shoots have more developed stomatal complexes, thicker cell walls with epicuticular wax, and depleted protein and lipid reserves.

In case of tobacco, the type of bud development in culture varies with the region from which the epidermal peel has been excised. Generally, the peel from floral branches produce only floral buds, and those from the basal part of the plant form vegetative buds. Peels taken from the middle region of the plant produce both types of buds in different proportions, depending upon their distance from the base of the plant. Formation of floral buds is also influenced by the physiological state of the donor plant. Floral buds are not likely to develop if the donor plant bears only flowers and no fruits. Similarly, floral bud differentiation is also greatly influenced by the chemical composition of the culture medium. Floral buds develop only in a medium containing kinetin and IAA in equimolecular concentrations and 2–3% sucrose. Increase or decrease in the hormonal combinations results in shifting the morphogenesis in favour of root, callus or vegetative bud formation. The flowers formed from buds originated from epidermal peels in cultures appear normal and fertile and also set seeds.

TOTIPOTENCY OF CROWN-GALL CELLS

The totipotency of crown-gall cells has been extensively used in plant genetic manipulations. It has been observed that the crown-gall cells, which show complete lack of organogenesis, have tremendous capacity of unlimited growth under *in vitro* conditions, independent of exogenous hormone supply. For instance, the bacterium *Agrobacterium tumefaciens* infects some plant species causing formation of crown-galls or tumours in the host tissue. The cells of these crown-galls are highly totipotent and can be differentiated into abnormal shoot buds and leaves in culture.

Brown (1959), who extensively worked in this particular area, used a nurse-culture technique to clone single cells of crown-galls in shake cultures which resulted in the development of abnormal shoots. Tips excised from these abnormal shoots were then grafted onto the cut ends of the stems of normal tobacco plants from which axillary buds had been removed. The successful grafts were again observed to have abnormal shoots. However, successive generations of repeated grafting resulted in the development of normal shoots. In this way, Brown and Wood (1976) obtained normal whole plants from crown-gall cells. Since the plants obtained from the crown-galls are fertile, the crown-gall system has been extensively used in plant genetic manipulation experiments.

A significant investigation on crown-gall cell culture reveals that explants excised from vegetative or reproductive organs of crown-galls, derived normal plants. When these explants were cultured, they revert to original crown-gall characteristics; but the F_1 progeny raised from seeds of such plants had lost the tumour trait. This shows that temporary loss of totipotency can be restored in crown-gall cells under *in vitro* conditions.

SIGNIFICANCE OF TOTIPOTENCY

The main objective in plant cell and tissue culture is the generation of plants from the totipotent cell. Although the process of differentiation is still mysterious in general, the expression of totipotent cell in culture has definitely provided a lot of information. On the other hand, the totipotency of somatic cell has been exploited in vegetative propagation of many economically, medicinally as well as agriculturally important plant species. Therefore, from fundamental to applied aspects of plant biology, cellular totipotency is highly imperative. Recent trends in plant cell culture technology include genetic modification of plants and production of homozygous diploid plants through haploid cell culture, somatic hybridization, mutation, etc. The success of all these studies depends upon the expression of totipotency. In many cases, successful and exciting results have been obtained.

Plant breeders, horticulturists and commercial plant growers are now more interested in plant cell culture only for the exploitation of totipotent cells in culture according to their desirable requirements. Totipotent cells within a bit of callus tissue can be stored in liquid nitrogen for a long period. Therefore, for germplasm preservation of endangered plant species, totipotency can be utilized successfully.

In totipotency, stem, leaf or root explants are successfully employed for the propagation of horticultural plants. The process of embryogenesis is considered as the main objective in maintaining a bank of germplasm of herbaceous and woody plants. The callus culture may be maintained for a longer time via regular subculture to fresh nutrient medium and preserved callus tissues may be transferred to other differentiation medium for the regeneration and growth of desirable plantlets. This technique is used to preserve endangered plants and plants whose seeds are very minute and difficult to germinate. The cultivation of orchid plants may prove successful. Most of the high coast orchids are hybrid in origin and they do not produce viable seeds. Explant tissues taken from shoot meristems readily proliferate when cultured in a liquid medium to form embryoids which can be further subdivided and cultivated to form more embryoids.

The genetic make-up of the cultured cell can be changed by introducing nucleic acid from different sources. The introduction of subcellular organelles is also possible via protoplast culture. The *in vitro* culture technique is of great importance in the improvement of crop and horticultural plants and also involves the production and propagation of haploid plants. The plants possess one half of the normal number of chromosomes found in the cells of growing vegetative cells. By applying the colchicine-mediated chromosome diploidization technique, if the chromosome number is doubled to form the normal diploid condition, there may be completely homozygous, true breeding plants in one generation.

REVIEW QUESTIONS

1. Define totipotency. What are the basics of totipotency?
2. Describe vascular bundle differentiation and organogenesis in cultured tissues.
3. Write notes on totipotency of epidermal and crown-gall cells.
4. What is the significance of totipotency?

REFERENCES

Guzzo, F., Barbara, B., Mariani, Loschiavo, P. and Terzi, M. (1994). Studies on the origin of totipotent cells in explants of *Daucus carota* L. *J. Exp. Bot.* 45: 1427–1432.

Hussey, G. (1977). *In vitro* propagation of *Gladiolus* by precocious axillary shoot formation. *Scientia Horticulturae.* 6: 287–96.

Kado, C.I. (1991). Molecular mechanisms in crown gall tumorigenesis. *Critical Reviews in Plant Sciences.* 10: 1–32.

Peterson, R.L. (1975). *The Development and Function of Roots.* Academic Press, Orlando.

Reinert, J., Bajaj, Y.P.S. and Zbell, B. (1977). Aspects of Organization- Organogenesis, Embryogenesis, Cyto-differentiation. In: Street, H.E. (ed.). *Plant Tissue and Cell Culture. Botanical Monographs.* 11. Blackwell, Oxford.

Skoog, F. and Miller, C.O. (1957). Chemical regulation of growth and organ formation in plant tissues cultured *in vitro. Sympos. Soc. Exp. Biol.* 11: 118–131.

Steward, F.C., Ammirato, P.V. and Mapes, M.O. (1970). Growth and development of totipotent cells. *Annal. Bot.* 34. 761–787.

7

MICROPROPAGATION: A SOURCE OF CLONAL REGENERATION

INTRODUCTION

The use of plant cell culture technology in vegetative propagation of plants has become the most widely used application of technology in agriculture, horticulture and forestry. In the last 35 years, plants have been regenerated from explants and/or callus of ornamental plants, food crops, vegetable and condiment plants, fruit and nut crops, medicinal plants and forest trees. These approaches have been used for the production of pathogen- and contaminant-free plants, *in vitro* conservation of plant genetic resource, asexual cell genetics, large-scale plant tissue culture and production of secondary metabolites. Besides these, somatic cell culture, single cell culture and protoplast culture technologies are fundamental to the modern approaches of plant genetic manipulation like genetic engineering of plants and for the manipulation of the plant cellular system for the production of useful products.

The successful expression of foreign genes, earlier in model systems such as tobacco and tomato, and recently in other crops, highlights the enormous potential for the integration of molecular biology and plant cell culture. But biological and technical problems still hamper the application of these technologies to important fruits, food and tree crops. There are limitations and constraints that still do not allow desired genetic manipulation of desired crops.

Plant tissue culture, also known as the *in vitro* culture or micro-culture of plants, is concerned with the growth and development of plants, plant cells, organs, or tissues in an aseptic, strictly controlled chemical and physical microenvironment. Regeneration of plants from cultured cells/tissues/organs/protoplasts is a unique property of higher plants, and it has become a critical component of plant biotechnology. Of all the terms which have been applied to the process, "micropropagation" is the term which best conveys the message of the tissue culture technique and it is the most widely used term. The prefix "micro" generally refers to the small size of the tissue taken for propagation, but could equally refer to the size of the plants that are produced.

According to Aitken-Christie and Connett (1992), micropropagation can be simply described as an *in vitro* method of producing clonal offspring identical to a superior stock plant. Micropropagation is, in fact, a complex multi-step process involving numerous different starting tissues and cell types, ranging from cotyledons and vegetative meristem to reproductive tissues, which can be induced to differentiate into plantlets, via either the organogenesis pathway or the embryogenesis pathway. The starting tissues, or explants, may be taken from either juvenile or mature/tested plant/tree of any age. Tissues from juvenile seedlings or seeds are frequently used in a new programme because they are easier to propagate and form the basis of development of suitable nutrient media and methods for a new species. When a breeding programme is in place, micropropagation can be carried out from genetically superior control-pollinated seed. However, it has been suggested that plantlets produced from the juvenile tissues of seedlings or seeds must be considered as untested clones, unless the tissues have been previously tested in a field clonal trial. There are only a few examples in woody plants/trees where the plantlets were produced from tested juvenile clones.

HISTORY

- Kotte (1922) in Germany and Robbins (1922) in the US postulated that true *in vitro* culture could be easily prepared by using meristematic cells such as those in root tips or buds.
- White (1934), working with tomato roots, obtained unlimited cultures with medium containing yeast extract.
- Ball (1946) showed that a part of shoot meristem was able to give a whole plant of *Tropaeolum majus*.
- The micropropagation is based on three important discoveries. The first critical finding was that cultured shoot meristems could give rise to virus-free plants (Morel and Martin, 1952). Another milestone was the elucidation of the role of cytokinins in shoot morphogenesis and in inhibition of apical dominance, which in turn release the axillary meristems from dormancy. The successful application of these principles to the *in vitro* multiplication of plants by this technology has been a key factor in the development of micropropagation into an important worldwide industry.
- Morel (1952–1955) demonstrated that certain virus infections could be eliminated by aseptic culture of stem tips of potato and dahlia.
- Steeves and Sussex (1966) successfully cultured the leaf primordia of *Osmunda* sp.
- Bornman (1976) reported the stimulating effect of auxin and gibberellin as xylem differentiation.
- Thorpe (1980) postulated that the endogenous auxin–cytokinin balance is the key factor in the initiation of organogenesis.

- Gupta *et al.* (1981) developed protocol for the clonal propagation of *Eucalyptus citriodora* and *Tectona grandis.*
- Kurtz *et al.* (1991) and Smith (1994) have described current methods of micropropagation and the various stages that lead to large-scale plant production.

METHODOLOGY

Micropropagation allows the production of large numbers of plants from small pieces of the stock plant in relatively short periods of time. Depending on the species in question, the original tissue piece may be taken from the shoot tip, leaf, lateral bud, stem or root tissue. In most cases, the original plant is not destroyed in the process. Once the explant is inoculated in nutrient culture medium, proliferation of lateral buds and adventitious shoots or differentiation of shoots directly from the explant begins. This results in tremendous increase in the number of shoots available for rooting. Rooted "plantlets" of many species have been successfully established and have been grown either in containers or in field plantings. The two most important findings from these experiments are as follows:

1. This methodology is a means of accelerated asexual propagation.
2. Plants produced by these techniques respond similarly to any self-rooted vegetatively propagated plant.

Micropropagation laboratories utilize the following basic methods to multiply plants *in vitro*:

1. Enhanced apical meristem axillary branching
2. Adventitious shoots
3. Somatic embryogenesis

The other plant production methods are based on differentiation of callus/suspension cultures through organogenesis and somatic embryogenesis. The enhanced axillary branching method is the most frequently used technique. Both terminal and axillary shoot meristems are used to initiate the culture. Similarly, the nodal shoot segment culture is another method. Since a callus intermediate is avoided in this method, there is least amount of possible genetic variation. Individual nodal selections with their associated axillary vegetative buds are excised upon subculture. By manipulation of the culture multiplication techniques and the culture environment, particularly the growth regulators in the culture medium, apical dominance is suppressed and axillary branching is induced. Cultures may be from either single shoots or shoot clumps and further multiplied to obtain the desired number of shoots. Although the initial multiplication rate is slow, it increases rapidly and reaches a steady plateau for several subcultures. The excised shoots may then be rooted *in vitro* or directly in soil/soilrite/vermiculite *ex vitro.*

The rates of culture multiplication and amplification are affected by several factors. In some plant systems the multiplying cultures sometimes exhibit metabolic and morphological derangements; however, in majority of plant systems, nodal sections (segments) are also used as explants, as these generate the least amount of possible genetic variation, because callus intermediate stage is avoided during this approach. Individual nodal sections with their associated axillary vegetative buds are excised upon subculture. The axillary bud grows and elongates, producing additional nodes for excision. One or two node sections can be excised and rooted. These techniques have several advantages but not all the species can be propagated successfully by these methods.

However, the modern techniques of DNA-based molecular marker analysis helps in the identification/characterization of desired clone even in juvenile stages of development. Such approaches are yet to be applied practically. Plantlets produced from tissues selected from mature elite plant (particularly trees/woody plants/perennials) have the advantage that the clones are likely to have been tested in plant/tree breeding programme and thus may be genetically superior. Alternatively, they may be selected from tested mature individual plant. The starting tissues can be either more juvenile (for example, stem sprouts on woody plants/trees) or more mature (for example, buds from the top of the tree). The micropropagation of mature plants is carried out to provide plants for clonal forestry or horticulture.

Adventitious shoots can be induced to differentiate *in vitro* directly from stem leaves, without forming intermediate callus. Similarly, induction of direct somatic embryo formation from the cultured explant is another preferred method of micropropagation. Plant regeneration through organogenesis and somatic embryogenesis from callus and suspension culture are other methods but these might lead to variations/ variants that are not desirable. Somatic embryogenesis is viewed as having the potential to be the most efficient plant regeneration method, since literally millions of plants could be produced from a relatively small amount of cultured cells or tissue.

Several factors contribute to the production of plantlets, their physiology and biochemistry *in vitro*. These are:

1. The chemical microenvironment—medium composition, medium pH and composition of vessel headspace.
2. The physical microenvironment—headspace temperature, incident light, vessel/ closure design, gel strength and other factors.
3. Facilities and aseptic methods.
4. Techniques used.

Medium components routinely include purified water, inorganic salts (macronutrients and micronutrients), carbohydrate sources, vitamins, hexitols, and growth regulators; they may also include amino acids, antibiotics, antioxidants and absorbents along with various natural complexes, which aid in the growth and development, and gelling agents.

The growth regulators which are instrumental in plant micropropagation include the auxins, cytokinins, gibberellins, ethylene and abscisic acid. However, the two classes of growth regulators with the most overwhelming influence on performance in cell culture are auxins and cytokinins. The cytokinins commonly used are BAP, kinetin, 2iP and zeatin, the latter two being natural cytokinins. Of these, BAP is the most active, the cheapest and the only one that can be autoclaved. Therefore, it is one of the most often used cytokinins, particularly in commercial micropropagation establishments. Lately, several synthetic cytokinins, other than BAP, have been tested. The best known of these is thidiazuron. This herbicide is a diphenylurea derivative which promotes the biosynthesis (or inhibits the breakdown) of endogenous purine (cytokinins). The pH of medium and composition of vessel headspace (humidity, gaseous composition of the headspace) are important factors. The headspace temperature, incident light, vessel/closure design and the strength of the gel of the culture media play a contributory role in the success of micropropagation.

STAGES IN STERILE MERISTEM CULTURE

In meristem culture, an LAF cabinet is used. The bench, roof and sides of the cabinet are swabbed with disinfectant. Latex gloves are preferred for carrying out the procedure. The gloves and bench surface are frequently swabbed during the experimental work. All instruments are sterilized and flamed before use. The general stages/steps in micropropagation technique are shown in Figure 7.1.

Stage 1 This stage is the initial stage of meristem culture. The stock plant, which is to be used as source of explant must be carefully monitored and grown under controlled conditions. Stock plants must be free from any type of disease. This practice is the base of the entire experiment, therefore selection of mother plant is a very important stage.

Stage 2 Juvenile tissue must be chosen to initiate cultures. It is always advisable to avoid sterilizing too many different genotypes at once. Problems of contamination and toxic effects of the sterilant increase with increased numbers of genotypes being treated simultaneously.

Stage 3 Cultured explants start responding at this stage and produce shoots. Condensation can be a problem in Petri dishes. This can be minimized by stacking the Petri dishes and rotating the position of each dish in the stack daily. The intensity of light reaching each culture does not appear to be critical at this stage, although light is necessary.

This stage is also known as multiplication stage of the cultures (Figure 7.2). Murashige and Skoog medium and Gambourg B_5 medium are the most frequently used media in the culture of axillary buds. They often need to be modified, mainly by varying the vitamins and growth regulators, depending on the species or variety being used. *In vitro* generated shoots can be harvested from the stage 1 and used as inoculum for this stage. Vitrification can be a serious problem during *in vitro* multiplication of cultures. In this condition, the shoots tend to elongate and look "glassy," the leaves

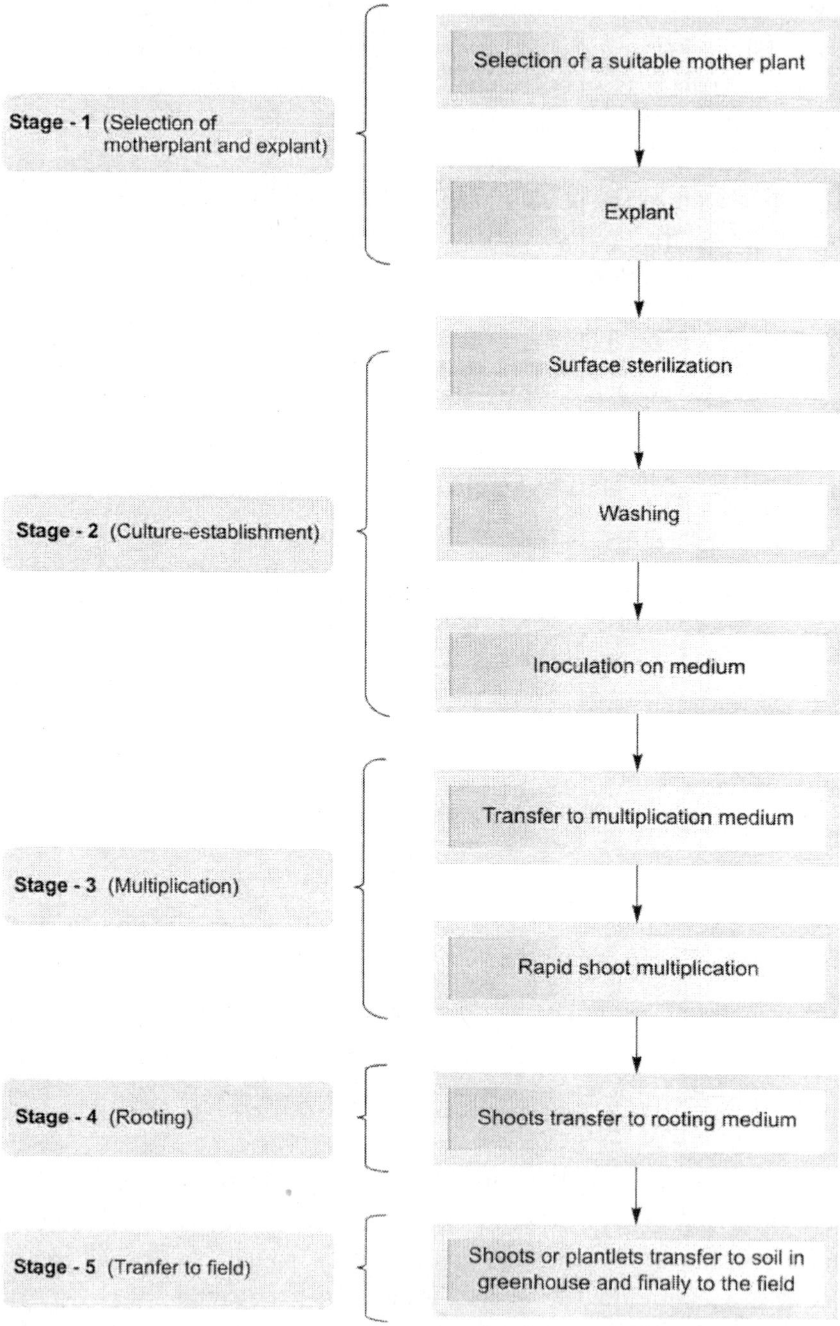

Figure 7.1 Stages of micropropagation

become reduced or malformed, and if not treated, the culture will die. The causes of vitrification are not fully understood, but the risk of vitrification is greatly reduced if there is adequate gaseous exchange between the culture and the external environment.

Stage 4 This stage is known as rooting stage. The elongated and proliferated shoots are to be cut here and transferred to root induction media, which contains auxins (IAA/IBA/NOA/NAA). The salt strength of the media used for rooting is generally half or one-fourth, which induce rooting from the *in vitro* generated shoots. Sometimes use of activated charcoal is also helpful in root initiation.

Figure 7.2 A culture showing regeneration of multiple shoots at multiplication stage (Plate 1.7)

Stage 5 Rooted shoots are carefully taken out and washed with sterile water to remove agar. This stage is also known as hardening stage. Rooting and transfer of regenerated plantlets to compost can be difficult. If the auxin level is too high, subsequent root development after transfer to compost is reduced. Some species can be transferred to compost without a rooting step. Some fruit tree species need a dark treatment as well as additions of auxin for successful rooting to take place. The transition from culture to compost is critical. High humidity must be maintained, but care must be taken not to waterlog the compost. If plants are transferred to the field, they have to be protected from wind damage, drought and birds.

SELECTION OF EXPLANTS FOR MICROPROPAGATION

Shoot apical meristem is generally used as explant for micropropagation. It lies in the "shoot tip" beyond the youngest leaf or first leaf primordium; it measures up to about 100 μm in diameter and 250 μm in length. Thus, a shoot tip of 100–500 μm would contain 1–3 leaf primordia, in addition to the apical meristem. In practice, shoot tips of up to 1 mm are used when the objective is virus elimination. Shoot-tip culture is widely used for rapid clonal propagation for which much larger (about 5–10 mm) explants are also used. In most cases, meristem cultures are essentially shoot–tip cultures. Nodal explants of various sizes are also commonly employed for rapid clonal propagation.

Moreover, when the objective is vegetative propagation, the size of shoot tip used for culture is not a limiting factor. But when the objective is to free the stock from a virus, it is essential that the apical meristem be excised along with a minimum of the surrounding tissue. The shoot tip may be cut into fine pieces to obtain more than one plantlet from each shoot tip. In species like cauliflower, pieces of curd (the inflorescence) are used in culture, while in plants having underground stems, shoot tips or tissue pieces bearing buds of such stems may be used. Generally, explants taken from actively growing plants at the beginning of the growing season are the most suitable explant sources in terms of *in vitro* response.

STERILIZATION OF EXPLANTS

The explant source (short piece of stem with apical bud/s) is sterilized by the following method:

1. The explant is dipped once in absolute ethanol for 30 seconds.
2. It is then treated with sterilant (for example, by using 5% v/v sodium hypochlorite with a drop of Tween 80). This is particularly important when the tissue surfaces are waxy or coated with epidermal hairs.
3. The explant is then rinsed 4–5 times with sterile distilled water thoroughly (Refer Chapter 4 for detailed discussion on explant sterilization).

EXCISION OF SHOOT TIPS

A dissection microscope with a magnification of at least 15X that must be mounted in an LAF cabinet is the primary requirement for initiating meristem culture. A piece of expanded polystyrene covered in white plastic film and taped to the microscope stage is ideal for holding sterilized stem segments at the dissection site, using sterilized pins. The tips of 12 gauge hypodermic needles are used to carry out the dissection.

After excision, the explant is inoculated directly onto the selected culture nutrient medium and then the culture vessel is closed. Media are commonly species-specific with regard to the formulation of the required growth regulators and their concentrations, and may also be specific for other organic and inorganic constituents. The specificity may even extend to the particular cultivar/race level with respect to the optimal growth.

ESTABLISHMENT AND MULTIPLICATION OF CULTURES

The inoculated meristem cultures are transferred to the growth room for the establishment of cultures. A growth room provides a controlled environment to incubate the cultures. Lighting at 4000 lux at culture level with a 16 hour photoperiod (provided by warm white fluorescent tubes) and a constant temperature of 25 ± 2°C provide optimal culture conditions.

If the explant is viable, then enlargement, development of chlorophyll and some elongation will be visible within 7–14 days of culture initiation. The developing young shoots are maintained *in vitro* until the internodes are sufficiently elongated to allow dissection into nodal explants. To prepare nodal explants (Figure 7.3), the regenerated young shoots are removed from the culture vessels under sterile conditions and cut into nodal segments. Each of these is separately transferred directly onto fresh nutrient medium to allow axillary bud outgrowth. Extension of these buds should be evident within 7–14 days of culture initiation.

Figure 7.3 Cultures showing bud breaking from nodal shoot segments (Plate 2.1)

CONTROL OF PHENOLICS IN CULTURE

In many species, phenolics leach into the medium from the cut surfaces of explants. These phenolics turn dark brown on oxidation and are detrimental to the cultures. This is very common in case of woody species, particularly when explants are taken from mature trees. In some species like mango, eucalyptus, phyllanthus, etc. control of phenolics is a major problem.

Browning of medium can be minimized by one of the following ways:

1. Frequent (every 3–7 days) subculture of explants on agar medium overcomes this problem in many species.
2. A brief period (usually 3–10 days) of culture in liquid medium is effective in many species, for example, apples, eucalyptus, etc.; this may remove not only polyphenols but also other inhibitory substances from the explants.
3. In difficult cases, an antioxidant like ascorbic acid (50–100 mg/l), cysteine-HCl (100 mg/l) or citric acid (50 mg/l) may be used to check the oxidation of polyphenols.
4. Adsorbents like activated charcoal (150–200 mg/l) or PVP (polyvinyl pyrrolidone) may be used to absorb the polyphenols secreted into the medium.
5. Sometimes, culture incubation in dark may be helpful, since light enhances oxidation of polyphenols.

ROOTING OF *IN VITRO*-GENERATED SHOOTS

In general, the rooting medium has a low salt content, for example, 1/2 or even 1/4th of the salts of the MS medium, and reduced sugar levels (usually 1.0 g/l). In some species like *Narcissus*, reduced salt level in nutrient composition proves to be essential for rooting, however, in contrast, in some species of *Narcissus* and strawberry, profuse rooting occurs on hormone-free basal medium.

But in general, most of the species require the addition of 0.1–1.0 mg/l NAA or IBA for rooting. In plants like citrus, however, a pulse treatment with an auxin (10 minute exposures with 100 mg/l NAA or IBA) gives optimum rooting.

In vitro grown multiple shoots are usually rooted in an agar medium (Figure 7.4), but the recent trend is to root them directly in vermiculite or potting mixture (*ex vitro* rooting). The cut ends of shoots are treated with a suitable auxin solution or powder mix and transplanted in pots and kept at high relative humidity and low light intensity. This approach, to some extent, saves the cost of technology because the rooting step is entirely eliminated through this technique. Furthermore, the soil acclimatization stages are also combined with rooting steps that further minimizes the culture duration because induction and development of roots generally takes about 10–15 days, depending on the plant species. Plantlets (Figure 7.5) with 0.5 to 1.0 cm long roots are usually transplanted into pots since longer roots tend to get damaged.

TRANSFER OF PLANTLETS TO GREENHOUSE

Rooted shoots are removed from the medium. The agar sticking to their roots is washed with sterile distilled water and they are transplanted into plastic pots (ranging from ordinary to specific design) containing

Figure 7.4 Cultures showing formation of roots from *in vitro*-formed young shoots on rooting media (Plate 2.2)

Figure 7.5 A picture showing *in vitro*-regenerated complete plantlets with root and shoot (Plate 3.7)

a suitable potting mixture (Figures 7.6 and 7.7). Plants are kept at high humidity (90% or more) and low light intensities initially. High humidity can be attained by one of the following methods:

1. Fog (water drops 10 μm or less)
2. Mist
3. Clear plastic to cover individual (plastic bags) or groups (plastic sheets) of plants

Figure 7.6 A picture showing hardening of tissue culture-raised *Aloe vera* plantlets in greenhouse (Plate 3.1)

Figure 7.7 A picture showing *in vitro*-generated plantlets under hardening conditions (in greenhouse) (Plate 3.2)

The potting mixture should not be too wet and water drops should not form on the plantlets. For this reason, fog is preferred over mist. The humidity is gradually decreased to the ambient level, after about 7–15 days and the light intensity is allowed to increase. The plants are finally exposed to greenhouse conditions (Figure 7.8).

Figure 7.8 A picture showing hardened plantlets in greenhouse ready to field transfer (Plate 3.3)

On the laboratory scale, individual plants may be covered with clear plastic bags and irrigated daily with 2–3 drops of tap water or 1/4 MS salts. After 7–10 days of transfer to the soil, the bags may be removed gradually over a period of another 7–10 days.

VITRIFICATION AND MORPHOLOGICAL VARIATIONS

During organogenesis, some shoots developed *in vitro* appear brittle, glassy and water-soaked; this is called vitrification or hyperhydracity. In many species, vitrification may be represented by symptoms not visible to the naked eye, for example, poorly developed vascular bundles, abnormal wax quality, abnormal functioning stomata, etc.

Remedy of Vitrification

Vitrification is the consequence of culture conditions, which leads to losses of plantlets. It may be overcome by the following methods:

1. Increasing the agar levels in the medium.
2. Cooling the bottom of culture vessels.
3. Adding desiccant ($CuSO_4$) into the medium.
4. Adding agar hydrolysates to the medium.
5. Adjusting the level of salts in the medium.

6. Using growth retardants.
7. Lowering cytokinin level.

Morphological variants may also arise during the multiplication state. Such variants occur in most cases, at a frequency comparable to that *in vivo*. But in commercial ventures using meristem cultures, visual selection is practised to eliminate variants and maintain the homogeneity of plantlets produced. Many commercial enterprises, therefore, prefer to multiply shoots for only about four cycles from an explant; after this, a fresh batch of cultures is initiated from field-tested plants.

MERICLONING

Micropropagation by meristem tip or axillary bud culture is also known as mericloning. The essence of meristem tip or axillary bud culture (Figure 7.9) is the excision of the organized apex of the shoot or axillary bud from a selected donor plant for subsequent *in vitro* culture. The conditions of culture are regulated to allow only for organized outgrowth of the apex directly into a shoot, without the intervention of any adventitious organs. It comprises the apical dome and a limited number of the youngest leaf primordia without any differentiated provascular or vascular tissues.

A major advantage of working with such a small explant is its potential to exclude pathogenic organisms present in the donor plant prior to *in vitro* culture. A second advantage is that since the genetic stability in meristem culture technique is inherent, plantlet production from an already differentiated apical meristem and propagation through adventitious meristems can be avoided. Shoot development directly from the meristem avoids callus tissue formation and adventitious organogenesis, ensuring that genetic instability and somaclonal variation are minimized.

However, if there is no requirement for such benefits, then the related technique of shoot tip culture may be more expedient for plant propagation. In this procedure, the explant is still a dissected shoot apex, but a much larger one, containing a relatively large number of developing leaf primordia. Typically, the explant is between 3–4 mm and 2 cm in length, and development *in vitro* is still regulated so as to be confined to the outgrowth of an organized shoot, without adventitious propagation. The axillary buds that develop on *in vitro*-grown plantlets derived from meristem-tip cultures may also be used as a secondary propagule. When the *in vitro*-formed plantlet expands internodes, it may be divided into segments, each containing a small leaf and an even smaller axillary bud. When these nodal explants are inoculated on fresh culture medium, the axillary bud will grow into a new plantlet and at each time such process can be repeated.

This technique adds a high propagation rate (Figure 7.10) to the original meristem culture technique, and together, these techniques form the basis of micropropagation.

If the culture conditions are designed appropriately, the excised meristem will develop and elongate in essentially the same manner as if it were still part of the original parent plant. The production of a shoot from the meristem does not rely on any callus tissue formation or adventitious organogenesis. It is possible that the callus tissue may develop on certain portions of the growing explant, particularly at the surface damaged by excision. The only acceptable situation under such circumstances is that the callus be sufficiently slow-growing and localized to be readily identified, so that it can be excised at the first available opportunity.

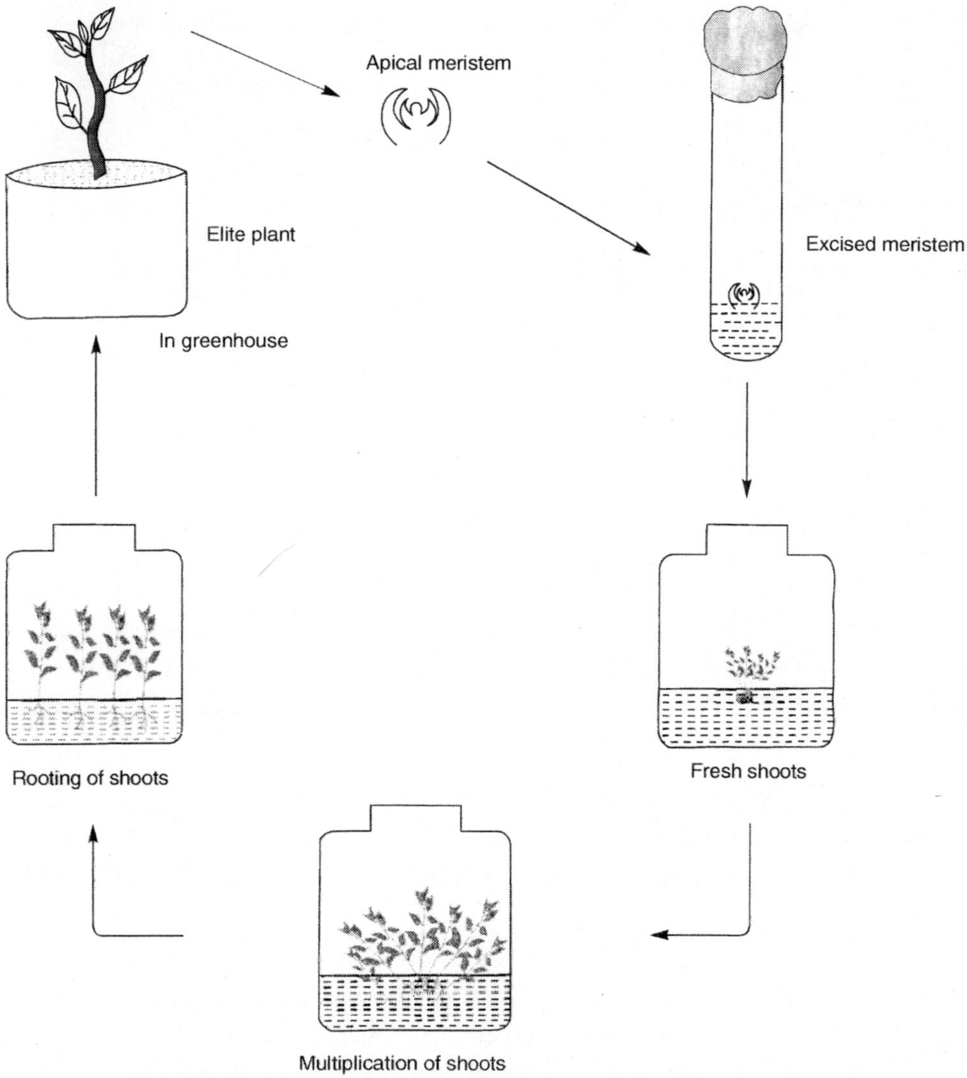

Figure 7.9 Shoot tip or meristem culture technique

Figure 7.10 Induction and regeneration of multiple shoots from various explants (Plate 2.6 & 2.7)

A further advantage of meristem culture is that the technique preserves the precise arrangement of cell layers necessary, if a chimeric plant genetic structure is to be maintained. In a typical chimera, the surface layers of the developing meristem are of differing genetic background, and it is their contribution in a particular arrangement to the plant organs that produces the desired characteristics, for example, the flower colour in some African violets. As long as the integrity of the meristem remains intact, and development is normal *in vitro,* the chimera pattern will be preserved. If, however, callus tissues were allowed to form, and shoot proliferation, subsequently, was from adventitious origins, there would be a risk that the chimera layers of the original explants may not all be represented in the specially required form in the adventitious shoots.

If a sufficiently small explant can be taken from an infected donor and raised successfully *in vitro,* there is a real possibility of the derived culture being pathogen-free. Such cultures, once screened and certified, can form the basis of a guaranteed disease-free stock for further propagation. The meristem-tip culture technique can be linked with heat therapy to improve the efficacy of disease elimination. If meristem culture alone is not successful in producing any virus-free plants, then temperature stress treatment of donor plants and/or the use of antiviral agents will have to be considered. There are no prescriptive, global methodologies for these treatments. Heat therapy relies on the growth of the donor plants at elevated temperatures, typically 30–37°C and may involve treatment for several weeks.

Antiviral chemicals can be used as additives in the culture medium and one of the most widely used antiviral compounds is virazole. This compound is a guanosine analogue with broad-spectrum activity against animal viruses, and it also appears to be active against plant virus, replication in whole plants. Increasing concentrations of ribavirin and increasing length of culture incubation in the presence of the compound, typically increase the effectiveness of virus elimination but causes slow growth, and phytotoxicity may be evident at high concentrations.

These cultures are often acceptable for international transport with respect to quarantine regulations. Again the meristem tip has to be sufficiently small and homogeneous in tissue type to provide a practical propagule for cryopreservation and other techniques of culture storage.

Procedure of Meristem Culture for Micropropagation

1. A suitable donor plant is selected from which apical meristem or axillary bud explants are procured. Also, the terminal portion of the stem and/or stem segments containing nodes are excised from the donor plant.
2. The mature and expanding foliage is removed to expose the terminal and axillary buds. The donor segments are cut into appropriate sizes and pre-sterilized, for example, by immersing in ethanol or industrial methylated spirit for 10–30 seconds.
3. The donor tissues are then sterilized by treating with mercuric chloride or sodium hypochlorite solution, the concentration and treatment durations depending upon the size and nature of the donor tissues. Treatment with a solution of mercuric chloride (0.1%) for 3–5 minutes would represent the typical protocol range. Domestic bleach may also be used at a concentration empirically determined for the tissues involved (0.5–10.0% solution with 15 minutes). A few drops of a detergent, such as Tween 80, are added to the sterilant solutions to improve wetting of the tissues, which is particularly important when the tissue surfaces are waxy or coated with epidermal hairs. However, caution is needed if domestic bleaches already containing high levels of detergent are to be used.
4. Following surface sterilization, the donor tissues are rinsed with sterile distilled water.
5. The tips of hypodermic needles are used to dissect away progressively smaller, developing leaves to expose the apical meristem of the bud.
6. After excision of the explant, it should be inoculated directly onto the selected growth medium.
7. The completed meristem culture is transferred and incubated to a culture room at 25 ± 2°C with illumination from warm white/cool white fluorescent lamps. A 12–16 hour photoperiod of intensities up to 4000 lux is optimal.
8. The developing multiple shoots should be maintained *in vitro* until the internodes are sufficiently elongated to allow dissection into nodal explants.

9. Culture-raised multiple shoots with nodal regions should be removed from their culture vessels, under sterile conditions, and separated into nodal segments. Each of these individual segments can be inoculated directly onto fresh growth medium of the original formulation, to allow axillary bud growth. Extension of this bud should be evident within 7–10 days of culture initiation.

Factors Affecting Meristem Culture

A number of factors are likely to affect meristem-tip or axillary bud culture. These are as follows:

1. Donor plants should be selected for general health and vigour. Donor tissues should be taken from young, preferably actively growing regions of the plant to ensure the best chance of success. Seasonally dormant plants are to be avoided.
2. In an attempt to eradicate surface pathogens, there is a possibility of oversterilizing the explants during surface sterilization, and so, it may damage the meristem tissues. Therefore, it is advised to use donor plants, which have been glasshouse-raised to minimize the infection problems.
3. It is only with practice that reproducible, high levels of culture success will be achieved, and early attempts at dissection will often result in damaged tissue. Hypodermic needles, used as dissecting tools, have to be discarded after 2–3 meristem excisions.
4. The presence of leaf primordia is essential for successful culture growth, and so removal of too small-sized explants may restrict success.
5. A consequence of dissection may be the production of toxic, oxidized polyphenolic compounds by the explant tissue. It interferes with growth of plant tissues in the culture.
6. When no previous reports of a successful medium can be found either for the plant in question or for a close relative, an empirical study becomes necessary. The modified medium of Murashige and Skoog (1962) has an inordinately wide application, and may be used with inorganic and organic constituents as a sensible first step in almost all cases. The energy source is typically sucrose at a 2–4% level. Most problems will be found in determining the appropriate type and concentration of the necessary growth regulators to ensure culture survival and the required pattern of development.

STRATEGY AGAINST CONTAMINATION DURING MICROPROPAGATION

1. Contamination by Viral Pathogens

The danger can be minimized by testing for infections by known viruses before initiating the culture. Once in culture, viral pathogens generally do not produce any symptoms, but are rapidly spread by axillary bud cultures. The smaller the piece of tissue used to initiate cultures, the lesser is the risk of transferring viral infection. To reduce this risk to a minimum, meristems can be used to initiate cultures.

2. Contamination by Bacterial and Fungal Pathogens

Theoretically, surface sterilization should remove most surface contaminants. Frequent scrutiny of cultures should allow healthy buds to be transferred to fresh medium if occasional contaminated buds are seen. Alternatively, inoculation of fewer explants in smaller Petri dishes reduces the problem, and infected plates can then be discarded. The use of antibiotics to suppress bacterial or fungal infections is generally ineffective.

APPLICATIONS OF MICROPROPAGATION

The most popular and widely commercialized global application of plant biotechnology is micropropagation. Due to the demand for high-quality planting stock of crops such as banana, sugarcane, cardamom, etc., micropropagation has become the most commercially exploited area of plant cell and tissue culture technology. Currently several genera and species encompassing the gamut of horticulturally important crops are being propagated in millions and at the commercial level (Table 7.1). The extensively micropropagated plants include a large number of foliage ornamentals, planting stocks of cut flowers. fruit-yielding and agro-forestry species. However, there are quite a few important tropical fruit and tree species such as mango, litchi, coconut, etc., which have remained recalcitrant.

Table 7.1 Some important examples of micropropagated plant

Type/nature of plants	Micropropagated plants
Medicinal plants	*Phyllanthus amarus, Withania somnifera, Catharanthus roseus, Atropa belladonna, Writia* sp., *Solanum* sp., *Artemisia annua, Urginea* sp., *Rauwolfia serpentina* sp., *Ipecacuanha* sp., *Aloe vera, Digitalis* sp., *Ephedra* sp., *Ferula* sp., *Adhatoda vasica, Cassia angustifolia, Glycyrrhiza glabra, Myrica esculenta,* etc.
Ornamental plants	*Adiantum* sp., *Anthurium* sp., *Ruscus* sp., *Asparagus* sp., *Allium* sp., *Asplenium* sp., *Begonia* sp., *Bougainvillea* sp., *Chrysanthemum* sp., *Freesia* sp., *Gladiolus* sp., *Hibiscus rosasinensis, Althaea rosea, Malva sylvestris,* carnations, *Rhododendron, Petunia,* African violet, orchids, tulips, lilies, *Dahlia* sp., *Ficus, Phlox* sp., *Dracaena,* etc.
Woody plants	*Eucalyptus* sp., *Tectona* sp., *Cedrus deodara, Pinus* sp., *Cassia* sp., *Shorea* sp., *Sequoia sempervirens, Camellia* sp., *Bauhinia* sp., *Capparis* sp., *Dalbergia* sp., *Acacia* sp., etc.
Vegetables/Condiments	*Beta vulgaris, Daucus carota, Raphanus sativus, Brassica rapa, Manihot esculenta, Solanum tuberosum, Allium cepa, Allium sativum, Spinacia oleracea, Amaranthus blitum, Lycopersicon esculentum, Cucumis sativus, Solanum melongena, Abelmoschus esculentus, Capsicum annuum,* etc.
Fruit plants	*Mangifera indica, Annona squamosa, Ananas comosus, Achras sapota, Citrus reticulata, Citrus sinensis, Musa paradisiaca, Psidium guajava, Carica papaya, Litchi chinensis, Punica granatum, Vitis vinifera, Aegle marmelos, Pyrus malus, Prunus persica, Citrullus lanatus, Morus alba,* etc.

Micropropagation offers several distinct advantages not possible with conventional propagation techniques. A single explant can be multiplied into several lakhs of plants in less than one year. With most species, the excision of the original tissue explant does not destroy the parent plant. Once established, actively dividing cultures are a continuous source of microcuttings which can result in plant production under greenhouse conditions without seasonal interruption. Using methods of micropropagation, the nurseries can rapidly introduce selected superior clones of ornamental plants in sufficient quantities to have an impact on the landscape plant market.

1. Production of Pathogen-free Plants

Plant meristem culture is uniquely suited for obtaining, maintaining, and mass propagation of specific pathogen-free plants. The concept behind indexing plants free of pests is closely allied to the concept of using tissue culture as a selection system. Plant tissues known to be free of the pathogen under consideration (viral, bacterial or fungal) are physically selected as explants for tissue culture. In most cases, the apical domes of rapidly elongating shoot tips are chosen. These are allowed to enlarge and proliferate under the sterile conditions of *in vitro* culture with the resulting plantlets tested for presence of the pathogen (a procedure called indexing).

Cultures which reveal the presence of the pathogen are destroyed, while those which are indexed free of pathogen are maintained as a stock of pathogen-free material. Procedures similar to these have been used successfully to obtain virus-free plants of a number of species and bacteria-free plants of species known to have certain leaf spot diseases. Obtaining pathogen-free nursery stock is only speculative, since little research documenting the transmission of viral, bacterial or fungal diseases through propagation of woody ornamentals is available. The main disadvantage of such a system is that the scope for physiological experimentation is restricted because the plants produced are generally juvenile in character, and there is a tendency for the amount of epicuticular waxes to be reduced, palisade cells to be smaller, mesophyll air sacs to be larger and stomatal physiology to be disrupted.

2. Preservation of Plant Genetic Resources

The need of a system for the conservation of plant genetic resources arises from the rapid changes occurring in modern agricultural practices. Primitive cultivars and wild relatives of crop plants constitute a pool of genetic diversity which is inevitable for future breeding programmes. These have already led to the replacement by new cultivars which encompass a much narrower range of genetic diversity. As a result, there is a very real danger of future breeding being impeded by the shrinking genetic bases of some crops. Hence, storage of this sort of irreplaceable breeding material or germplasm (gene combinations available for breeding) and establishment of a centralized gene bank are the practical ways to solve these problems.

Conventionally, germplasms are stored in the form of seeds because they occupy a relatively smaller space and can be stored for many years. But there are a number of important species, in particular, root and tuber crops, which are normally propagated vegetatively. These include potato, sweet potato, yam and cassava. So, the conventional preservation method is not applicable to vegetatively propagated plants.

On the other hand, the cost of maintaining a large proportion of the available genotype of these plants in nurseries or fields is very high and risk of plant loss due to disease or environmental hazards is very real. It is now possible with modern cell culture technique to provide a germplasm storage procedure which uniquely combines the possibilities of disease elimination and rapid clonal multiplication. Cryopreservation technique has already been shown to be successful with a range of cell cultures (for example, carrot cells, tissues and meristem tips) from a number of crop plants, including asparagus, tomato and potato. Plant tissue and meristems of such plants have been successfully recovered from liquid nitrogen and grown into normal, mature fertile plants.

DISADVANTAGES OF MICROPROPAGATION

Micropropagation may appear to be the perfect means of multiplying plants, but it has certain disadvantages. These are as follows:

1. The greatest limitation is the cost of production. It is very expensive and can have a labour cost of more than 70%.
2. All plants cannot be successfully regenerated through tissue culture technique, often because the proper medium for growth is not known.
3. Some plants produce secondary metabolic chemicals (for example, phenolic compounds) that stunt or kill the explants during culture.
4. Sometimes, plants or cultivars do not come true-to-type after culture; this is often dependent on the type of explant material utilized during the initial phase or the result of the age of the cell or propagule line.
5. Some plants are very difficult to disinfect, and elimination of the fungal contaminants is not achieved.

A thorough understanding of production costs serves as a meaningful competitive advantage in the micropropagation industry. It is important that cost-effective procedures be defined for plant species under micropropagation. Several general strategies can be employed to reduce current production costs. They include:

1. *In vitro* biological optimization.
2. Elimination of production stages or repeated subcultures.
3. Automation and mechanization of the process.

Mechanization of the process would eliminate most of the labour cost, but this has proven difficult so far despite active attempts to develop this technology.

REVIEW QUESTIONS

1. Define micropropagation and describe the different levels of success in cloning.
2. Explain the process of micropropagation by meristem-tip culture.
3. What are the factors that affect micropropagation?
4. What are the advantages and problems in meristem-tip culture?
5. Describe the limitations and precautions of micropropagation technique.

REFERENCES

Aitken-Christie, J. and Connett, M. (1992). Micropropagation of forest trees. In: Kurata, K. and Kozai, T. (eds.). *Transplant Production System.* Kluwer Academic Publishers, The Netherlands, pp. 163–194.

Altman, A. and Loberant, B. (1998). Micropropagation: clonal plant propagation *in vitro*. In: Altman, A. (ed.). *Agricultural Biotechnology.* Marcel Dekker, Inc., New York. pp. 19–42.

Boulay, M. (1985). Some practical aspects and applications of the micropropagation of forest trees. Proc. International Symp. on "In Vitro Propagation of Forest Tree species", Bologna, (May 9–10, 1984) A.R.F. Bologna. pp. 53–81.

Debergh, P.C. and Read, P.E. (1991). Micropropagation. In: Debergh, P.C. and Zimmerman, R.H. (eds.). *Micropropagation: Technology and Application.* Kluwer Academic Publishers, The Netherlands. pp. 1–13.

Debergh, P.C. and Zimmerman. (1991). *Micropropagation: Technology and Application.* Kluwer Academic Publishers, The Netherlands. pp. 484.

Deora, N.S. and Shekhawat, N.S. (1995). Micropropagation of *Capparis decidua* (Forsk.) Edgew-a tree of arid horticulture. Plant Cell Reports. 15: 278–281.

Kurtz, S.L., Hartman, R.D. and Irwin, I.Y.E. (1991). Current methods of commercial micropropagation. In: Vasil, I. K. (ed.). *Scale-up and Automation in Plant Propagation: Cell Culture and Somatic Cell Genetics of Plants.* Vol. 8, pp. 7–34.

Rathore, T.S., Deora, N.S. and Shekhawat, N.S. (1992). Cloning of *Maytenus emarginata* (Willd.) Ding Hou- a tree of the Indian Desert, through tissue culture. *Plant Cell Reports.* 11: 449–451.

Shekhawat, M.S. and Dixit, A.K. (2007). Micropropagation of *Phyllanthus amarus* Schum and Thom. through nodal shoot segment culture. *Int. J. of Plant Sci.* 2, 2, 79–84.

Shekhawat, M.S. and Dixit, A.K. (2008). Micropropagation of *Eucalyptus camaldulensis* Dehn. Clones selected for the Tsunami affected areas. *Int. J. of Plant Sci.* 2, 3, 31–35.

Shekhawat, N.S. (1997). Application of Biotechnologies at field site for adoption by rain-fed farming communities emphasizing participatory approaches-Micropropagation. In: Proc. Regional Training-cum-workshop on application of Biotechnologies to rainfed farming system, including bioindexing emphasizing participatory approach at community level. Sponsored by FAO/UNDP/UNIDO/MSS Research Foundation, Madras.

Shekhawat, N.S., Singh, R.P., Choudhary, N., Yadav, J., Shekhawat, M.S., Rathore, J.S. and Arya, V. (1999). Micropropagation: Achievements and limitations with special reference to plants of arid areas. In: Proc. National Symp. *"Role of Plant Tissue Culture in Biodiversity Conservation and Economic Development"*. GBPIHE&D, Kosi-Katarmal. p. 3.

8

CALLUS: INDUCTION AND DIFFERENTIATION

INTRODUCTION

The intact plant represents a highly organized, coordinate system in which correlative factors operate in an integrated way in the development of the whole plant. When its organs such as leaves, root tips, shoot tips and floral parts suffer from injuries, the normal function of the tissue system is hampered and the coordination between the cells operate in another way to protect and heal the injured portion. The vascular tissue or the tissues beneath the injury become meristematic and produce a mass of undifferentiated parenchyma cells which fill up the injured portion. This undifferentiated mass of cells forms a type of tissue that is known as wound callus tissue.

In basic plant biology, callus cells are those cells that cover a plant wound and protect the wounded tissues from microbial infection. However, in plant biotechnology, a callus is an unorganized mass of undifferentiated cells initiated from explant, in response to *in vitro* culture. Callus culture is mainly concerned with the initiation and continued proliferation of undifferentiated parenchyma cells from parent tissue on a defined nutrient media. Such cultures may be maintained for extended periods by frequent subcultures at 2–3 week intervals. This is a convenient approach for long-term maintenance of calli.

In vivo, callus is frequently formed as a result of wounding at the cut edge of a root or stem, following invasion by microorganisms or damage resulting from insect feeding. Its formation is controlled by endogenous auxin and cytokinin. By incorporation of these plant growth regulators into a growth medium, callus can be induced to form *in vitro* on explants of parent tissue.

A callus culture is usually sustained on gel media, much in the same manner as bacteria are grown. Nutrient media consist of agar and the usual mix of macronutrients along with micronutrients in specific proportions. For plant cells, a medium enriched with nitrogen, phosphorus and potassium is especially important. Water is provided as a constituent of both the gel and liquid media. In the early years of plant cell culture technology, dicots were used to a large extent in the initiation of a callus. But successful culture of excised tissues from mosses, ferns and gymnosperms were on the anvil, though considered less important.

Since the majority of plant cell cultures are photosynthetically incompetent and hence heterotrophic, it is necessary to supply them with a carbon source, usually in the form of sugar, for example, sucrose or glucose. Plant cells in culture also require the same macro and micronutrients that a whole plant may need, including mineral salts (usually supplied in the medium), amino acids, vitamins and plant growth regulators. The presence of plant growth regulators is generally essential to promote growth but the nature and required concentration varies from species to species. Typically, an auxin, such as indole 3-acetic acid (IAA), 2,4-dichlorophenoxyacetic acid (2,4-D) or naphthalene acetic acid (NAA) and a cytokinin, such as kinetin, benzylaminopurine (BAP) or zeatin are required either singly or more commonly in combination. There are numerous plant cell culture media documented in literature which facilitate callus formation. The most commonly used is that of Murashige and Skoog's (MS) medium. However, if there is no callus formation after incubation, the use of an alternative nutrient mixture should be investigated.

To raise the callus tissue, one must have a clear understanding of some basic concepts of the *in vitro* regeneration technique. In general, a cell or tissue excised from any part of the plant, when inoculated on a suitable medium under aseptic conditions, is able to dedifferentiate and multiply. This results in the formation of an amorphous mass of cells, which can be induced to redifferentiate on an appropriate medium in order to develop shoot buds or embryoids. These later on develop into young plantlets, which eventually give rise to whole plant regeneration on transfer to root induction nutrient media (Figure 8.1).

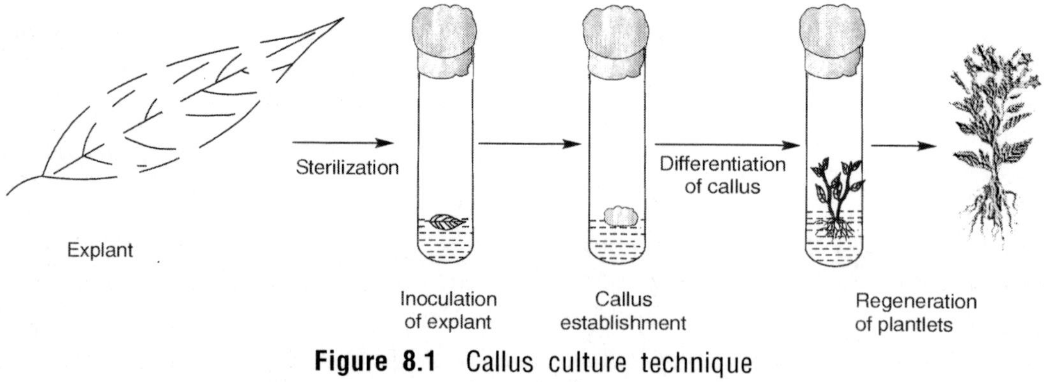

Figure 8.1 Callus culture technique

HISTORY

The history of callus culture can be summarized as follows:

- Gautheret (1934–1937) attempted to achieve *in vitro* induction of callus tissue. He first succeeded in promoting the development of callus tissue from excised cambial tissue of *Salix capraea* and other woody species.
- Nobecourt (1939) cultured tap root explants of *Daucus carota* on semi-solid agar medium. He was able to maintain the culture viability by simply transferring a part of the callus to a fresh medium at regular intervals.
- Caplin and Steward (1948) also worked on *Daucus carota* and obtained the growth of differentiated non-cambial cells on medium containing coconut milk.
- Skoog (1954 & 55) was able to achieve the callus tissue from cut pieces of the stem of *Nicotiana tabacum* in auxin-containing medium.
- Skoog and Miller (1957) experimented with the concept of hormonal control of organ formation from callus. They formulated that equal concentration of auxin and kinetin induced the continuous growth of callus tissue.

SOURCE TISSUE FOR CALLUS INDUCTION

A wide range of plant organs and specialized tissues can be used to initiate callus formation. However, sterile seedling and non-woody or immature stem tissue tend to facilitate ease of callus initiation. Seeds should be checked for viability before raising the seedling and excision of stem tissue for culture initiation should be devoid of senescence tendencies. Typical explants for callus initiation are segments of root or stem, pieces of leaf lamina, immature inflorescence, etc.

When an explant is cultured *in vitro* in a medium supplemented with proper growth substances, it absorbs growth substances and nutrients. Moreover, it grows by cell division asynchronously, enlarges in size, forms irregular cellular masses around the cut edge and forms an unorganized mass of callus tissue.

PREPARATION OF EXPLANT TISSUES

The growth rate of plant cells in culture is slow as compared to microbial growth. Since the employed plant growth medium is nutritionally rich, it is very important to maintain complete sterility, that is, free from microbial contamination, because microorganisms can grow in the supplied plant growth medium. Manipulations of tissue cultures should, therefore, be carried out using standard microbiological techniques in a sterile area.

Plants grown either in field or in greenhouse are much prone to contamination by microorganisms like bacteria, fungi, etc. Excised plant parts should necessarily be sterilized by chemical treatments prior to explant inoculation. Disinfection of the explants is a pre-requisite for *in vitro* culture initiation. The selection of sterilizing agent depends on the sensitivity of the plant tissue and the extent of surface microflora.

STERILIZATION OF EXPLANTS AND INOCULATION

Explants are first washed with liquid detergent (generally 5.0% "Teepol"). Further, explants are surface sterilized by the most commonly used sterilizing agents such as mercuric chloride or sodium hypochlorite for a limited time. After surface sterilization, the explants are repeatedly rinsed with autoclaved distilled water. Explant inoculation is carried out under aseptic conditions inside an LAF cabinet. The work bench must be wiped with alcohol intermittently during culture. It would be always ideal to wash the internal surfaces of the cabinet with alcohol at the commencement of work sessions.

Now, the explants are kept on a sterilized Petri dish containing a sterilized filter paper. The explant material is cut into a series of transverse slices by a sterilized sharp scalpel. The size of the explant is a critical factor in the induction of callus tissue from nodal segments in general and leaf base tissues in particular. Explants below a critical size and cell number may not respond at all to culture conditions. However, in general, callus induction is normally not dependent on the size or shape of the initial explants. It has also been observed that an explant consisting of uniform parenchyma cells may not be conducive for callus induction, in comparison to one that contains several types of cells. Presence of vascular tissue in explants is favourable to callus growth from parenchyma or cambial cells. The explants are finally inoculated aseptically on a suitable agar gelled nutrient medium.

FACTORS INFLUENCING CALLUS CULTURE

For plant cell culture, a medium enriched with nitrogen, phosphorus and potassium is required. Solidified agar medium or semi-solid nutrient medium, after preparation and sterilization by autoclaving at 121°C with 15 psi pressure for 15 minutes, is used for the induction of callus tissue. For healthy callus growth, usually both auxin and cytokinin are required. Incubation of culture under controlled physical conditions such as temperature, light and humidity is indispensable for the proper initiation of callus tissue.

The suitable temperature for *in vitro* callus induction and growth is usually 25 ± 2°C and approximately 2000 to 5000 lux artificial light intensity is needed. Furthermore, generally 60% to 75% relative humidity is to be maintained in the culture room. However, some plant tissues do not require light, and callus formation is possible even in dark. Once the growth of the callus tissue is well-established, fragments of the callus tissue can be removed and transferred directly to fresh nutrient medium to continue growth. In this manner, callus cultures can be continuously maintained by serial subcultures.

The explants are incubated under optimal conditions and observation data is collected regularly in terms of frequency and kind of explant responses. Initial response often takes place along the cut edge of the explant in contact with the agar or at the axillary bud removal site. Once sufficient callus growth

Figure 8.2 A culture showing friable callus with appearance of greenish structures all over the callus surface in a dicot plant (Plate 3.4)

is achieved (Figure 8.2), a part of it is carefully removed from the explant and transferred to a fresh medium. It is possible to maintain the callus line by subculture at regular intervals, for example, every 2–4 weeks, depending on the growth rate and chemical discharge from the callus to the culture medium. The well-developed callus (Figure 8.3) must be subdivided and a fragment of it can be transferred to fresh medium for further proliferation and regeneration. Care should be taken not to reculture senescent tissue and also, the subcultured callus tissue must be of a sufficient size to be able to maintain growth. A typical subculture cycle varies from 3–4 weeks for a healthy callus culture, however, if the callus exhibits rapid growth, the growth measurements should be made every 2–3 weeks, since it is important that the growth is not inhibited by nutrient depletion. On the other hand, if the callus growth rate is very slow, it is not advisable to extend the subculture cycle beyond 6 weeks. Further delay of subculture may cause gradual dry out of agar from the medium and this will affect the growth rate of callus cells.

When callus tissue is being transferred from one medium formulation to another, it is important to allow sufficient time for the cells to adapt to the new medium. This is probably necessary for two reasons:

1. The culture may survive the first transfer into "new" medium because of the carryover of nutrients from the old medium and it is necessary to reduce earlier nutritional effect before a meaningful measurement of growth is undertaken.
2. Secondly, transferring cells into "new" medium can often cause a drop in culture viability during the first few transfers. It is possible that with due course of time, these cells may adapt to the new medium, where cell viability will gradually be retrieved.

Figure 8.3 A culture showing induction and regeneration of friable callus (Plate 2.5)

When transferring a known weight of callus during growth experiments, it is important to ensure that the weight transferred consists of a single piece of callus rather than several pieces of calli. It is likely that the availability of nutrients to several small pieces will be higher than to a single large piece of callus. This may affect the apparent growth rate of the callus tissue. In most cases, it is possible to enhance the friability of callus tissue by either increasing the auxin or decreasing the cytokinin concentration in the growth medium.

STAGES IN CALLUS GROWTH

A callus tissue, during its growth and development, proceeds through various stages. A typical callus develops by the following four major interacting processes: 1) Growth (Induction) 2) Regressive change (Division) 3) Differentiation and 4) Pattern formation.

1. Induction

In the beginning, explant cells are inactive and they do not increase in size. The addition of plant growth regulators at this stage works as a triggering factor which targets a specific set of genes. These events increase cellular RNA content. Many cytological changes occur and these have been observed to explain the quiescent cell transforming to active cell.

The cells of the storage parenchyma of *Helianthus tuberosum,* when excised for callus culture, results in synchronous division at the cut edges. This division has been greatly facilitated by constitutional changes before cell division. After excision, the initial cell division cycle falls into G_1, S, G_2 and mitosis phases. The G_2 phase is short and the DNA synthesis phase is very quickly followed by mitosis. After excision and during first cell cycle, the accumulation of DNA, RNA and total protein is discontinuous. The oxygen uptake increases until DNA synthesis starts, while RNA accumulation takes place and it occurs in the following squences:

i. About 12% increase in the total RNA has been seen after excision of explants.
ii. Six hours after excision, 30 to 40% increase in RNA accumulation has been noticed.
iii. Last increase occurs just before the start of the mitosis phase.

The pattern of total protein accumulation is same as for total RNA accumulation. A marked change can be observed in the protein composition of quiescent cells and in the cells that are about to divide. The enzyme activity pattern in the cell is also changed. All enzymes increase in activity after DNA replication begins. Moreover, DNA synthesis proceeds during most of the cell cycle, however, RNA accumulation is not linear and periodic enzyme synthesis is the feature of the second cycle after excision.

2. Division

This phase includes active synthesis of cells and decrease in cell size is initiated due to the occurrence of division. A regressive change results in this phase. As the metabolic changes take place and synthesis of metabolic substances occur, the entire process culminates in cell division. Not all the cells in a piece of excised tissue respond uniformly during this phase of callus culture. Only the cells in the peripheral or outer regions are induced to divide. Thus, the excised tissue consists of actively dividing peripheral cells surrounding the non-dividing core cells. There are also some cells in the periphery callus, which fail to divide, but they are not similar to the core cells. The non-dividing cells of the periphery accumulate substantial amounts of protein for preparation for later division.

It is interesting to note that the cells which are potent to divide can synthesize and accumulate RNA. The initiation of division in the periphery is related to a number of interacting factors—wound response at the cut surface, greater availability of oxygen, more rapid release of CO_2 and rapid release of a volatile inhibitor.

3. Differentiation

Differentiation phase starts only after the division phase proceeds throughout the outer region of the callus, leaving a small core of undivided cells within. The formation of differentiated structures begins to supersede the regressive changes. In regressive change, it is observed that there is an increase in cell number as the division of cells proceeds swiftly. The rise in cell number does not keep pace with increase in fresh weight. The average cell size reduces as the rise of cell number falls later on.

Dedifferentiation is presented in such a manner that initially RNA and other constituents of the dividing cell diminish with succeeding division. At some stage, it will be observed that the division and accumulation of cellular constituents are stopped and the decrease in the size of the cell is arrested. Thus, the cell is small and non-vacuolated like a typical cell found in the apex of the root and which reverts to the meristematic state. RNA level per cell declines and leaves only DNA to retain its constancy. The phenomenon of decreasing cell size is interesting and it is perhaps concerned with enzyme activity involved in DNA replication which remains constant during the initial period of division.

4. Pattern Formation

At the end of the division phase, morphogenic changes take place in the callus. Peripheral cell division slows down and is supplemented by secondary growth formation. On the radial rows of cells of the wounded cambium, there appears to be a new pattern of cell differentiation. It may be superimposed upon wounded cambium. This new course is marked by increasing cellular differentiation. This is reflected by the appearance of nodular structures resembling vascular bundles. The meristematic nodules become the growth centres and produce expanded parenchymatous cells from their periphery. These cells form the frothy proliferation and expand considerably, forming large inflated cells. This cell expansion is balanced by cell division continuing throughout this phase so that a more or less mean cell size is established.

Depending on the type of explants, either enlargement in size or the swelling of explants is followed by rupture of the tissue within few days of inoculation. This change indicates the response of explants to callus formation and is followed by the appearance of little irregular cellular masses around the cut edges or from the ruptured surface. Thus, the initial formation of cellular mass, particularly at the cut end, may be due to injury during excision. Some endogenous growth substance oozes out through the injured tissue at the cut end and stimulates cell division which is simultaneously induced by the exogenously supplied hormones. There is another explanation that both endogenous product and exogenous hormones make a threshold level and their interaction results in the formation of unorganized cellular growth at the cut end.

Whatever may be the fact, it is generally accepted that exogenous hormones play a major role in the formation of callus tissue in culture. Both auxin and cytokinin are required for indefinite growth and cell division in callus culture. Sometimes, 2,4-D alone is sufficient for callus formation. Regarding the action of auxin and cytokinin, there is a general belief that auxin is required for cell growth and cytokinin for cell division.

MORPHOLOGY OF CALLUS TISSUE

The texture and physical properties of the callus tissue are different from normal plant tissue. Callus tissue proliferates as an amorphous mass of cells having no regular shape. So, it is very difficult to describe its external morphology. Apparently, callus tissue originated from any explant tissue and excised from any plant source looks alike that is, homogenous mass of cells; however, these can be distinguished on the basis of texture, colouration, hormone requirements, etc. Even callus tissues initiated from explants of the same plant species may show considerable variations among themselves. On the basis of texture, some callus tissues are compact and hard while others are friable and disintegrate freely. The compact hard callus may be nodular in form and grows as lumps and does not give suspensions, while the friable soft callus is suitable for suspension culture, and mechanical agitation leads to a dispersion of the tissue.

The internal structures of these two types of calli have demonstrated spectacular differences. In friable types of callus, large numbers of organized meristematic centres are separated by large number of undifferentiated cells. Non-friable types of calli exhibit less differentiated cells with much large portion of vacuolated cells. The compact hard callus has higher amount of cell polysaccharides but a decreased percentage of cellulose compared to pectin substances. The greater amount of cellulose will increase the rigidity of the cells. The soft callus shows lesser amount of cellulose.

Points to remember in callus culture

1. If the stem tissue available is too narrow to be cut into half lengthwise, then a cut should simply be made along the length of the explant and the cut side has to be placed on the surface of the agar.
2. It is preferable to initiate and maintain callus lines in either continuous light or continuous dark, so that if required, the effect of photoperiod can be easily investigated.
3. The morphology of callus cultures can be controlled by the manipulation of the plant growth regulators in the medium. Increasing auxin concentration in nutrient medium will increase the friability of the callus, which is generally required to be used to initiate suspension cultures.
4. Similarly, differentiation of callus can be induced or halted by manipulating the ratio of auxin to cytokinin in the medium. For example, a high ratio of cytokinin to auxin can induce the formation of shoots and a high ratio of auxin to cytokinin can induce the formation of roots in general and dicot callus tissue in particular.

HABITUATION OF CALLUS TISSUE

Generally, callus tissue needs growth hormones in the nutrient medium in order to grow as long as they are maintained through serial subcultures. But it has been observed that the callus tissue in some plant species may become habituated after prolonged culture. This means that the callus tissue is able to grow on a standard maintenance medium or basal medium which is devoid of growth hormones. This property is known as habituation and the callus tissue is known as habituated callus tissue.

This is possible because the cells in callus tissue, after repeated subcultures, develop the capacity to synthesize adequate amount of auxins and/or cytokinins. This probably accounts for their independence of exogenously supplied hormones. Even the plant tumour tissue can be grown directly on the simple basal medium without addition of any hormone, just like the habituated callus.

APPLICATIONS OF CALLUS CULTURE

1. In plant cell culture, callus consists of somatic undifferentiated cells originated from mature, fully differentiated plant tissues. A callus need not necessarily be genetically homogeneous because a callus is often made from structural tissue, not individual cells. Nevertheless, callus cells are often considered similar enough for standard scientific analyses to be performed on a single subject. For example, an experiment may be carried out where one half of a callus can be given a specific treatment as the experimental group, while the other half undergoes a similar but non-active treatment as the control group.
2. Plant cell callus may be made to differentiate into the specialized tissues of a plant or even a whole plant with the addition of a number of hormones or enzymes. This is an inherent ability known as totipotency.
3. Callus tissue is of particular use in micropropagation where it can be used to grow genetically almost identical copies of plants with desirable characteristics.
4. The propagation of elite or endangered plant species may be achieved and this way germplasm conservation of rare species is possible.
5. The whole plant can be regenerated in large numbers from callus tissue through manipulation of the nutrient and hormonal constituents in the culture medium. This phenomenon is known as plant regeneration.
6. Somatic embryos can be developed directly from the somatic cells of explant tissue and these embryos directly give rise to the complete plant.
7. Callus culture may be a good source of genetic or karyotypic variability, so it can be used to produce genetically variable cells and tissues.
8. Callus cultures are useful in the production of valuable secondary metabolites.
9. Callus can be a prime source to initiate cell suspension culture in liquid medium.
10. It is also the suitable material for protoplast isolation, which is employed for somatic hybridization and transgenic crop experimentation and production.
11. Callus is often the suitable target for specific DNA insertion in gene delivery technique.

REVIEW QUESTIONS

1. What is callus? Define callus culture and how it can be induced from explant tissue?
2. What are the processes in callus culture establishments?
3. Explain the process of habituation and its significance.
4. Write notes on applications of callus culture.
5. Write notes on the following:
   i. Induction
   ii. Differentiation
   iii. Morphology of callus tissue

REFERENCES

Aitchison, P.A., Macleod, A.J. and Yeoman, M.M. (1977). Growth patterns in tissue (callus) cultures. In: Street, H.E. (ed.). *Botanical Monographs*. 11. Oxford, Blackwell.

Halder, T. and Gadgil,V.N. (1982). Shoot bud differentiation in long-term callus cultures of *Momordica* and *Cucumis*. *Indian J. Experimental Biology*. 20: 780–782.

Henshaw, G.G., O'hara, J.F. and Webb, K.J. (1982). Morphogenetic studies in plant tissue cultures. In: Yeoman, M.M. and Truman, D.E.S. (eds.). *Differentiation in vitro: British Society for Cell Biology Symposium* 4. Cambridge University Press. pp. 231–251.

Holmberg, N. and Bulow, L. (1998). Improving stress tolerance in plants by gene transfer. *Trends in Plant Science*. 3: 61–66.

Moreno, V., Garcia-Sogo, M., Granell, I., Garcia-Sogo, B. and Roig, L.A. (1985). Plant regeneration from calli of melo (*Cucumis melo* L. cv. Amarillo Oro). *Plant Cell, Tissue and Organ Culture*. 5: 139–146.

Vikrant and Rashid, A. (2001). Direct as well as indirect somatic embryogenesis from immature (unemerged) inflorescence of a minor millet *Paspalum scrobiculatum* L. *Euphytica*. 120:167–172.

Vikrant and Rashid, A. (2001). Comparative study of somatic embryogenesis from immature and mature embryos and organogenesis from leaf-base of *Triticale*. *Plant Cell, Tissue and Organ Culture*. 67:33–38.

Yu, H.J., Oh, S.K., Oh, M.H., Choi, D.W., Kwou, Y.M. and Kim, S.G. (1997). Plant regeneration from callus cultures of *Lithospermum erythrorhizon*. *Plant Cell Reports*.16: 261–266.

9

CELL SUSPENSION CULTURE

INTRODUCTION

In multicellular eukaryotes, all cells are derived from a single, uninucleate zygotic cell, which by the process of controlled division, forms millions of cells of the body. Each and every cell receives the same genetic composition as originally present in the zygote. Thus, studies on single cell culture play a critical role in understanding the mechanism behind the inter-relationships among the various kinds of cells in a multicellular system. Moreover, knowledge of a single cell culture and its different developmental pathways of differentiation offer an opportunity for researchers to establish the links between cell communication and cell behaviour.

Cell culture is a method of growing isolated cells aseptically, obtained by any method and culturing it on a nutrient medium for its growth. If culture initiation and maintenance of cell growth and differentiation is carried out in an agitated liquid nutrient medium, it is known as cell suspension culture. Liquid cultures must be constantly agitated, generally by a gyratory shaker at 100–150 rpm (revolutions per minute), to facilitate aeration and dissociation of cell clumps. Suspension cultures grow much faster than callus cultures and require subculture every week. This allows a more accurate determination of the nutritional requirements of cells and is amenable to scaling up for a large-scale production of cells and even somatic embryos (SEs).

The basic reason behind the development of cell suspension culture technique is the elimination of disadvantages faced during callus culture establishment. Moreover, cellular events in callus culture are possibly different and therefore, it is relatively not simple to analyse the exact mechanism of cell differentiation in an unorganized mass of cells.

HISTORY

- Suspension culture of plant cells was first established in agitated liquid media by Caplin and Steward in 1949.
- Muir *et al.* (1954) reported that the fragments of callus of *Nicotiana tabacum* could be cultured in cell suspension.
- Steward and Shentz (1956) obtained large number of plantlets from the carrot root explant's suspension cultures.
- Tulecke and Nickell (1959) envisaged the production of mass amounts of plant tissue by submerged culture and its practical use for the study of biosynthesis of secondary compounds.

INOCULUMS FOR CELL SUSPENSION CULTURE

Callus is a mass of unorganized and undifferentiated cells. It is generally used to achieve an ideal cell suspension (Figure 9.1). Friable callus is transferred to agitated liquid medium where it breaks up anddisperses.

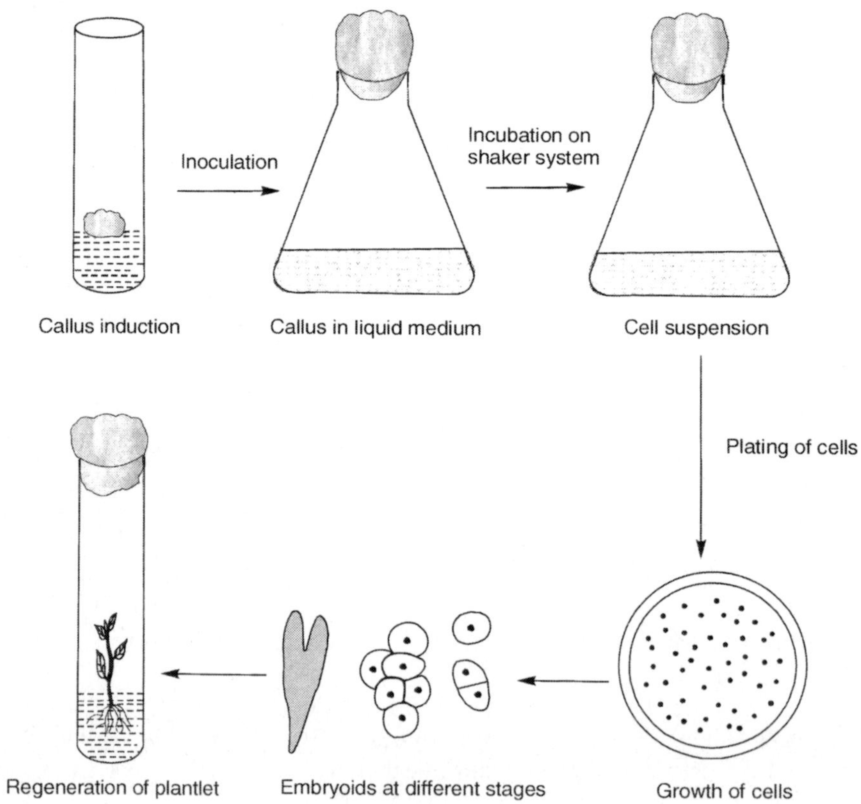

Figure 9.1 Stages of cell suspension culture and regeneration of plantlets

After discarding the larger parts of callus, only small clumps of cells or small cell aggregates are transferred to a fresh medium and after three weeks of subculture, actively growing cell suspension is produced. This suspension is then propagated by regular subculture of an aliquot to fresh medium. In principle, suspension culture should consist of single cells which are physiologically and biochemically uniform. Movement of cells in relation to nutrient medium facilitates gaseous exchange, removes any polarity of the cells due to gravity and eliminates the nutrient gradients within the medium and at the surface of the cells.

Also, suspension cultures can be initiated directly from the explant into a liquid medium, which can be continuously agitated on a moving rotatory shaker. As cell division starts on the explant surface tissues, cells gradually shed away and disperse directly into the medium. An increase in the concentration of auxins or addition of very low concentration of cellulase and pectinase enzymes in the liquid medium is also effective in segregating and dispersing the cells. The cells in suspension may vary in shape and size (Figure 9.2). They may be oval, round, elongated, etc. Suspension culture cells are generally thin-walled.

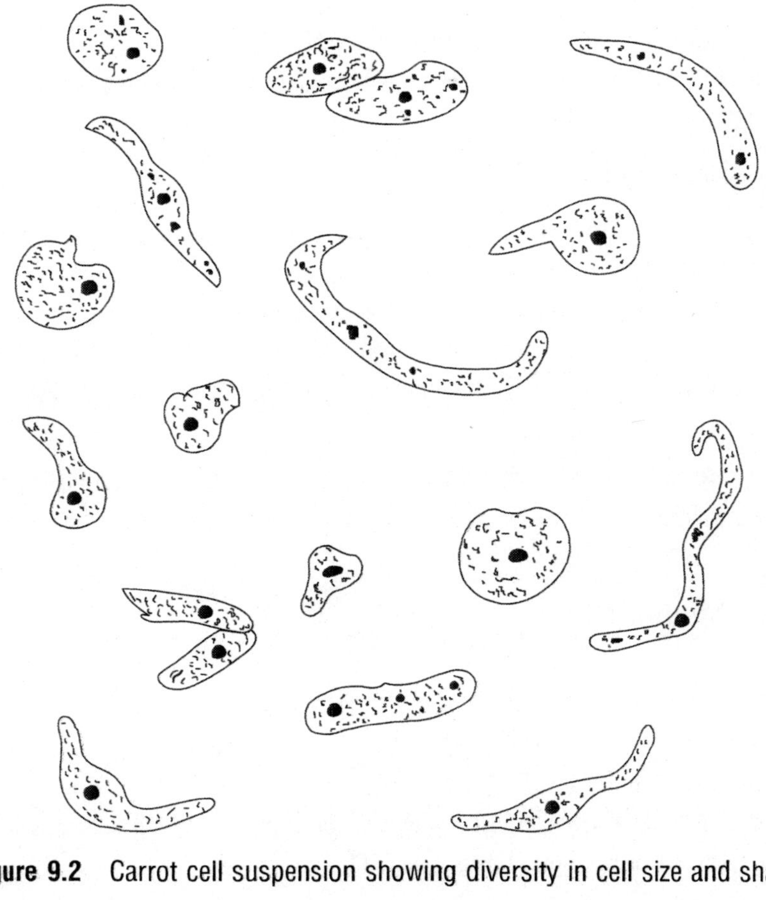

Figure 9.2 Carrot cell suspension showing diversity in cell size and shape

Use of liquid medium in suspension cultures allows easy and extensive scaling up by employing bioreactors, although a limited scaling up can be achieved by using larger culture flasks (usually a 50-ml flask having 50 ml of medium). A bioreactor is a culture vessel, generally with a large volume (ranging from 1 to 1000 litres) and having provisions for aeration, stirring to achieve mixing of cells in media, contamination control and replacement of used medium. Bioreactors used for culture of plant cells may be of following four types:

1. Batch bioreactors
2. Continuous bioreactors
3. Multistage bioreactors
4. Immobilized cell bioreactors

The first three bioreactors are called as stirred tank reactors.

ISOLATION OF CELLS

Suspension culture with single cells can be obtained from intact plant organs either mechanically (grinding the tissue followed by cleaning, filtration and centrifugation) or enzymatically (treating excised and peeled leaves with macerozyme).

1. Mechanical Method

Rossini (1969) made large-scale mechanical isolation of parenchymatous cells from the leaves of *Calystegia sepium* and *Arachis hypogaea*. The leaves are initially surface sterilized with 95% ethanol followed by brief rinsing in 0.1% solution of $HgCl_2$ and finally washed with sterile distilled water. Then, leaves are cut into pieces and 2.0 g of the leaf material is taken and ground with 10 ml of culture medium. Homogenates are filtered through metal Tyler filters. The filtrate is further centrifuged and the suspended cells are cultured in liquid medium.

The advantage of this method is that the possible harmful effect of chemicals and enzymes are eliminated and the process does not involve plasmolysis, which may not be beneficial for plant cell isolation.

2. Enzymatic Method

Originally this method was described by Tekebe *et al.*, in 1965 and was modified by Evans and Cooking in1975. In this method the surface sterilized leaf is utilized and the lower epidermis is peeled off with sterilized forceps. About 2.0 g of peeled leaf pieces are treated with an enzyme solution (0.5% macerozyme, 0.8% mannitol and 1.0% potassium dextran sulphate). After fixed time duration, the enzyme with pieces of the leaf part is filtered with a vacuum pump. The flasks containing filtrates are incubated at 25°C for 2 hours in a rotatory flask. The enzyme solution should be changed every

30 minutes. Finally, these cells are washed and cultured. Thus, this method gives us true single type of cells instead of mixed type cells.

TYPES OF SUSPENSION CULTURES

The medium used for suspension culture is usually the same as for callus growth except that agar is not included. A change or manipulation of auxin/cytokinin ratio may be desirable to achieve cell suspension. The suspension culture needs to be regularly agitated to avoid clumping of cells and to provide aeration. A shaking speed of 30 to 130 rpm is satisfactory for this purpose. Based on the method of culture, there are three types of suspension cultures:

1. Batch culture
2. Continuous culture
3. Immobilized cell culture

1. Batch Culture

In this method, the same media and all the constituent cells produced are retained in the culture vessel till the end of the culture process. Here cells grow in a finite volume of nutrient medium. Culture flasks of various sizes (100–250 ml) are used as culture vessels. Each vessel contains 20–75 ml of culture medium for culture initiation and routine growth of the cells. These types of cultures are most commonly maintained in conical flasks, incubated on orbital platform shakers at the speed of 80 to 120 rpm.

The cell biomass of batch culture exhibits a typical sigmoidal curve (Figure 9.3), having a lag phase, a logarithmic phase (log phase) and a stationary phase. The lag phase is the period where the cells adjust themselves to the nutrient medium and synthesize everything necessary prior to cell division. This is followed by very rapid cell division leading to logarithmic increase in cell number. This phase is called as logarithmic phase. As nutrients are depleted and some of the cells of the culture begin to show senescent characteristics, the rate of cell division within the culture declines and it passes through the stationary phase. The lag phase duration depends mainly on the inoculum size and the growth phase of the culture from which inoculum is taken. The log phase lasts for about 3–4 cell generations, and the duration of cell generation may vary from 25–45 hours depending upon the plant species. The stationary phase is forced on the culture by depletion of the nutrients and possibly due to accumulation of cellular wastes. If the culture is kept at stationary phase for a longer period, the cells may lyse and die.

Batch cultures can be maintained on frequent subculture at regular intervals. In every case, the cultures are continuously propagated by taking a small aliquot (20 ml, 40 ml or 60 ml) of the cell suspension and transferring it to a fresh medium. Batch cultures are properly maintained by regular subculture. They are used for initiation of cell suspensions, which may be used for cloning, cell selection or as seed cultures for scaling up and for continuous cultures. Batch cultures are much more convenient than continuous cultures and, as a result, are routinely used.

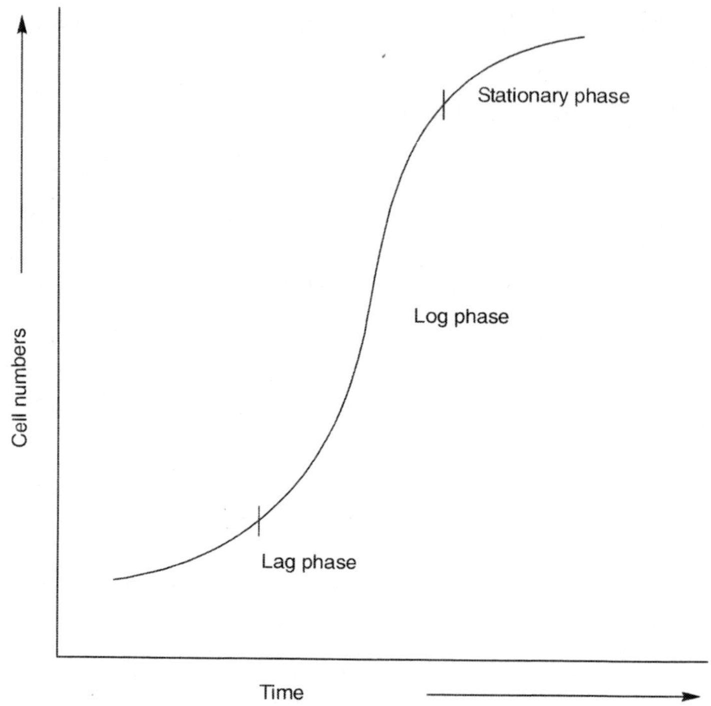

Figure 9.3 Sigmoidal curve for cell numbers in a batch culture

The growth of cells is controlled by the culture medium; this type of culture is not suitable for cell growth and metabolism because there is a constant change in cell density and nutritional status of the medium. The steady-state growth period may not be achieved for a longer period of time as the relative cell concentrations of metabolites and enzymes are constant.

2. Continuous Culture

In this method, the cell population is maintained in a steady state by regularly replacing the same quantity of spent medium by the fresh liquid medium to stabilize the physiological state of the growing cells. Normally, the liquid medium need not be changed until the depletion of some nutrient medium components. Also, if the cells are kept in the same medium for a longer period, the medium is changed. In continuous culture system, nutrient depletion does not occur due to continuous flow of nutrient medium and the cells always remain in the steady state of active growth phase. Continuous culture systems are of the following types:

i. ***Closed type continuous culture*** In this culture, cells are separated out from the used medium that has to be replaced, and added back to the culture. So, the cell biomass keeps on increasing with the age of the culture.

ii. ***Open type continuous culture*** In this culture, both cells and used medium are taken out from cultures and replaced by providing equal volume of fresh medium. The replaced medium and cells are adjusted so that cultures remain at sub-maximal growth indefinitely. The open cultures are of either chemostat type or turbidostat type.

In chemostats, culture vessels are generally cylindrical or circular in shape and possess inlet and outlet pores for aeration and introduction and removal of cells and medium. The inflow of fresh sterilized medium, which is pumped in at a constant rate into the vessel is balanced by the displacement of an equal quantity of used medium and cells. Thus, in a steady-state condition, the density, growth rate, chemical composition and metabolic activity of the cells remain constant. In the turbidostat, cells are allowed to grow up to a pre-selected turbidity when a predetermined volume of the culture is replaced by fresh, normal culture medium. The changes of turbidity of the culture medium can be measured by the changes of optical density of the medium.

3. Immobilized Cell Culture

Immobilization is an effective technique for the industrial production of some valuable chemicals. It avoids the wastage of cells during the recovery of products. Plant cell and cell clumps may be entrapped in a suitable material, for example, agarose, polyacrylamide, polyurethane fibre, calcium alginate, etc. or in membranes or stainless steel screens. The gel arrests the free movement of cells and the leakage of cells into the reaction mixture. Liquid medium is continuously run through the gel to provide nutrients and aeration to cells. Immobilization of cells changes their cellular or physiological condition in comparison to suspension culture cells. Immobilization induces the division of cells entrapped in the gel. Hence, the biosynthetic activity of the cells remains constant.

An immobilization system, which could maintain viable cells over an extended period of time and release the bulk of the product into the extracellular medium in a stable form, could dramatically reduce the costs of phytochemical production in plant cell culture. However, an immobilized system also has some limitations as listed below.

i. Immobilization is normally limited to cases where production is decoupled from cell growth.

ii. The initial biomass must be grown in suspension cultures.

iii. Secretion of product into the extracellular medium is imperative.

iv. At the site where secretion occurs, there may be chances of extracellular degradation of the products.

v. When gel entrapment is used, the gel matrix introduces an additional diffusion barrier.

SUSPENSION CULTURE GROWTH/SUBCULTURE

After a period of time, it becomes necessary to transfer cells and tissues to fresh media mainly due to nutrient depletion and medium drying. This is particularly true of tissue and cell cultures, where a part of tissue is used to inoculate new culture flasks and this is known as subculture.

In general, suspension cultures need to be subcultured every 3–4 days. Plant cell and tissue cultures may be maintained indefinitely by serial subculture to fresh nutrient medium. The subculture step should be done prior to the time of their maximum growth. The inoculum volume should be 20–25% of the fresh medium volume; in any case, the initial cell density of the fresh culture (just after inoculation) should be 5×10^4 cells/ml or higher; otherwise, the cells may fail to divide. When the cell number in suspension cultures is plotted against the time of incubation, a sigmoidal growth curve is obtained. It depicts that culture initially passes through a lag phase, followed by a brief exponential growth phase and a decline in growth after three to four cell generations, indicating the stationary phase. Cell suspensions are clonally maintained by the routine transfer (subculture) of cells in the early stationary phase to a fresh medium. During the incubation period, the biomass of the suspension cultures increases due to cell division and cell enlargement.

For a subculture, the flask containing the suspension culture is allowed to stand for a few seconds to enable the large colonies to settle down. A pipette or a syringe with an orifice, fine enough to hold aggregates of two to four cells, or only single cell, is used. The suspension is taken from the upper part of the culture and transferred to a fresh medium. The incubation period from culture initiation to the stationary phase is determined primarily by:

1. Initial cell density
2. Duration of lag phase
3. Growth rate of cell line

While initiating a new suspension culture, in order to achieve maximum growth, it is necessary to determine optimal cell density proportionate to the volume of the culture medium.

CELL CULTURE SYNCHRONIZATION

Suspension culture in which all the constituent cells are at the same stage of cell cycle is called synchronous culture. It is very useful to study gene regulations, cell metabolism and biochemical pathways. Cells in a suspension culture vary greatly in size and shape, but have the same nuclear status and DNA content. Moreover, the cell cycle time varies considerably within individual cells. Therefore, cell cultures are mostly asynchronous. It is essential to manipulate the growth conditions of an asynchronous culture in order to achieve a higher degree of synchronization. A synchronous culture is one in which the majority of cells proceed through each cell cycle phase (G_1, S, G_2 and M) simultaneously. Synchronization is expressed as percentage synchrony of cells in suspension cultures. Synchronization can be achieved by physical or chemical methods.

1. ***Physical methods*** These methods include selection of cells of similar size from the culture after temperature shock (low temperature). Physical properties of cells (size of individual cell, small cell aggregates) or environmental growth conditions (light, temperature, etc.) are factors that are generally monitored.
2. ***Chemical methods*** These methods are grouped into three categories:
   i. ***Inhibition*** It involves inhibition of the progression of events temporarily in the cell cycle using a biochemical inhibitor and then releasing the block. Inhibitors of DNA synthesis (aminouracil, 5-fluorodeoxyuracil(FUDR), thymidine, etc.) are used to achieve synchrony of cell division.

   ii. ***Starvation*** Cells are kept temporarily on starvation of nutrients (suspension cultures deprived of an essential growth compound and resupplying) necessary for cell division, so that cells are arrested at G_1 or G_2 stage. These arrested cells are supplied with the nutrient to obtain synchronous cell division.

   iii. ***Mitotic arrest*** Arresting of cells at metaphase stage by adding colchicine to the suspension culture is called mitotic arrest.

MEASUREMENT OF CELL GROWTH

Cell number is estimated only when information obtained justifies the efforts. In contrast, packed cell volume of suspension cultures is easily determined by pipetting a known volume into a 15 ml graduated centrifuge tube, spinning at 200 g for 5 minutes and then reading the volume of the cell pellet, which is expressed as per ml cells of culture. Fresh and dry weights of culture are the most commonly used measures of growth of both suspension and callus cultures. In the case of callus cultures, the cell mass is placed on a pre-weighed dry filter paper or nylon filter and weighed to determine fresh weight. Cells from suspension cultures are filtered onto a filter paper or nylon filter, washed with distilled water, excess water removed under vacuum and weighed along with the filter; the filter is pre-weighed in wet condition.

For dry weight determination, the cells and the filter are dried in an oven at 60°C for 12 hours and weighed; the filter is pre-weighed in dry condition. Fresh and dry weight of the cells may be expressed either as per ml (suspension culture) or per culture. Assessment of growth in suspension cultures is accomplished by cell counting, packed cell volumes, and fresh or dry weight increase of cells and cell colonies. Cell count is a relatively more accurate measure adopted to determine the growth of cultures. Determination of cell number is a simple but tedious procedure since suspension cultures invariably carry cell colonies of various sizes.

Methods of Cell Growth Measurement

Several methods are used for the measurement of cell growth in suspension cultures. Some important methods are given below:

1. ***Cell packing method*** A 15 ml of culture suspension is transferred to a graduated measuring tube and centrifuged for a few minutes. The cells get settled at the bottom of the tube. The total volume of packed cells is measured against the reading of the graduated measuring tube. Then, the concentration of cells in the culture is expressed as a percentage of the total volume of the suspension.
2. ***Direct cell count method*** Cell number is the most useful method to measure cell growth. This measurement is applicable only to suspension cultures, and cell aggregates must be treated with pectinase, so as to dissociate them into single cells before counting the cell number in a haemocytometer. Cell count data obtained from haemocytometer is multiplied by a factor $\times 10^3$ and the result can be expressed in terms of cell number per unit volume of culture. By comparing the cell numbers at the beginning of culture and after certain days of incubation, the growth can be measured.
3. ***Protein measurement technique*** Protein is extracted from a known volume of cell suspension, and is measured by using Lowry's method. The protein content is expressed in mg/ml. The protein content of the filtrate should not exceed one mg/ml. It indicates the maximum concentration of cells in the suspension. Immediately, the suspension should be subcultured for further use.
4. ***Direct weight method*** Fresh and dry weight of cultures are the most commonly used methods to measure the growth of suspension cultures. A known volume of cell suspension is filtered through Whatmann filter number 1. The filter is dried and weighed. Its original weight is subtracted from the final weight. The weight of cells is expressed in g/litre.

CELL VIABILITY TESTS

The growth of suspension cultures depends on the number of viable cells present in the culture. For proper maintenance of suspension cultures, all the cells should be in the viable condition. Viability of cells in suspension can be assessed by viewing a sample of the suspension under a microscope, after staining or without staining. Cell viability can be determined by any one of the following approaches:

1. Phase contrast microscopy
2. Staining with 2, 3,5-triphenyltetrazolium chloride (TTC)
3. Staining with fluorescein diacetate (FDA)
4. Staining with Evan's blue

Live cells show cytoplasmic streaming and a well-defined healthy nucleus which are easily observable with a phase contrast microscope or even a light microscope. Cell masses can be stained with 1–2% solution of TTC, which is reduced by living cells to formazan that yields red colour. Formazan can be extracted and measured with a spectrophotometer to give a quantitative estimation of viability, but it is not suitable for single or few cells.

Apart from observations of protoplasmic streaming and presence of an intact nucleus and staining with TTC solution, cell viability is also assessed by the use of vital stains such as fluorescein diacetate (FDA). FDA dissolves in 5.0 ml of acetone and is added to cell population at 0.01% of the final concentration. A dead cell fluoresces red and living cells green under UV light. Evan's blue, used at a concentration of 0.01%, is specific for dead cells. The solution stains these dead cells blue but viable cells remain unstained.

Evan's blue is not taken up by live cells while it freely enters into damaged dead cells. Therefore, such cells that take up the stain are dead cells. Evan's blue is usually applied in conjunction with FDA.

SUSPENSION CULTURE AND SECONDARY METABOLITE PRODUCTION

Plant cell suspension cultures are mostly used for the biochemical investigation of cell physiology, growth, metabolism, and for large- or medium-scale production of secondary metabolites. For such purposes, normally suspension cultures are used which are propagated in Erlenmeyer flasks on a gyratory shaker and are maintained by regular subculturing after short intervals (usually 1 or 2 weeks).

For such use, callus cultures are also maintained under a mineral oil layer that reduces the subculture frequency. However, from the callus phase, it takes additional time to establish a new suspension culture again, and this new suspension may exhibit changed biochemical traits.

STORAGE OF CELL SUSPENSION CULTURE

For storage purposes, several strategies have been developed, all aiming at reduction of labour and costs of maintenance, while preserving all properties of the cells. Long-term conservation of suspension cultures is usually successful by cryopreservation. However, it requires special and expensive equipment and is not suitable for routine work, as reactivation of cells requires prolonged incubation and has to pass through a callus phase. Therefore, for practical work there is a need for an easy and a time-saving method that allows the storage of actually investigated suspension cultures for a relatively longer time.

For storage, 100 ml of the cell suspension from the early exponential growth phase is transferred to a 650 ml screw-capped plastic bottle with membranes, which permits sterile air to pass through. Depending on the volume of cell suspension to be maintained, smaller 250 ml bottles with 43 ml cell suspension can also be used. However, the surface-to-volume ratio should remain the same as with big bottles. These bottles are supplied as tissue culture flasks that prevent contamination. Alternatively, wide-mouthed Erlenmeyer flasks with new cotton plugs can also be used. The volume of the added cell suspension in the vessels should lead to a high surface-to-volume ratio in order to assure sufficient oxygen supply for the cells. The suspension cultures are stored without shaking in the dark and at a reduced temperature of 10°C. To prolong cell survival, it is recommended to add some stabilizers, such as 0.01% activated charcoal and 1.0% gelatin, or double to triple concentration of sugar.

REGROWTH TEST

For regrowth, an aliquot of 5 ml of the stored cell suspension is transferred with a sterile pipette from the plastic bottles to 100 ml Erlenmeyer flasks containing 25 ml of the medium. The cells are cultured in these flasks for 1–2 weeks at 24°C on a rotatory shaker (100 rpm) under continuous light (600 lux).

To determine regrowth on solid media 1.0 ml of cell suspension is placed on a Petri dish (60 mm diameter). For a periodic viability assay of the preserved cultures, the ability of regrowth is determined. After appropriate intervals of storage (every 2–3 weeks or longer, depending on the sensitivity of the cell suspension), aliquots of the stored cells and the control samples (grown under normal conditions) are recultivated in equal amount of liquid media. After a definite period of growth (about one week when the cells are in exponential growth phase), the cells are harvested by suction filtration. The dry weight of the regrown cells is determined, which is taken as a measure of the inoculum quantity (the cell density developed from the surviving cells during storage). The resultant cell mass from the stored cells is compared after different storage intervals and with the control samples.

STABILITY TEST

For stability checking by high pressure liquid chromatography (HPLC), 2 g (fresh weight) of the harvested cells are extracted in boiling methanol for 1 hour. Twenty ml of the resulting extracts are analysed by reverse phase HPLC. The result of the chromatographic procedure is a characteristic "fingerprint" of cell metabolites. Methanol is used as an organic solvent. Substances are eluted by a standard gradient system, which runs from 0% methanol to 100% methanol within 30 minutes. The majority of cell cultures show most characteristic "fingerprints" when a slightly acidic solvent system is used. Comparability of such "fingerprints" has to be assured by extracting cells from the same growth stage of the cell culture.

APPLICATIONS OF CELL SUSPENSION CULTURE

Cell suspension cultures are capable of contributing significant information about cell biochemistry, physiology and metabolic pathways at the level of cells. Some of the main applications are as follows:

1. ***Mutant selection*** Suspension cultures can be utilized for crop improvement through *in vitro* mutant selection. The frequency of mutation can be increased several-fold through mutagenic treatments, and millions of cells can be screened. Since selection of mutations are exercised at cellular level, no chimeras are obtained.

   If haploid cells are used for cell culture, the problem of dominance does not permit the expression of mutation in F_1 generation. For example, tobacco protoplasts selected for resistance to methionin sulfoximine (MSO) should enhance resistance to *Pseudomonas tabacii* also, even though MSO is not the phytotoxin of *P. tabacii*. Cell cultures of *Nicotiana* are also selected, for resistance against several herbicides like amirol, sulfometuron methyl, chlorsulfuron and sodium chlorate. The selected mutant cells are grown into whole plants which are then multiplied and tested for the mutant phenotype.

2. ***Production of secondary metabolites*** Cell cultures are used as a source for extraction of cellular enzymes and many secondary metabolites. Plants are important sources for a wide variety of chemicals used for a variety of purposes including pharmacy, medicine and industry. The yield and quality of the product is more consistent in cell cultures because it is not influenced by the environment. Again, the production schedule can be predicted and controlled in the laboratory in suspension. These secondary metabolites are alkaloids, steroids, phenolics, terpenoids and a variety of flavours, fumes, agrochemicals, etc. as listed in Table 9.1.

Table 9.1 The secondary metabolites produced by plant species using cell suspension culture techniques

Plant species	Products	Functions
Chrysanthemum cinerariaefolium	Pyrethrin	Insecticide
Cinchona officinalis	Quinine	Antimalarial
Catharanthus roseus	Vincristine	Antileukaemic
	Ajmalthine	Antiarrhythmic
	Ajmalicine	Tranquilizer
	Serpentine	Anti-epileptic
Dioscorea deltoidea	Diosgenin	Antifertility
Datura stramonium	Scopolamine	Antihypertension
Digitalis lanata	Digoxin	Antifertility
	Reserpine	Hypotensive
Jasminum sp.	Jasmine	Perfume
Papaver somniferum	Morphine	Analgesic, sedative
Nicotiana tabacum	Nicotine	Ganglion blocker
	Ubiquinone-10	Cardiovascular agent
Atropa belladonna	Atropine	Blocking of cholinergic
Morinda citrifolia	Anthraquinones	Laxative, dye
Lithospermum erythrorhizon	Shikonines	Dye, pharmaceutical
Glycyrrhiza glabra	Glycyrrhizin	Expectorant

(*Contd.*)

Table 9.1 (Continued)

Plant species	Products	Functions
Stevia rebaudiana	Stevioside	Sweetener
Coptis japonica	Berberine	Antibacterial
Thaumatococcus daniellii	Thaumatin	Sweetener
Papaver bracteatum	Codeine	Analgesic
Cephalotaxus harringtonia	Cephalotaxine	Antitumour

3. ***Biotransformation*** In this technique, low-cost precursors are used as a substrate and are transferred into value-added high cost products. Suspension cultures of *Digitalis lanata* can convert digitoxin or methyl digitoxin into medically important digoxin or methyldigoxin respectively. Cell cultures of *Stevia rebaudiana* and *Digitalis purpurea* can convert stevio into steviobiocid and stevioside, which are 100 times sweeter than cane sugar.
4. ***Single cell proteins*** Single cell cultures are also used for production of single cell proteins (SCP). SCP means any microbial biomass from unicellular bacteria, fungi or algae, which can be used as food or feed additives.

   In India, the Central Food Technology Research Institute (CFTRI), Mysore, has conducted research on the use of the blue-green algae, *Spirulina*, as a supplement in diet. This alga is cultured as suspension culture, dried, powdered and then used in the form of tablets.
5. ***Making clones*** Cell suspension cultures are used to regenerate identical clones from single cells and helps to understand cloning studies.
6. ***Production of somatic embryos*** Suspension culture forms important tools for the study of mode of tissue differentiation, such as somatic embryogenesis.

REVIEW QUESTIONS

1. What is cell suspension culture? Describe the inoculums to initiate suspension culture.
2. Give an outline of the methods of cell suspension culture.
3. What are the advantages of suspension cultures over callus cultures?
4. How can we isolate cells from intact explants?
5. Describe the methods of cell growth and measurement in suspension culture.
6. Write notes on the applications of cell suspension culture.
7. Write short notes on storage, regrowth and stability checking of cell suspensions.

REFERENCES

Bergmann, D. (1959). A new technique for isolating and cloning cells of higher plants. *Nature*. 184: 648–649.

Constabel, F. and Tyler, R.T. (1994). Cell culture for production of secondary metabolites. In: Vasil, I. K. and Thorpe, T.A. (eds.). *Plant Cell and Tissue Culture*. Kluwer Academic Publishers, The Netherlands, pp. 271–292.

Gamborg, O.C., Miller, R.A. and Ojima, K. (1968). Nutrient requirements of suspension cultures of soybean root cells. *Experimental Cell Research*. 50: 151–158.

Street, H.E. (1977). Cell (suspension) Culture-Techniques. In: Street, H.E. (ed.). *Botanical Monographs*. 11. Blackwell, Oxford.

Torry, J.G. and Reinert, J. (1961). Suspension culture of higher plant cells in synthetic media. *Plant Physiol.* 36: 483–491.

Vasil, V. and Hildebrandt, A.C. (1965). Differentiation of tobacco plants from single isolated cells in micro cultures. *Science*. 150: 889–890.

10

SINGLE CELL CULTURE: TECHNOLOGY AND APPLICATIONS

INTRODUCTION

Plant cells are destined to grow almost continuously if supplied with the appropriate nutrients and conditions even after their removal from parent tissues. This process when carried out in a laboratory at *in vitro* level is called cell culture. The culture process allows single cells to act as independent units, much like a microorganism, to grow in the culture. The cells divide, increase in size and continue to grow and divide until limited by some culture variable such as nutrient depletion.

Plant cells have been cultured to produce many ingredients needed for food and pharmaceutical industries. Tremendous progress has been made in recent years in understanding the basics of plant metabolism and in the development of bioprocesses as well as in the design and operation of large- scale bioreactors for plant cell culture. A wide range of food ingredients including flavours, colourants, essential oils, sweeteners and antioxidants have been produced through cell culture. Japan has so far been the most successful country in the world to carry out plant cell culture on a commercial scale. Ginseng products, derived from cell suspension cultures of *Panax ginseng* are used as additives in wine; tonic drinks and herbal liquors are being produced in Japan since 1990.

WHAT IS SINGLE CELL CULTURE?

Single cell culture is a method of growing isolated single cell, aseptically on a nutrient medium, under controlled conditions (Figure 10.1). Establishment of a single cell culture provides an excellent opportunity to investigate the properties and potentialities of plant cells. Such studies contribute to our understanding of the interrelationships and complementary influences of cells in multicellular organisms. Several researchers have successfully isolated single cell and even raised complete plants from single cell cultures. The advantage of single cell culture over callus culture or cell suspension culture is that single cell culture system is an ideal system for studying cell metabolism, the effects of various exogenous substances on cellular responses and to obtain single cell clones. Using cell cultures in studies designed to describe the pathways of cellular metabolism was another aspect that initially attracted the attention of plant biologists. It was soon realized that single cell systems have great potential for crop improvement. Free cells in cultures permit quick administration and withdrawal of diverse chemicals and substances, thereby making them easy targets for mutant selection.

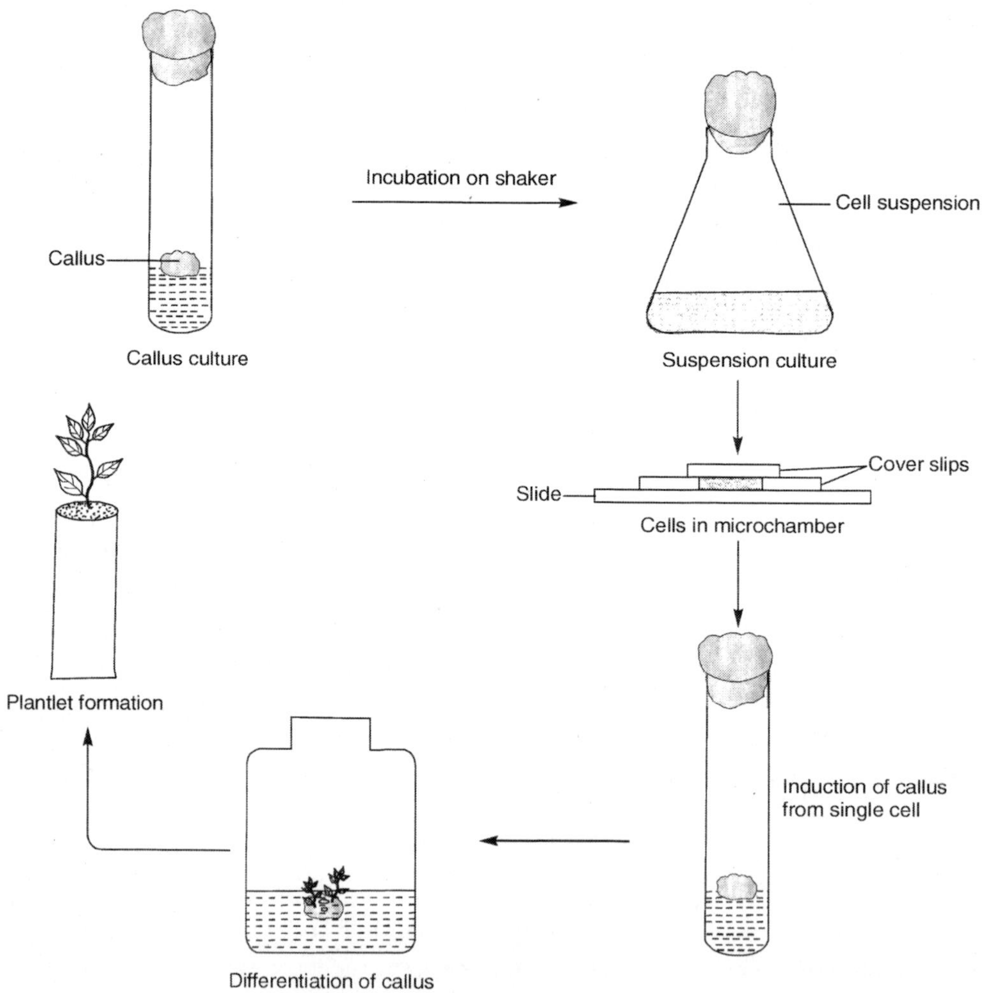

Figure 10.1 Single cell culture and regeneration of plantlets

Single cell cultures normally contain cells of one type, although mixed cultures, especially of bacteria, are common in food sciences and waste water treatment related studies. The cells in culture may be genetically identical (homogeneous population) or may show some genetic variation (heterogeneous population). A homogeneous population of cells derived from a single parental cell is called a clone. Therefore, all cells within a clonal population are genetically identical. Apart from this, the individual cells within a population of cultured cells invariably show cytogenetic and metabolic variations depending on the stage of the growth cycle and culture conditions. Such variability, termed spatial heterogeneity, has been the subject of much interest since differences between cells in their karyotype and the ability to accumulate secondary metabolites are manifested during morphogenesis in the clones regenerated from single cells.

A clone of cells consists of all the cells derived through mitosis from a single cell, and the process of obtaining a clone is called cloning. Therefore, all the cells of a clone are expected to be identical copies of each other in their genotype and karyotype, except for the changes that may arise afresh during and after cloning. Cloning is based on single cells separated from tissues and cultured in a manner to allow separate recovery of the cell mass derived from them. In this way, the cell line selection technique can be usefully applied to produce high-yielding cultures as well as plants with superior agronomic traits.

HISTORY

Although the technique of single cell culture was initiated by Haberlandt as early as 1902, these cells were induced to divide in culture only after several decades. During the last few decades, the progress in this area has been so spectacular that cultured cells are being used in all areas of biology for a variety of purposes.

- Muir (1954) described the paper raft nurse technique for the culture of single cells.
- Later in 1960, Hildebrandt attempted to culture single cells of hybrid of tobacco plant using micro-chamber technique.
- Kohlenbach in1966 isolated the single cells from the leaf mesophyll of *Macleaya cordata* and cultured them to produce embryoids.
- Ball and Joshi (1968) isolated single cells from the leaves of *Arachis* sp.
- Bergmann (1968) grew single cells by the use of Petri dish plating technique.
- Takebe (1968) isolated metabolically active single cells from mesophyll tissues by the use of enzymatic method from tobacco leaves.
- Rossini in1969 described a method for the large-scale mechanical isolation of single cells from the mesophyll tissue of *Calystegia sepium*.
- Zenk (1978) reported manifold increase of anthroquinone in cell cultures of *Morinda citrifolia*. Similarly Roller (1978) also mentioned the synthesis of relatively high amounts of serpentine and ajmalicine by the cell and callus cultures of *Catharanthus roseus*.

ISOLATION OF SINGLE CELLS

The most suitable material for the isolation of single cells is the leaf tissue, since it has a more or less homogeneous population of cells which are good candidates for raising defined and controlled large-scale cell cultures. Significantly, single cells can be isolated either from cultured tissues or intact plant organs, the former being more convenient than the latter. Isolation of single cell from cultured tissue is possible by initiation of culture from desired and organized tissue on nutrient media. Furthermore, single cells can be isolated using mechanical or enzymatic methods.

Isolation of Single Cells from Intact Tissues

Single cells from the intact plant tissue (leaf, stem, root, cladode, etc.) are isolated either mechanically or enzymatically.

1. ***Mechanical method*** Earlier, mechanical isolation involved tearing or chopping the surface sterilized explant to expose the cells followed by scraping of the cells with a fine scalpel to liberate the single cells. However, only few living cells were obtained even after lot of time, effort and precaution. The mechanical procedure involves mild maceration of 10 g leaves in 40 ml of the grinding medium (20 μ mol sucrose, 10 μ mol $MgCl_2$, 20 μ mol tris-HCl buffer at pH 7.8) with a mortar and pestle. The homogenate is passed through two layers of muslin cloth and the cells thus released are washed by centrifugation. This method is now widely used for the large-scale mechanical isolation of viable cells. The advantage of this method is that harmful effects of chemicals and enzymes are eliminated and this method needs no steps for plasmolysis which may not be beneficial for plant cell isolation.
2. ***Enzymatic method*** Enzymatic method is a more efficient way for the large-scale isolation of free cells from the surface sterilized explants. It involves dissolving the intercellular cementing material (that is pectin) by pectinase or macerozyme treatment as described in Chapter 9.

Cultured tissues are also widely used to obtain single cells. Partial vacuum may be used to facilitate the entry of enzyme into the tissue but the medium solution must be changed during the treatment. Enzyme treatment tends to weaken cell walls; therefore, a suitable osmoticum (mannitol) is added to the enzyme and culture solutions.

Isolation of Single Cells from Cultured Tissues

Single cells can also be isolated from the established friable callus tissue and cell suspension culture. Mechanically, single cells are carefully isolated from cell suspension or friable callus with a needle or fine glass capillary. Alternatively, the friable tissue is transferred to liquid medium which is then continuously agitated by a shaker. Agitation of liquid medium breaks and dispenses the single cells and cell clumps in the medium. As a result, it makes a cell suspension. The cell suspension is first filtered to remove cell clumps and the filtrate is then centrifuged to collect the single cells in the form of a pellet.

CELL CULTURE SYNCHRONIZATION

A synchronous culture is one in which the majority of cells proceed simultaneously through each cell cycle phase (G_1, S, G_2 and M). Synchronization is expressed as a percentage synchrony of cells in suspension cultures. Cells in suspension cultures vary greatly in size, shape, DNA and other nuclear content. Moreover, the cell cycle time varies considerably within individual cells. Therefore, cell cultures are mostly asynchronous. It is essential to manipulate the growth conditions of an asynchronous culture in order to achieve a higher degree of synchronization. Synchronization can be achieved by physical or chemical methods (as described in Chapter 9).

METHODS OF SINGLE CELL CULTURE

Free cells isolated either from plant organs (mesophyll tissue) or cell suspensions are grown as single cells under *in vitro* conditions using a suitable agar gelled medium or liquid medium. Several techniques have been developed to culture single cells (Figure 10.2).

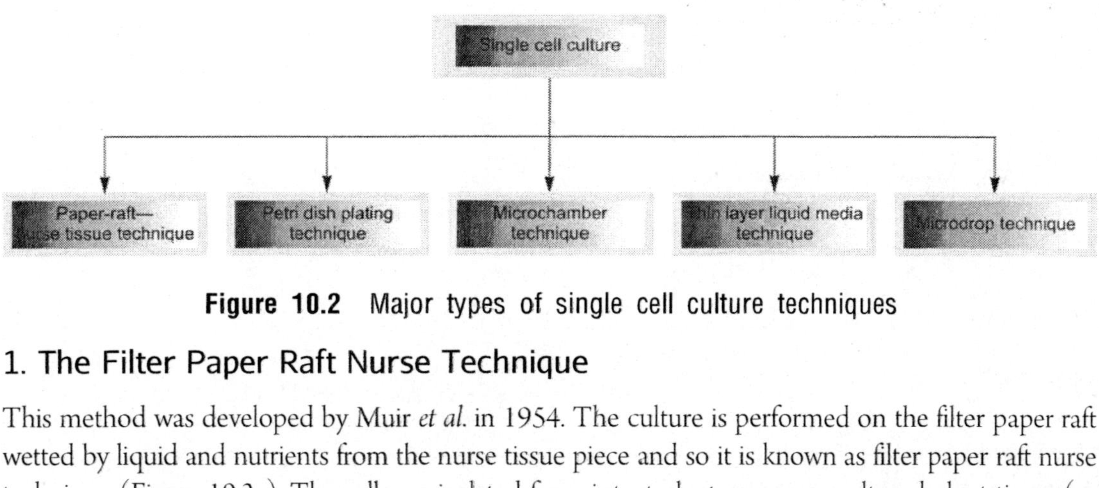

Figure 10.2 Major types of single cell culture techniques

1. The Filter Paper Raft Nurse Technique

This method was developed by Muir *et al.* in 1954. The culture is performed on the filter paper raft wetted by liquid and nutrients from the nurse tissue piece and so it is known as filter paper raft nurse technique (Figure 10.3a). The cells are isolated from intact plant organs or cultured plant tissue (as described earlier), with the help of a micropipette. Few days before cell isolation, sterile 8 mm × 8 mm squares of filter paper are placed aseptically on the upper surface of the actively growing callus tissue of the same or different species. The filter paper is wetted by soaking in the water and nutrient from the callus tissue. The isolated single cell is placed aseptically on the wet filter paper raft.

The whole culture system is incubated for 15 hours under cool white light (3,000 lux) or under continuous darkness at 25°C. The single cell divides and redivides and ultimately forms a small cell colony. When the cell colony reaches an appropriate size, it is transferred to a fresh medium where it gives rise to the callus tissue. The callus tissue, on which the single cell is growing, is called the nurse tissue. Actually the callus tissue supports the cell not only with the nutrients from the culture medium but also something that is critical for cell division. The single cell absorbs nutrients through the filter paper. The nutrients actually diffuse upward from the culture medium through the callus tissue and filter paper to the single cell. A callus tissue originating from a single cell is known as single cell clone.

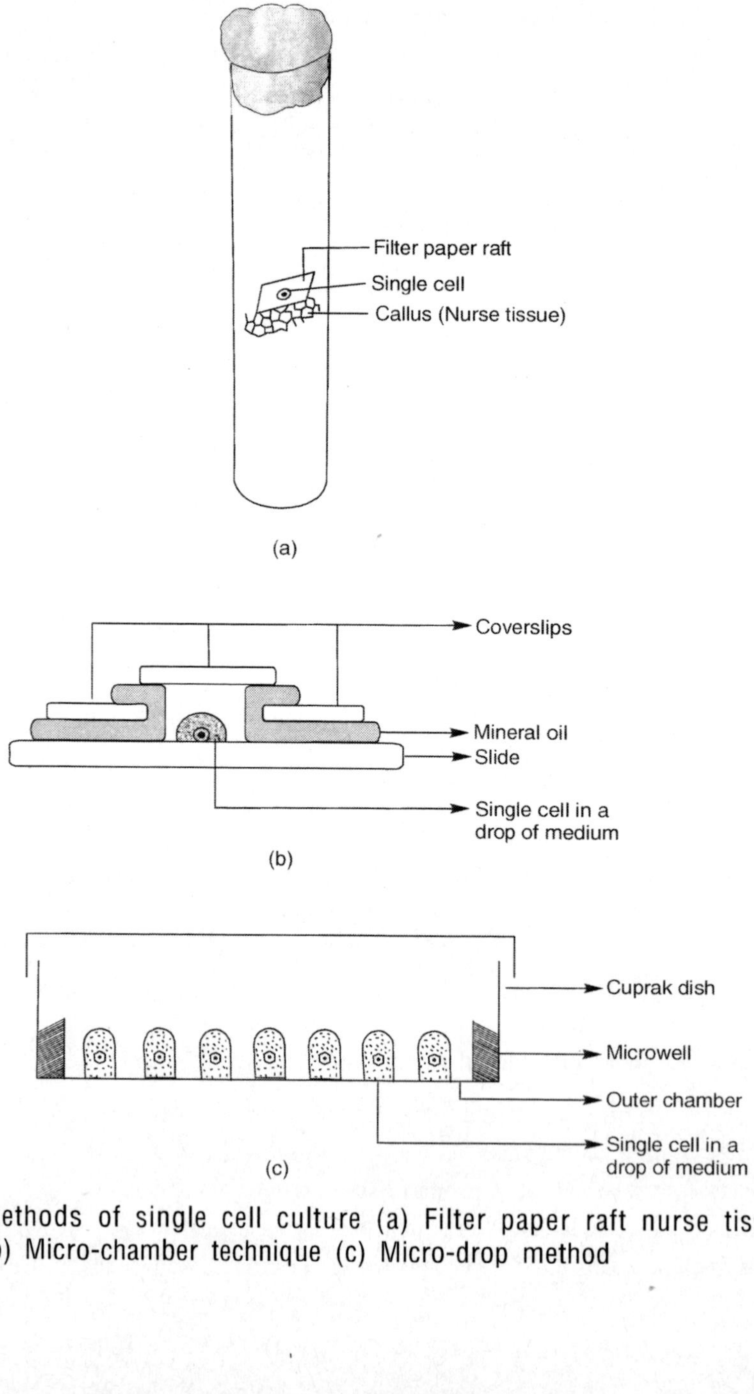

Figure 10.3 Methods of single cell culture (a) Filter paper raft nurse tissue technique (b) Micro-chamber technique (c) Micro-drop method

2. Micro-chamber technique

This method was developed by A.C. Hildebrandt to culture single cells of tobacco plant in 1960. A micro-chamber can be created either by using a microscope slide and coverslip or by a cavity slide. Single cells are suspended in conditioned medium, and a drop of medium having single cells is placed in the micro-chamber, which remains covered by a coverslip and ringed with sterile mineral oil. One drop of mineral oil is then placed on either side of the ringed culture drop and a cover glass is placed on each oil drop. A third cover glass is then placed on the culture drop bridging the two cover glasses (Figure 10.3b). As a result, a micro-chamber is formed enclosing the single cell aseptically within the mineral oil. The oil prevents water loss from the chamber but permits gaseous exchange.

In case of cavity slide, the drop of culture is placed on a coverslip which is then inverted onto the slide cavity. Micro-chambers allow microscopic observation, and they can be kept in a Petri dish for incubation. The micro-chamber slide is now incubated by placing it in a Petri dish. The cover glass is removed as soon as the cell colony becomes visible to the naked eye and the tissue is subcultured by transferring to fresh liquid or a semisolid medium.

3. Micro-drop method

A specially designed dish (Cuprak dish), having a chamber (incorporating several numbered microwells) is employed. Micro-drops of 0.25–0.5 ml are distributed in the microwells and the dish is sealed with paraffin. Cell density in the medium is so adjusted as to give, on an average, one cell per droplet as shown in Figure 10.3c. This method has been successfully used for protoplast culture and it should work with single cell culture as well.

4. Bergmann's Plating Technique

This method was developed by Bergmann in 1962. The cells to be cultured are suspended in a liquid medium at a cell density that is twice the desired density in the plate. Sterilized agar medium is kept melted in a water bath at 35°C. Equal volumes of the liquid and agar media are mixed thoroughly and quickly spread into a thin layer in a Petri dish. The dish is cooled to solidify the medium and then sealed with paraffin. Finally, the dishes are incubated in the dark or diffused light at 25°C. The cells remain embedded in the soft agar medium and are observed under a microscope. When macroscopic colonies develop, they are isolated and cultured separately.

5. The Thin Layer Liquid Medium Culture

Cells can be plated in a thin layer of liquid medium in a Petri dish to allow adequate aeration. Since cells are not in a fixed position, it is not possible to follow up individual cells during culture. This technique is common for protoplast culture.

Precautions

The following points must be taken into consideration during cell culture:

1. While growing single cells in the dark, microscopic observations have to be made to know the minimal level of light required, because light has a detrimental effect on cell proliferation.
2. In addition, either a conditioned medium or a suitably enriched medium should be used, since standard tissue culture media are unsuitable.

Some highly enriched synthetic media permit the culture of as few as 25 to 50 cells/ml, while some media supplemented with casamino acids and coconut milk support cell division at 1–2 cells/ml. It is postulated that some biochemicals essential for cell division leach into the culture when they are subcultured, and achieved the lag phase. Cells begin to divide only when equilibrium is established for these metabolites between the medium and the cells. This happens much later at lower cell densities, and below a critical cell density, it may not be achieved. For this reason, conditioned or specially enriched media is required for culture of cells at low densities.

VIABILITY TESTS FOR CULTURED CELLS

The growth of cell cultures is largely dependent on the viability of the cells. Cells' viability can be assessed by microscopic examination of untreated cells or of stained samples.

Viability tests that measure "vital metabolic functions" can be grouped into three categories:

1. Observations of cytoplasmic streaming.
2. Measurement of membrane integrity, such as electrolyte leakage, plasmolysis and stain exclusion or retention.
3. Measurement of biochemical activity, such as protein synthesis, tetrazolium chloride reduction, DNA and RNA synthesis, fluorescein diacetate staining and ATP content.

Phase contrast microscopy is used mainly to observe cytoplasmic streaming and presence of a healthy nucleus in cell. Several viability tests should be used to reduce interpretation error. The validity of the viability test should be established for each culture system studied. Validity would mean that the assay results correlate with the viability of the cultured tissue. It can be validated that the mode of action of the assay is the same in the plant system as in other systems reported in the literature by comparing the assay results from actively growing samples with samples killed by freezing and thawing.

The viability test should be conducted, once a tissue has been removed from the experimental situation and placed back into conditions of optimal growth.

2,3,5-triphenyltetrazolium Chloride (TTC) Reduction

The reduction of tetrazolium salt test is used to measure efficiency of cell's respiration. This test is carried out as follows:

1. Place 10–30 mg of freshly cultured cells in a centrifuge tube with 1 ml of the 23.9 mM TTC solution.
2. Incubate the mixture at 28°C for 6 hours in the dark. The formation of the water-insoluble formazan (alcohol soluble) in a positive reaction will yield red pigmented cells.
3. Centrifuge the solution for two minutes and remove the TTC solution followed by further removal of supernatant carefully with a Pasteur pipette.
4. Add 0.5 ml of 95% ethanol to the tube, and pulverize the tissue with a plastic pistil that fits the centrifuge tube. Add to the tube an additional 0.5 ml of 95% ethanol and set covered at room temperature for 4 hours to completely solubilize the formazan.
5. Centrifuge the resultant material for 5 minutes (pellet should be white after complete extraction).
6. Determine the absorbance of the supernatant spectrophotometrically at 485 nm.

Precautions

i. The TTC solution should be stored frozen or at 4°C in the dark.

ii. If the solution has a red tinge to it, the solution must not be used.

iii. Since TTC reduction requires mitochondrial activity, factors known to slow down or inhibit mitochondrial activity that is, low temperature, also affect TTC reduction. Consequently, the assays must be conducted under optimal growth conditions in order to achieve an effective viability test for TTC reduction.

Fluorescein Diacetate (FDA) Staining

The fluorescein diacetate test helps to locate the living cells under UV light by the accumulation of fluorescein in the cytoplasm of intact cells. Evan's blue stains the dead and damaged cells. The procedure is as follows:

1. Dilute 0.5 ml of FDA stock with 24.5 ml of culture medium (0.01% w/v solution).
2. On a glass slide, mix one drop of cell suspension and one drop of diluted FDA.
3. Wait for at least 5 minutes before viewing under a microscope to allow fluorescence to develop.
4. View the slide under the microscope and count fluorescing cells as viable cells and non-fluorescing cells as dead cells.

APPLICATIONS OF SINGLE CELL CULTURE

1. This is the technique by which we can produce true cell clones. Almost identical plants could be regenerated from the callus tissue derived from the single cell clones.
2. Many cell lines resistant to different cytotoxic substances have been isolated by direct selection method. This is useful for selecting cells with phenotypes that have the advantage of growing under selected conditions, toxic to the wild type cells.
3. The occurrence of high degree of spontaneous variability in the cultured tissue and its exploitation through single cell culture are very important in relation to crop improvement programmes.
4. The plant cells in culture show the synthesis of secondary compounds in higher amounts than the parent plants (*Morinda citrifolia, Catharanthus roseus,* etc). However, there are some reports where cell cultures have failed to produce secondary metabolites in higher amounts. The genetic and epigenetic instability of the cells, inadequate knowledge of the cultured conditions and the production of secondary metabolites affecting the growth phases, etc. are considered to be the causes of poor performance of cell cultures in the production of secondary metabolites. If these problems are solved, cell culture could become a practical method for the industrial production of some important natural compounds.
5. The selection of mutants is possible through single cell culture technique and it is important in crop plant improvement. The frequency of certain phenotypes could be increased several-fold by treating the cells with mutagens.
6. In biotransformation method, the cellular conversion of an exogenously supplied substance, which is not available in the cell, takes place and at the end, new compounds are obtained. There are reports of such biotransformation in cell culture of *Stevia rebaudiana* and *Digitalis purpurea.* In these species, addition of sterioside can transform glyolyrae steroil to steviobioside. A glycon stervioil is not sweet whereas the glycoside steviobioside can be used as a sweetening agent.
7. One of the major problems of mutation breeding in higher plants is the formation of chimeras following the mutagenic treatment of multicellular organisms. In this respect, single cell culture is an additional and supporting technology. Isolated single cells can be handled as a microbial system for the treatment of mutagens and for mutant selection. In practice, single cells are grown on a medium containing the mutagenic compounds and the proliferating cell lines are isolated. The mutant nature of the selected cell lines can be confirmed by regenerating the plants and comparing their phenotypes with a normal plant. Many cell lines resistant to amino acid analogues, antibiotics, herbicides, fungal toxins, etc. have been selected by single cell culture.
8. To study the complex interrelationships and complementary influences of different cells in multicellular organism, scientists have attempted to culture the isolated cells in order to see whether each and every cell retains the ability to behave like a zygote and finally give rise to a complete plant.

ADVANTAGES OF CELL CULTURE PLANTING MATERIAL

1. True to the type of mother plant.
2. Uniform growth, increases yield.
3. Early maturity of crop—maximum land use is possible in low land areas.
4. Round-the-year planting is possible as samplings are made available throughout the year.
5. Two successive ratoons are possible in a short duration which minimizes cost of cultivation.
6. No staggered harvesting.
7. 95– 98% plants bear bunches.
8. New varieties can be introduced and multiplied in a short duration.
9. The major advantage of using cell culture is the consistency and reproducibility of results that can be obtained from using a batch of clonal cells.

LIMITATIONS OF SINGLE CELL CULTURE

After a period of continuous growth, cell characteristics may change and become quite different from those found in the starting population. Cells can also adapt to different culture environments (for example, different nutrients, temperatures, salt concentrations, etc.) by varying the activities of their enzymes.

REVIEW QUESTIONS

1. Describe the techniques used for isolation of single cells from plant tissues.
2. How will you isolate and culture single cells?
3. What are the methods to test the viability of the cultured cells?
4. Write short notes on the following:
    i. The filter paper raft nurse technique
    ii. Micro-chamber technique
    iii. Bergmann's plating technique
    iv. Micro-drop method
5. What are the applications of single cell culture?

REFERENCES

Fujita, Y., Takahashi, S. and Yamada, Y. (1985). Selection of cell lines with high productivity of shikonin derivatives by Protoplast culture of *Lithospermum erythrorhizon* cells. *Agric. Biol. Chem.* 49, 1755–1759.

Muir, W. H., Hildebrandt, A. C. and Riker, A.J. (1954). Plant tissue cultures produced from single isolated cells. *Science.* 119, 877–878.

Thomas, E. and Davey, M.R. (1975). *From single cells to plants.* Wykeham Publ. (London) Ltd., London and Winchester.

Vasil, V. and Hildebrandt, A.C. (1965). Differentiation of tobacco plants from single isolated cells in micro cultures. *Science.* 150, 889–890.

11

EMBRYO CULTURE

INTRODUCTION

Plant embryogenesis is the process that produces a plant embryo from a fertilized ovule by asymmetric cell division and the differentiation of undifferentiated cells into tissues and organs. Embryogenesis occurs naturally as a result of sexual fertilization resulting in the formation of zygotic embryos. It occurs during seed development, when the single-celled zygote undergoes a programmed pattern of cell division resulting in a mature embryo. The embryo, along with other cells from the mother plant, develops into the seed for the next generation, which grows into a new plant after germination. A similar process continues during the plant's life within the meristems of the stems, leaf bases and roots.

The fertilized egg follows a predetermined mode of development and gives rise to an embryo, which has an inherent potential to form a complete plant. A division of the fertilized egg, which is a highly polarized and isolated cell, gives rise to a small apical cell and a large basal cell. Based on the plane of division of the apical cell, different types of embryos are formed, which further form plumule or primordial shoot and radical or primordial root at each pole. This reflects the possession of a basic organization of embryonic cells in the formation of an adult plant. The variations in the developmental pattern of embryo during early embryogeny are common to monocotyledons and dicotyledons. The zygote undergoes embryogenesis and thus an embryo is formed inside the seed.

The entire process of embryogenesis has two distinct phases:

1. Morphogenetic events, which form the basic cellular pattern for the development of the shoot–root body and the primary tissue layers; it also programmes the regions of meristematic tissue formation.

2. Post-embryonic development, which involves the maturation and differentiation of cells, possible by cell growth and the storage of macromolecules.

If any of these two phases is not suitable, the embryo development would be incomplete. In such cases, the researchers need to go for embryo culture to get transplantable seedlings. Embryogenesis involves cell growth and division, cell differentiation and programmed cellular death. The zygotic embryo is formed following double fertilization of the ovule, giving rise to two distinct structures: 1) the plant embryo, and 2) the endosperm. These distinguished structures together go on to develop into a seed. Seeds may also develop without fertilization, which is referred to as apomixis. Plant somatic cells can also be induced to form embryos without fertilization during *in vitro* culture technique; such embryos that are originated from somatic tissues, instead of germinative tissues, are called somatic embryos. The plant breeders usually rescue inherently weak, immature or hybrid embryos to prevent degeneration. The successful regeneration of plants from the cultured zygotic embryos largely depends upon the maturation stage of the developing zygotic embryos and the composition of the nutrient medium.

Embryo culture represents the earliest technique to obtain viable offspring following interspecific and intergeneric hybridizations where routine fertilization failed to produce a well-defined and full-term embryo. Most extensive use of embryo culture technique has been made for making interspecific and intergeneric crosses within the tribe *Triticeae* of the family *Poaceae.*

METHODOLOGY

The embryos of different developmental stages formed within the female gametophyte through sexual process can be isolated aseptically without any damage, or dissected embryo from the bulk of maternal tissues of ovule, seed or capsule (Figure 11.1). They are cultured *in vitro* under aseptic and controlled physical conditions in glass vials/tubes containing nutrient medium to grow directly into plantlets.

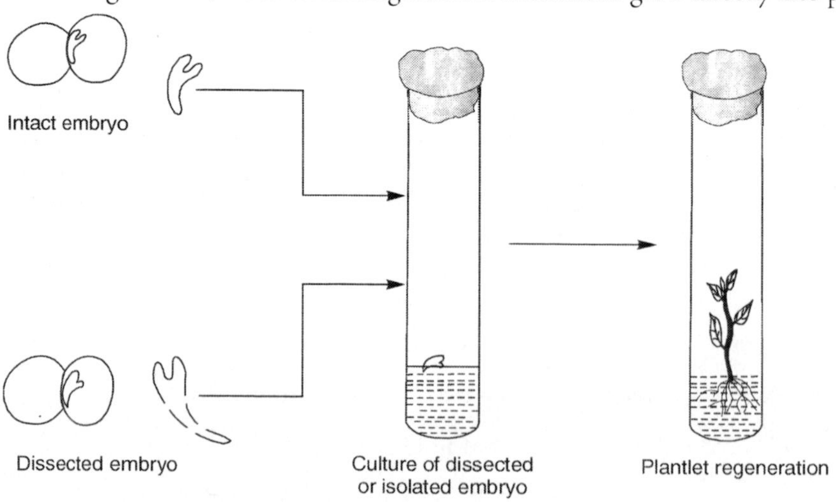

Figure 11.1 Embryo culture techniques

These techniques are popular approaches for raising hybrids from a number of incompatible crosses. Embryo culture allows one to investigate the factors that influence the embryonic growth under controlled conditions. Zygotic embryo can be used in artificial culture to investigate the factors that influence the embryonic growth.

The young or immature embryos are removed from developing seeds and inoculated on a suitable medium to obtain plantlets (Figure 11.2). The young embryos in culture generally do not complete their natural course of development and germinate prematurely. Sometimes, embryos from mature seeds may also be used for embryo culture. So far, it has not been possible to culture embryos before a certain stage of development (for example, before the globular stage, particularly in case of cereals and millets). However, in some plant species, even very young embryos of few cells have been successfully cultured. Young embryos need to be isolated with their intact suspensor as they provide gibberellins to the developing embryo.

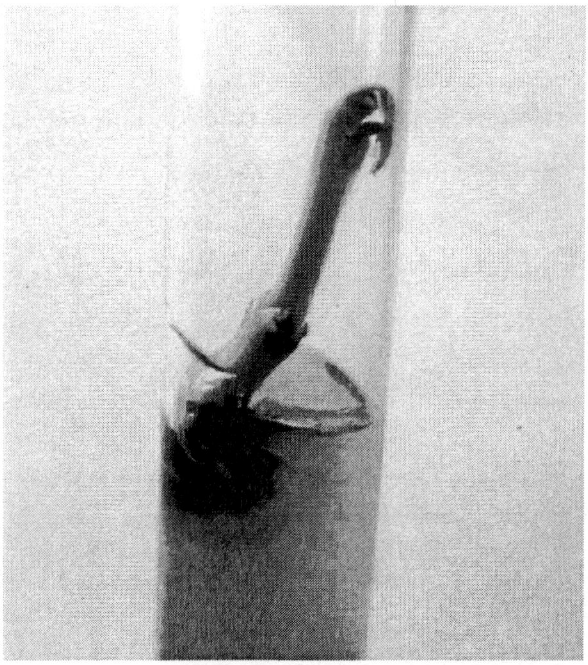

Figure 11.2 Germinating embryo on MS medium (Plate 4.7)

HISTORY

- Hanning (1904) was the first to excise mature embryos from the seeds of *Raphanus sativus* and cultured them in aseptic conditions.
- Dubard *et al.* (1913) stated about the role of endosperm in the development of embryos.

- Knudson (1923) was the first person to demonstrate the germination of orchid embryos into plantlets *in vitro* in the absence of any fungus.
- Laibach (1929) demonstrated for the first time successful embryo culture to obtain an interspecific cross between *Linum perenne* and *L. austriacum*.
- Van Overbeek (1942) showed that the mature embryo can be grown on a simple nutrient medium but proembryo grows only on nutrient-rich medium.
- Lee (1955) cultured various segments of *Lupin* embryos and observed that hypocotyl segments give rise to hypocotyls; the radicular segments give rise to radicle only, while the segments taken from just below the plumule give rise only to callus.
- Bajaj (1966) demonstrated that culture of *Dendrophthoe falcata* embryos, with their radicular cut ends gave rise to holdfast with haustoria and with cotyledons cut, gave rise to callus of undifferentiated tissues, whereas, cultures of plumular region with cotyledon, gave rise to a complete plantlet.
- Kruse (1974) analysed the importance of endosperm transplant in his experiment of embryo culture.
- Monnier (1978) described a new culture method which allows complete development of embryo in two different media without moving the embryo from its original position.
- De Lautour (1980) observed a problem of rescuing hybrids where embryo abortion occurs at a very early stage of development.

IN VIVO/IN VITRO EMBRYOGENY

The zygotic embryos develop through the proembryo stage, globular heart-shaped stage, torpedo-shaped stage, cotyledonary stage and coleoptilary stage. A fully developed embryo undergoes a period of maturation, during which the embryo becomes hardy. In general, up to a certain stage of embryogenesis, for example, globular stage in *Capsella,* the embryo is heterotrophic as it takes some part of its nutrition from the endosperm. Beyond this stage, the embryo becomes autotrophic and is able to synthesize its biochemical needs from simple nutrients like salt and sugar. In general, as the embryo grows older, its nutritional requirements become simpler.

Extensive studies have shown that embryos are quite sensitive to the salt solutions of culture media. Further, embryo survival and its growth need varied requirements. Salt solutions are significant for growth but not compulsory for survival and vice-versa. *Capsella* embryos show improved survival when iron level of the MS medium is lowered to two–thirds of the normal, and $CaCl_2$ concentration is doubled. In addition, increased potassium and micronutrient concentration promotes embryo growth without affecting survival. Sucrose is the best carbon source; 8–10% sucrose is generally used, which approximates the osmotic potential of the young embryo sac. High osmolarity of the medium prevents precocious germination of embryos.

Mannitol at 120 g/l is proved to be suitable for the growth and development of heart-shaped embryos. Generally, plant growth regulators are not used since embryos are usually autotrophic in this respect. ABA checks precocious germination, and promotes orderly embryo development and maturation. The best recommended medium should contain 180 g/l sucrose, high Ca^{+}, no Fe^{2+}, low salts and should be rich in amino acids.

Embryo cultures are kept at temperatures ranging from 25–30°C; but in some species of *Datura*, higher temperature may be more significant. Light does not appear to be a critical factor in several cases, however in barley, it suppresses precocious germination.

VARIOUS APPROACHES IN EMBRYO CULTURE

There are five different types of embryo culture. These are as follows:

1. ***Immature embryo culture*** Embryo before the cotyledonary stage is said to be an immature embryo. This stage of embryo is also known as proembryo; it is the early developmental stages of the embryo that precede cotyledon initiation. Globular and heart-shaped stages of zygotic embryo are appropriate for this cultures. The embryos are taken from immature or unripe fruits.

2. ***Inviable embryo culture*** This technique is useful in many interspecific and intergeneric breeding experiments, where sometimes inviable embryos may develop due to unsuccessful crosses. As a result, the non-viable seeds do not germinate normally, but it is now possible to raise hybrid plants by culture of the abortive embryos *in vitro.*

3. ***Mature embryo culture*** The aim of this study is to analyse the various parameters of embryonic growth and the metabolic and biochemical aspects of dormancy and germination.

4. ***Undifferentiated embryo culture*** This is a very useful approach to culture the undifferentiated embryos, particularly of orchids, because each orchid fruit develops thousands of small seeds which contain morphologically undifferentiated embryos. These embryos are a spherical mass of tissues lacking both radicle and plumule. There is even no storage tissue in the seeds and the seed coat is reduced to a membranous structure. Thus, the entire seed of orchid contains undifferentiated embryos which are treated as explants for embryo culture. In nature, these seeds germinate only in association with proper fungus or else they perish.

5. ***Adventitious embryo culture*** In citrus, besides the zygotic embryo produced from egg cell, some additional embryos are produced from nucellar tissue in polyembryonic seeds. These additional abortive embryos can be exploited in culture for clonal propagation.

EMBRYO NURSE ENDOSPERM TECHNIQUE

Often very young embryos may prove to be difficult to culture directly on the medium. In such cases, embryos may be placed on to or implanted into developing endosperms that are in culture; this technique is known as embryo nurse endosperm technique. Using this approach, it has been possible to obtain hybrids, which could not be otherwise obtained, for example, *Hordeum* × *Secale* hybrid, interspecific hybrids in *Trifolium,* etc.

In some interspecific crosses, flowers may drop much before the embryos as explants reach a certain size to permit their isolation. In such cases, isolated ovules, the whole ovaries or even the flowers can be cultured for a period of time, so that the embryos become large enough to be isolated. Interspecific hybrids of *Gossypium*, such as *G. arboreum* × *G. hirsutum*, have been recovered on culture of ovules, 3 days after pollination.

It is relatively easy to obtain pathogen-free embryos, since the embryo is lodged in the sterile environment of the ovule or seed or capsule or fruit, therefore, surface sterilization of the embryos as such is not required. Thus, the entire seed or fruit containing the ovules are surface sterilized and the embryos are then aseptically isolated from the surrounding tissues. In case of very small embryos like orchid seeds, the entire seed or ovule is cultured because the seed contains morphologically undifferentiated, spherical embryos having no functional storage tissue like endosperm. Each fruit of an orchid contains thousands and thousands of tiny seeds, so a large quantity of sample can be cultivated easily just after excision of surface-sterilized fruits.

As far as nutritional requirements of embryos are concerned, the younger the embryo, the more complex is the nutritional requirement, whereas mature embryos require relatively simpler nutrient composition. Thus, the mature embryo can be grown in an inorganic salt medium supplemented with a carbon source while relatively young embryos require in addition, different combination of vitamins, amino acids, growth hormones, etc.

SIGNIFICANCE OF EMBRYO CULTURE

1. The degeneration of embryos is observed in seeds obtained from the interspecific crosses such as *Lilium speciosum album* × *L. aurotum*. In this case, the microsurgical technique of embryo culture is useful in raising full hybrid plants.
2. The degeneration of endosperm is observed in the fruit obtained from the plants of the crosses between *Hordeum vulgare* and *H. bulbosum*. As a result, embryos die. Such embryos can be rescued by this technique.
3. To produce disease-resistant varieties of *Lycopersicon esculentum*, a series of crosses of this variety with *L. peruvianum* was attempted. The seeds formed showed abnormal development and no viability was observed. The embryo of such plants are cultured to avoid seed formation and disease-resistent plantlets can be generated.

4. Hybridization between *Corcorus capsularis* and *C. olitorius* is a major aim of jute breeders but their attempts have failed due to early abscission of pollinated flowers, resulting in a low fruit set. Fruits are associated with premature abortion of embryos. The problem is overcome by adopting embryo culture.
5. Dormant seeds fail to germinate under apparently suitable conditions. In dormant seed, the enclosed embryo is not able to grow. This temporary suspension of growth is definitely due to some internal condition of seeds or due to some environmental factors. Prolonged seed dormancy presents a special problem to plant growers. So, several methods have been devised to break the dormancy of seeds. But, in some seeds, dormancy cannot be broken by any conventional method. Hence, embryo culture can be successfully used to bypass the traditional treatments to overcome seed dormancy and to accelerate germination in certain types of seed. For example, *Lactuca sativa, Nepenthes insignis, Phacelia tanacetifolia* and *Citrullus colocynthis.*
6. The breeding cycle may be shortened by embryo culture method. Rose normally takes a whole year for flowering, but it has been possible to produce two generations by embryo culture technique.
7. The nature of sterility in some economic fruits like banana, kachoo seeds, etc., which never germinate in nature, could be overcome by embryo culture.
8. It is useful in obtaining rare hybrids of some crop plants. The breeding programme fails in some cases due to the presence of some barriers such as embryos or endosperm not maturing normally; hence, embryo culture works as an alternative technique for crop breeding programme.
9. *Melilotus officinalis* has high coumarin content which is harmful to cattle. By employing embryo culture technique, the hybrid plants obtained by the crossing between *M. officinalis* and *M. alba* (with low coumarin), some success was achieved in producing harmless varieties.
10. Conventional hybridization technique is an important method for the study of evolutionary relationship, speciation and for determining the taxonomic position of plants. By hybridizing the different species of *Datura* and culture of their abortive embryos in some cases, some interesting results have been obtained about their controversial taxonomic position and evolutionary relationship.
11. It is useful to determine the factors that regulate the growth of the primordial organs of the seedling plant and to study the metabolic and biochemical aspects of dormancy and germination.

EMBRYO RESCUE

Abortion of embryos at one or the other stage of development is a characteristic feature of distant hybridization. Currently embryo rescue holds great promise not only for effecting wide crosses, but also for obtaining plants from inherently weak embryos, for obtaining haploid plants as well as for shortening the breeding cycle.

One of the classic examples of successful application of embryo rescue techniques in plant nematology is the production of tomato cultivar, resistant to root knot nematode. Cultivated tomato species (*Lycopersicon esculentum*) is highly susceptible to *Meloidogyne* sp. But the related wild species *L. peruvianum* has a high degree of resistance against these nematodes. A cross between the two, yielded a hybrid in which the endosperm did not develop and therefore, the embryo was aborted at an early stage. This problem was solved by the application of embryo culture. This provided a mechanism to transfer nematode resistance gene from *L. peruvianum* into cultivated species of tomato.

Thus, embryo culture is an additional and alternative technology of embryo rescue and production of biotic/abiotic stress-resistant plants, particularly nematode resistance in tomato where general hybridization experiments have failed.

The hybrids raised through embryo culture have been utilized for the following purposes:

1. Phylogenetic studies and genome analysis.
2. To raise synthetic crops like *Triticale* by producing amphiploids from the hybrids.
3. Transfer of useful agronomic traits from wild genera and species to the cultivated crops.
4. Production of haploids through distant hybridization followed by elimination of the chromosomes of one parent in the hybrid embryo culture, for example, hybridization of barley and wheat with *Hordeum bulbosum* leading to the production of haploid barley and haploid wheat respectively. Haploid wheat plants have also been successfully obtained through culture of hybrid embryos from wheat × maize.

REVIEW QUESTIONS

1. Define embryo culture and embryo rescue, and discuss their applications.
2. Write down the different categories of embryo culture.
3. Explain embryo nurse endosperm culture technique.
4. What is the the significance of embryo culture?
5. Write a short note on embryo rescue.

REFERENCES

Raghavan, V. (1966). Nutrition, growth and morphogenesis of plant embryos. *Biol. Rev.* 41, 1–58.

Shekhawat, N.S., Mughal, M.H., Johri, B.M. and Srivastava, P.S. (1998). Indian contribution of plant tissue culture and organ culture. In: Srivastava, P. S. (ed.). *Plant Tissue Culture and Molecular Biology: Application and Prospects.* Narosa Publishing House, New Delhi. pp. 751–811.

Smith, R.H. (2000). *Plant tissue culture: Techniques and Experiments.* Academic Press, Tokyo.

Van Overbeek, J., Conklin, M.E. and Blakeslee, A.F. (1941). Factors in coconut milk essential for growth and development of very young *Datura* embryos. *Science.* 94, 350–351.

12

SOMATIC EMBRYO: INDUCTION AND REGENERATION

INTRODUCTION

In plants, totipotency of the fully differentiated and mature somatic cell has been well-established. It is now known that the nucleus of every living somatic cell contains all the genetic information necessary to direct the development of the entire plant.

A plant or an animal reproduces naturally through the development of zygotic embryos. An embryo is composed of actively growing cells and the term is normally used to describe the early formation of tissue in the first stages of growth. It can refer to different stages of the sporophyte and gametophyte plant, including the growth of embryos in seedlings. It can also refer to 1) meristematic tissues, which are in a persistently embryonic state, and 2) the growth of new buds on stem. Moreover, formation of the embryo begins with the division of the fertilized eggs or zygote within the embryo sac of the ovule. Through an orderly progression of divisions, the embryo eventually differentiates, matures and develops into the young seedling.

In both gymnosperms and angiosperms, the young plant contained in the seed, begins as a developing egg cell, formed after fertilization (sometimes without fertilization, in a process called apomixis) and becomes a plant embryo. This embryonic potential also occurs in the buds that form on stems. These buds have specific tissue that has differentiated partially but not grown completely into embryo-like structures. These can remain in a resting or dormant state for a longer period or when conditions are dry, and when suitable growth conditions commence, they start growing into stem and leaves. Therefore, these buds are said to be in an embryonic state.

In plant system, embryogenesis can also be induced *in vitro* from somatic cell instead of zygotic cell and these embryos are called somatic embryos. Alternatively, the plants can be derived from a single somatic cell or a group of somatic cells. This regeneration process, which differs from the natural pathway, is called somatic embryogenesis. All the plantlets thus produced have the same genetic make-up. Combined with genetic engineering, regeneration through somatic embryogenesis provides an efficient means or tools of producing a large number of elite or transgenic plants.

Induction of somatic embryo and its development is analogous to its zygotic counterpart (Figure 12.1). The embryo development process is characterized by a series of morphological changes which correspond to its maturation. Somatic embryos are very different from buds. Embryos that have a bipolar structure radicle-plumule are without a vascular connection with the mother tissue, and for this reason, can be easily separated from the tissue. In addition, the initial leaves of the embryo have characteristics that are typical of cotyledons.

Somatic embryos are formed from non-germinative or somatic cells that are not normally determined to be involved in the development of embryos, that is, ordinary plant tissue. No endosperm or seed coat is formed around a somatic embryo. Cells derived from competent source tissue are cultured to form an undifferentiated mass of cells called a callus. Plant growth regulators in the nutrient culture medium can be manipulated to induce callus formation and subsequently changed to induce embryo formation from the callus. The ratio of different plant growth regulators are required to induce callus or embryo formation and it varies with the explant type or source of explant. Significantly, the embryogenic callus can be subcultured routinely on a solid medium composed of agar and nutrients. Culturing on solid medium is advantageous over liquid medium culture.

ZYGOTIC EMBRYOGENESIS VS SOMATIC EMBRYOGENESIS

Zygotic and somatic embryos share a number of characteristics of developmental stages; however, the very early steps in their development are not well-correlated. Following fertilization, the zygote undergoes an asymmetrical cell division that gives rise to a small apical cell that becomes the embryo and a large basal cell (called the suspensor) that functions to provide nutrients from the endosperm to the growing embryo. However, the origin of somatic embryo is generally a single cell, which divides to form a group of meristematic cells. Usually, this multicellular group becomes isolated by breaking cytoplasmic connections with the other cells around it and subsequently by cutinization of the outer walls of this differentiating cell mass.

Asymmetrical cell division also seems to be an important factor in the development of somatic embryos. However, while failure to form the suspensor cell is lethal to the formation of zygotic embryo, it is not lethal for somatic embryogenesis. From the eight-cell stage (octant) in the zygotic embryo, "embryo patterning" is apparent; however, somatic embryos at this stage may be quite variable; therefore, zygotic and somatic embryos become most comparable from the globular stage. In the globular stage,

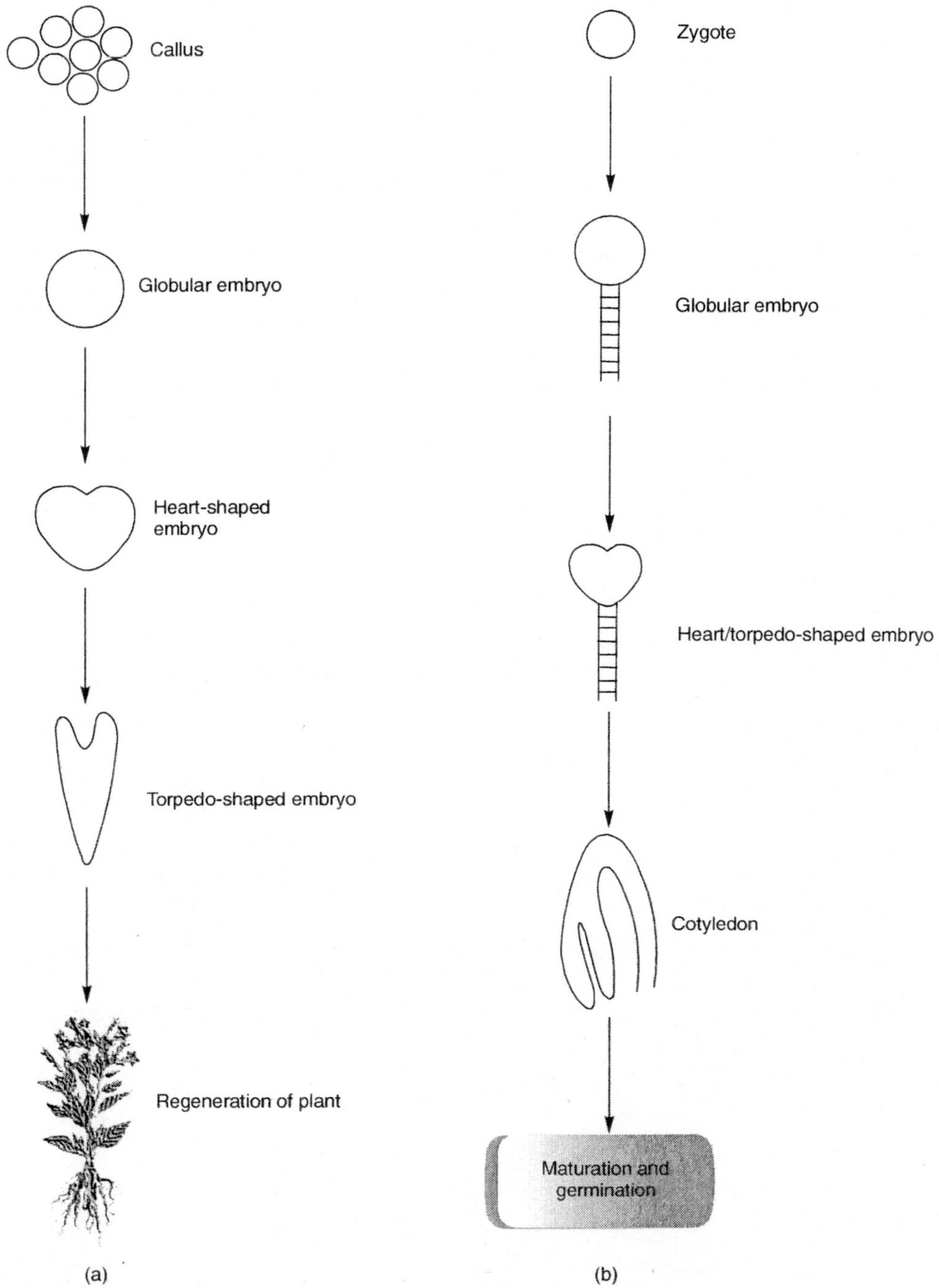

Figure 12.1 Comparison of (a) somatic and (b) zygotic embryogenesis

the embryo develops into a radial pattern through a series of cell divisions, with the outer layer of cells differentiating into the "protoderm". The globular embryo can be thought of as two layers of inner cells with distinct developmental fates; the apical layer will go on to produce cotyledons and shoot meristem, while the lower layer will produce the hypocotyl and root meristem. Moreover, bilateral symmetry is apparent from the heart stage; provascular cells will also differentiate at this stage. In the subsequent torpedo and cotyledonary stages of embryogenesis, the embryo completes its growth by elongating and enlarging. In a dicot embryo, the hypophysis, which is the uppermost cell of the suspensor, differentiates to form a part of the root cap.

Moreover, the development of somatic embryos closely resembles that of zygotic embryos. (Figure 12.2). The embryos of the first recognizable stage; the globular stage, generally grow out of the small cell clusters selected for somatic embryogenesis within 5 to 7 days after cells are shifted to auxin- free medium. In some cases, a small suspensorlike region can be seen, although this is more easily visualized in embryos developing on solid media than in liquid cultures. After 2 to 3 days of isodiametric growth, the globular stage is followed by an oblong stage which signals the shift from isodiametric to bilaterally symmetrical growth and the beginning of the heart stage.

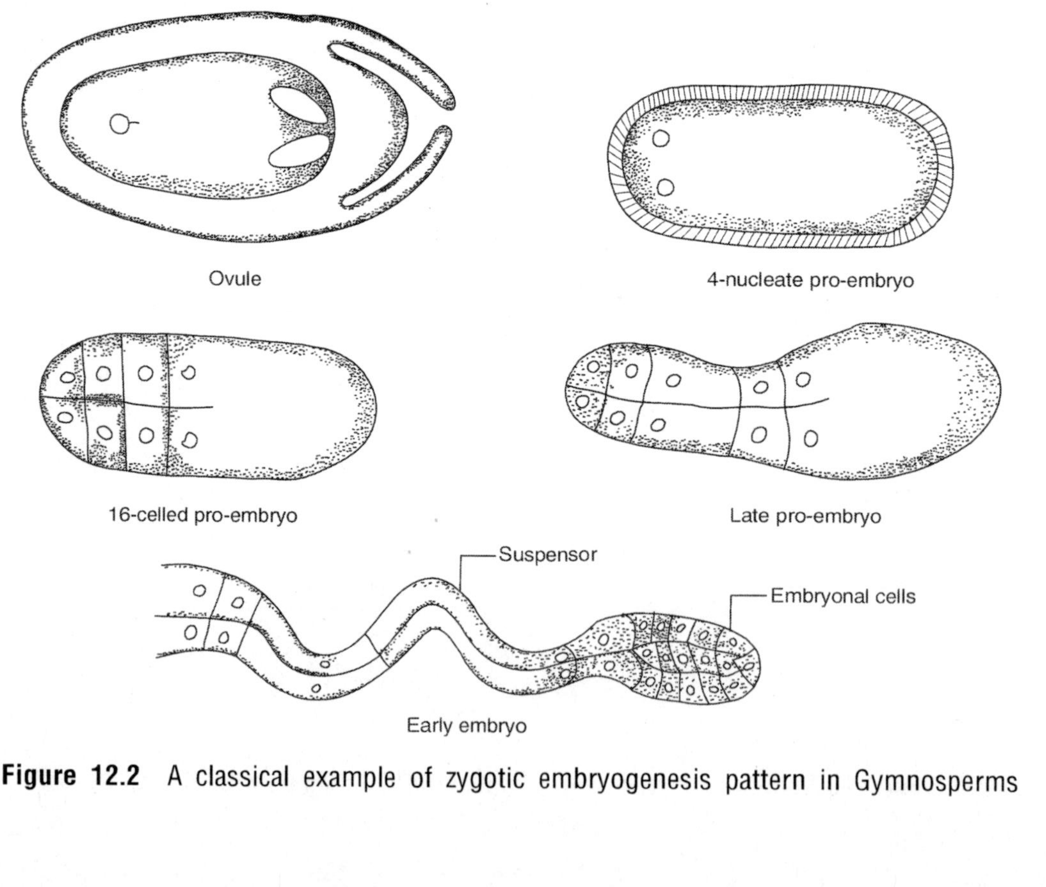

Figure 12.2 A classical example of zygotic embryogenesis pattern in Gymnosperms

The globular to heart transition is clearly marked by the outgrowth of the two cotyledons, the elongation of the hypocotyl, and the beginning of the development of the radicle. These processes continue through the torpedo and plantlet stages, and by 2.5 to 3 weeks after induction, plantlets can be identified that contain green cotyledons, elongated hypocotyls, and developed radicles with clearly differentiated root hairs. These plantlets will continue to grow in liquid culture or they can be transplanted to solid medium for regeneration into whole plants.

The similarity between zygotic and somatic embryogenesis is both striking and remarkable. The somatic embryos develop completely irrespective of the physical constraints and the genetic content of the maternal tissue. The fact that structurally and developmentally normal embryos can develop from somatic cells demonstrates that the genetic programme for embryogenesis and its elaboration are totally contained within the cell and can function completely in the absence of gene products from the maternal environment. Although the hormonal content of callus growth medium may somehow mimic some natural signal to initiate embryogenesis, it is clear that the morphology and size of the various embryogenic stages are completely intrinsic in the embryogenic programme and are not controlled by any environmental information or spatial limitations.

Sometimes embryos are formed by the unfertilized egg and such embryos are called parthenogenic embryos. Also, any cell of the female gametophyte or the sporophytic tissue around the embryo sac may give rise to an embryo and such embryos are called non-zygotic embryos.

Immature ovaries, embryos, placental tissue, leaf discs, petiole, glumes, stem hypocotyls, roots, floral buds, bulb leaf-disc, pedicel, epidermal peels, seed embryo, endosperm, cambium, etc. can be utilized as explant sources from which somatic embryos can be formed.

PRINCIPLE OF SOMATIC EMBRYOGENESIS

Somatic plant cells are not terminally differentiated and can therefore regain totipotency and initiate embryo development under appropriate conditions. Although this phenomenon is well known for more than 50 years, the details are still mysterious as we do not know why certain genotypes, explants or cells are more amenable for somatic embryogenesis than the others. It is also not known why so many different conditions are used to initiate somatic embryogenesis and which are the key molecular steps that are common in all cases. Recent progress in plant molecular and developmental biology now allows us to establish new hypotheses in the transition of somatic cells to the embryogenic state (Figure 12.3).

Somatic embryos can differentiate directly either from the explant without any intervening callus phase or indirectly after a callus phase. Direct somatic embryogenesis is most likely to occur in explants such as microspores (pollen grains), immature inflorescence, immature embryos and young seedlings. In the case of other explants, calli are first induced from the explants and then they are induced to produce somatic embryos.

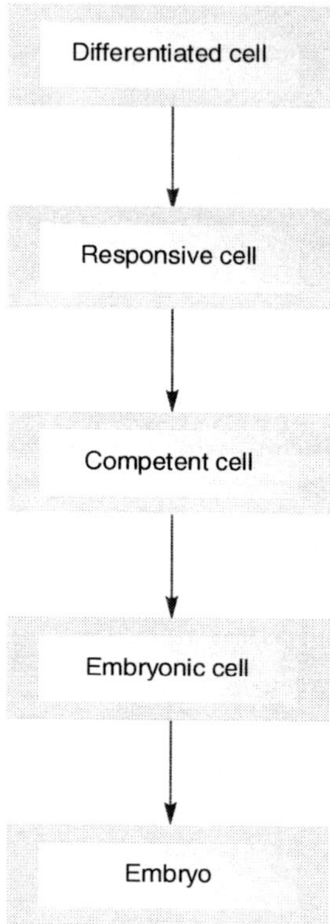

Figure 12.3 Induction of somatic embryogenesis in differentiated plant cells

Direct Somatic Embryogenesis

According to Conger *et al.* (1938), in some explants, somatic embryogenesis occurs directly in the absence of any callus production from pre-embryonic determined cells (PEDCs). These are already programmed for embryo differentiation. For example, the mechanically isolated leaf mesophyll of orchard grass is cultured on a medium supplemented with a synthetic auxin 2,4-D alternative, dicamba (3, 6-dichloro-O-anisic acid). The culture, after a few days, is subcultured on the same medium but without the growth hormone, dicamba; the formation of plantlet is observed without any callus formation. In this process, the cells of the explants undergo direct embryogenesis from PEDCs.

Indirect Somatic Embryogenesis

This is the most common pathway of somatic embryo formation. The cells of explants first undergo callus proliferation and then embryoids develop within the callus tissue from induced embryogenic cells. Two different nutrient media are required for this mode of somatic embryogenesis. One medium is needed for the initiation of embryonic cells and the other for the subsequent development of these cells into embryoids. The first medium is also known as induction medium. In majority of cases irrespective of dicot or monocot tissue, induction medium must be supplemented with auxin. However, somatic embryogenesis can be initiated in the second medium by removing the hormone or lowering its concentration.

The mode of somatic embryogenesis, whether direct or indirect, solely depends upon the endogenous and exogenous level of hormones in explant at the time of culture initiation. Moreover, a set of genes are involved in the mechanism of somatic embryogenesis. Some auxin-responsive genes are found to be responsible for somatic embryogenesis.

For direct embryogenesis, auxin works as a triggering factor. On application of auxin to the explant tissue, the genetic programme at the cellular level of the tissue gets altered and, therefore, normal somatic cell becomes embryogenic. Direct embryogenesis is not a normal process as it requires subculture and the medium is a complicated one. As indirect embryogenesis occurs in most of the plants and as the somatic cells are induced under culture condition, this method is normal. For indirect embryogenesis, cells are initially incompetent and, therefore, need to be made competent by treatment with specific growth regulators during culture incubation and finally these cells achieve competency which leads to induction of somatic embryos.

Embryoids are generally initiated in callus tissue from the superficial clumps of cells associated with enlarged vacuolated cells that do not take part in embryogenesis. The embryogenic cells are generally characterized by dense cytoplasmic contents, large starch grains, and a relatively large nucleus with a darkly stained nucleolus. It is interesting to note that the process of somatic embryogenesis is a short-term process and the potentiality of the somatic cell is decreased with the increasing duration of the culture. In long- term culture, the cells lose certain biochemical properties of their own and their potentiality to form embryo. High concentration of auxin treatment for a very short period (one minute) or pulse treatment is good enough for explant cells to trigger the induction of somatic embryos in rare cases.

INDUCTION OF SOMATIC EMBRYOS

Tissue culture techniques make it possible to grow excised plant tissues into somatic embryos. Plant regeneration via somatic embryogenesis includes five steps:

1. Initiation of embryogenic cultures by inoculation of the primary explant on nutrient medium supplemented with plant growth regulators (PGRs), such as auxins and cytokinins.
2. Proliferation of embryogenic cultures on solidified or in liquid medium supplemented with PGRs, in a similar fashion to initiation.
3. Pre-maturation of somatic embryos in medium lacking PGRs; this inhibits proliferation and stimulates somatic embryo formation and early development.

4. Maturation of somatic embryos by subculture on medium supplemented with ABA and/or having reduced osmotic potential.
5. Regeneration of plants on medium lacking PGRs.

Generally, embryogenesis occurs after two successive cultures. The first one involves culturing the explant tissue (primary culture) on a medium with auxin where there is a loss of differentiation of explant (embryogenic induction). The second (secondary culture) consists of the subculturing of the dedifferentiated tissue on an auxin-free medium where the embryogenic capacity is expressed.

Primary Culture

The immature explants (for example, stems) of the plant are sterilized by plunging them into the sterilant (for example, calcium hypochlorite) solution for 15 minutes. They are then washed thrice in sterile water. The ends of the stems that were bleached by hypochlorite are discarded and the remaining stem is cut into segments (each of 0.5 cm). Each segment is inoculated either horizontally or vertically on the nutrient medium in a tube. Nutrient medium for primary culture mainly consists of auxins, especially 2,4-D in higher concentration (0.5 to 1.0 mg/l).

After one to two weeks in culture, the surface of the explant becomes rough and the ends of the explant segment form nodulelike structures (Figure 12.4), which are proembryonic masses. The explant tissue grows very slowly, and only after one month of culture initiation, it becomes sufficiently large enough to be transferred. The embryonic tissue in the healthy state seems to be a friable mass of cells while actual explants and necrotic tissues appear brown in colour.

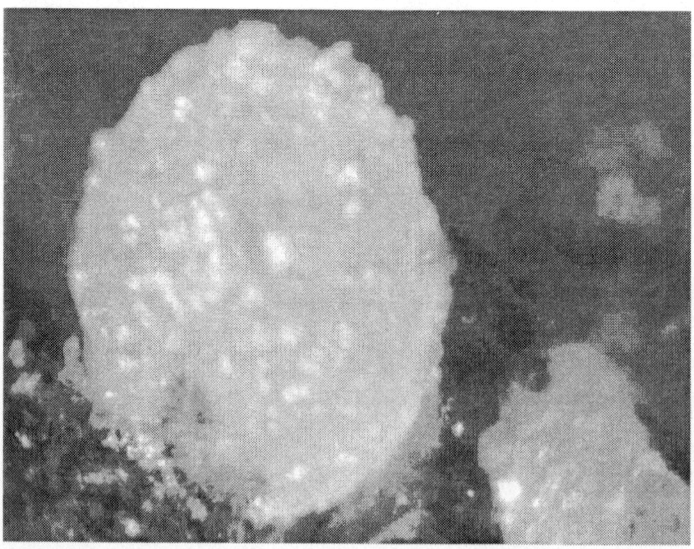

Figure 12.4 Proembryonic (nodular) mass of cells from explant (Plate 4.1)

Secondary Culture

The healthy callus is cut into small pieces and inoculated into a liquid medium. About 2.5 to 3.5 g callus is required to establish 50 ml of suspension culture. The medium is the same as described for primary culture, except that the auxin is omitted. After the transfer of healthy callus tissue that bears protuberances to the new medium, the nodules grow into embryos after 2–3 weeks. The tissue consists of an intertwining of embryos, sometimes aberrant with green blades. In the callus, all the intermediary stages from embryos to typical leaves are visible. For observation, the callus can be put into a Petri dish with a small quantity of medium and dissociated with needles. Hundreds of embryos are apparent that show the recognizable steps of embryogenesis such as globular, heart-shaped, torpedo-shaped, mature and germinating embryos.

The mature embryos can be transferred to a filter-paper bridge dipped into a liquid medium. The nutrient solution of Murashige and Skoog at half strength and sucrose (5.0 g/l) is necessary for somatic embryo germination and its growth. The tubes are placed on a well-lit shelf in the growth chamber for embryogenesis in callus. Numerous intertwining somatic embryos appear among abnormal embryos, which resemble leaves (for example, wild carrot). When the roots are sufficiently developed, the plantlets are transferred into pots in a soil mixture. The pots must be covered by plastic for a few days to prevent desiccation of the plants.

Mature somatic embryos are called "somatic seedlings". Somatic seedlings are planted in soil and can be grown in a greenhouse until they are ready to be transplanted to the field.

DEVELOPMENTAL PATTERNS OF SOMATIC EMBRYOS

Somatic embryos generally originate from single cells, which divide to form a group of meristematic cells. Usually, this multicellular group becomes isolated by breaking the cytoplasmic connections with the neighbouring cells, and subsequently cutinization of the outer walls of this differentiating cell mass occurs. Due to the continuous mitotic activity, the embryonic tissue increases in size and becomes an embryo. The cells of the meristematic mass continue to divide to give rise to globular (round ball-shaped), heart-shaped, torpedo and cotyledonary stages (Figures 12.5,12.6 and 12.7). A well-defined protoderm and an organized root and shoot apices develop in the torpedo-shaped embryo. Because of the growth of the protoderm, edges grow into cotyledon-like structures, leading to the formation of the cotyledonary-stage embryo. The apical poles become dome-shaped structures to form apical meristems to lead apical growth. In general, the essential features of somatic embryo development, especially after the globular stage is comparable to those of zygotic embryos. Cotyledonary embryos are produced after four weeks of incubation. At this stage, the embryo accumulates plenty of proteins and lipids.

Somatic embryos are bipolar structures in that they have a radicle and a plumule. The radicular end is always oriented towards the centre of the callus, while the plumular end always sticks out from the cell mass. In contrast, a shoot bud is monopolar as it does not have a radicular end. In many somatic embryos, the radicle is suppressed so that they often do not produce roots; in such cases, roots have to be regenerated from the shoots produced by germinating somatic embryos. In case of monocot, generally embryogenic cultures (Figure 12.8) can be regenerated into plantlets (Figure 12.9) either on media supplemented with low 2,4-D or on auxin-free media.

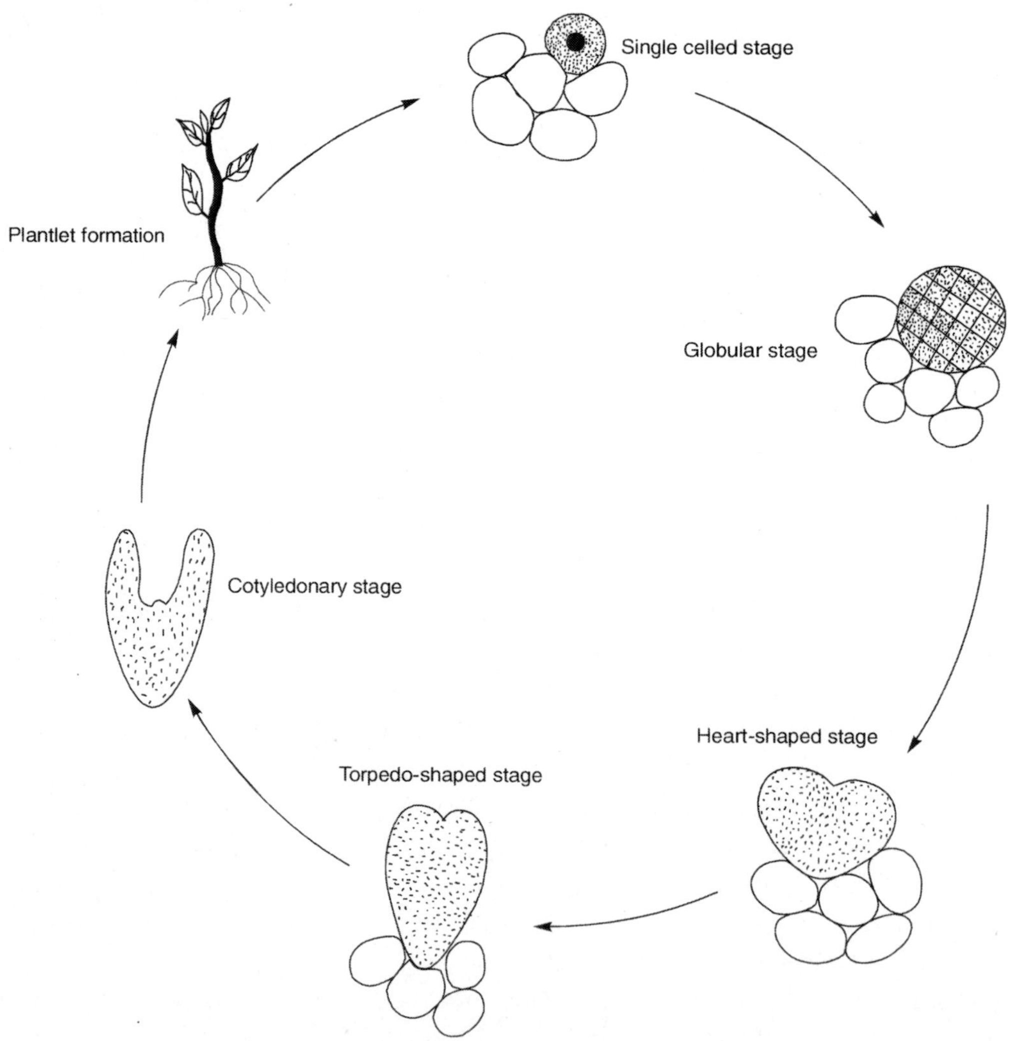

Figure 12.5 Stages of somatic embryogenesis

Figure 12.6 A callus showing differentiation of globular-stage and heart-shaped somatic embryos (Plate 4.2)

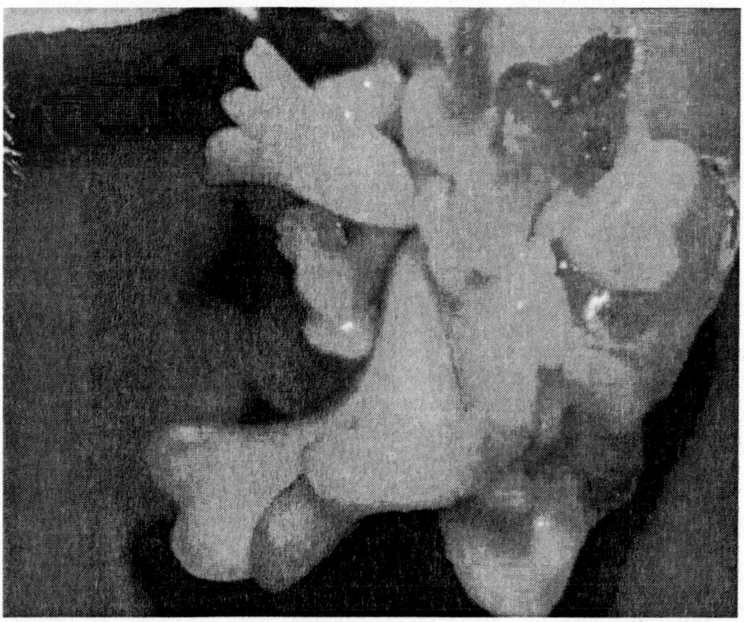

Figure 12.7 A culture showing various stages of somatic embryogenesis (heart-shaped and torpedo stage of somatic embryos) in a dicot plant (Plate 4.3)

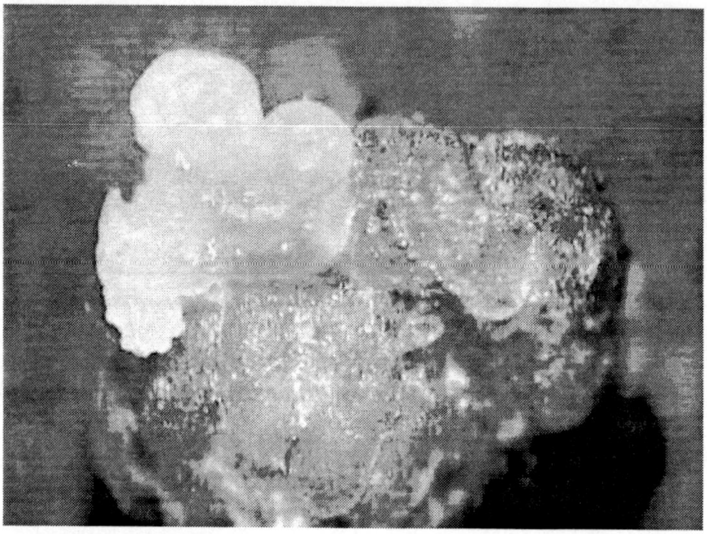

Figure 12.8 A compact callus with somatic embryos in a monocot crop (Plate 4.4)

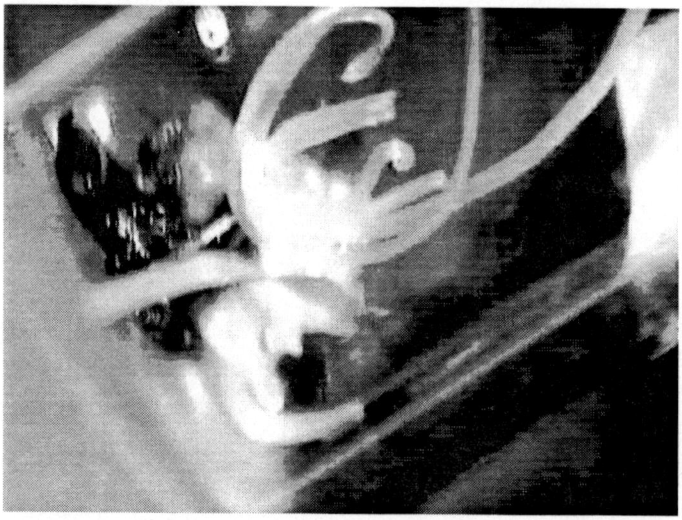

Figure 12.9 Regeneration of somatic embryos into plantlets in a monocot plant (Plate 4.5)

Significantly, studies of Dudits *et al.* (1991) have highlighted the utility of alfalfa microcallus cells as an alternative system, particularly for studying the induction of embryo development from cultured cells. Somatic embryos are induced from cultured callus cells by a relatively simple manipulation of the culturing conditions, as summarized in Figure 12.10. In carrot, this generally involves 1) the establishment of a callus cell line from small hypocotyl pieces cut from sterile germinated individual seeds, 2) the selection of an embryogenic subpopulation of the cultured cells through sieving or gradient fractionation, 3) the removal of auxin from the culture medium, and 4) the dilution of the cells to a relatively low density. The overall embryogenic potential of a culture is highest when it is relatively young and resides primarily within a subpopulation of the culture that has been termed "proembryogenic masses" (PEMs).

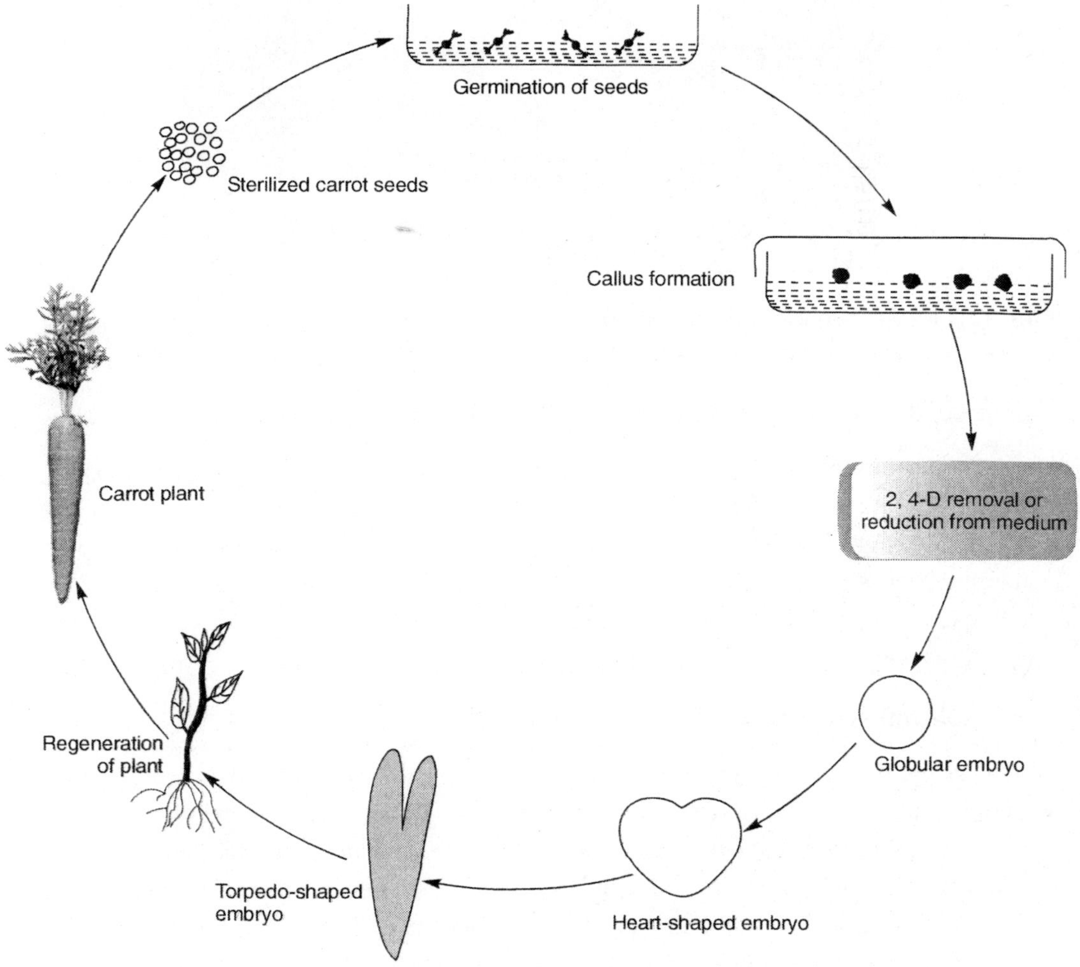

Figure 12.10 Culture of carrot tissue and somatic embryogenesis

Somatic embryos often show abnormal developmental features such as three or more cotyledons, bell-shaped cotyledon, larger size, etc. These problems are often overcome by the presence of ABA or a suitable concentration of mannitol. Somatic embryos induced directly or via callus formation are further subjected to regenerate into plantlets (Figure 12.11) followed by their gradual acclimatization in greenhouse.

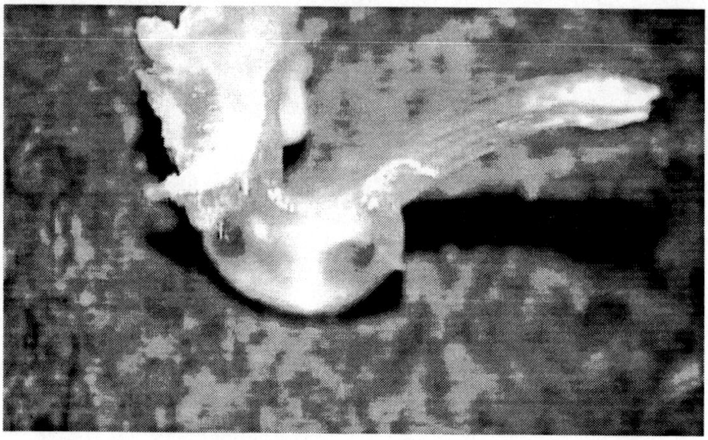

Figure 12.11 Germination of a somatic embryo showing emergence of coleoptile and coleorrhiza (Plate 4.6)

GENE EXPRESSION DURING SOMATIC EMBRYOGENESIS

Gene expression during somatic embryogenesis can be evaluated either by isolating genes expressed in somatic embryos and subsequently identifying the function of those genes or by studying the expression of a variety of other genes isolated from non-embryonic tissues in the hope that they may also play some role in embryogenesis. A number of genes have been identified that are enhanced in expression in developing embryos, and severa l of these are being used to analyse mechanisms of gene regulation during embryogenesis.

Genes Isolated from Somatic Embryos

The dramatic transition from unorganized callus cell growth to somatic embryo development, coupled with measurements that indicate active RNA synthesis during this transition, suggests that a substantial re-programming of gene expression occurs, presumably at the transcriptional level. Early studies showed that few changes were apparent in the abundant proteins that were being synthesized in somatic embryos compared to callus cells. These results were optimistically interpreted as showing that at least a few changes in gene expression could be observed by protein analysis, and it was speculated that even more changes may have occurred in the production of less abundant proteins or mRNAs. Several groups took a similar approach in trying to identify "embryo-enhanced genes" from carrot somatic embryos. The basic experimental strategy relies on a comparison between genes and proteins being expressed in somatic embryos versus callus cells.

In addition to these general differential screening approaches, other experimental strategies have allowed the isolation of a small number of additional genes that are upregulated in somatic embryos. For example, embryogenic cultures secrete some unique proteins. Sterk *et al.* (1991) isolated genes encoding two extracellular proteins (EP), EP2 and EP3 that are enhanced during embryogenesis. Experiments designed to isolate post-transcriptionally controlled genes yielded two other genes, EF1-α, and ATP2. Each of the classes of proteins that have been identified as being upregulated during embryogenesis has added to our understanding of both the process of somatic embryogenesis as well as the relationship between somatic and zygotic embryogenesis.

Late Embryogenesis Abundant (LEA) Gene Expression in Somatic Embryos

Several genes that are preferentially expressed in somatic embryos appear to have characteristics of a class of proteins called LEA proteins. LEA proteins are hydrophilic proteins that are abundantly expressed late in zygotic embryogenesis in many plant species such as cotton, barley, rice, oilseed rape and wheat. The timing of their expression in embryogenesis and their ABA inducibility has led to the suggestion that in zygotic embryos, they play a role in protecting the embryo during desiccation. The LEA genes isolated from carrot somatic embryos include representatives of groups of LEA proteins, which are characterized by significant homology in either amino acid composition or general protein structure.

In carrot, all of the LEA genes that have been analysed to show detectable expression in callus cells (for example, ECP31 and ECP40), are highly expressed in the PEMs of callus cell cultures, and most of the LEA transcripts increase significantly in abundance in somatic embryos at the heart stage. Indeed, it has been found that the LEAs are most abundant and differentially expressed mRNAs in somatic embryos, as compared to callus cells which probably explains why so many LEA genes have been isolated through differential screening approaches using carrot somatic embryos. All of the carrot LEA genes tested can be induced by ABA treatment of callus cells and somatic embryos; however, it appears that only DC3, a group 3 LEA, are induced in non-embryonic tissues by either ABA or desiccation stress, whereas the other LEAs that have been tested (DC8, DC59, ECP31 and ECP40) cannot be induced in non-embryonic cells.

In addition to the analysis of regulatory mechanisms governing the expression of LEA genes in embryogenesis, an investigation of the distribution and timing of EMB-1 LEA gene expression in somatic and zygotic embryos of carrot has been performed. *In situ* hybridization analysis showed that EMB-1 mRNA begins to accumulate uniformly in both somatic and zygotic globular embryos, which is significantly earlier in development than the time of maximum production of LEA proteins. As development proceeds to the heart stage in both zygotic and somatic embryos, EMB-1 mRNA accumulates to higher levels and begins to show a polarity of distribution, with hybridization predominantly over the peripheral regions of the embryo. In mature zygotic embryos, very high levels of

EMB-1 mRNA are seen, primarily associated with the procambium and shoot and root apical meristems. In plantlet-stage somatic embryos, EMB-1 also accumulates predominantly in the meristematic cells, but at lower levels. There is no detectable EMB-1 mRNA in the endosperm, developing seed coat, or developing carpels or fruit of zygotically produced seeds, and EMB-1 mRNA does not accumulate in any cell type of young plants.

The apparent difference in levels of accumulation of this LEA mRNA in somatic and zygotic embryos supports the idea that zygotic embryos experience a secondary signal (possibly a pulse of ABA) from the maternal environment that could not only enhance LEA gene expression, but could also signal the beginning of the dormancy programme.

Somatic embryos provide a useful backdrop for analysis of the intrinsic hormonal influences on embryo induction, morphogenesis, and maturation preceding desiccation and dormancy. If the entire process is hormonally regulated, the activation of the entire signal transduction pathway must be triggered by the embryo itself rather than from any maternal interaction, because this is totally lacking in developing somatic embryos.

FACTORS AFFECTING SOMATIC EMBRYOGENESIS

Somatic embryogenesis is influenced by several factors. The most important chemical factors included in the medium are auxin and reduced nitrogen. Besides chemical factors, some physical factors also influence the process of somatic embryogenesis.

Physical Factors

1. ***Activated charcoal*** The medium supplemented with activated charcoal has facilitated embryogenesis in several cultures. It has been suggested that charcoal absorbs a wide variety of inhibitory substances of somatic embryogenesis from the medium.
2. ***Oxygen*** High concentration of potassium and low dissolved O_2 levels promote somatic embryo regeneration in some species. During subculture, the dissolved oxygen helps in this process. Even for cell suspension, the presence of oxygen is essential.
3. ***Light and temperature*** These two factors are also important in somatic embryogenesis. High light intensity has been shown to be inhibitory to organogenesis and plantlet formation. It has been observed that the callus grows better at 18–25°C and shoot bud differentiation is very good at higher temperature (above 29°C).

Chemical Factors

1. ***Plant growth regulators*** In most cases, 2,4-dichlorophenoxyacetic acid (2,4-D) is essential for somatic embryogenesis. The auxin causes dedifferentiation of a proportion of cells of the explants, which begin to divide. Somatic embryogenesis in carrot, coffee and alfalfa are the classical examples.

Somatic embryogenesis is a two-step process. Cells first develop into a callus tissue in the medium containing the 2,4-D (0.5 to 1.0 mg/l). When such callus tissue is transferred to the same medium with a very low level of auxin (0.1 to 0.5 mg/l), embryoids are formed.

If the callus tissue is maintained continuously in the medium containing auxin, formed embryoids would follow recallusing pathway instead of proceeding into further stages of somatic embryo development. In the continuous presence of auxin, the embryogenic clumps grow and break up into smaller cell masses, which again produce embryonic clumps. But only when the auxin is removed or reduced, further development of globular stage of somatic embryos is possible. Each somatic embryo is believed to develop from a single superficial cell. The ability to regenerate somatic embryos is acquired by cells during dedifferentiation in response to high auxin treatment. Some glycoproteins produced by totipotent cell masses are secreted into the medium. When these proteins are added into the culture medium, they speed up the process of acquisition of totipotency. A class of proteins, called arabinogalactan proteins, induces somatic embryo regeneration from undifferentiated carrot cells, indicating their role in the process.

Hormone requirements for the two phases are drastically different. In most cases, somatic embryos begin to germinate immediately after the cotyledonary stage; this is called somatic embryo conversion. But often the plantlets obtained are rather weak. Somatic embryo maturation is essential and can be achieved by culturing them on a high sucrose medium or in the presence of a suitable concentration of ABA.

In some species, somatic embryos are produced in response to a cytokinin (BAP). BAP induces somatic embryogenesis from hypocotyls of young *Trifolium* seedling. But somatic embryogenesis can be possible on immature cotyledons of these explants only when 2,4-D is used along with BAP in the medium. It seems that cytokinins are effective in somatic embryo regeneration from embryogenic cells of young zygotic embryos, while auxins are effective on differentiated cells of both zygotic embryo tissues and somatic tissues. Other plant growth regulators like gibberellins have no positive effects on somatic embryogenesis.

2. ***Nitrogen*** The use of reduced nitrogen (NH_4^+) shows effective results in somatic embryogenesis. In carrot culture, the addition of NH_4Cl to the embryogenic medium having KNO_3 produces the near optimal number of embryoids. The presence of a low level of NH_4^+ in combination with NO_3^- is required for somatic embryo regeneration. In carrot, NH_4^+ is essential during somatic embryo induction, while somatic embryo development occurs on a medium containing NO_3^- as the sole nitrogen source.

3. ***The genotype of explant source*** The explant genotype of alfalfa, wheat, maize, rice, etc. has a marked influence on somatic embryo regeneration. In case of alfalfa, individual genes affecting somatic embryogenesis regeneration have been identified. Variation for regeneration ability is mainly additive and highly heritable in maize, rice and wheat.

4. ***The source of carbon*** In alfalfa, use of maltose as carbon source improves both somatic embryo induction and maturation when compared to those on sucrose.
5. ***Role of amino acids*** Glutamine is the most important amino acid but others such as serine and proline, can be added along with glutamine at a lower concentration (100 mg/l). It has been reported that tryptophan induces production of somatic embryos, particularly in monocots.
6. ***Role of other adjuvants*** Casein hydrolysate (800 mg/l) can promote somatic embryo formation in some cases. Similarly, coconut milk (20%) has been proven beneficial. Addition of silver nitrate in the culture medium has proven to be a good inducing agent for the generation of somatic embryos by inhibiting the ethylene level produced by explant tissues. Sometimes, abscissic acid also promotes embryo formation from the cotyledon explants.

Choice of the Source Plant

It is preferable to work with some families of plants known to produce somatic embryos in culture, especially members of the Solanaceae, Umbelliferae and Ranunculaceae families. Nevertheless, in these families, some species successfully produce embryogenic callus, whereas others do not. Even in the same species, different cultivars have shown variable potentialities for somatic embryogenesis.

In carrot, any part of the plant taken at any stage of development successfully produces somatic embryos in culture. However, for other species, only certain regions of the plant may respond in culture. Tissues that are younger and less differentiated can be more efficient for embryogenesis than mature tissues. For instance, immature zygotic embryos can develop into a callus from which multiple somatic embryos may be differentiated whereas mature zygotic embryos normally germinate into young plant without showing any sign of somatic embryo induction. In general, reproductive tissue has proven to be an excellent source of embryogenic material. However, zygotic embryo size is more critical and significant in terms of induction of somatic embryos. Zygotic embryos in too early stages (less than 1.0 mm) prove to be nonresponsive; in contrast, size larger than 2.0 mm does not support somatic embryogenesis tendencies (Figure 12.12), but it supports the germination of embryo without callus formation.

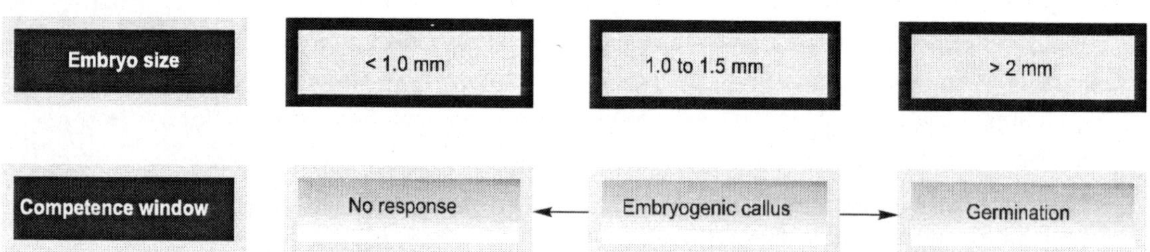

Figure 12.12 Growth stage of embryo and its *in vitro* response

SIGNIFICANCE OF SOMATIC EMBRYOGENESIS

Somatic embryogenesis has several practical applications in clonal propagation of genetically uniform plant materials, elimination of viruses, and provision of target tissue for genetic transformation and development of synthetic seeds. The significance of somatic embryogenesis is discussed below:

1. Somatic embryogenesis is used to quickly produce many genetically identical plantlets. These plantlets can be genetically modified to enhance certain tree characteristics, such as increased growth rate, disease resistance and improved fibre quality. In future, such trees may be used to improve wood quality for industry or to populate a deforested area.
2. Elite plant genes may also be preserved for long term via somatic embryo storage. This is accomplished by deep-freezing the somatic embryo, or by drying them to create "artificial seeds". Somatic embryos need to be preserved until they are tested and approved for safety. They are allowed to be grown and planted only after being approved. Preserved somatic embryos also help to protect endangered and commercially valuable tree species.
3. Trees grown from somatic embryogenesis are of early maturing types.
4. The high volume of multiplication of embryonic propagules is the most commercially attractive application of *in vitro* somatic embryogenesis.
5. The mass production of adventitious embryos in cell culture is still regarded by many as the ideal propagation system. The adventitious embryo is a bipolar structure that develops directly into a complete plantlet and there is no need for a separate rooting phase as with shoot culture.
6. Somatic embryos have advantages for mass propagation in comparison to both conventional clonal propagation method and micropropagation. The somatic embryogenesis method is very fast as compared to other alternative methods it has very high multiplication rates. Unlimited number of embryos can be generated from a single explant.
7. Another importance of somatic embryogenesis is that, for many species, both growth of the embryogenic tissue and development of the somatic embryos can be carried out in liquid medium, making possible the handling of enormous numbers of embryos at a time. It helps for rapid bulking of reproductive propagules without much labour.
8. Somatic embryos have no food reserves, but suitable nutrients could be packaged by coating or encapsulation to form a kind of artificial seeds. These artificial seeds produce the plantlets directly in the field as the natural seeds do.
9. Plants derived from asexual embryos may in some cases be free of viral and other pathogens. For example, *Citrus* plants propagated from embryogenic callus of nucellar origin are free of viruses.

Although somatic embryogenic cultures can be successfully generated, the subsequent stages of somatic seedling development are not reliable enough for all plants to make the process economically sustainable. Researchers are trying in various ways as a means to develop somatic seedlings in an economically feasible way, with the goal of standardizing the method for applications in industry and use in commercial endeavours.

Researchers are also looking for identification of tree-specific genes. This will help to understand how the different tree species grow, how genes behave, and in what way the tree species can be improved by genetic modification. Scientists are also identifying the possible impacts of genetically modified (GM) plants in the environment. This can be achieved by breeding GM plants with natural plants and furthermore, by studying the effects of their offspring on the surrounding plant communities.

SUSTAINABLE DEVELOPMENT AND SOMATIC EMBRYOGENESIS

Sustainable development is made possible through careful research and assessment programmes to preserve the environment and promote the biological diversity of a forest. The development of improved and standardized protocols would make somatic embryogenesis more economically viable. Educating both industry and the public about somatic embryogenesis and genetic engineering would promote social responsibility through informed decision-making when producing or buying forest products. In the future, somatic embryogenesis could play a key role in both improved forest conservation methods and increased forest productivity. Conservation may be carried out at many levels, from somatic embryogenesis-aided reforestation efforts to the preservation of endangered tree species by artificial seed production and by deep-freezing somatic embryos.

REVIEW QUESTIONS

1. What is somatic embryogenesis? Discuss the principles of somatic embryogenesis.
2. What is the role of auxins and cytokinins in somatic embryogenesis?
3. Write notes on the significance and applications of somatic embryogenesis.
4. Describe the important factors affecting somatic embryogenesis.
5. Write a note on gene expression during somatic embryogenesis.
6. Explain the stages and process of somatic embryogenesis.

REFERENCES

Burza, W. and Malepszy, S. (1995). *In vitro* culture of *Cucumis sativus* L. XVIII. Plant from protoplasts through direct embryogenesis. *Plant Cell, Tissue and Organ Culture.* 41: 259–266.

Dirks, R. and Buggenum, M. (1989). *In vitro* plant regeneration from leaf and cotyledon explants of *Cucumis melo* L. *Plant Cell Reports.* 7: 626–627.

Gupta, P.K., Timmis, R., Pullan, G., Yancey, M., Kreitinger, M., Carlson, W. and Carpenter, C. (1991). Development of embryogenic system for automated propagation of forest trees. In: Vasil, I. K. (ed.). *Scale-up and Automation in Plant Propagation.* Academic Press, Inc., London. pp. 76–94.

Gupta, P. K., Pulman, G., Timmis, R., Krietinger, M., Carlson, W.C., Grob, J. and Wetty, E. (1993). Forestry in the 21st century. The biotechnology of somatic embryogenesis. *Biotechnology.* 11: 454–459.

Muralidharan, E.M. and Mascarenhas, A.F. (1995). Somatic embryogenesis in *Eucalyptus.* In: Jain, S., Gupta, P. and Newton, R. (eds.) *Somatic Embryogenesis in Woody Plants.* Vol. 2. Kluwer Academic Publishers, The Netherlands. pp. 23–40.

Nomura, K. and Komamine, A. (1999). Physiological and morphological aspects of somatic embryogenesis. In: Soh, W.Y and Bhojwani, S.S. (eds.). *Morphogenesis in Plant Tissue Cultures.* Kluwer Academic Publishers, The Netherlands. pp. 115–131.

Punja, Z.K., Abbas, N., Sarmento, G.G. and Tang, F.A. (1990). Regeneration of *Cucumis sativus* var. *Hardwickii, C. melo,* and *C. metuliferous* from explants through somatic embryogenesis and organogenesis. *Plant Cell, Tissue and Organ Culture.* 21: 93–102.

Rangaswamy, N.S. (1986). Somatic embryogenesis in angiosperm cell tissue and organ cultures. *Proc. Indian Acad. Sci.* (*Plant Sciences*). 96: 247–271.

Wang, T.L. and Cuming, A. (1996). *Embryogenesis—the Generation of a Plant.* Bios Scientific Publishers, New Delhi.

Vikrant and Rashid, A. (2002). Somatic embryogenesis from immature and mature embryos of a minor millet *Paspalum scrobiculatum* L. *Plant Cell, Tissue and Organ Culture.* 69: 71–77.

Vikrant and Rashid, A. (2002). Induction of multiple-shoots by thidiazuron from caryopsis cultures of minor millet (*Paspalum scrobiculatum* L.) and its effect on regeneration of embryogenic callus cultures. *Plant Cell Reports.* 21: 9–13.

Williams, E.G. and Maheswaran, G. (1986). Somatic embryogenesis: Factors influencing coordinated behaviour of cells as an embryogenic group. *Ann. Bot.* 57: 443–462.

13

HAPLOID PRODUCTION-I (ANDROGENESIS)

INTRODUCTION

Haploids are defined as sporophytes with gametophytic chromosome number, and these have been produced in many plant species using a variety of methods. Although the significance of haploids in genetics and plant breeding has been recognized for a long time, with the advent new approaches in plant biotechnology, it has received renewed emphasis. Now, the production of haploids has become an important and integral part of crop biotechnology programmes.

Angiosperms, like other green plants, have two phases in their life cycle—diploid sporophytic phase and haploid gametophytic phase. The plant body, which usually develops from the seed, is differentiated into root, stem and leaves. This is the dominant phase in the life cycle of higher plants. Flowers are the reproductive organs and contain haploid microspores (pollen) and megaspores, which are formed in the stamens and carpels, respectively, by meiosis. The microspore gives rise to the male gametes. Therefore, the gametophytic phase in higher plants is very inconspicuous and extremely reduced, in comparison to lower plant groups, where the gametophytic plant body is the dominant phase of life cycle.

Spontaneous production of haploids usually occurs through the process of embryo development from an unfertilized egg under *in vivo* conditions. Each pollen possesses a unique genome where every gene is present as a single set. Their phenotype remains unmasked by gene dominance effects; production of homozygous diploid by doubling the chromosome number of haploids *in vitro* makes a pure line in a single step. Such homozygous pure line is of great importance in plant improvement.

The stamen is typically a slender organ and its distal fertile part is known as anther. Anther consists of the microsporangia, which contain numerous pollen grains. The microspore is the mother cell of the male gametophyte. The egg cell produced within the ovule is very difficult to separate from the complex tissue integration. Thus, pollen is more suitable material than egg cell for the production of haploid plants. With reference to higher plants, haploids may be obtained as sporophytes with gametophytic chromosome constitution. In nature, their frequency is extremely low.

For the first time, to study meiosis, anthers were cultured *in vitro* by a Japanese cytologist, Shimakura (1934). But only much later, were anthers brought into the culture tube by Guha and Maheswari (1964) with the objective to understand the regeneration potential of germinative tissues under *in vitro* controlled conditions. They came across many embryo-like structures in the cultured anther of *Datura innoxia,* which they called embryoids. Later, it was demonstrated that these embryoids were originated from pollen grains through a series of divisions and therefore, the totipotent nature of pollen grains was confirmed. Moreover, Beurgin and Nitsch (1967) obtained complete haploid plantlets from the anther culture of *Nicotiana tabacum.*

Production of haploids is possible by methods such as delayed pollination, irradiation of pollen, temperature shock treatment, colchicine treatment and distant hybridization; however, the most important methods currently being used in biotechnology programme include:

1. Anther or pollen culture and ovule (or ovary) culture.
2. Chromosome elimination following interspecific hybridization (also called bulbosum technique).

One of the very common methods for production of haploids is through the culture of anthers or microspores on artificial culture media. This leads to the growth of microspores into sporophytes. After the initial reports of successful production of haploids from anther culture in *Datura* by Guha and Maheshwari in 1964, haploid production from anther was obtained in more than 250 species belonging to 34 families of angiosperms by Maheshwari *et a1.* in 1980. These include a wide variety of economically important species. This number is increasing every year.

Furthermore, induction of somatic embryos directly from the pollen grains of cultured anthers have been achieved in many plant species, for example, *Datura, Atropa, Brassica campestris, B. napus, Nicotiana tabacum, N. rustica, Petunia axillaris,* etc. In these cases, the plants obtained from germination of somatic embryos are generally haploid, but some polyploids are also produced. However, in many other species like *Oryza sativa, Hordeum vulgare, Triticum aestivum, Triticale,* etc., pollen grains produce callus from which plantlets could be regenerated under suitable culture conditions. Haploid plantlets have been regenerated from pollen grains of about 200 species of over 50 genera and 25 families. Simultaneously, some of the basic factors that have been considered significant and proved to be critical for successful anther and microspore culture are 1) genotype of donor plant, 2) physiological status of donor plant, 3) the developmental stage of anther/microspore, 4) the mode and dose of explant pre-treatment, 5) the chemical nature of nutrient medium, 6) the incubation conditions during culture, etc. These factors influence the frequency of androgenesis.

Anthers are cultured more often than microspores, since the extraction and culture methods for microspores are somewhat difficult and have been successful only in a few species (such as *Datura innoxia, Nicotiana sylvestris, N. tabacum, Oryza sativa,* etc.).

ANTHER CULTURE

Anther culture is a technique by which the developing anthers at a suitable stage are excised aseptically from unopened flower buds. These anthers are cultured on a nutrient medium, where the pollen grains develop into either callus or embryoids that finally give rise to haploid plantlets as shown in Figure 13.1. The technique of anther culture was first developed by Guha and Maheshwari in 1964. Some of the advantages, which make anther culture a valuable method for obtaining haploid plants are:

1. The technique is fairly simple.
2. It is easy to induce cell division in the immature pollen cells in some species.
3. A large proportion of the anthers used in culture respond well (induction frequency is high).
4. Haploid plants can be produced in large numbers very quickly.

In experiments using *Datura innoxia,* haploid induction frequency is almost 100% and a yield of more than one thousand plantlets or calluses have occurred under optimal conditions from one anther. Success rate can be determined within 24 hours of culture initiation as cells begin to divide. The haploid embryoids or the callus tissue can be seen as the anther dehisces in cultures. But there is always a fair possibility that the diploid somatic cells of the anther will also respond to culture conditions and so produce unwanted diploid callus. In order to avoid this problem, free pollen isolated from the anther can be grown in an appropriate nutrient medium.

In either case, flower buds are brought to an LAF cabinet and sterilized using appropriate chemical treatment before their dissection. While removing anthers from the flower buds, care is exercised to avoid injury, because injury leads to development of callus, giving a mixture of diploids, haploids and aneuploids. The anthers are generally cultured on solid agar medium, where they may directly give rise to embryoids or may lead to callus formation, before differentiation. The embryoids develop into haploid plantlets, which are further treated with colchicine to get diploid homozygous plants.

Significantly, microspore culture may be preferred over anther culture, although the degree of success is low in this case.

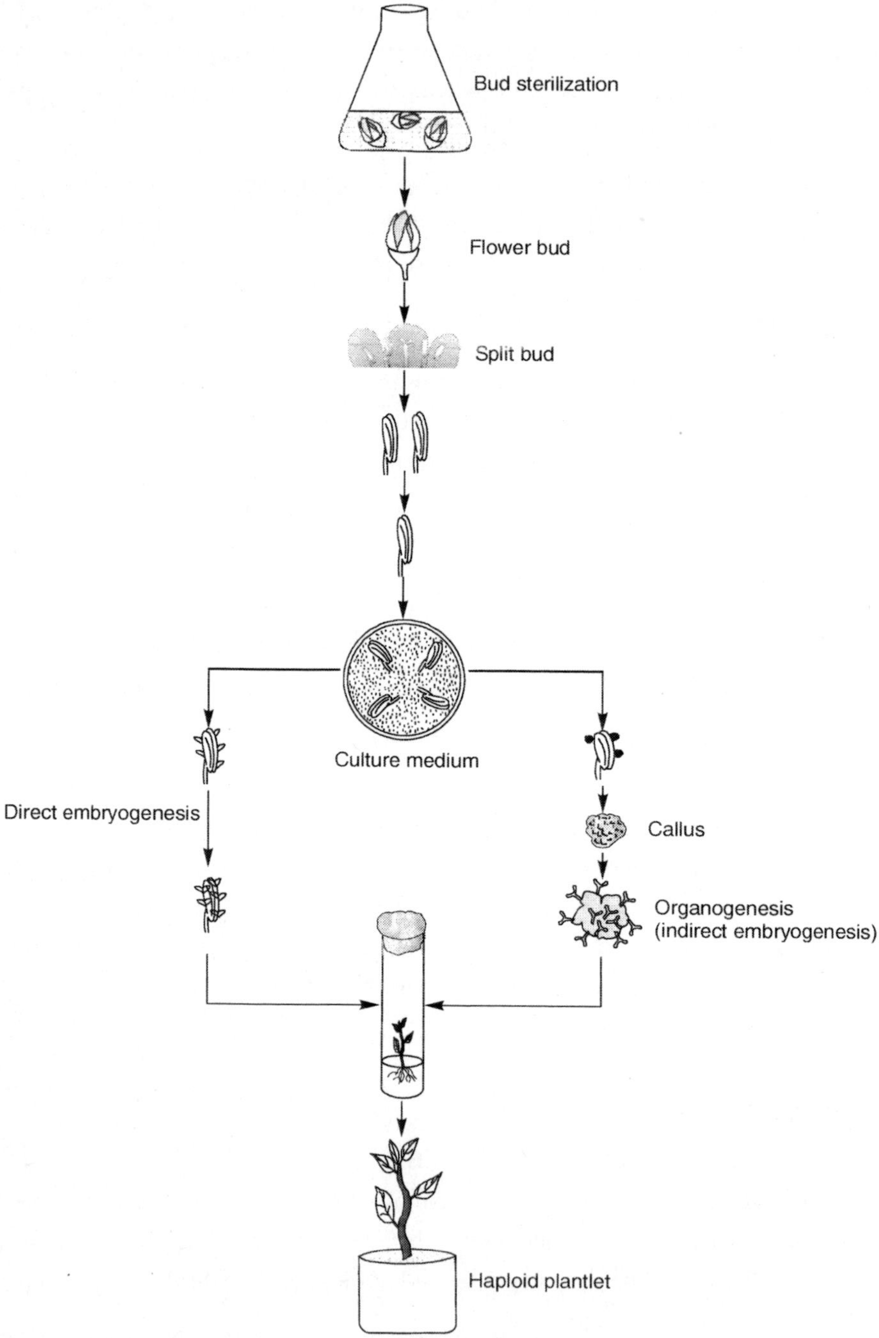

Figure 13.1 Anther culture technique

Source of Anther

The success of androgenesis depends upon factors like the age of the donor plant, kinds and durations of anther pretreatment, the environmental conditions, etc. The frequency of androgenic response has been observed relatively more in anthers of short-day plants, than the anthers of long day plants, and tobacco, mustard, etc., exhibit this type of androgenic response. Pretreatment of anthers causes effective increase in response to androgenesis. The pretreatment methods, such as cold treatment of anthers prior to detachment from flower bud, showed pronounced activity in multiplication of embryonic cells.

Furthermore, in order to achieve maximal androgenic response, the experimental plants from which floral buds are to be excised for haploid production should be of young age, because with increasing age of the source plant, the androgenic responses decline and abnormalities appear frequently.

The process of androgenesis can be completed via two modes:

1. ***Direct androgenesis*** The pollen grains behave like a zygote and undergo repeated divisions to form embryoids, which ultimately give rise to plantlets.
2. ***Indirect androgenesis*** The pollen grains first undergo the formation of an undifferentiated mass by repeated divisions to form a callus tissue which finally redifferentiates and forms haploid plantlets.

Nutrient Medium for Androgenesis

The medium requirement may vary with the species, the genotype, the age of source plant and anthers, and the conditions under which the source plants are grown. Miller's (1963) medium is used for cereal crops with some modifications. The nutritional requirements of excised anther is very much similar to those of isolated pollens. The pollens can be induced for androgenesis by keeping them in sucrose solution for 24 hours. A basal medium, as formulated by Murashige and Skoog (1962), is the ideal medium for *Nicotiana tabacum*. Addition of activated charcoal to agar medium is advocated, since charcoal is thought to absorb inhibitory compounds present in trace elements in the culture medium and also ensures a greater survival of the induced pollen grains. The iron in the medium also plays a very important role for the induction of haploids. Potato extract, coconut milk and growth regulators like auxins and cytokinins are also used for anther and pollen culture because of their stimulatory effects on androgenesis.

In *Datura,* sucrose is most essential for androgenesis but there are some reports where pollen could develop embryos even in the complete absence of sucrose (as in tobacco). The addition of additives like ascorbic acid and glutathione, adenine sulphate and glucose as substitutes of sucrose in the basal medium has proved to be stimulatory for androgenesis.

Pathways of *In vitro* Androgenesis

Under normal conditions, the microspores mature into pollen grains *in situ* but in culture, their morphogenesis is altered. Depending on the composition of the medium, development may lead either to the formation of embryoids and/or plantlets or a mass of parenchymatous callus. Based on the few initial divisions in the microspores, four pathways have been identified as leading to *in vitro* androgenesis (Figure 13.2).

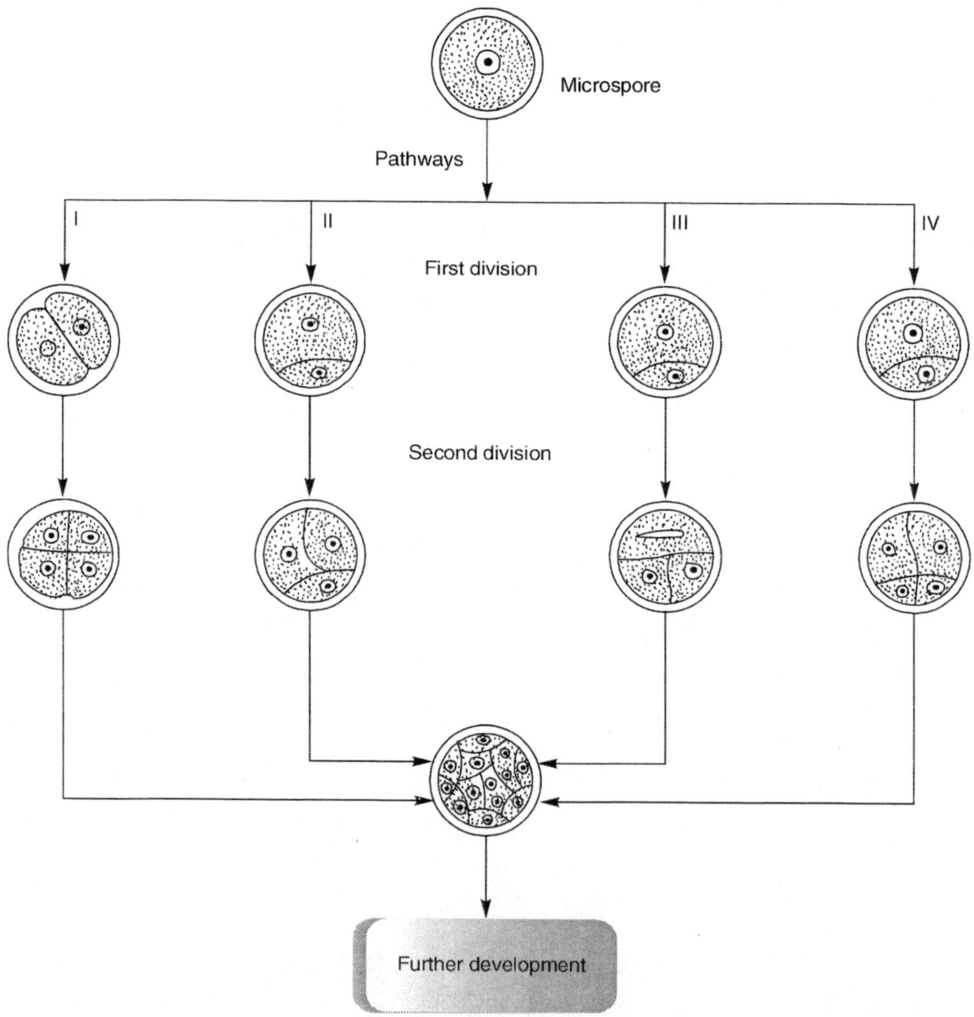

Figure 13.2 Pattern of early divisions in microspore during androgenesis

1. ***Pathway I*** In *Datura innoxia,* the uninucleate pollen grain may divide symmetrically to yield two equal daughter cells, both of which undergo further divisions. The microspores divide by an equal division and two identical daughter cells contribute to sporophyte development. Vegetative and generative cells are not distinctly formed in this pathway.
2. ***Pathway II*** In this case, the mononucleate pollen divides unequally. The division of uninucleate microspores is unequal, resulting in the formation of a vegetative and a generative cell. The sporophyte or plantlet arises through further divisions in the vegetative cell while the generative cell, either does not divide or does so only once or twice before degenerating (for example, *N. tabacum,* barley, wheat, etc.).

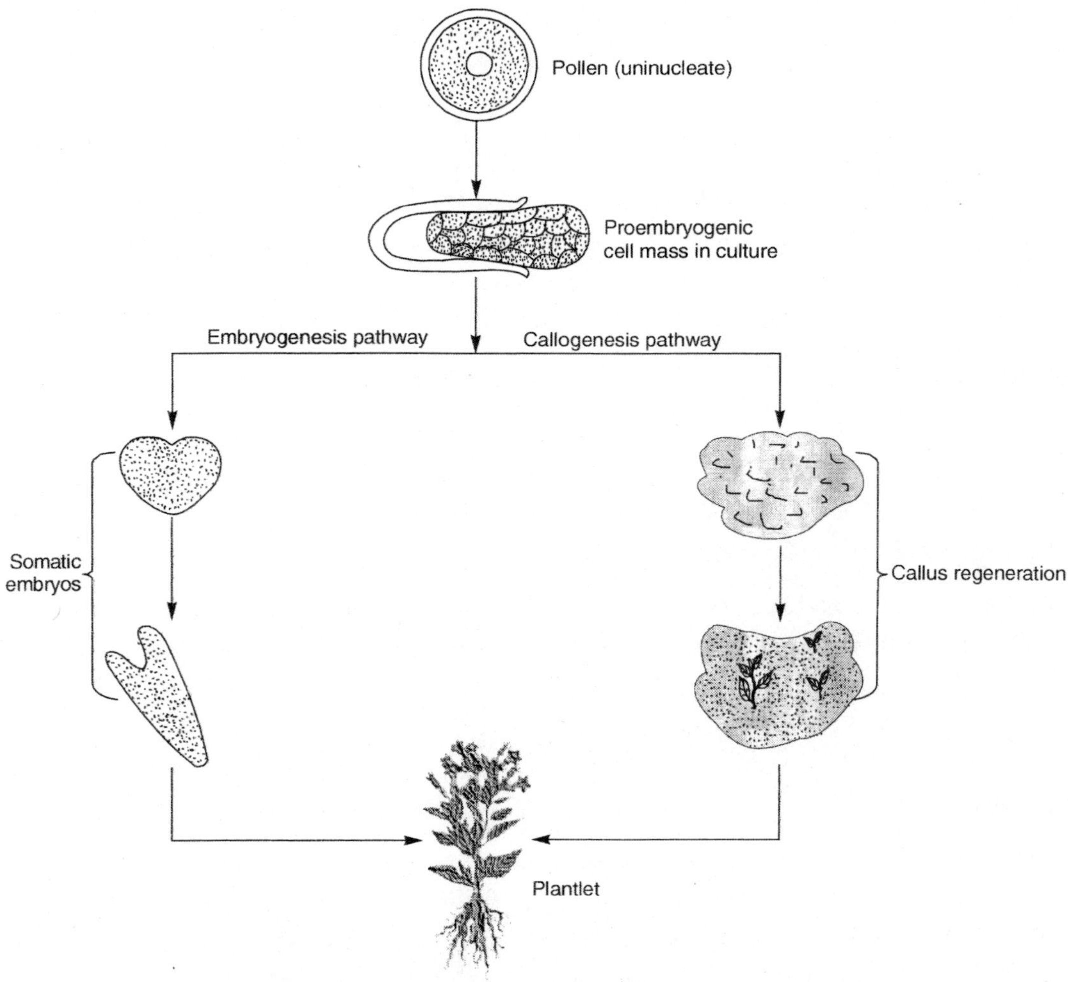

Figure 13.3 Pathways of pollen differentiation in culture

3. ***Pathway III*** The uninucleate microspore undergoes a normal unequal division but the pollen embryos are predominantly formed from the vegetative cell alone (for example, *Hyoscyamus niger*). The generative cell, either does not divide at all or does so only to a limited extent.
4. ***Pathway IV*** In some species, the uninucleate pollen grains divide unequally, producing generative and vegetative cells but both these cells divide repeatedly to contribute to the developing embryo. (for example, *Datura metel, Atropa belladonna,* etc.).

Irrespective of the above early patterns of microspore division, the embryogenic pollen grains ultimately become multicellular and burst open, gradually assuming the form of a globular embryo. This is followed by the normal stages of post-globular embryogeny until the development of an embryoid takes place. Alternatively, the multicellular mass liberated from the bursting pollen grain proliferates to form callus which may later differentiate into whole plants either on the same medium or on a modified medium.

It is sometimes possible to obtain androgenic haploids via embryo formation or through callusing (Figure 13.3) from anthers of the same species (*Oryza sativa*) by manipulating the constituents of the medium. In general, many microspore culture procedures involve a heat treatment which causes the microspores to undergo equal mitosis and essentially behave like egg cells and develop into an embryo.

Anther Culture Technique

The experimental plants being used for anther or pollen culture should be grown under controlled conditions of light, temperature and humidity. This is helpful in drawing a possible correlation between the developmental stage of anther and the morphological characteristics of the flower such as bud length and emergence of sepals and petals.

Unopened flower buds are sterilized with 70% ethanol for 10 minutes, followed by 20% sodium hypochlorite for 10 minutes, after which sepals and petals are removed aseptically. Contact between the dissecting instruments and the anthers should be avoided. With a pair of sterilized forceps, the stamens are taken out and kept on a sterilized Petri dish. The filaments are excised from the anther and the anthers are inoculated on solidified basal medium. The remaining anthers from the flower buds are excised to avoid injury.

The culture plates/tubes with the inoculated anthers are kept into the growth/incubation/culture chamber. Anther cultures are generally maintained in alternating periods of light (12–18 hours; 5000–10000 lux m^2) at 28°C and darkness (6–12 hours) at 22°C, but the optimum condition varies with the plant species. If the material is *N. tabacum,* the anther generally opens after 4–5 weeks at 25°C. The wall of responsive anthers turns brown and within 3–8 weeks, these anthers burst and open due to the induction and development of callus or embryoids. Further, callus differentiation takes place by production of shoot buds followed by young shoot formation. However, in case of embryoid development from anther, these embryoids germinate and develop into young plantlets. Moreover, when young shoots are 3–5 cm long, they are transferred to a medium conducive to good root development. Some stress treatments are also helpful in induction of androgenesis (Table 13.1).

Table 13.1 Some stress treatments that induce androgenesis

Plant	Stress treatment	Stage of responsive pollen grain
Brassica napus	Heat treatment (32°C for 8 hrs), gamma rays, colchicine	Late unicellular to early binucleate pollen grains
Hordeum vulgare	Stress by mannitol, calcium and ABA	
Nicotiana tabacum	Glutamine and sugar starvation (then transferred to high glucose medium)	Young bicellular pollen grains
Triticum aestivum, Oryza sativa and other millets	Cold treatment (3–24 days at 4–10°C)	

The rooted plantlets are again transferred to sterilized potting mixture in small pots and kept in greenhouse for their further growth.

Detection of Competent Anthers

To identify the correct stage of microspore (pollen) development within the anther, it is necessary to examine the microspores as follows:

1. The anthers are collected from flower buds at various stages of development.
2. They are placed on a clean slide, and a drop of 0.8% aceto-orcein stain in 40% acetic acid is added. (1.0% orcein in 50% acetic acid or 1.0 g orcein in 50 ml glacial acetic acid is used. It is then boiled in a flask with a funnel on it, and immediately removed from heat. This solution is cooled and then 50 ml distilled water is added).
3. The tissue is gently macerated with a scalpel.
4. The large pieces of anther wall are pushed to the side using the scalpel.
5. The slide is then gently heated. Care is taken not to scorch it.
6. The slide is allowed to sit for 5–10 minutes.
7. A coverslip is placed over the material.

8. Then the slide is placed on a paper towel, and the towel is folded over the slide; the coverslip is gently and evenly pressed.
9. The slide is then placed under the microscope and observed.

Limitations of Anther Culture

1. In some species, the majority of the regenerated plants probably turn out to be non-haploid.
2. In cereals, very few green plants are obtained and most of the plants are either albinos or green-albino chimeras.
3. It is tedious to remove the anthers without causing damage.
4. Sometimes, a particular orientation is necessary to achieve a desired response.

POLLEN CULTURE

The pollen grain or microspore, when cultured on nutrient medium, produces haploid plants without giving rise to male gametophyte. This process is known as pollen culture. The pollen grains are taken out from the anther aseptically and cultured in liquid medium separately. Thus, the process is also called pollen suspension culture.

Sometimes the anther wall produces some growth-inhibitory substances in the medium. Such types of problems can be minimized by pollen cultures.

Pollen Competency and Differentiation

Pollen stage varies according to the species. In *Datura* and tobacco, the suitable stage is just before or soon after the first pollen mitosis, while the early binucleate stage is the most appropriate for species like *Atropa belladona* and *N. sylvestris,* and it is absolutely essential for *N. knightiana*. Significantly, in cereals, the best stage appears to be the early or mid-uninucleate stage, that is, before the first pollen mitosis, while in *Brassica,* trinucleate pollen grains stage is the best stage. The beginning of starch accumulation in pollen grains of tobacco marks the end of their embryonic potential.

An anther or pollen has two distinct morphologies that have the potential to develop into two different modes of generations: the gametophyte and the sporophyte. This is referred to as pollen dimorphism. The pollen capable of forming haploids (embryogenic pollen) is small with a cytoplasm that stains less intensely with acetocarmine than gametophytic pollen. Such pollen is usually low in quantity and accounts for a very low frequency of haploid formation. Exposure of plants to low temperatures before flowering increases the frequency of embryogenic pollen.

The pollen mother cell is diploid and has a determinate structure. It undergoes cell reorganization for differentiating into microspore (haploid) cells which are indeterminate structures. Determination into either androgenic (gametophytic) or embryogenic (sporophytic) grains is possibly a function

of differential gene expression. Gametophytic pollen germinates in the presence of metabolic inhibitors in contrast to embryogenic pollen, which does not germinate because of sensitivity to these antimetabolites.

Pollen Culture Techniques

There are several pollen culture techniques that result in the formation of new plantlet of only gametophytic origin, without any heterogeneity. All the techniques are based on the mode of isolation of pollen grains from anther and further culture on either semi-solid medium or liquid medium. Some of techniques are described here:

1. ***Nurse culture technique*** This culture technique was introduced by Sharp *et al.* (1972). All major steps like selection of flower bud, sterilization and excision of anther without filament are the same as described in anther culture. The intact anthers are placed horizontally on the top of the semi-solid medium within a conical flask. A small filter paper disc is kept over the intact anther and about 10 pollen grains in suspension are then placed on the filter paper disc. Hence, the intact anthers are considered as the nurse tissue. Sometimes, callus tissue derived from any tissue of the plant is used as nurse tissue. After 2–4 weeks, colonies of parenchymatous cells appear on the disc. In the absence of nurse anthers, the pollen does not form colonies.
2. ***Nitsch's technique*** This method was developed by Nitsch in 1974. This culture technique is considered to be an ideal method for pollen culture. The experimental plants are kept in controlled conditions and anthers from sterilized buds are placed in a small sterile beaker containing 20 ml of liquid Nitsch medium. By pressing the anthers with the help of a sterilized glass rod, the pollen grains are squeezed out. The homogenized anthers are then filtered through a nylon sieve to remove the anther tissue debris. The filtrate is then centrifuged at low speed (500–800 rpm) for 5 minutes. The pellet of pollen is suspended in fresh liquid medium and washed twice by repeated centrifugation and resuspension in fresh medium. Pollens are finally mixed with a measured volume of liquid Nitsch medium. About 2.5 ml of pollen suspension is pipetted off and is suspended in a 5 cm Petri dish. Pollen can also be grown by plating it on soft agar medium. These Petri dishes are incubated at 25 ± 2°C under 500 lux light intensity for 16 hours. Young embryoids are produced within one month. These embryoids ultimately develop into haploid plantlets, which, on maturity, can be transferred to soil for further development.
3. ***Float culture technique*** This technique was developed by Sunderland and Roberts in 1979. In this method, the excised anthers are floated on a liquid medium and anthers dehisce pollen grains that develop into haploid plantlets. The cultured Petri dish is incubated at 27–30°C under 500 lux light intensity. After a few days, the anthers dehisce and pollen grains are discharged into the liquid medium. These pollen grains germinate to give rise to haploid plantlets.

Protocol of Pollen Culture

A general protocol of pollen culture technique is described below:

1. Flower buds are collected from the plants for culture.
2. The buds are placed in a Petri dish and the length of each bud is measured using a scale. Buds of corolla at proper length are chosen, as the pollen in them would have usually completed the first mitosis.
3. The buds are chilled for 1–2 hours at 7 to 8°C in a refrigerator.
4. They are then surface-sterilized in a Petri dish containing a suitable sterilizing agent, for instance 0.1% solution of mercuric chloride.
5. Using sterile double-distilled water, the buds are rinsed 3–4 times in a sterile LAF cabinet.
6. Using forceps and a dissecting needle, the buds are carefully teased open and the anthers are taken out and kept in a sterile Petri dish.
7. The anthers dissected out from each bud are grouped separately on removal.
8. One anther from each group is then taken and squashed in acetocarmine stain in order to determine the suitable stage of pollen development. The pollen should be in the state of first pollen division.
9. For each culture, anthers from flower buds are placed in 5 ml of liquid medium in a Petri dish.
10. At regular intervals of 6, 10 and 14 days of culture, the anthers are removed from the culture and discarded. This ensures the dehiscence of all anthers and release of pollen would have occurred into the culture medium.
11. The dishes are then sealed with parafilm and incubated at 28°C in darkness for the first 14 days of culture.
12. After 14 days, the cultures are transferred to an illuminated growth chamber (1000 lux, 12 hours day length, at 25 ± 2°C).
13. The growth of haploid embryos to plantlets, developed from the released pollen, can be observed.

The basic approach of microspore isolation and culture can be followed (Figure 13.4) which is applicable in routine practice. The nurse culture technique is used for raising haploid tissue clones from isolated pollen grains. Intact microspores are mounted on a filter paper disk which is then placed over an intact anther or callus derived from the floral parts. The extract of cultured anthers has equal nursing effects on the growth and development of microspores and may be used in place of nurse callus or anther. A purely synthetic medium developed for raising haploid plants from isolated pollen grains of tobacco and datura may also be suitable for other systems.

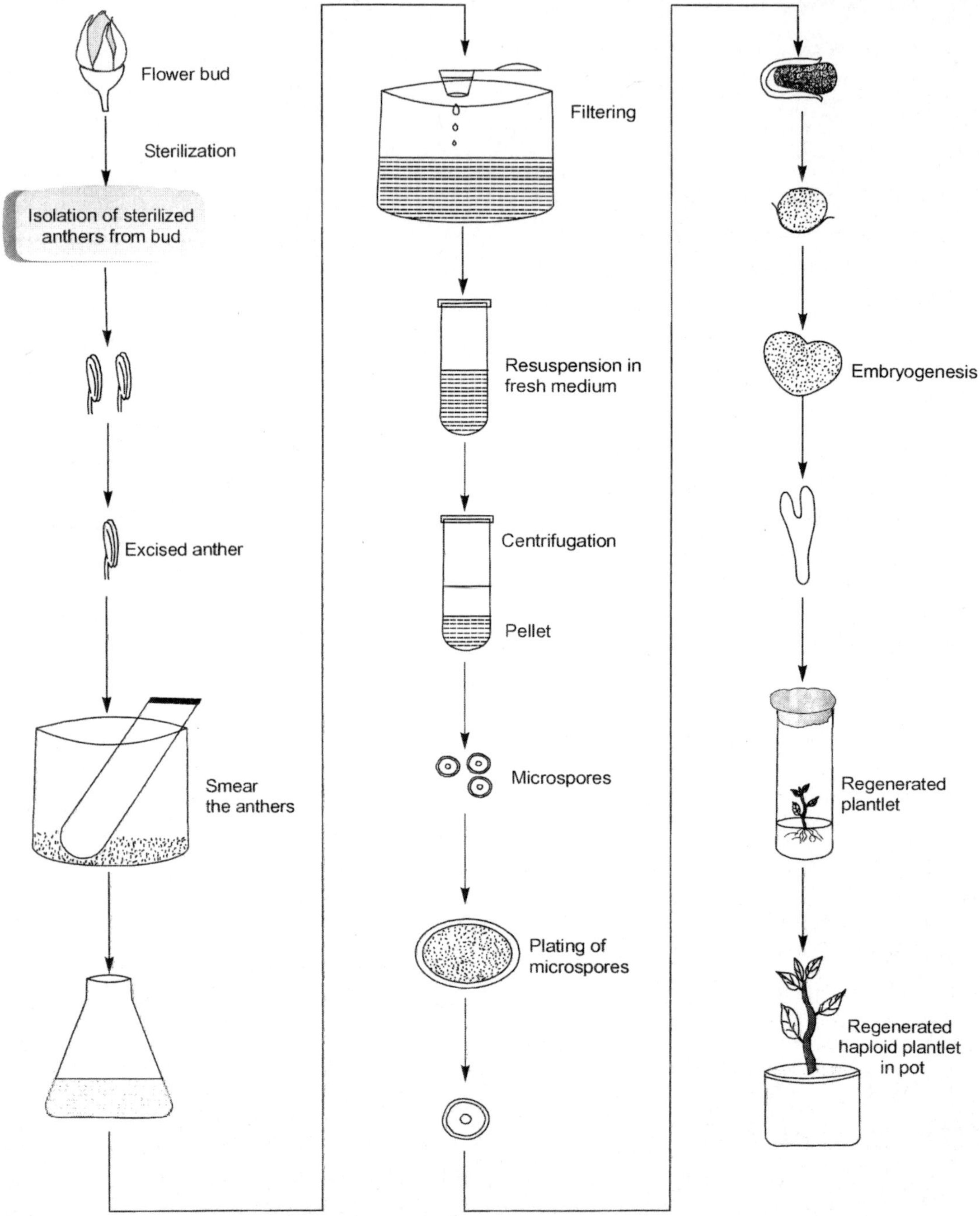

Figure 13.4 Microspore isolation and culture technique

HAPLOID PRODUCTION THROUGH DISTANT HYBRIDIZATION

The production of haploid plants is not only achieved by pollen or anther culture. There are other alternatives too. Selective elimination of chromosome of one species during interspecific crosses of two different species of the same plant and retaining one chromosome of one species, is used to obtain haploids in barley and also in wheat.

1. ***Barley haploids through crosses with Hordeum bulbosum*** A condition similar to somatic reduction occurs, when, in an interspecific or intergeneric hybrid, chromosomes of one parental genome are preferentially eliminated. This method has been extensively utilized for production of haploids in cereals. The production of haploids in barley in this manner involves interspecific hybridization of *Hordeum vulgare* ($2n = 14$) with *H. bulbosum* ($2n = 14$), a cross-pollinated and self-incompatible species. Most progeny (95%) are barley haploids, by this method. A cross between $2x$ *H. vulgare* and $4x$ *H. bulbosum* mainly produces triploid hybrid plantlets.

   Cytological examination of early embryo and endosperm in (VV × BB or BB × VV) hybrids shows gradual elimination of chromosomes.
2. ***Wheat haploids through crosses with Hordeum bulbosum*** Barclay in 1975 demonstrated that the haploids in hexaploid wheat (var. Chinese Spring) can also be produced through this method in hybrids of wheat with *H. bulbosum* (both $2x$ and $4x$).

DOUBLE HAPLOID PRODUCTION

Haploid plants obtained either from anther or ovule culture may grow normally under *in vitro* conditions up to the flowering stage but viable gametes are not formed due to the absence of one set of homologous chromosomes and, consequently, there is no seed set. The only mechanism for perpetuating the haploids is by duplicating the chromosome complement in order to obtain homozygous diploids. In pollen-derived plants, duplication of chromosomes may occur spontaneously in cultures but due to the small percentage of such double haploids, it is necessary to diploidize the haploids by chemical means.

A simple procedure designed to achieve diploidization involves immersion of very young haploids in a filter-sterilized solution of colchicine (0.4%) for 2–4 days, followed by their transfer to the culture medium for further growth. In this procedure, chromosome or gene instabilities are minimal in comparison to other methods of colchicine or chemical treatment. In nature, spontaneous duplication of chromosome in haploid, plants has been observed but its frequency is very low. Thus, artificial techniques are being used by scientists. These are as follows:

1. ***Colchicine treatment*** Colchicine has been utilized widely as a spindle inhibitor to induce chromosome duplication and to produce polyploid plants. This method has been employed for obtaining homozygous diploid plants from haploid culture. The young pollens, while still enclosed within the anther, are treated with 0.4% colchicine solution for 24–48 hours. Young haploids are

immersed in 0.4% colchicine solution for 2–4 days. Treated plantlets are replanted in the medium after thorough washing. In case of mature haploid plantlets, 4% colchicine–lanoline paste may be applied to the axils of the leaves.

2. ***Endomitosis*** Endomitosis refers to chromosome duplication without nuclear division. Explants taken from haploid plants are cultured in a suitable medium to induce callus. The tendency of such cell is to produce diploid homozygous cells by endomitosis. The initial callus may have some diploid cells but their frequency would increase in repeated subcultures. New large numbers of homozygous diploid plantlets can be obtained by this method.
3. ***Fusion of pollen nuclei*** It refers to spontaneous fusion of two similar nuclei of the cultured microspore. This is possible in *Brassica.*

The most important use of haploids often advocated is their use in the production of homozygous line, which may be directly used as cultivars or may be used in breeding programmes. A doubled haploid "Marglobe tomato" has been used commercially. To obtain fertile homozygous diploids, various diploidization techniques in plant breeding have been described.

MENTOR POLLEN

Various techniques have been employed to treat pollen so as to alter its fertilizing ability. By means of these treatments, it is possible to achieve hybridization between closely related species or genera, and also self-pollination of normally cross-pollinated plants. The pollen that has been exposed to treatments with the foregoing objective is called mentor pollen. Its action is on pollination with normally incompatible pollen, that is, pollen that is viable but is unable to set seed on its own pistil (self-incompatibility) or the stigmas of another species (interspecific incompatibility). Thus, mentor pollen may be defined as compatible pollen that has been treated in several ways to block its fertilizing power, but retaining its ability to stimulate incompatible pollen to effect fertilization. It may or may not be capable of undergoing germination.

Mentor pollen technology has already become an important tool in plant biotechnology, and has the following advantages:

1. It permits transfer of particular gene required in a breeding strategy directly through the vehicle of seed.
2. Its effect can be quantified.
3. It helps to study the pollen–stigma recognition process.

FACTORS AFFECTING HAPLOID PRODUCTION

There are many factors which affect the production of viable haploid plantlets by anther or pollen culture. Some of these are as follows:

1. Genotype of the Donor Plant and Anther Wall Factor

The genotype of the donor plant plays a significant role in determining the frequency of haploid plant production. There are some differences in anther culture response among the genera, species, subspecies and varieties. In cereals haploid plantlet formation is by and large controlled by many genes. For example, out of 18 lines of *Arabidopsis thaliana*, only three cultivars responded during anther culture. Anther wall factors also support pollen embryo development. Histological studies support this view. As induction of the pollen into embryoids occurs most easily within the anther, the anther wall seems to provide a nursing effect. There are two different thoughts regarding the role of the anther wall:

i. It may have a stimulatory effect on the growth of pollen embryos (probably due to the presence of enhanced levels of some amino acids such as glutamine and serine).
ii. It may release some inhibitory substances into the culture medium, thereby, blocking the growth of more pollen into embryos.

Weatherhead and Henshaw in 1979 observed that pollen grains of one cultivar of tobacco successfully developed into plantlets even if transferred into the anthers of another cultivar. Subsequent work has conclusively proved that anther wall serves as a source of nutrition for developing embryoids. Even the extract of anthers is known to stimulate development of embryoids. Recent studies have shown that for successful development of pollen grains into embryoids, their contact with the tapetum is essential. Occasionally, embryoids develop suspensorlike structures attached to the anther wall, which further substantiates the nutritive role of the anther wall. It has been suggested that wall factors are involved in the induction of embryogenic divisions in pollens.

2. Composition of the Medium

The culture medium also plays a vital role, since the requirements vary with the genotype and probably the age of the anther as well as conditions under which donor plants are grown. The medium should contain the correct amount and proportion of inorganic nutrients to satisfy the nutritional as well as physiological needs of the many plant cells in culture.

3. Growth Regulators

In addition to basal salts and vitamins, PGRs in the medium are critical factors for embryo or callus formation. Cytokinins (for example, kinetin) are necessary for induction of pollen embryos in many species of Solanaceae family, except tobacco. Auxins, in particular 2,4-D, greatly promote the formation of pollen callus in cereals. For regeneration of plants from pollen calli, a cytokinin and lower concentration of auxin are often necessary.

4. Source of Carbohydrate

Sucrose has been considered the most effective carbohydrate source which cannot be substituted by other disaccharides. However, glucose can be used in anther culture in some cases but fructose is far less effective. The concentration of sucrose also plays an important role in induction of pollen haploid plants.

5. Activated Charcoal

Activated charcoal is also added to the culture medium. It helps in two ways:

i. The removal of inhibitors from the agar medium.
ii. The adsorption of 5-hydroxymethylfurfural, a product of sucrose dehydration during autoclaving, assumed to be an inhibitor of growth in anther cultures.

6. Physiological Status of the Plant

The physiological state of the parent plant plays a key role in haploid production. Success in haploid induction is in part dependent on knowledge of the physiology of the pollen-yielding plant. In various plant species, it has been shown that the frequency of androgenesis is higher if anthers are harvested at the beginning of the flowering period and response declines with the increase of plant age. This may be due to deterioration in the general condition of the plants, especially during seed set. The lower frequency of induction of haploids in anthers taken from older plants may also be associated with a decline in pollen viability.

Seasonal variations, physical treatment, and application of hormones and salts to the plant also alter its physiological status, which is reflected in a change in anther response.

7. Temperature and Light

Temperature and light are two physical factors which play an important role in the culture of anthers. Frequency of haploid formation and growth of plantlets are generally better in light. A temperature in the range of 23–28°C is usually suitable for divisions in pollen grains, but it has been observed that chilling of anthers before inoculation, increases the number of pollen embryoids. The stimulatory effect of chilling treatment is indirect; it increases pollen viability, thereby enhancing embryoid formation. For example, in *Nicotiana,* only 21% of the anthers produced embryoids when cultures were maintained at 22°C. But if the buds were pretreated at 5°C for 25 hours prior to anther culture, 57% of anthers yielded embryoids. On the other hand, high-temperature treatment of anther cultures of *Brassica napus* (30°C for 14–21 days) before transferring them to normal temperature (25°C) resulted in more embryoid production.

8. Pretreatment Method

Certain physical and chemical treatments given to flowers, buds or anthers prior to culture can be highly conducive to the development of pollen into plants. The most significant is cold treatment.

9. Developmental Stage of Anther/Pollen Grains

The developmental stage of pollen greatly influences the fate of the microspore. Androgenesis occurs when a microspore or pollen is induced to shift from a gametophytic pathway to a sporophytic pathway by embryo formation.

The success of androgenesis depends to a large extent upon the stage of pollen development at which anthers are cultured. This is critical for the induction of division and varies with species. For example, in *Nicotiana tabacum* and *Datura innoxia*, the best results are obtained if pollens are cultured just before or soon after first mitotic division. In *Petunia, Brassica* and *Atropa belladonna*, the two-nucleate stage of the pollen grain responds better in cultures, however, in many cereals multinucleate stage was found to be the most responsive. Usually very young or very late stages of pollen development do not give suitable results. In *Lycopersicon esculentum*, however, the ideal stage for androgenesis is when microspore mother cells are in meiosis I.

10. Other Factors

Certain organic supplements added to the culture medium often enhance the growth of anther cultures. Some of these include:

i. The hydrolysed products of proteins such as casein (found in milk), nucleic acids and others.
ii. Coconut milk obtained from tender coconuts is often added to tissue culture media. It contains a complex mixture of nucleic acids, sugars, growth hormones and some vitamins.

SOMATIC REDUCTION AND CHROMOSOME ELIMINATION

Cases are known where either spontaneously or due to specific treatments, the chromosome number was reduced to half in the somatic tissues, a phenomenon described as somatic reduction or reductional mitosis. Swaminathan and Singh (1958) induced a haploid branch on a watermelon by irradiation of the seed used. This must have resulted in the reduction of chromosome number in the somatic tissue through an unknown mechanism (perhaps due to spindle organizer abnormalities).

Similarly, colchicine treatment has also been found to lead to the reduction in chromosome number. Franzke and his associates found that in *Sorghum vulgare*, somatic tetraploid ($2n = 4x$) cells responded to colchicine treatment and gave rise to diploid cells, which took over the growing point completely, thus giving rise to diploid individuals.

In general, attempts to induce somatic reduction in other species by treating seeds or seedlings with colchicine have failed or met with only partial success, as in flax. There are also a number of other chemicals like chloramphenicol and parafluorophenylalanine (PFPA, an amino acid analogue) which have sometimes been successfully used for production of haploids in a number of materials.

MONOPLOID PRODUCTION

Monoploids are defined as sporophytes with the basic gametic chromosome number. They have a single genome or chromosome set (for example, $2n = x = 7$ in *Hordeum vulgare*). Monoploids are generally sterile but by doubling the chromosomes, homozygous diploid fertile plants are produced. Monoploids can be produced by interspecific hybridization, followed by chromosome elimination. The hybridization method consists of crossing cultivated barley, *Hordeum vulgare* ($2n = 2x = 14$), with the wild, diploid, cross-pollinating and perennial *H. bulbosum* ($2n = 2x = 14$). This technique, the Bulbosum method, involves the fertilization of the female gamete of barley by the *H. bulbosum* gamete.

During the formation of the embryo, the chromosomes of *H. bulbosum* having the barley genome in the embryos are eliminated. These are cultured *in vitro* as immature embryos. Plantlets from these monoploid embryos can be made to give fertile flowers bearing homozygous offspring, following an efficient chromosome doubling technique. There are two ways by which monoploids can be induced artificially.

1. Based on the male gamete (microspore)
2. Based on the female gamete (megaspore)

Potentially, the first route, via anther or microspore culture, has the advantage because there are far more potential monoploids per spike in the form of male gametophytes than female gametophytes. Both methods are based on embryogenesis and the development of plants from monoploid embryos is followed by chromosome doubling to obtain homozygous diploids.

Monoploids as Tools in Plant Breeding

The advantage of monoploids as tools in plant breeding or genetics becomes more apparent when their direct application is visualized.

1. They provide the quickest possible way to complete homozygosis.
2. They may serve to recover recessives.
3. Linkage data can be obtained directly by sampling gametes as monoploids.
4. Doubled monoploids give an immediate product of stable recombinants from species crosses.
5. Monoploids can be used to determine homology within a genome and in between the genomes.
6. They are ideal for the study of mutation frequencies and spectra.

7. They provide an ideal system for fundamental cell biological problems (that is, biosynthesis) in cell and protoplast culture.
8. Monoploid cells as protoplasts provide unique material for gene transfer, host–pathogen reactions and cytoplasmic and/or chromosomal incompatibility.
9. One of the main advantages of using monoploids for breeding purposes is that completely homozygous lines are produced directly from gametes of F_1 hybrids or from later (advance) selections. This allows for a direct fixation of quantitative characters.
10. For practical plant breeders, this saves time and the desired product is readily recognizable.
11. Monoploid protoplasts are powerful tools in plant modification and somatic hybridization.

DETECTION AND IDENTIFICATION OF HAPLOIDS

Stomata size, pollen size and pollen abortion can be the general criteria to help initial screening, although in each case cytological confirmation is essential. A number of methods have been developed for the detection of haploids in a population of diploids. They are:

1. Use of genetic markers—the most effective method.
2. Phenotypic effects—can also be used for identification of haploids.

The vegetative and floral parts and the cell size in haploids are reduced relative to diploids. In maize, out of the 45 plantlets selected out of 10,000 plantlets, on the basis of size (that is, thin coleoptiles, etc.), five of them were found to be monoploids showing some success in screening for haploidy.

APPLICATIONS OF ANTHER AND POLLEN CULTURE

1. Anther and pollen cultures are used to produce haploid plants. In haploid plants, as a chromosome contains dominant or recessive genes, only one set of homologous chromosomes present is derived from pollen grains. All the recessive genes will be expressed phenotypically in the progeny as there is no masking effect of recessive gene by dominant gene, so it is easy to identify recessive characters.
2. The selection of superior plants among haploids is possible from F_1 hybrids through anther culture. This is known as hybrid sorting and it means selection of recombinant superior gametes. The frequency of superior gametes is higher than the frequency of corresponding superior plants in F_2 generation, and haploid breeding reduces significantly the time required for the development of a new variety. Hybrid sorting has been successfully utilized in many countries for the development of new varieties in several crops (for example, tobacco cultivar breed is resistant to bacterial wilt).
3. A comparison of haploid method of breeding with conventional methods such as pedigree, bulk, single seed descent, suggests that the time period required for haploid breeding is only 4–5 years, as against 10 years for conventional methods.

4. Cryopreservation of haploid cell, pollen embryoids and haploid meristem tips at super low temperature in liquid nitrogen (–196°C), offers a novel approach for the long-term preservation of genetic stability and to establish a haploid germplasm bank. These materials are more suitable for cryogenic studies.
5. Haploids can be successfully used in genetic engineering for gene transfer. Haploid tissues of *Arabidopsis* and *Lycopersicon* have been used for the transfer and expression of three genes from *E. coli.*
6. Homozygous plants obtained through anther culture have been employed for the selection of breeding lines of *N. tabacum* with high alkaloid content.
7. Nullisomic plants of tobacco have been produced by doubling the chromosome complement of nulli-haploids derived from cultured anthers of monosomic plants.
8. Anther and pollen culture techniques allow a greater survival rate of embryos.
9. Homozygous super males (YY) are produced from cultured anthers from male plant of *Asparagus.* When such males are crossed with pistillate plants, the entire progeny consists of males.
10. Anther-derived haploids have been extensively used in protoplast culture and somatic hybridization experiments.
11. The haploid technique in cereal breeding is a quick method to convert heterozygotes to homozygotes.

SIGNIFICANCE OF HAPLOIDS IN MUTATION RESEARCH

Use of haploids in mutation research is also well-established and provides the following.

1. Due to the presence of only one set of chromosomes, even the recessive mutations are immediately expressed in haploids.
2. Pollen can be plated on solid medium like single cells, and therefore, much larger population can be screened in the laboratory, without growing the plants in the field.

There are some of the common haploids that have been directly employed in the areas of mutational research study:

1. When *N. tabacum* is cultured from microspore, the plantlet just emerges out and then if gamma rays are applied, the plantlet will show aberrant phenotype having abnormal leaves and flowers. The whole plant treatment will show the appearance of sectorial chimeras. If the flower bud containing stage-4 anthers is irradiated and cultured, the plantlet will show totally aberrant plants.
2. In raising isogenic pure lines, which generally take 7–8 years by conventional methods in cross-pollinating crops, culturing of F_1 plants produces haploids in single generation.

3. The resistance acquired from bromodeoxyuridine-resistant lines (by using chemical in the medium to allow limited growth) is inherited as a single Mendelian factor. Cell lines resistant to an antibiotic have been isolated from haploid callus of *Petunia hybrida* without prior mutagenic treatment. Cultures raised from pollen derived haploids and single cell systems are used for selection of cell lines resistant to environmental stresses.
4. In *Ginkgo*, arginine-requiring strains from pollen culture have been obtained. With the availability of the technique, a large number of biochemical mutants have been isolated in a number of plant species using haploid cell. Cell lines, tissues and complete plants resistant to streptomycin, 5-bromodeoxyuridine and methionine sulfoximine have been obtained.
5. Nitrate reductase mutants have also been reported in *Nicotiana tabacum*.

USES OF HAPLOIDS

The important uses of haploids are described below:

1. As a result of haploid induction followed by chromosome doubling, homozygosity can be achieved in the quickest possible way making genetic and breeding research much easier. The genetic segregation is simplified in homozygotes, and recessive genes are not masked by dominant ones.
2. Homozygosity is still more important for those plants which have a very long juvenile phase, such as fruit trees, bulbous plants and forest trees. Even though homozygosity can be achieved through self-pollination, it is a long process in such plants. As a result of homozygosity, lines which produce pure F_1 hybrids are made possible.
3. In polyploid plants haploid induction helps to work at a low ploidy level. Monoploids have an advantage for the breeder in that the recessive mutations are different from others and can easily be identified.
4. By haploid induction followed by chromosome doubling, it is possible to obtain exclusively male plants. An important example of this is *Asparagus officinalis,* in which male plants have a higher productivity and yield earlier in the season than female plants.
5. If haploids are produced from anthers of male *Asparagus* plants, these are either "X" or "Y"; chromosome doubling of Y results in super-male plants YY, which can subsequently be vegetatively propagated.
6. When homozygotes are artificially made into diploids *in vivo* with the use of colchicine, problems may arise which are not usually found *in vitro. In vitro* haploids often double spontaneously, directly giving rise to homozygotes. It is much easier to work with haploid protoplasts rather than diploid ones for somatic hybridization.

Uses of Haploids in Plant Breeding

Among the different uses of haploids, their use in crop improvement is considered to be the most significant and stands out as the most important reason for emphasis on haploid research.

1. *Commercial crops where haploids have been used* include alfalfa, asparagus, barley, broccoli, citrus, clover, coffee corn, cucumber grape, kale, peanut, pepper, potato, rapeseed, etc.
2. *Use of haploids in production of disomic inheritance of homozygous line* is the most important use of haploids, which has often been advocated. The production of homozygous lines may be directly used as cultivars or may be used in breeding programmes. For instance, a doubled haploid, "Marglobe tomato" has been used commercially. Similarly, doubled haploids were used as inbred lines in hybrid maize programme.

LIMITATIONS IN HAPLOID PRODUCTION

Although haploids have been raised mainly from anther culture of a large number of species, this technique has still not been proved successful for all genotypes of crop species. A number of problems are encountered with the induction of haploids. The following are some of the widely recognized limitations:

1. Anthers often fail to grow *in vitro* or the initial growth is followed by abortion of the embryos.
2. The tissue or callus developing from the anther generally comprises a chimera of diploid, tetraploid and haploid cells.
3. Selective cell division must take place in the haploid microspores, concomitantly restricting proliferation of unwanted diploid and polyploid tissues. However, selective cell division is often impossible.
4. Formation of albinos in anther cultures, especially with cereals cannot be avoided. Also the loss of plants due to albinism is not reduced.
5. The technique of inducing haploids *in vitro* is not economically viable due to low success rate. The cost–benefit ratio in haploid breeding is often not favourable, thus discouraging the use of haploid breeding despite its obvious advantages.
6. Haploids will express recessive deleterious traits, and deleterious mutations may arise during anther culture.
7. Callus derived from anther or pollen in a medium supplemented with growth regulators is usually detrimental for haploid production.
8. It is difficult to select a haploid from a mixture of haploids and higher ploidy levels since the polyploids outgrow the haploids. Different ploidy levels may be available so that haploid status

may need to be confirmed cytologically; alternatively, pollen culture may need to be used, which is more expensive and gives relatively low success rate.

9. Doubling of haploids is time-consuming and may not always result in the production of a homozygote. Double haploids sometimes exhibit segregation in their progeny.
10. The anther culture technique works well on plants belonging to only a few families, like Solanaceae, Cruciferae, etc.
11. The frequency of embryoids obtained is often very low.

REVIEW QUESTIONS

1. Describe the different methods of production of haploid plants, and their various applications for crop improvement.
2. Discuss the achievements, merits and demerits of haploid plants.
3. Discuss in detail about the use of haploids in plant breeding programmes.
4. What are mentor pollen and anther squash techniques?
5. How are homozygous diploids raised?
6. What are the factors affecting anther and pollen culture?
7. Explain the suitable stages of pollen for culture.
8. Write short notes on:
   i. Nurse culture technique
   ii. Nitsch's culture technique
   iii. Float culture technique

REFERENCES

Bajaj, Y.P.S. (1978). Effect of super low temperature on excised anthers and pollen embryos of *Atropa, Nicotiana* and *Petunia*. *Phytomorphology*. 28: 171–176.

Knox, R.B *et al.* (1987). Mentor pollen techniques. *Internat. Rev. Cytol.* 107: 315–332.

Nitsch, J.P. (1972). Haploid plants from pollen. *Z. Pflanzen-zuecht.* 67: 3–18.

Prakash, J. and Giles, K.L. (1987). Induction and growth of androgenic haploids. *Internat. Rev. Cytol.* 107:273–292.

Sangwan, R.S. and Sangwan-Norreel, B.S. (1987). Biochemical cytology of pollen embryogenesis. *Internat. Rev. Cytol.* 107: 221–272.

Sharp, W.R., Raskin, R.S. and Somner, H.E. (1972). The use of nurse culture in the development of haploid clones of tomato. *Planta*. 104: 357–361.

14

HAPLOID PRODUCTION-II (*IN VITRO* POLLINATION FERTILIZATION AND GYNOGENESIS)

INTRODUCTION

Sexual reproduction in higher plants consists of a series of complex processes including pollination, double fertilization, embryo differentiation and seed formation. Development of methods for pollination and fertilization under *in vitro* controlled conditions has greatly facilitated the study of these processes. In addition, manipulation of reproductive processes is also possible by these methods. Thus, plants have been obtained following *in vitro* self-pollination with self-incompatible species. Hybrid plants also have been obtained from interspecific and intergeneric crosses. Utilization of *in vitro* methods for wide hybridization has also provided a method for induction of parthenogenesis and development of haploid plants.

When pollen is applied to the stigma of cultured ovary under *in vitro* conditions or directly to ovule in culture with or without placental tissue, it is called *in vitro* pollination. Ovaries are collected from emasculated flowers, usually 1–2 days after anthesis and these are cultured by the following ways:

1. Culture of intact ovary.
2. Culture of ovary without ovarian wall (with exposed placenta).
3. Culture of entire placenta.
4. Culture of placental pieces along with attached ovules.

Pollination is an important step in fruit formation. Seed and fruit development through *in vitro* pollination of exposed ovules is known as test-tube fertilization or *in vitro* fertilization. Seed development following stigmatic pollination of cultured whole pistils has been referred to as *in vitro* pollination.

There are more terminologies in this respect. The application of pollen to excised ovules, the application of pollen to ovules attached to the placenta and application of pollen to the stigma are referred to as *in vitro* ovule pollination, *in vitro* placental pollination and *in vitro* stigmatic pollination, respectively.

In species like maize, small pieces of cob (bearing about 10 ovaries) may be employed in culture, usually 2–6 days after silk emergence. Generally, ovaries/ovules are cultured along with the pedicel. In some dicot species, it may be desirable to retain the calyx; while in monocots like barley, the glumes need to be retained since their removal is deleterious to embryo development. Wetting of ovules and stigma should be avoided because it may reduce pollen germination and interfere with normal pollen tube growth.

The female reproductive structure of a complete flower is the carpel, which consists of ovary, style and stigma. Seed and fruit development take place in the ovary while the stigma and style are the receptive organs for the development of pollen grains and formation of male gametophyte prior to fertilization. Fruit physiology is of great value to horticulturists and seed physiology is equally important to agriculturists.

IN VITRO POLLINATION

Success of *in vitro* pollination depends upon two basic controlling factors:

1. Selection of correct developmental stage of pollen grain and ovule.
2. Selection of suitable nutrient medium.

The nutrient medium used for culture initiation should support both pollen germination and pollen tube growth, and also the development of fertilized ovules into mature seeds. It is advisable to determine the nutritional and hormonal requirements for the excised fertilized ovules of the species, to be set as the female parent, prior to *in vitro* pollination in a new species. Nitsch's (1951) medium supplemented with vitamins and sucrose (5%) is the most commonly used nutrient formulation for *in vitro* pollination.

In some species like maize, a higher sucrose level (7%) is found to be optimal. Though the information on hormone requirements is scanty, in some cases, IAA (up to 10 μg/l) and kinetin (0.1 μg/l) are found to improve seed set significantly. The cultures are usually kept at 25 ± 2°C; however, in plants such as *Narcissus*, 15°C proves to be the best in terms of seed set. In general, light is not necessary; the cultures are usually incubated either in diffuse light or more often in dark.

Pollen grains are collected from surface-sterilized anthers that are kept in a sterile Petri dish so as to obtain contamination-free pollen. The pollen grains are usually deposited directly to ovules, placenta or stigma depending on the type of culture (Figure 14.1). In most cases, culture of ovules along with the placenta gives the best seed set.

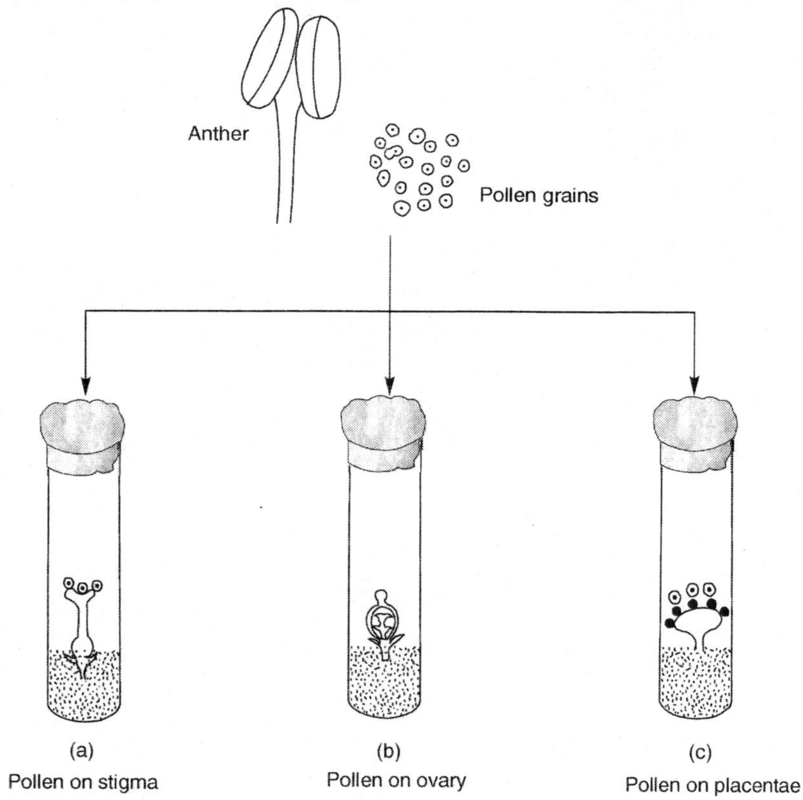

Figure 14.1 Methods of in vitro pollination

In vitro pollination has been proved to be a very successful technology in many species, including some intergeneric crosses, for example, *Melandrium album* × *Viscaria vulgaris, M. album* × *Minuartia laricifolia, M. album* × *Dianthus serotinus, N. tabacum* × *Petunia hybrida,* etc. However, in most interspecific combinations, only nongerminable seeds are obtained. More surprisingly, seeds obtained from self-pollination in many species, like *Antirrhinum majus* are also non-germinable.

Ovaries excised from pollinated flowers are able to grow on nutrient media supplemented with simple minerals salts, sugar and vitamins; however, ovaries from unpollinated flowers fail to grow on such medium. An auxin may replace the pollination stimulus. In apomictic plants, though there is no fertilization, the stimulus alone enables the growth of the ovary and formation of seeds. This technique is useful in breeding and hybridization programmes, particularly when zone of incompatibility lies in the stigma, style or ovary. In self-incompatible species, pollen tube does not enter the ovary, although the pollen germination is good during self-pollination. In this case, self-placental pollination causes fertilization and seed set occurs regularly. In *Petunia axillaries* and *Petunia hybrida,* the two self-incompatible species show development of viable seeds on placental pollination by self-pollination. So, self-incompatibility can be overcome by this process.

Likewise, cross-incompatibility can also be overcome by *in vitro* pollination technique. Another importance of this technique is the production of haploids through parthenogenesis.

IN VITRO FERTILIZATION

Research on overcoming self- and cross-incompatibility has been undertaken by several scientists, from the time the test-tube fertilization technique was developed by Kanta *et al.* (1962). Some of these studies have proven to be successful in several species. It has been reported that self-incompatibility of *Petunia hybrida* can be overcome by *in vitro* fertilization. However, the number of seeds formed after fertilization *in vitro* was relatively less and very variable. This is caused by various factors such as the technique of pollen application and the composition of nutrient medium on the development of fertilized ovules.

Success of *in vitro* fertilization (measured by production of viable seeds) is dependent upon the following basic controlling factors:

1. Age of the explants, particularly that of the ovule.
2. Adequate pollen germination.
3. Successful microsporogenesis.
4. Successful pollen tube entry into ovules.

In vitro Fertilization Technique (Protocol for *Trifolium repens*)

In vitro fertilization technique in case of *Trifolium repens* involves the following steps:

1. Plants are grown in greenhouse for a period of 8–12 weeks, with supplementary lighting to facilitate vigorous vegetative growth.
2. Flowering occurs after 12–16 weeks of vegetative growth in the greenhouse.
3. Modified B_5 culture medium can be utilized for this technique.
4. Florets that are to provide pollen as the male partner of the cross are selected from the lower part of the flower head. Moreover, the florets in which the keel petal is almost fully open are collected.
5. The florets are dissected with the help of a clean scalpel, tweezers and microscope by making an incision at the base of the floret in the sepals and through the petals.
6. The floret is then opened to reveal the anthers connected near the base in a tube. This tube may be excised and removed with anthers attached. Remarkably, anthers would have dehisced, showing the golden coloured pollen.

7. *T. repens* is self-incompatible, and to achieve fertilization and seed formation, it is necessary to collect florets of the female partner and pollen of the male partner from different plants. Florets to be used as the female partner in the cross are selected at a more immature stage.
8. In these florets, anther dehiscence will occur in the succeeding 24-hour period. If possible, floret stalks are excised, which are attached to each floret. The florets are transferred to a small plastic basket for surface sterilization, which is performed by immersion of florets for 20 minutes in a solution of sterilant (calcium/sodium hypochlorite or mercuric chloride).
9. The florets are agitated occasionally during this period. The hypochlorite solution is decanted and the florets are rinsed thrice in sterile water (5 minutes/rinse). Aseptic conditions are provided during this and subsequent stages. The florets are removed from the baskets and transferred to culture dishes.
10. Pollination of each floret is performed separately. For pollination, each floret is transferred to the lid of the Petri dish and an incision is made carefully, along the length of the floret. The stigma and style may then be gently pushed out through the incision. It is not necessary to remove anthers, since *T. repens* is self-incompatible. The stigma should bear a slightly moist appearance.
11. Pollen is transferred to the stigma by rubbing anthers on the surface. Pollen should be applied liberally so as to cover the stigmatic surface. The stalk of the pollinated floret is inserted into the culture medium. The lid is replaced on the culture dish, and transferred to the culture room.
12. After two weeks of culture, petals gradually turn brown and finally get withered, but sepals, flower stalk and ovary remain green and turgid. Now, the pollinated floret is transferred to the lid of the culture dish and the fertilized ovules are aseptically dissected out. Care should be taken not to make deep incisions that will damage the immature seeds. The number of immature seeds present are recorded, thereafter, the testa is carefully excised and removed (outer covering) from each seed, since, this is essential to promote seed germination. These testa-less seeds are transferred to fresh medium.
13. Germination of seeds begins after seven days. Furthermore, the seedlings are cultured for a further 3–4 weeks. The resultant plantlets are transferred to half strength medium for a further 3–4 weeks, prior to its transfer to the greenhouse.

Applications of *In Vitro* Fertilization

The technique of *in vitro* fertilization is commonly applicable for:

1. Production of hybrids.
2. Overcoming sexual self-incompatibility (homomorphic incompatibility).
3. Induction of haploid plants.
4. Studying pollen physiology and fertilization.

OVULE CULTURE

Ovule is a megasporangium, attached with placenta inside the ovary by its funiculus, and covered by integument. An ovule contains a female gametophyte or megaspore, which after fertilization forms a single cell zygote and ultimately leads to embryo, which possesses root and shoot primordia. Haploids have also been successfully produced from cultured female gametophytes. Success in this connection was first achieved with some gymnosperms like *Zamia, Ephedra* and some cycads. Although initial attempts for the production of haploids from cultured ovules in angiosperms were unsuccessful, recently haploids were obtained from cultured ovules in a few crop species like barley, wheat and tobacco. With more efforts in the future, haploids from ovule culture can be produced in a number of plant species. Further research is, therefore, necessary in the following aspects:

1. The events leading to induction of haploidy in female gametophytes.
2. The factors which control the *in vitro* development of proembryo into the fully organized plants.
3. The differences in the growth patterns of *in vitro* development of unfertilized female gametophytes (in ovules) and the male gametophytes (in pollen).

For many difficult interspecific/intergeneric crosses (involving members of the families, Malvaceae, Fabaceae, Cruciferae, Solanaceae, etc.), the ovules, after fertilization, have been successfully cultured to obtain mature embryos/seeds. However, ovule culture has been tried only in those cases where the embryo aborts very early, and embryo culture is not possible due to difficulty of its excision at a very early stage. In some cases, the medium may need to be supplemented with some fruit/vegetable juice to accelerate the initial growth.

In this technique, the ovules are aseptically isolated from ovary and grown on chemically defined nutrient medium under laboratory conditions. Ovule culture is successfully employed in the production of haploid plants in the plant species such as *Allium cepa, Helianthus annuus, Solanum tuberosum, Beta vulgaris, Cucumis melo, Brassica oleracea, Petunia axillaris, Melandrium album, Hevea brasiliensis, Gerbera jamesonii, etc.*

In all these plants, diploid plants were made from haploids via *in vitro* colchicine treatment.

Applications of Ovule Culture

1. Ovule culture is a boon for the plant breeders in obtaining seedlings from crosses which are normally unsuccessful because of abortive embryos. It has been observed that in several interspecific crosses, the hybrid embryo of *Abelmoschus* fails to develop beyond the heart or torpedo stage of embryo. Viable hybrids have been obtained by culture of the fertilized ovule and some of the hybrids which have been produced by ovule culture include

- *A. esculentus* × *A. ficulneus*
- *A. esculentus* × *A. moschatus*
- *A. tuberculatus* × *A. moschatus*
- *Brassica chinensis* × *B. pekinensis*
- *Lolium perenne* × *Festuca rubra*

2. Haploid plants can also be obtained by ovule culture *in vitro.* Uchimiya *et al.* in 1971, cultured unfertilized ovules of *Solanum melongena* and produced callus on a medium containing IAA and kinetin. Finally, plantlets are produced on the plant regeneration medium. It is an important attempt to obtain haploid plants from an alternative source rather than ovary, anther or pollen culture.
3. Polyembryony can be induced artificially via ovule culture and therefore, this technique holds a great potential. It has been observed in the nucellus of monoembryonic ovule of *Citrus,* where adventive embryos are induced in cultures.

OVARY CULTURE (GYNOGENESIS)

The knowledge of fruit physiology is of great value to horticulturists in improving the quality of fruits. When the effect of various chemicals on fruit growth is studied using entire plants, it remains questionable whether the chemical tested is acting in its original form or is getting modified in the plant system before reaching the ovary. Therefore, ovary culture provides an alternative technique for the production of haploids and also information about the basic mechanisms connected with fruit physiology. Moreover, when an unfertilized ovary is cultured under *in vitro* conditions, in order to obtain haploid plants from egg cell or other haploid cells of the embryo sac, it is called ovary culture, and the process is termed as gynogenesis.

The first report of gynogenesis in case of barley was demonstrated by San Noem in 1976. Subsequently, success has been obtained with many species, such as, wheat, rice, maize, tobacco, petunia, gerbera, sunflower, sugarbeet, etc. However, the rate of success varies considerably with different species and is markedly influenced by explants genotype since some cultivars do not respond at all. In rice, *japonica* types are far more responsive than *indica* cultivars. In most cases, the optimum stage for ovary culture is the mature embryo sac stage, but in rice, ovaries at free nuclear embryo sac stage is most responsive.

Ovary culture technique was developed by Nitsch (1951) who grew detached ovaries of *Cucumis anguria, Fragaria* sp., *Lycopersicon esculentum, Nicotiana tabacum* and *Phaseolus vulgaris* on synthetic medium. The ovaries of *Cucumis* and *Lycopersicon,* excised from pollinated flowers, developed into mature fruit containing viable seeds (Figure 14.2). Applewhite *et al.* (1997) reported the demonstration of *in vitro* development of red, ripe fruits of tomato with normal flavors from pre-anthesis flower buds. On MS medium (3% sucrose, pH 5.8) supplemented with 1 μM BAP and 10 μM IBA, 100% of the buds formed red berries. However, the size of the *in vitro* developed fruits was smaller than those formed in nature.

The reduced fruit size *in vitro* is due to the reduction in both cell division and cell enlargement. It is interesting to note that the DNA content per cell increases 15-fold during the growth of tomato fruits on the plants as compared to only 2–3 folds in cultured fruits.

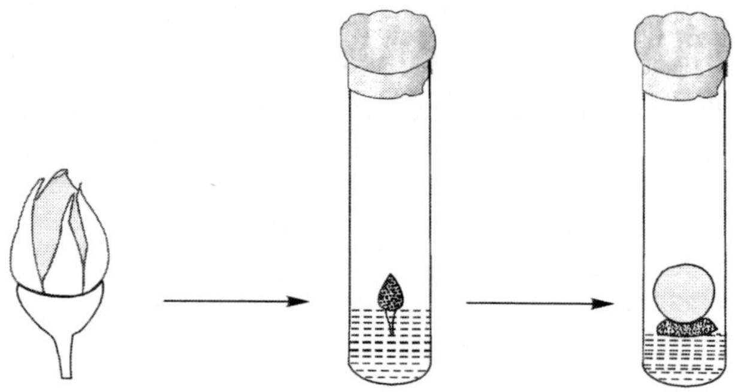

Figure 14.2 Ovary culture technique

These differences in fruit growth *in vivo* and *in vitro* could be due to insufficient supply of carbohydrates under *in vitro* conditions. Until this bottleneck is removed, the *in vitro* system cannot be recommended as a model for fruit development. The largest fruits formed in ovary cultures of *Linaria maroccanna* and *Tropaeolum* were smaller than their natural fruits.

Maheshwari (1958) succeeded in rearing the ovaries of *Iberis amara* excised from flowers a day after pollination. At this stage, the ovules contained a zygote and a few endosperm nuclei. On a simple medium containing mineral salts and sugar, the growth of ovaries was good but the embryos remained smaller than those formed in nature. With the addition of vitamin B to the above medium, normal healthy fruits were obtained. *In vitro* formed fruits were even larger than the *in vivo* formed fruits, when IAA was added to the mineral salts–sugar–vitamins medium.

In culture, normal development of ovaries, excised at the zygote or two-celled embryo stage, has also been reported in *Anethum graveolens* and *Hyoscyamus niger.* On a medium supplemented with coconut milk, the ovaries of *Anethum* surpassed the size of natural fruits. Significantly, ovaries excised from pollinated flowers are able to grow on simple mineral salts–sugar–vitamins medium; however, ovaries from unpollinated flowers fail to grow on such a medium. An auxin can replace the pollination stimulus.

Most apomictic plants require pollination stimulus for the growth of ovary and seeds. *Aera tomentosa* is an obligate apomictic plant. Puri (1974) cultured ovaries, flowers and portions of spikes of these apomicts and studied the effect of IAA, casein hydrolysate, coconut milk, and yeast extract on the growth of the embryos. The best response was noted when portions of spikes were cultured instead of individual flowers or ovaries. In the presence of IAA and yeast extract the seed set was comparable to that in nature. The development of endosperm and embryo was also fairly normal.

Generally, culture of whole flowers, ovary and ovules attached to placenta respond better, but in gerbera and sunflower, isolated ovules show relatively good response. Significantly, cold pretreatment (24–48 hours at 4°C in sunflower and 24 hours at 7°C in rice) of the inflorescence prior to ovary culture has been proved highly effective, and it enhances gynogenesis. Moreover, plant growth regulators are crucial in gynogenesis and at higher levels, they may induce callusing of somatic tissues and even suppress gynogenesis. Sometimes, hormone requirement seems to depend on species type such as in sunflower, hormone free medium is the best and even a low level of 2-methyl 4-chlorophenoxyacetic acid induces callus and somatic embryo formation.

Ovaries/ovules are generally cultured in light, but in some species, for example, sunflower and rice, dark incubation favours gynogenesis and minimizes somatic callusing. Generally, the whole process of gynogenesis completes in two stages and each stage may have distinct requirements. These two stages are induction phase and regeneration phase.

1. During induction phase, ovaries are floated on a liquid medium having low auxin and kept in dark.
2. In regeneration phase, they are transferred on to an agar medium with higher auxin concentration and incubated in light.

Haploid plants generally originate from egg cell in most of the species (*in vitro* parthenogenesis) but in some species, such as, rice, they arise chiefly from synergids and in *Allium tuberosum* even antipodals produce haploid plant (*in vitro* apogamic). As in anther culture, androgenesis may occur either via embryogenesis or through plantlet regeneration from callus induction. In rice, when cultured on 2-methyl 4-chlorophenoxyacetic acid, there appears a small amount of protocorm like callus formation from which shoots and roots are regenerated, while picloram promotes embryo induction and germination. In contrast, sugarbeet usually shows direct embryo development, while in sunflower, embryos regenerate following a callus phase. In general, regeneration of plantlets from a callus phase appears to be easier than direct embryogenesis.

Ovary Culture Technique

The general technique of ovary culture is described as follows:

1. The experimental plant is kept in controlled condition for few days and a normal sized flower is taken for culture.
2. The flower is surface-sterilized with 5% sodium hypochlorite for ten minutes and then washed with sterile distilled water 3–4 times.

3. Using aseptic method, the pistil is kept on a Petri dish after removing sepals, petals and stamens. In the Petsri dish only fully grown mature pistil should be taken.
4. Then the excised ovary is inoculated on the basal medium aseptically.
5. The culture is incubated at 25°C for 16 hours photoperiod with 2000 lux light intensity. If the ovary is unpollinated, it will not grow on a simple medium containing mineral salts and sucrose. With the addition of B vitamins to the medium, normal healthy fruits can be obtained.

In vitro rooting of the pedicel and a limited growth of ovaries on a culture medium were very frequently seen in ovary cultures. Although, technique for growing excised ovaries on culture medium was developed later in several plant materials including tobacco, tomato, beans etc., subsequently mature fruits containing seeds could be successfully obtained from ovaries excised from pollinated flowers. In India, at Delhi University, healthy fruits from excised ovary culture were successfully obtained in several plants including *Iberis amara.*

Seed Development in Cultured Ovary

Satisfactory development of fruits is no indication of normal development of seeds in it. Nitsch (1951) stated that in ovary cultures of tobacco and tomato, although fruit development was fairly good, seeds failed to develop on the basal nutrient medium. In gherkin also, most of the *in vitro* formed fruits showed only a few seeds.

In *Tropaeolum* ovary cultures on basal medium, the embryos showed some growth and differentiation but were invariably smaller than those formed *in vivo*. However, in some plants such as *Hyoscyamus, Iberis* and *Linaria,* embryo and endosperm development was normal on the basal medium. In *Anethum,* normal development of endosperm and embryo occurred only when ovaries were cultured on a medium enriched with coconut milk.

Applications of Ovary Culture

Besides being useful for understanding the physiology of fruit development, ovary culture has been used as an aid to produce haploids and rare hybrids.

1. Hess and Wagner in 1975 raised parthenogenic haploids of *Mimulus luteus* cv. *tigrinus grandiflorus* by pollinating their exposed ovules in ovary cultures. Anther cultures of this plant did not form haploids. Parthenogenic development of haploid plants in the cultures of unfertilized ovaries has also been reported in many species as shown in Table 14.1.
2. Ovary is the most important part of the carpel as it contains the ovules which develop into seeds. Thus the knowledge of fruit physiology depends on the growth and development of ovary. This knowledge is essential to horticulturists in improving the quality of fruits. Cultured ovaries grow in limited form and in many cases, the general development pattern is the same as that in nature.

3. Ovary culture has been successfully employed to obtain hybrids (interspecific) between sexually incompatible parents *Brassica campestris* and *B. oleracea*. It has got an immense value in plant breeding where problems of sterility pose a barrier to normal sexual reproduction.
4. Ovary culture has been successfully employed in the production of haploid plants in several economically valuable plants. These haploids have been used in crop improvement programmes. Examples are listed in Table 14.1.

Table 14.1 Haploid plants raised through *in vitro* gynogenesis

Plant Name	Family
Beta vulgaris	Chenopodiaceae
Gerbera jamesonii Helianthus annuus	Asteraceae
Hordeum vulgare	Poaceae
Oryza sativa	Poaceae
Triticum aestivum	Poaceae
Zea mays	Poaceae
Lilium tuberosum	Liliaceae
L. davidii	Liliaceae
Allium cepa	Liliaceae
Nicotiana tabacum	Solanaceae
N. rustica	Solanaceae
Petunia axillaris	Solanaceae
Solanum tuberosum	Solanaceae
Brassica oleracea	Brassicaceae
Cucurbita pepo	Cucurbitaceae

Limitations of Ovary Culture

1. So far it has been successful only in about 43 species.
2. The frequency of responding ovaries (1–5%) and the number of plantlets/ovaries (1–2) are quite low.
3. Anther culture is preferred over ovary culture. Therefore, only in those cases where anther culture fails (for example, sugarbeet), ovary culture technique can be used.

REVIEW QUESTIONS

1. What is *in vitro* pollination? Discuss the *in vitro* pollination technique in detail.
2. Define *in vitro* fertilization and discuss a protocol of this technique.
3. Write notes on the significance of *in vitro* pollination and fertilization in crop improvement.
4. What is ovary culture? Write notes on ovary culture technique.
5. Describe the applications and limitations of ovary culture.

REFERENCES

Kasha, K.S. and. Reinbergs , E. (1980). Achievements with haploids in barley research and breeding. In: Davies, D.R. and Hopwood, D.A. (Ed.). *The Plant Genome*. John Innes Charity, Norwich. pp. 215–230.

Morrison, R.A. and Evans, D.A. (1988). Haploid plants from tissue culture: New plant varieties in shortened time frame. *Biotechnology*. 6: 684–690.

Nitzsche, W. and Wenzel, G. (1977). *Haploids in Plant Breeding*. Paul Parey, Berlin.

15

ENDOSPERM AND NUCELLUS CULTURE

INTRODUCTION

In angiosperms, double fertilization and triple fusion give rise to embryo and endosperm, while in gymnosperms, the enlarged female gametophyte forms the endosperm. In angiosperms, the endosperm surrounds the embryo and is rich in nutrients and growth substances. It is a genetically unique triploid tissue that arises from the fusion of one male gamete with two polar nuclei. However, in gymnosperms, the endosperm is a haploid tissue. Most of the angiospermic families are known to possess triploid endosperm in their developing seeds. On the basis of presence or absence of endosperm, seeds can be classified into two groups:

1. ***Endospermic seeds*** Seeds which have endosperm up to the germination stage are called endospermic seeds, e.g., cereals, coconut, coffee, magnolia, etc.
2. ***Non-endospermic seeds*** Seeds which have endosperms at the beginning but are consumed during maturity are called non-endospermic seeds, e.g., legumes, cucurbits, etc.

ENDOSPERM CULTURE OR PRODUCTION OF TRIPLOIDS

Endosperm culture under *in vitro* controlled conditions is significant and unique due to the following two reasons:

1. Its function of supplying nutrition to the developing embryo.
2. It being triploid in its chromosome constitution.

Triploid plants are useful for the production of seedless fruits (for example, apple, banana, watermelon, etc.) and for production of trisomics for cytogenetic studies. Moreover, lack of vascular elements and any degree of differentiation also makes it a system for morphogenetic studies.

Generally, these triploids are routinely obtained by crossing colchicine-induced tetraploids with diploids followed by rescuing the triploid embryos. However, there may be strong crossability barriers, making it difficult to produce triploids from $4x \times 2x$ crosses, so that endosperm culture may be used as an alternative method for triploid production.

HISTORY

- *In vitro* culture of the endosperm tissue began in the 1930s. Lamp and Mills cultured *Zea mays*, but the material did not respond properly. In the process of culture of endosperm tissues of *Hordeum* and *Triticum*, organogenesis could not be achieved.
- Rangaswami and Rao (1963) reported their success in mature endosperm culture of *Santalum album* up to the callusing stage.
- Nakano *et al.* (1975) successfully achieved callus organogenesis on immature endosperm culture of rice.
- Lakshmi Sita *et al.* (1980) reported successful regeneration of sandalwood endosperm.
- Ho Sun and Chu (1981) reported plantlet regeneration in *Hordeum* from endospermic tissues.
- Thomas *et al.* (2000) reported production of triploid plants of mulberry through endosperm culture.
- The failure of earlier workers to induce organogenesis and embryogenesis may be either due to the improper selection of the explants stage or improper nutrient media treatment.

ENDOSPERM CULTURE TECHNIQUE

Extensive work on endosperm culture was initially done at Delhi University, where successful culture of endosperm was demonstrated initially in some parasitic angiosperm families (Santalaceae and Loranthaceae). In almost all the parasitic angiosperms, endosperm shows a tendency of direct differentiation of organs without prior callusing, whereas in autotrophic taxa, the endosperm usually forms callus tissue followed by differentiation of shoot buds, roots or embryos. The technique of endosperm culture involves the following basic steps:

1. The immature seeds are dissected under aseptic conditions and endosperms, along with embryos, are excised; sometimes mature seeds can also be used.
2. The excised endosperms are cultured on a suitable medium and embryos are removed after initial growth.

3. The initial callus phase is followed by embryogenesis or shoot bud differentiation.
4. The shoots and roots may subsequently develop, and complete triploid plants can be established for further use.

NUTRIENT MEDIUM AND PHYSICAL FACTORS

The use of MS medium containing a high level of nitrogen, helps enormously in promoting the response of endosperm culture. The culture of both immature and mature endosperm requires selection of seeds of a precise developmental stage; in addition, supply of an undefined nitrogen source, that is, yeast extract, casein hydrolysate and coconut milk helps in the success of culture.

In most of the cases, growth hormones are essential in one or the other combinations to promote regeneration. Various physical factors also have significant effect on growth of the endosperm culture; pH 7.0 was found to be highly effective for fresh weight increase. The maximum growth of the endosperm culture occurs between 24 and 27°C; 12–16 hours diffuse photo-period support callusing as well as regeneration.

The successes of more than 40 species of angiosperms have thus established the totipotency of the endosperm. Some examples of plants regenerated from endosperm cultures are listed in Table 15.1.

Table 15.1 Regenerated plantlets (triploid) from endosperm culture

Monocotyledon plants	Dicotyledon plants
Zea mays	*Acacia nilotica*
Oryza sativa	*Santalum album*
Asparagus officinalis	*Actinidia chinensis*
Hordeum vulgare	*Annona squamosa*
	Citrus grandis
	Citrus sinensis
	Codiaeum variegatum
	Emblica officinalis
	Juglans regia
	Lyceum chinensis
	Morus alba

FACTORS AFFECTING ENDOSPERM CULTURE

1. Isolation of uninjured cellular endosperm is a prerequisite for endosperm culture.
2. Selection of a particular age of responding endosperm tissue is a very important factor and it may vary from cellular endosperm to mature endosperm (7 days in rice and more than 10 days in maize, wheat, barley, etc.).
3. The embryo–endosperm relationship is another factor for proliferation of endosperm tissue *in vitro.*

APPLICATIONS OF ENDOSPERM CULTURE

1. Triploid plants have more vigorous vegetative growth than their diploid counterparts. Hence, in plants where the vegetative parts are economically useful, triploids are useful.
2. Seed sterility is the prime character of triploid plants. Triploidy is not desirable where seeds are important for agriculture. But there are some plants, especially some tropical fruits, where seedlessness caused by triploidy, may have some commercial value. Triploid production in tropical fruits by the conventional method is laborious and time-consuming. It is as an alternative method for crop improvement.

NUCELLUS CULTURE

Nucellus from pollinated flowers can also be used for micropropagation, particularly in cases where adventive embryos develop from nucellar cells (e.g., *Citrus*). Nucellus culture has been utilized to study the factors responsible for the formation of adventive embryos. Normally, the embryo originates from the zygote, but sometimes in plants such as citrus and mango, the embryos arise adventitiously from cells of the nucellus or integument.

Rangaswamy (1958), studied for the first time, nucellar embryony *in vitro*. He excised the nucellus from unpollinated ovules of *Citrus microcarpa*. It is a natural polyembryonic species and it was cultured on a modified White medium supplemented with casein hydrolysate. The nucellus proliferated into a callus mass and differentiated into embryo-like structures. These structures, termed as pseudobulbils, eventually developed into plants. Rangan *et al.* (1968) demonstrated that it is possible to induce nucellar embryogenesis in monoembryonic citrus species as well.

The plantlets obtained from the nucellar tissue are of parental type and thus may be important for maintaining purity of horticultural stocks. In *Citrus,* these may be better than those obtained from cuttings due to the following reasons:

1. Nucellar saplings have tap root, while those from cuttings have lateral roots.
2. The vigour lost due to repeated propagation by cuttings is restored in nucellar plantlets.
3. Nucellar saplings are disease-free.

NUCELLUS CULTURE TECHNIQUE

In the procedure of nucellus culture for direct embryogenesis in citrus, immature green fruits, 2.5 to 3.0 cm in diameter, are harvested from trees after 100–120 days of pollination. After proper sterilization, a shallow incision around the circumference of the fruit is made and ovules are excised. With the aid of a microscope, the zygotic embryos along with nucellus are rolled away from the integument. The micropylar end is pierced, the embryo is pushed and the entire nucellus is placed on the media.

The callus may then give rise to pseudobulbils, followed by differentiation into embryoids, which eventually develop into plantlets. Nucellus from unfertilized ovules may also develop embryoids, when grown under suitable culture conditions. These embryoids can be excised and grown on medium supplemented with GA_3, and plantlets can be obtained.

Adventive embryos were successfully initiated in nucellus cultures of *C. grandis* on MS medium supplemented with either malt extract (500 mg/l) or combination of adenine sulphate (25 mg/l) + NAA (0.5 mg/l) + orange juice (5.0%). In these species, nucellar explants did not produce callus or pseudobulbils, but they gave rise directly to highly organized multiple embryos.

Pollination and fertilization are generally considered essential prerequisites for the induction of nucellar embryogenesis. However, in later studies, it was shown that embryogenesis could be induced *in vitro* from nucellus of unfertilized ovules of *C. sinensis*. MS medium is generally adequate for both poly and monoembryonic varieties. A high sucrose concentration of 5.0% is generally recommended. Nutrient medium should be supplemented with complex substances rich in amino acids, such as casein hydrolysate. Malt extract can be replaced by combinations of adenine sulphate and kinetin.

APPLICATIONS OF NUCELLUS CULTURE

1. The adventive embryos are of considerable importance to the horticulturists. They are genetically uniform and reproduce the characters of the maternal parent without inheriting the variations brought about by gametic fusion.
2. In contrast to clones established from cuttings or by conventional methods of propagation, plants derived from nucellar embryos are free from most viruses. Moreover, many of the desirable plant vigour and fruiting characteristics associated with juvenility are restored in trees established from nucellar sapling.

3. The formation of nucellar embryos from unpollinated ovules is also of considerable interest in that it can be employed to import seedless citrus varieties without the risk of introducing new viral diseases, thereby avoiding the necessity of subjecting vegetative material to long periods of quarantine.

REVIEW QUESTIONS

1. What are triploids? Explain the endosperm culture technique.
2. Write a note on nucellus culture technique.
3. What are the applications of endosperm and nucellus cultures?

REFERENCES

Bhojwani, S.S. and Razdan, M.K. (1983). *Plant Tissue Culture: Theory and Practice.* Elsevier, Amsterdam.

Cocking, E.C. (1979). Parasexual reproduction in flowering plants. *New Zealand Jour. Bot.* 17:665–671.

Johari, B.M. (1982). *Experimental Embryology of Vascular Plants.* Springer-Verlag, Heidelberg.

Raghavan, V. (1966). Nutrition, growth and morphogenesis of plant embryos. *Biol. Rev.* 41: 1–58.

16

PROTOPLAST TECHNOLOGY– ISOLATION AND REGENERATION OF PROTOPLAST

INTRODUCTION

All organisms, be it plants or animals, reproduce sexually by the fusion of male and female gametes, each containing a single set of chromosomes from either parent. This is possible only in the compatible and related species. Wide crosses are not common without human intervention. However, plant protoplast culture provides a unique system, wherein each protoplast has the potential to give rise to a whole plant and two or more protoplasts can be induced to produce a hybrid (fusion between two nuclei and cytoplasmic contents) or a cybrid (fusion between cytoplasms). This technique is used to overcome sexual incompatibility barriers through the production of unique somatic hybrids involving vegetative cells.

The most significant development, after the successful establishment of plant cell culture, is the isolation, culture, manipulation and regeneration of protoplasts. The presence of inert cell wall is the only barrier between the external environment and the interior of the cell. The cell possesses some problems to conduct a variety of experimental manipulations which are targeted to study about cellular events. Such problems can be overcome only if the cell wall of the developing cell is removed experimentally.

A protoplast can be defined as a plant cell without cell wall but with cell membrane. Every plant cell possesses a definite cellulosic cell wall and the protoplast lies within the cell wall except some reproductive cells and the free floating cells in some fruit endosperm (coconut milk). Therefore, protoplast of plant cell consists of plasma membrane and every cellular component is contained within it. Experimentally produced protoplasts are known as isolated protoplasts. According to Torrey and Landgren, "The isolated protoplasts are the cells with their walls stripped off and removed from the proximity of their neighbouring cell." According to Vasil, "The protoplast is a part of plant cell which lies within the cell wall and can be plasmolysed and which can be isolated further after removing the cell wall by mechanical or enzymatic procedure."

Isolated protoplast is only a naked plant cell surrounded by plasma membrane (Figure 16.1), which is potentially capable of cell wall regeneration, cell division, growth and plant regeneration in culture. Since, in plant cell, the plasma membrane is protected by a rigid cellulose wall, it has been relatively difficult to handle plant cells. Fortunately, plant scientists have made intelligent attempts in this direction to remove the wall of plant cell and culture the naked cell on nutrient medium.

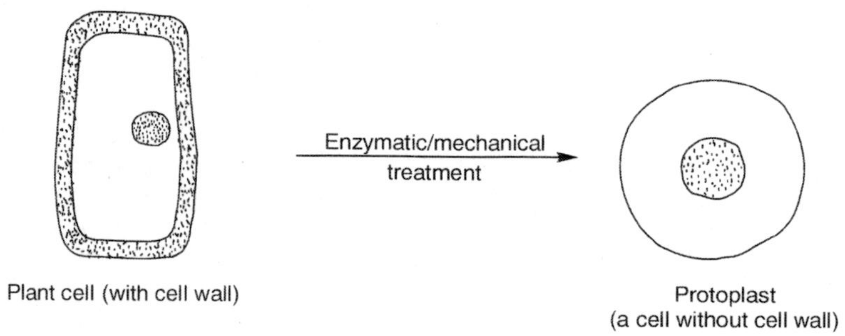

Figure 16.1 A plant cell and a protoplast

Historically, Hanstein for the first time used the term "protoplast" in 1880. The isolation of protoplasts from plant cells was first achieved by Klercker by microsurgery on plasmolysed cells in 1892. With refinements in the technique, protoplasts were isolated in large numbers by enzymatic removal of cell wall. In 1960, it was demonstrated by E.C. Cocking that the naked cells without their cell walls, called the protoplasts, can be obtained through enzymatic degradation of cell walls. This led to significant developments in the field of somatic cell genetics in higher plants.

A major reason for the rapidly expanding interest in protoplast culture is its potential use in plant cell genetics and especially in cell fusion and transfer of genetic information by DNA uptake and organelle implantation. Lack of cell wall barriers offers significant advantages for the introduction of genetically engineered foreign DNA into the naked cells. The essential ingredients of the technique include isolation of protoplast, culture of protoplasts, introduction of foreign DNA into protoplasts, raising whole transgenic plants from genetically manipulated protoplast culture and fusion of protoplasts leading to somatic hybridization.

The plant species, conditions under which the plants are grown, plant age, method of protoplast isolation, nutrient manipulations and culture durations are often critical for sustained division of protoplasts. Therefore, there are no standard methods for the isolation and culture of protoplasts. Considerable success has been achieved over the past two decades, since a number of techniques have been employed for the culture of protoplasts of numerous crop species.

Since 1960, when the first successful protoplast isolation took place, significant progress was also made towards improving the technology. Attempts were made to isolate protoplasts from several crop species and protoplast-based plant regeneration systems were made available for a great number of species. The improvements that have occurred include modification of protoplast isolation procedures, media composition, preconditioning of protoplast donor tissues, utilization of conditioned media or feeder cells and manipulation of culture environment.

PROPERTIES OF PROTOPLASTS

Freshly isolated protoplasts show the following characteristics in plasmolyticum solution:

1. They are highly fragile in structure.
2. They appear rounded in shape.
3. They show high pinocytic activity (which can be utilized to introduce foreign DNA, cell organelles, viruses, bacteria, etc. into the protoplasts).
4. They have a negative surface charge from –10 to –30 mV.

The freshly isolated protoplasts do not fuse and they repel each other due to similar charges. But in the presence of some fusion-inducing chemicals in the medium, the protoplasts isolated from distantly related plant groups can be induced to fuse. Such fusion may lead to the formation of hybrid protoplast. Protoplast fusion is a highly traumatic event because it takes place when the molecular distance between the fusing protoplasts is 10 Å or less.

SOURCE OF PROTOPLASTS

Protoplasts can be isolated from almost all plant parts like leaves, fruits, roots, stem, root nodules, tubers, pollen and pollen mother cells, endosperms, cells of callus tissues, cells from cell suspension cultures, etc. Furthermore, protoplasts can also be obtained from petals and reproductive parts of flower but the conventional and the most suitable source of protoplast are leaf mesophyll and cells from liquid suspension cultures. Protoplast yield and viability are profoundly influenced by the growing condition of the source plant. However, the prevailing conditions of light, photoperiod, humidity, temperature, nutrition and watering are also other contributory factors.

Aseptically grown plantlets are a good source to procure sterile leaves, stems, roots, etc. for the isolation of protoplast but cell suspension cultures may also provide a more reliable source for obtaining constant quality and supply of protoplast. It is necessary, however, to establish and maintain the cells at maximum growth rates and utilize the cell at the early log phase.

ISOLATION OF PROTOPLASTS

The success of a protoplast culture system primarily lies with the consistent yields of a large population of uniform and highly viable protoplasts. Several protoplast isolation and purification protocols have been published to optimize the yield and reproducibility. These procedures are often elaborate and labour-intensive. Also, these protocols involve too many explant or protoplast handling steps, and require extended exposure of explant for digestion treatment. Further, the efficacy of such protocols or that of enzyme combinations used therein could be limited to a few plant species only. These restrictions have to be overcome by improving the existing conditions and methods. A number of commercial cellulases and pectinases which allow easy protoplast release are available. By manipulating the source and concentrations of these isolation enzymes, protoplast procurement could be possible from most of the tissues; however, generalizations cannot be made particularly for protoplast isolation from monocot tissues.

All plant parts can be utilized to isolate protoplast but leaf mesophyll and cells from liquid suspension cultures are the most suitable materials. The essential step in the isolation of protoplast is the removal of the cell wall without damaging the cell. The plant cell is an osmotic system. The wall exerts inward pressure upon the enclosed protoplast; likewise, the protoplast also exerts equal and opposite pressure upon the cell wall. The enzymes and techniques used for isolation of protoplasts have a strong influence on their subsequent behaviour and development. Methods with too many steps involved often result in the introduction of cell contamination at some stage or the other.

There are two methods of isolation of protoplasts, namely, mechanical method and enzymatic method. The earliest attempts for the isolation of protoplasts made by Klercker in 1892 used the mechanical method which involved the preliminary plasmolysis of the cells and subsequently dissection of the cell walls to release the protoplasts.

1. ***Mechanical method*** The tissues are kept in a suitable plasmolyticum and cut with a fine knife, so that protoplasts are released from the cells cut through the cell wall. Then the tissue is again deplasmolysed. This method is suitable for isolation of protoplasts from vacuolated cells (Figure 16.2). This method gives poor yield of protoplasts and is not suitable for isolating protoplasts from meristematic and less vacuolated cells. This operation has to be done carefully and small pieces of tissue should be cut under a microscope using a microscalpel.
2. ***Enzymatic method*** A considerably more efficient way of liberation of protoplast is to digest the cell wall around them, using cell wall-degrading enzymes. This method was developed by Cocking in 1960, and it makes use of cell wall-degrading enzymes such as cellulase, hemicellulase and pectinase. These digestive enzymes could be applied to tissue of excised plant parts as well as to callus cultures

or cell suspension. The cells and tissues are soaked in a high concentration of the enzyme for several hours. The isolation of protoplast is carried out under sterile conditions. Important enzymes used for the isolation of protoplasts are listed in Table 16.1.

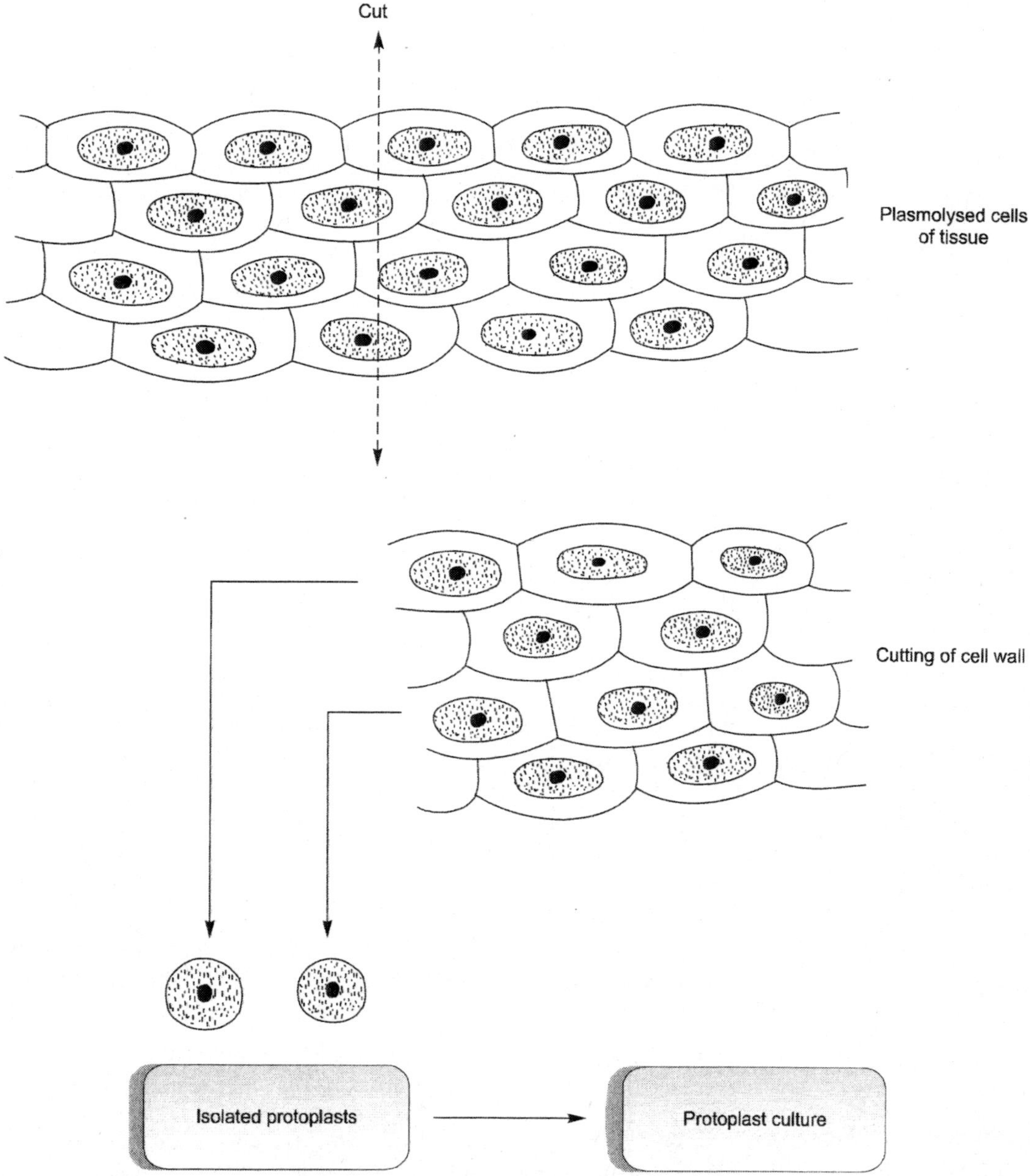

Figure 16.2 Mechanical isolation of protoplast

Table 16.1 Enzymes used for isolation of protoplasts

Enzymes	Source	Manufacturing company
Pectin-degrading enzymes		
Pectinase	*Aspergillus niger*	Sigma (C. St. Louis, US)
Pectinol AC, Pectolyase Y 23	*Aspergillus niger*	Seishim Pharma., Tokyo, Japan
Macerozyme R-10	*Rhizobium* sp.	Kinki Yakult Biochem. Ltd., Japan
PATE (Peptic acid acetyl transferase)	*A. japonicus*	Farbwerke-Hoechst, FRG, San Diego
Cellulose-degrading enzymes		
Cellulysin	*Trichoderma reesei*	Calbrochem-Behring, San Diego, US
Driselase	*A. niger* and *Irpex lacteus*	Kyowa Hakko Kogya Co. Ltd, Japan
Onozuka cellulase P-1500	*Trichoderma viride*	Kinki, Yakult Nishinomia, Japan
Hemicellulose-degrading enzymes		
Hemicellulase	*A. niger*	Sigma London Chemical Co. Poole, Dorset, UK
Rhozyme HP 150	*A. niger*	Rohm and Haas, Philadelphia, US

Onozuka enzyme-1500 is successfully used to release protoplast from a variety of tissues. Cellulase (0.2–2.0%), commonly used with hemicellulase (0.1–0.5%) and/or pectinase and driselase, are efficient enough to give a high yield of protoplasts in 1–2 hours with cell cultures derived from different species.

Isolation of Protoplasts from Leaves

Young cell suspensions are practically ideal for the isolation of protoplasts in large quantities. However, the most commonly used organs are leaves (Figure 16.3) which can be employed for isolation of protoplasts with the following steps:

Sterilization of leaves Young fully exposed leaves from the upper part of 10-week-old plants growing in greenhouse are detached and washed thoroughly with tap water. The leaves are dipped in 70% ethanol for one minute, followed by dipping into 0.4–0.5% sodium hypochlorite solution for 30 minutes. Further, treated leaves are washed 4–5 times with sterile distilled water to remove adhered traces of hypochlorite. The whole process of sterilization is done inside the LAF cabinet.

Peeling of the epidermis The impermeable lower epidermis of the surface-sterilized leaves is peeled off as completely as possible with the help of a sterilized forceps. In cereals, where it is difficult to peel off

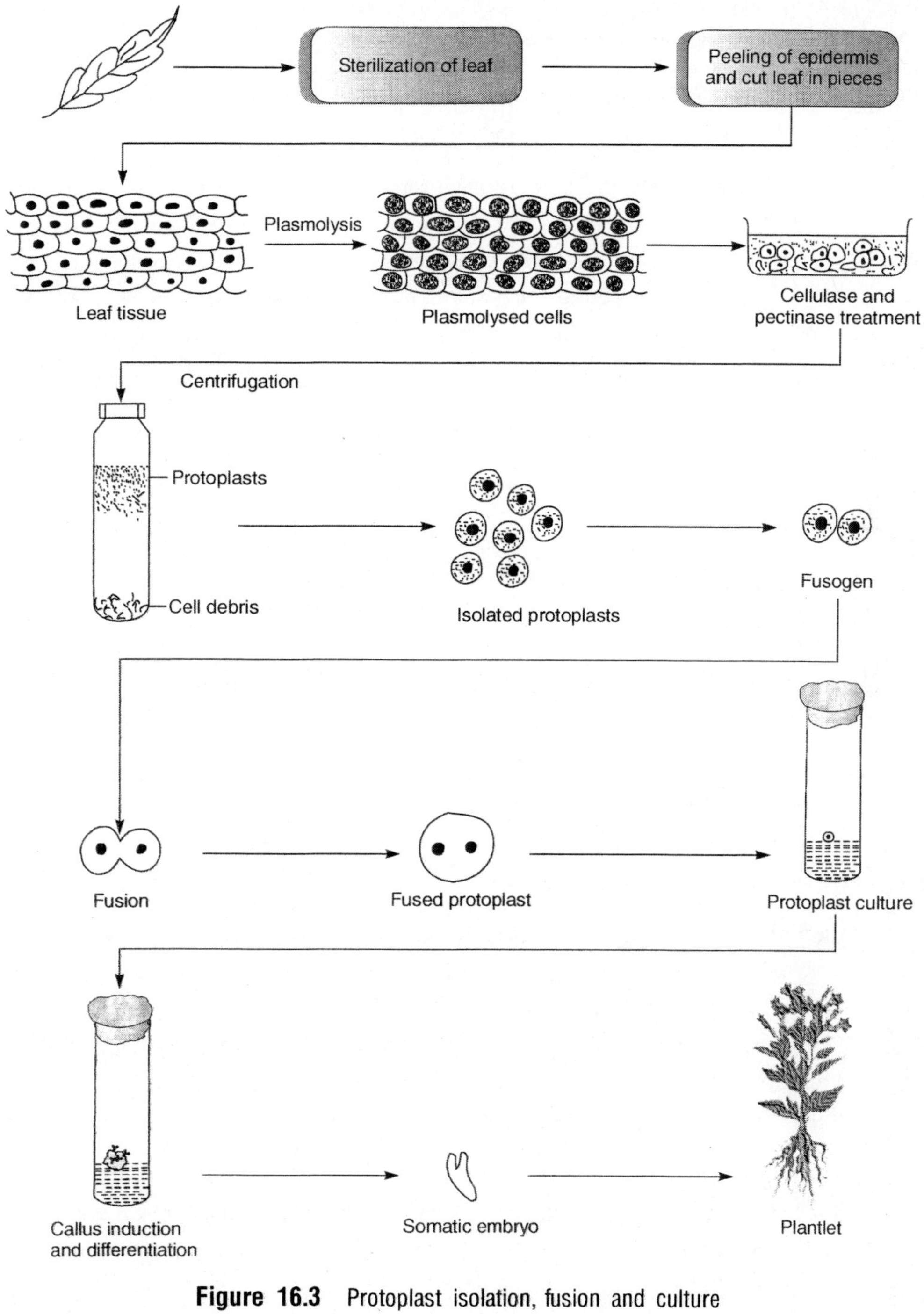

Figure 16.3 Protoplast isolation, fusion and culture

the epidermis, leaves are cut in large strips. It is sufficient enough to allow entry of the enzyme mixture into the cells. The peeling operation is relatively easier for the leaves carrying flaccid cell.

Enzymatic treatment The protoplasts from these peeled-off leaves are isolated either by direct enzyme treatment method or sequential method.

Direct enzyme (single-step) method This method is useful for the isolation of protoplasts from spongy and palisade parenchyma in leaves. Peeled leaf pieces are kept with their lower surface down in 25 ml solution of 13% mannitol and inorganic salts for one hour. After one hour, the above solution is pipetted out with the help of a Pasteur pipette and 25 ml solution of enzyme mixture (0.5% macerozyme + 2% Onozuka cellulase in 13% sorbitol at pH 5.4) is added. Leaf pieces in the enzyme solution are incubated in the dark at 24–26°C for 16–18 hours. The Petri plates are shaken well to facilitate the release of the protoplast. Then the larger pieces of leaf peels are pushed to one side with a sterile forceps by keeping the Petri dish at an angle of 15°. When protoplast settles down in the Petri dish, this mixture is transferred to 15 ml centrifuge tubes and centrifuged at 100 g for 5 minutes.

The protoplasts form a pellet in the centrifuge tube. The supernatant is pipetted off and the pellet is resuspended in 21% sucrose solution. It is again centrifuged for 5 minutes at 200 g. The viable protoplasts will float on the top surface of the sucrose solution, while the remaining cells and debris will sink to the bottom of the tube. The viable protoplasts are collected from the surface and resuspended again in 13% sorbitol. After 2–3 centrifugations, the protoplasts are suspended in measured volume of NT liquid medium (Nagata and Tabeke, 1971).

Sequential enzyme (two-step) method In this method, cells are first isolated using macerozyme (A) and then treated with cellulase (B) to isolate protoplasts. The two enzyme mixtures (A and B) are used one after the other to isolate protoplasts from palisade tissues. The protoplasts isolated this way are better for culture studies.

Step I: Isolation of cells The peeled leaf segments, kept in a Petri dish containing an enzyme mixture A (0.5% macerozyme + 0.3% potassium dextran sulphate in 13% mannitol at pH 5.8), are vacuum-filtered for 5 minutes followed by transfer to a water bath at 25°C and then subjected to slow shaking. After 15 minutes, the mixture A is replaced by fresh enzyme mixture A and leaf segments are incubated for one more hour. The mixture is filtered using a nylon mesh, centrifuged at 100 g for one minute and 2–3 times with 13% mannitol to get a pure sample of isolated cells.

Step II: Isolation of protoplasts The isolated cells are then incubated with enzyme mixture B, containing 2% cellulase in a 13% solution of mannitol at pH 5.4 for 90 minutes at 30°C. After incubation, the mixture is centrifuged at 100 g for one minute, so that the protoplasts form a pellet, which is cleaned 2–3 times to get a pure sample of isolated protoplasts with 20–25% sucrose solution.

Leaves of *Nicotiana tabacum* are highly standardized material for protoplast isolation. Currently, the single step method (direct method) is very popular and this procedure is followed as routine work in most of the laboratories.

Isolation of Protoplast from Cells in Suspension Cultures

Cell suspension cultures are the most suitable material for isolation of protoplasts in good quantity. A new suspension does not yield many protoplasts until it has been subcultured at least twice. Suspension cultures are generally harvested at their early exponential growth phase. Large cell clumps are filtered and actively growing free cells in suspension are first treated with colchicine and some chelating chemicals to prevent the formation of aggregates. Secondly, inclusion of cellulase (0.2%) prior to enzyme digestion would soften the cell wall, thus quickening the process. After filtration, the solution is added to cells. The Petri dish containing cells and enzyme is placed on the rotating gyratory shaker (40–50 rpm) for 4–6 hours.

Incubation mixture consists of protoplasts with cell debris. The protoplast suspension is filtered to remove the larger debris. The filtrate is centrifuged at 80 g for 5–10 minutes so as to sediment the protoplast and then the supernatant is pipetted off. The pellet is resuspended in 21% sucrose solution. The protoplast suspension is again centrifuged at 100 g for 5–6 minutes. Collected protoplasts are washed with 13% mannitol by repeated centrifugation and finally protoplasts are resuspended in liquid culture medium.

The yield of protoplasts can be measured by hemocytometer. B_5 medium, supplemented with 2,4-D (1.0 mg/l), kinetin (2.0 mg/l) and casein hydrolysate (1.0 mg/l) has been proved to be the best medium for the culture of protoplasts isolated from cell suspension culture of *Daucus carota.*

Protoplast isolation protocol for rose cells Protoplasts are isolated from actively growing cell suspension in exponential phase (7–8 days after subculture) in fresh MS medium containing 5 μM each of 6-benzylaminopurine (BAP) and 2,4-dichlorophenoxy acetic acid (2,4-D). The cell culture medium is decanted and the cells are plasmolysed in a plasmolysing solution (protoplast washing salt solution with 13% mannitol) for one hour. Later on, this solution is replaced by an enzyme mixture (cellulase R10, 1.0–5.0%; macerozyme R10, 0.25–1.0%; and driselase, 0.25–1.0%) and incubated at 25°C in dark on a shaker (60 rpm) for 14–16 hours. After incubation, protoplasts and partly digested cells are passed through a sieve (90 μm pore size) with undigested tissues and finally rinsed with 10 ml of 13% mannitol solution.

The filtrate containing protoplasts is centrifuged (80 g) for 10 minutes and the supernatant is discarded. For purification, protoplasts are suspended in high-density sucrose solution (21% sucrose) and centrifuged (100 g) for 10 minutes. Protoplasts free of debris are carefully removed from the interface of the solutions and resuspended in known volume of MMM550 ($MgSO_4 \cdot 7H_2O$, 1.69 g/l; $MgCl_2 \cdot 6H_2O$, 2.3 g/l, MES 1.95 g/l; and mannitol, 130 g/l). Protoplasts could be counted at this stage.

The protoplasts are cultured at a plating density of 4×10^5 protoplasts/ml in modified MS medium. The first division is observed on day 1 and 8–10 celled colony stage is observed within a week. In two weeks, 30–40 celled colonies are formed and it could be transferred to regenerating medium. A plating efficiency of 71.4% could be recorded after 10 days of culture.

Isolation of Protoplasts from Pollen Grains (PG)/Pollen Mother Cells (PMC)/Pollen Tetrad (PT)

Protoplasts obtained from mature PG, PMC or PT could be used in the production of haploid protoplasts. Haploid protoplasts are used in the study of somatic hybridization, mutation, pollen ontogeny, cell modification and production of haploid plantlets. Haploids contain uniform chromosome number and hence fusion of two haploids form a normal diploid, which is another procedure of hybridization instead of doubling of the chromosome number of the new sterile hybrid.

The cell wall of PMC and PT is made up of callose whereas the exine of mature pollen is coated with sporopollanins, which is very difficult to digest. Bhojwani and Cocking used the snail enzyme complex halicase, to degrade the callose wall enclosing PMC and PT. The contents of anthers are squeezed out during incubation of enzymatic treatment for 30 minutes followed by enzyme complex replacement by 10% sucrose, where the PMC and PT are released and settled in sucrose solution. Once protoplasts are settled down, they are reset by washing. Finally, the protoplasts are cultured in nutrient medium.

VIABILITY OF PROTOPLASTS

The modern technique to test the viability of protoplast is as follows:

Dyes like fluorescein diacetate (FDA 0.01%), phenosafranine and calcofluor white (CFW) can be used to test the viability of protoplasts. Since the FDA accumulates within the plasma membrane, the viable protoplasts fluoresce green/white. Practically, it should be examined within 10 minutes after the treatment, otherwise, beyond that FDA dissociates. Similarly, phenosafranine (0.1%) can stain and detect only dead protoplasts, whereas viable protoplasts remain unstained even after 2 hours of exposure to phenosafranine stain. However, calcofluor white detects the onset of cell wall formation around the plasma membrane of viable protoplasts in the form of a ring of fluorescence.

ISOLATION OF SUBPROTOPLASTS

Subprotoplasts are fragments derived from protoplasts and do not contain the entire constituents of plant cell/protoplasts. They may include the following:

1. *Cytoplast* containing cytoplasm of a cell and lacking a nucleus used for cybrid production.
2. *Karyoplasm*, also known as mini-protoplasts, containing a nucleus surrounded by some cytoplasm.
3. *Micro-protoplasts* containing only a few of all chromosomes, and a fraction of the cytoplasm.

Cytoplasts can be prepared by one of the following methods:

1. X-ray treatment or laser beam microsurgery of freshly isolated protoplast (X-ray and laser inactivates the nucleus).
2. In tomato pericarp, cytoplasts occur spontaneously at the time of ripening of fruits.
3. Sometimes shaking of protoplasts may lead to separation of buds; these separated buds without any nucleus, are known as cytoplasts.

The size of sub-protoplasts of green algae *Bryopsis plumosa* were mostly 30–40 μm in diameter. The survival of sub-protoplasts was partially dependent on their size. No sub-protoplast less than 10 μm in size can survive to form secondary membrane.

PROTOPLAST CULTURE METHODS

A number of techniques have been developed over the years for the culture of plant protoplasts (Figure 16.4). These include the use of feeder layer, microchambers, hanging drop culture and multi-drop array technique, depending on the objectives of the experiment. However, the most common technique for protoplast culture is agar embedding method. In this method, the protoplast suspension is mixed with equal volume of melted 1.6% agarified medium at 30–35°C and the protoplast-agar mixed medium is poured into a Petri dish. The medium is dispersed throughout the dish gently and quickly by slightly rotating the Petri dish. Embedding of protoplasts on the top surface layer of solid medium is known as plating of protoplasts. These can be handled easily and the agar medium is used as a support to the protoplasts.

An improvement of this method is the use of alginate as the gelling agent; alginate can then be dissolved in osmotically adjusted sodium citrate solution and the protoplast-derived micro-colonies could be easily recovered without affecting the growth. However, another improvement in protoplast regeneration efficiency has been possible by the use of a combination of both semi-solidified blocks/thin layers and liquid media.

1. *Agar embedded culture* In this method, protoplasts may be allowed to form a new cell wall in liquid culture medium before embedding. In the original method of Nagata and Takebe (1971), the isolated protoplasts are mixed with 0.1% agar culture medium and maintained at 37°C. Small amount of liquid agar–protoplast mixture is then poured into sterile Petri plates for subculturing.

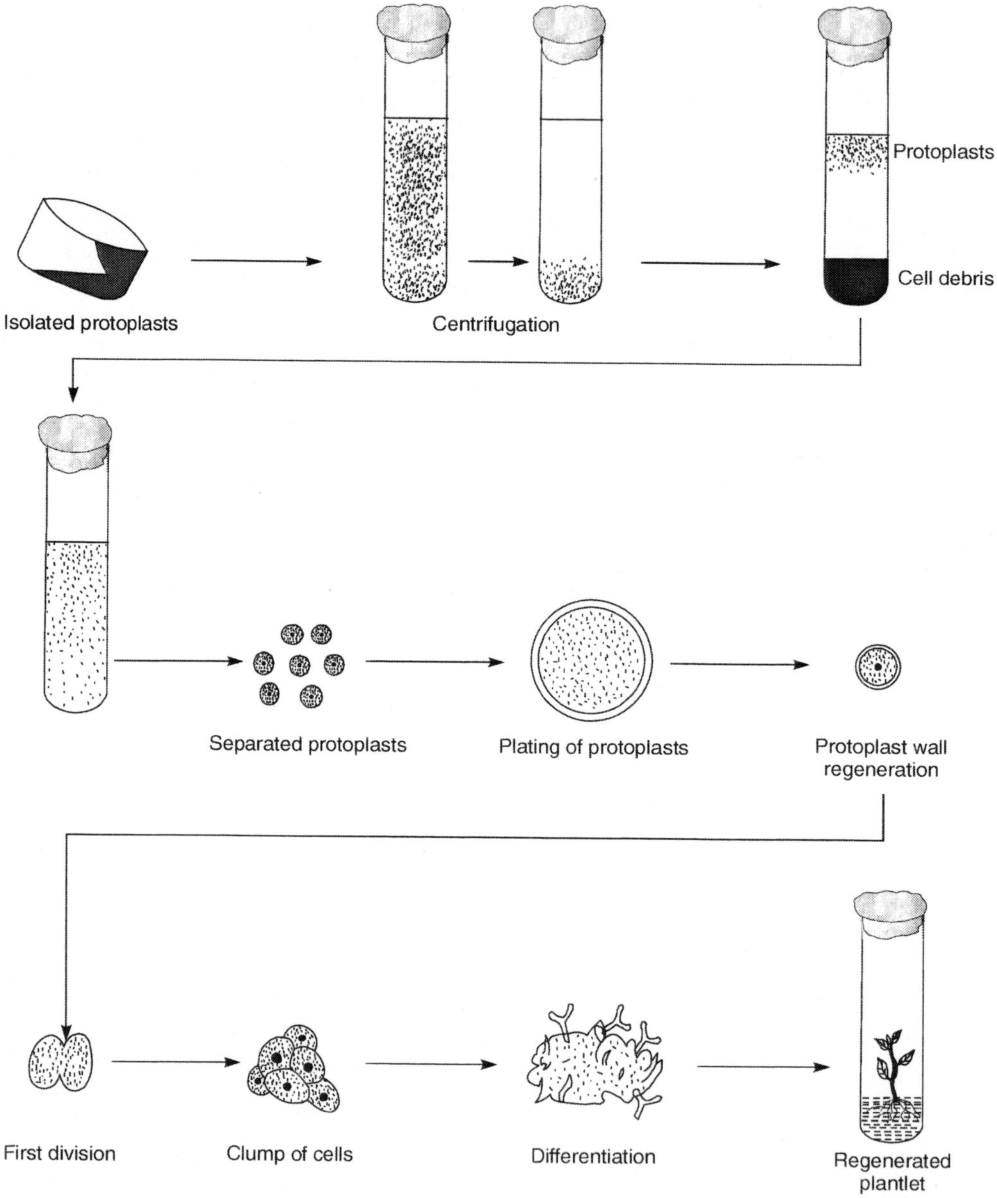

Figure 16.4 Protoplast culture and plant regeneration

2. *Droplet culture* The suspending protoplasts in the liquid culture media are transferred to Petri dishes in the form of droplets with the help of a micropipette. This method enables the subsequent microscopic examination of protoplast development.

3. *Co-culture* It means culture of fast growing protoplasts along with the less fast growing protoplasts and these are plated on the solid agar media. The fast growing protoplasts provide some growth factors which induce the growth and development of the desirable slow growing protoplasts.
4. *Bead culture* In this method, protoplast suspension is mixed with alginate, difco agar, carrageenan, etc., and then small beads are made by dripping the mixture into liquid medium. Finally, the beads in liquid medium are put on a shaker. This technique saves the protoplasts from mechanical breakage and increases the stability, aeration and viability, because the protoplasts are embedded in the beads.
5. *Hanging drop method* This method was developed by Kao, Keller and Miller in 1970. Culture of protoplasts in an inverted liquid droplet is known as hanging droplet method. Each droplet contains very small numbers of protoplast. This method is used to avoid evaporation of the media while observing protoplasts. The drop is placed in the depression of a specially prepared hanging drop slide, which is then sealed with a coverslip and oil, or the drops can be put on the lid of a Petri dish. A very thin layer of mannitol solution is generally kept on the lower part of the Petri dish to provide a humid condition inside the Petri dish as well as to prevent the desiccation of the droplets.
6. *Feeder layer technique* In this method, one protoplast layer contains mitotically blocked protoplasts, by low doses of X-ray treatment. The inactivated protoplast layer is called the feeder layer. This improves the growth and development of desired normal protoplasts on the upper layer.
7. *Thin alginate layer (TAL) technique* This technique is an improvement over all the methods described above. It encompasses several positive points over earlier methods and was used for *Nicotiana* mesophyll protoplasts. In this method, squares of polypropylene mesh (2.0 × 2.0 mm grid) are cut so that it exactly fits into 60 mm Petri dish. Protoplasts at twice the required density in MMM550 are suspended in equal volumes of 2.8% alginic acid to give a final alginate concentration of 1.4%. For one grid preparation, 625 μl of this mixture is placed on an agar layer (20 mM $CaCl_2$), and a polypropylene mesh (10 × 10 meshes; 2.0 × 2.0 mm mesh size) is placed over it.

   After one hour, the grid is slide gently towards one side and taken out. This is then placed upside down in a 60 mm Petri dish containing 10 ml of protoplast culture medium. Excess of calcium chloride is removed by washing twice with protoplast culture medium (10 ml) and finally cultured in 2.0 ml of the same medium. These are then incubated in dark/diffused light at 25°C. The protoplasts divide to form colonies. When colonies of 10–20-celled stage are formed, these are transferred to the regeneration medium.

Despite the fact that TAL technique has certain advantages like cell tracking, convenience of transfer to fresh medium and reduced release of toxic substances to the medium, it also has certain limitations. These are:

i. This process is time-consuming and costly.

ii. When the grid is transferred to the regeneration medium, callus or differentiated tissue sometimes folds upwards and ultimately slips out of the grid, thereby losing contact with the medium.

iii. The alginate layer is thick, hence protoplasts in multilayer are difficult to trace.

iv. It uses a specific polypropylene mesh, which is not easily available.

Moreover, all the methods described above suffer from one or the other limitations; however, an improvement over TAL technique, recently proposed, is the extra-thin alginate film (ETAF) technique.

8. *Extra-thin alginate film technique* When compared to TAL, this technique is simpler to perform, less time-consuming and economical. The advantages of this technique can be enumerated as follows:

   i. It is less expensive than TAL technique as the agar component of $CaCl_2$ is eliminated (25 ml of gelled $CaCl_2$ medium is poured into Petri plates where two grids can be prepared. For 25 ml of medium, 200 mg of agar is used, therefore for one grid 100 mg of agar is required).

   ii. Only 150 μl of $CaCl_2$ solution is required instead of 25 ml of $CaCl_2$ medium. Similarly, for one cover glass, only 50 μl of alginate is required.

   iii. Microslides and coverglasses are less expensive, can be reused and are easily available everywhere than the polypropylene grid.

   iv. Just one ml of protoplast culture medium is required is for each washing, compared to 10 ml required in TAL technique.

   v. This technique also saves time as complexation time is only 5 minutes as compared to one hour in TAL technique.

   vi. Two washings of 15 minutes each are required instead of two washings for 30 minutes. Preparation of grid (sizing, autoclaving and drying) consumes a lot of time, whereas microslide and coverglass can be sterilized after giving a dip in alcohol and by subsequent flaming.

   vii. Protoplasts are evenly spread and easy to track due to ultra thin layer.

viii. While using TAL technique, callus or differentiated tissue sometimes folds upwards and ultimately slips out of the grid thereby losing contact with the medium. However, in the ETAF technique, none of these limitations prevail.

ix. While performing fusion experiments on coverglass, the protoplasts can be easily embedded using this technique.

Thus, this technique overcomes the limitations of all the protoplast culture techniques described so far.

ROLE OF HAEMOGLOBIN IN PLANT PROTOPLAST CULTURE

The inability to supply adequate amounts of oxygen to cells can limit the success of *in vitro* plant cultures. A recent method of improving oxygenation of plant cell or protoplast cultures involves the use of perfluorocarbons (PFCs). These compounds are remarkable for their ability to carry and release extremely large amounts of oxygen as compared with other liquids.

The use of PFCs and haemoglobin provide new options for enhancing division of cultured plant cells. The relative effectiveness of oxygenated PFC and haemoglobin separately or in combination, may be species-specific. Scientists have also compared the ability of bovine haemoglobin to enhance division of *Petunia hybrida* protoplasts in the presence or absence of the copolymer surfactant, Pluronic F-68. It is a compound that has been shown to promote growth of cultured plant cells, tissues and organs.

Haemoglobin with or without Pluronic F-68, provides a further option for enhancing the growth of cultured plant protoplasts and protoplast-derived cells under static and, perhaps, agitated conditions. This technology could readily be scaled up to fermenter systems and may be effective in stimulating growth of cultured plant tissues and organs, which are now employed extensively as a source of economically important plant products.

ACTION OF MACEROZYME/PECTINASE AND CELLULASE ON PLANT CELL WALL

Pectin present in the middle lamella of the cell walls functions as an intercellular cementing material. Its structure consisting of two carbohydrate residues—D-galacturonic acid residues bound together by alpha (1–4) linkage and L-rhamnase residues interrupted with (1–2) linkages. Pectinase or macerozyme, by either complete or partial hydrolysis of pectin, loses the cell mass from tissues and exposes the cellulosic layer to cellulase enzyme for further action as cellulose is made up of 10000 units of D-glucose bound together with β(1–4) glycosidic links. This linkage of cellulose is hydrolysed by cellulase is enzyme complex and finally, cellulose is converted into glucose.

PROTOPLAST REGENERATION

New cell walls are regenerated over the plasma membrane with the deposition of cellulose microfibrils and pectins immediately (depending on the species, it takes 10 minutes to 1 day) on culture of the protoplasts in a suitable medium. Cell wall formation can be detected by plasmolysis and by the use of calcofluor white (CFW). The CFW binds to the linked glycosides in the newly synthesized cell wall. Cell wall regeneration is observed by a ring of white fluorescence around the plasmalemma. The cellulose is deposited either between the plasmalemma and the multilamellar wall material or directly on the plasmalemma. The nature of biosynthesis of the cell wall depends on the plant material and the system of protoplast cultures.

Electron microscopic studies and freeze etching studies have revealed much about the structure and progressive development of cell wall around the protoplast in culture medium. Cell wall regeneration is a prerequisite for nuclear and cell division. After the formation of cell wall, the walled cells expand and divide into two cells. In soybean protoplast culture, the first report indicates that nuclear divisions in protoplasts are followed by cytokinesis. Cell division stages can also be stained using CFW. After wall formation, the reconstituted cells show considerable increase in size and the first cell division usually takes place within 5–7 days of culture initiation. Subsequent divisions give rise to small cell colonies. After 2–3 weeks, small colonies are formed which can be transferred to an osmoticum-free medium to develop a callus. The small callus mass can be subcultured at regular intervals and used for differentiation.

The ultimate goal of protoplast culture is the regeneration of plant from the single protoplast. In order to regenerate protoplast, callus may be induced to undergo organogenic differentiation, which ultimately leads to whole plant regeneration. It is generally noted that plant regeneration occurs very easily in members of the family Solanaceae and rarely observed in Convolvulaceae, Ranunculaceae, Umbelliferae, Cruciferae and Asparagaceae. However, recently, considerable success of protoplast regeneration has been reported in both monocot and dicot plants as shown in Table 16.2.

Table 16.2 Plant species in which plantlets have been regenerated from protoplast culture

Plant species	Family	Source of protoplast
Nicotiana tabacum	Solanaceae	Leaf and cell cultures
Brassica napus	Brassicaceae	Leaf
Datura innoxia	Solanaceae	Leaf
Daucus carota	Umbelliferae	Cell cultures
Solanum tuberosum	Solanaceae	Leaf
Asparagus officinalis	Liliaceae	Cladodes
Petunia hybrida	Solanaceae	Leaf
Citrus sinensis	Rutaceae	Nucellus cells and callus

APPLICATIONS OF PROTOPLAST CULTURE

The isolation, culture and fusion of protoplasts are the most fascinating areas of research. The freshly isolated protoplasts with totipotent nature are very important because of the far-reaching applications in studies of fundamental and applied research in experimental biology, somatic cell genetics, somatic hybridization, genetic engineering and crop improvement.

The following are the major applications of protoplast technology:

1. *Gene transformation* The introduction of exogenous DNA into experimental protoplast brings some genetic modification and produces new plant varieties. The study on genetic diversity is possible by using the DNA virus as vector; it is quite possible to insert foreign DNA into the viral genome. So, DNA virus is a suitable vector for gene transfer. The genetic manipulations of plant protoplast can be studied in detail in crown gall-causing bacteria (*Agrobacterium tumefaciens*). The autonomous growth of the tumour cell is caused by the integration of a piece of DNA from a Ti plasmid.

   This naturally transformed system can be used in experiments to transform plant protoplasts into hormone-independent cells. In order to give the plant a self-fertilizing property, this technique can be used to transfer the nitrogen-fixing *nif* gene from the bacterial genome of *Rhizobium* sp. to the protoplasts non-leguminous crops and cereals, in which this particular bacterial species is unable to cause an infection.
2. *Transplantation study* Incorporation of bacteria, isolated nucleus, yeast, blue-green algae, etc., into protoplast has been attempted in order to study the symbiotic association with higher plants and nuclear cytoplasmic inheritance with foreign and genetic information. Bacterial cell uptake by plant protoplasts has been investigated with species of *Spirulina, Rhizobium, Azotobacter,* etc. Introduction of *Anabaena* sp. and *Gleocapsa* sp. into protoplasts has also been studied. Nuclear transplantation establishes different hybrid varieties of cultivated crops.
3. *Virus uptake* Protoplasts can be directly inoculated with any pathogenic virus in the medium. The process of uptake of virus particles, their replication inside the protoplasts and their mode of action at the molecular and cellular level are made possible with the aid of protoplasts.
4. The effect of various environmental factors on the osmotic behaviour can be studied using plant protoplasts.
5. This is useful for the study of morphogenesis because, under optimum conditions, protoplast regenerates its own cell wall and becomes a walled cell.
6. Cell division followed by plant regeneration may occur from the cell system either through organogenesis or embryogenesis, which can be studied by this technique.
7. Plasma membrane can be isolated and studied, when newly released protoplasts are placed in hypotonic solution.

8. In future, this technique will be one of the most frequently used research tools for *in vitro* technologists, physiologists, pathologists, molecular biologists, microbiologists, etc.
9. Protoplast is a very convenient material for isolation of cell organelles like chloroplast, mitochondria, nuclei, etc.
10. Transplantation of chloroplasts, microorganisms, nuclei and chromosomes are possible in protoplasts.

REVIEW QUESTIONS

1. What is protoplast and what are its sources?
2. Describe the methods for isolation and purification of protoplasts.
3. What are the enzymes used for the isolation of protoplasts?
4. How is the isolation of protoplasts from cell suspension culture different from that of an intact explant?
5. How can protoplasts be isolated from pollen grain/pollen tetrads?
6. How can subprotoplast be generated?
7. What are the methods of protoplast culture? Which method is most suitable and why?
8. Describe any protoplast isolation protocol.
9. Write short notes on cell wall regeneration and cell division in protoplast culture.
10. What are the important applications of protoplast technology?

REFERENCES

Burza, W. and Malepszy, S. (1995). *In vitro* culture of *Cucumis sativus* L. XVIII. Plant from protoplasts through direct embryogenesis. *Plant Cell, Tissue and Organ Culture*. 41: 259–266.

De, K.K. (2004). *Plant tissue culture*. New Central Book Agency (P) Ltd., Calcutta.

Kao, K.N., Keller, W.A., and Miller, R.A. (1970). Cell division in newly formed cells from protoplasts of soybean. *Exp. Cell. Res*. 62: 338–340.

Nagata, T. and Takebe, J. (1971). Plating of isolated tobacco mesophyll protoplasts on agar medium. *Planta*. 99: 12–20.

Takebe, I., Labib, C. and Melchers, G. (1971). Regeneration of whole plants from isolated mesophyll protoplasts of tobacco. *Naturwissen*. 58: 318–320.

Teulieres, C. and Boudet, A.M. (1991). Isolation of protoplasts from different Eucalyptus species and preliminary studies on regeneration. *Plant Cell, Tissue and Organ Culture*. 25: 133–140.

17

PROTOPLAST TECHNOLOGY–SOMATIC HYBRIDIZATION AND CYBRIDIZATION

INTRODUCTION

In conventional breeding, successful crosses is generally effective only between closely related species but not with widely differing genera and species. Sexual hybridization was a total failure for distantly related plant species as well as sexually incompatible plant species in the conventional process. Therefore, if a system could be developed for transferring genetic information between widely different plant species, it would provide the basis for a technology to overcome the limitations of conventional sexual hybridization. One of the most important practical uses of protoplast culture is to develop methods for facilitating somatic cell fusions and genetic engineering in higher plants.

The process of somatic hybridization involves the fusion of isolated protoplasts under *in vitro* conditions to produce hybrid protoplasts, and growing of the hybrid protoplasts into whole transplantable plants. Such hybrids are known as somatic hybrids. Although somatic hybridization was successfully achieved first in animals, its significance was realized only later in plants. The hybrid cells can be induced to regenerate into complete plants. In plants, somatic hybridization through cell fusion is possible only if the cell wall, which forms a barrier between the cell and external environment, is removed, leaving the cell with intact plasma membrane, but without cell wall. Plasma membrane is not a rigid surface but a membrane which is capable of developing buds and invaginations. The lack of cell wall allows the plasma membrane of two or more protoplasts to come into intimate contact.

Thus, somatic hybridization is defined as the *in vitro* fusion of plant protoplasts, derived either from the somatic cells of the same plant or from two genetically different plants. When the protoplasts from vegetative and gametic cells are fused, such fusion is called somato-gametic hybridization.

Protoplast fusion as a possible plant breeding tool has been used to make crosses within species (intraspecific), between species (interspecific) and also between genera (intergeneric). It also facilitates transfer of cytoplasmic elements. This method permits gene flow from one species to another, which is generally not possible by sexual means. Protoplast fusion has been proved to be helpful in several members of Gramineae, Solanaceae and Cruciferae to overcome sexual barriers to produce plants with resistance to diseases or tolerance to selective herbicides, to produce somatic hybrids having efficient organelle system and to widen the genetic base (Glimelius *et al.*, 1991).

The hybrid protoplast contains genome and organelles of the two parent cells. Chloroplast and mitochondria of any one of the parents degenerate due to sorting during mitotic cell division. So, the hybrid protoplast has a new combination of genes with some new traits.

Furthermore, hybrid protoplasts are grown *in vitro* into a plantlet via either embryogenesis or organogenesis and this technique involves the following steps:

1. Isolation of protoplasts of desired species/varieties.
2. Fusion of protoplasts isolated from the different species.
3. Regeneration of cell wall by the hybrid cells.
4. Fusion of nuclei.
5. Cell division of hybrid cells and their growth to produce callus.
6. Identification and selection of hybrid callus.
7. Induction of organogenesis or embryogenesis in the callus tissues derived from hybrid cell.
8. Formation of young plantlets from regenerated shoots followed by complete plant production.

ISOLATION OF PROTOPLAST

Protoplasts have been isolated from virtually all plant parts, but leaf mesophyll tissue has been proven to be the most preferred tissue for the isolation of well-defined protoplasts. In general, fully expanded leaves are surface sterilized, then their lower epidermis is peeled off with a pair of forceps and the peeled areas are cut into small pieces with a scalpel and further, the pieces are suspended in the enzyme mixture. If the epidermis cannot be peeled off, the leaf may be cut into pieces and directly treated with the enzyme mixture; vacuum infiltration may also be used to facilitate the entry of enzymes into the tissues. In some cases, protoplasts may also be isolated from callus cultures or cell suspension cultures.

After the incubation period, protoplasts are washed with a suitable washing medium in order to remove the enzymes and the cell wall debris. Furthermore, these protoplasts can be cultured on a suitable nutrient medium in order to achieve hybrid plant regeneration.

Generally, MS and B_5 nutrient media, with minor modifications are used for protoplast culture. These media are supplemented with a suitable osmoticum along with an auxin and a cytokinin; however, their types and concentrations are mainly dependent on the plant species or specific tissues. After 7–10 days of culture, protoplasts regenerate cell wall and the osmolarity of medium is gradually reduced to that of normal medium. Protoplasts are very sensitive to light; therefore, these are incubated in diffuse light or in dark for the first 4–7 days.

METHODS OF PROTOPLAST FUSION

A number of strategies have been followed to induce fusion between protoplasts of different strains/species. Protoplasts of desired strains/species are mixed in almost equal proportion; generally these are mixed while still suspended in the enzyme mixture. Various methods have been developed for inducing fusion between protoplasts; however, the fusion frequency remains low. The mode of protoplast fusion can be grouped into three major categories:

1. Spontaneous fusion
2. Induced fusion
3. Mechanical fusion

Spontaneous Fusion

This type of fusion is strictly observed among the protoplasts of the same species. During isolation of protoplasts for cultures, enzymatic degradation of cell walls is effected and some of the protoplasts, in close proximity to each other, may undergo fusion through their plasmodesmata to form homokaryocytes. It never yields hybrid protoplasts. The occurrence of multinucleate fusion bodies is more frequent, particularly when protoplasts are isolated from actively dividing cells. The spontaneous fusion of meiocyte derived protoplasts is used to produce intergeneric fusion, as between *Lilium longiflorum* and *Trillium kamatschticum*.

Spontaneous fusion seldom occurs in protoplast population, if they are completely separated. This type of fusion between two or more adjoining somatic protoplasts is practically not very useful. There seems to be a correlation between the size of the leaf and the percentage of protoplasts undergoing spontaneous fusion. Protoplasts from young leaves are more likely to undergo such fusion.

Induced Fusion

Fusion of freely isolated protoplasts from different sources, with the help of fusion-inducing chemical agents is known as induced fusion. In general, isolated protoplasts do not fuse with each other because the surface of the protoplasts carries negative charges around the plasmalemma and thus repel one another. So, this type of protoplast needs a fusion-inducing agent (fusogen), which reduces the electronegativity of the isolated protoplasts and allows them to fuse with each other.

This technique has the possibility and ability to combine different genotypes beyond the limits imposed by the sexual process. Fusion can be induced by employing chemical fusogens such as sodium nitrate, higher concentration of calcium ions, polyethylene glycol (PEG), dextran and others. The use of PEG causes agglutination of the protoplasts and brings about fusion.

Mechanical Fusion

Induction of protoplast fusion by pushing together two protoplasts is called mechanical fusion. In this method, protoplasts of two different plant varieties of desired characteristics are sucked in a micromanipulator or perfusion micropipette. This micropipette is partially blocked within one millimetre of the tip by a sealed glass rod. The rapid flow of protoplast suspension through the partially sealed mouth creates a mechanical force. This force induces adhesion of adjacent protoplasts highly. As a result, the two or many protoplasts may fuse together and form hybrids. This kind of fusion is not dependent upon the presence of any fusogen and has been commonly observed in soybean, *Arachis* and *Vinca rosea*. Occasional fusion of protoplasts has been observed by this technique. However, protoplasts are likely to get injured by this procedure.

MAJOR FUSOGENS

Fusion can be induced by employing chemical agents called fusogens such as, sodium nitrate, high calcium ions, polyethylene glycol (PEG), dextran, etc. Some of the commonly used fusogens which have been employed in protoplast fusion technology are described as follows.

Polyethylene Glycol (PEG)

PEG, as a fusion-inducing agent, was first introduced by Kao and Michayluk in 1974. This chemical gives a high frequency of fusion with reproducible results and involves low toxicity. This method can be used for fusion of protoplasts from unrelated plant taxa (for example, soybean–tobacco, soybean–maize and soybean–barley), from unrelated animal taxa and also between those from animal and plant cells. PEG causes rapid and tight agglutination of protoplasts which leads to its widespread use in protoplasmic fusion. The most effective molecular size ranges between PEG 1540 (MW 1300–1600 daltons), PEG 4000 (MW 3000–7000 daltons) and PEG 6000 (MW 6000–7500 daltons). Demonized PEG promotes better heterokaryon formation than the commercial grades. Also, on the basis of requirements of protoplasts quantity during fusion treatment, protoplast fusion method is conveniently grouped into two categories:

1. When protoplasts are available in sufficient quantities, one ml of the culture medium with suspended protoplasts is added to one ml of 56% solution of PEG and the tube is shaken for 10 seconds. The protoplasts are allowed to sediment for 15 minutes followed by washing with liquid growth medium and are finally examined for successful agglutination and fusion.

2. If the protoplasts are in micro-quantities, drop cultures can be used. This method involves the following steps:
   i. Two types of protoplasts are mixed together in equal quantities and 4–6 micro-drops (100 µl each) are kept in a small Petri plate and allowed to settle for 5 to 10 minutes at room temperature.
   ii. Two to three micro-drops of PEG are added from the periphery in each Petri plate, followed by further addition of PEG from the centre of each Petri plate and left for 20 minutes.
   iii. One ml of 0.7 M mannitol solution is added to dilute the PEG solution. One side of the Petri plate is slightly raised and the protoplast clumps adhering to the surface of the plate are washed with 10 ml of 0.7 M mannitol.
   iv. Finally, PEG and mannitol residues are removed and protoplasts are observed under the inverted microscope.

PEG manipulation and protoplast fusion The performance or frequency of PEG-mediated protoplast fusion can be enhanced several-fold by the following modifications:

1. PEG + 15% dimethyl sulfoxide (DMSO) is more effective than PEG alone.
2. Addition of concanavalin A (Con A) to PEG increases fusion frequency of protoplasts.
3. Polyvinyl alcohol (PVA) and PEG, being non-toxic surfactants, promote cell adhesion–fusion activity.

Sodium Nitrate/Potassium Nitrate

Kustor, in 1909, for the first time, used sodium nitrate for the fusion of onion protoplasts. Further in 1937, Michel demonstrated the use of potassium nitrate in protoplast fusion. These chemicals favour the fusion tendencies in between the protoplasts in substantial frequencies and, therefore, could be used in inter- and intraspecific fusions. The method requires 5 ml of 5.5% sodium nitrate in 10% sucrose solution as an aggregation mixture. The suspended protoplasts are kept at 35°C for 5 minutes and then centrifuged at 200 g for 5 minutes. The pellet is kept for incubation at 30°C for 30 minutes to fuse the protoplasts. Finally, the protoplasts are washed with liquid culture medium by repeated centrifugation. Protoplasts isolated from young immature parts of the plant give better results in the fusion experiments than the protoplasts derived from mature cells.

Carlson *et al.* used $NaNO_3$ to fuse the protoplasts of *Nicotiana glauca* and *N. langsdorffii* and released the first somatic hybrid in 1972.

Calcium and High pH

This method was developed by Keller and Milcher in 1973, to fuse *Nicotiana* protoplasts. The method involves centrifugation of equal quantities of protoplasts in 0.05 M $CaCl_2 \cdot 2H_2O$ in 0.4 M mannitol with pH 10.5 for 30 minutes at 50 g, after which the tubes are placed in water bath (at 37°C) for 40 minutes. About 20–50% protoplasts are responsive in fusion technique. This protoplast fusion approach is succesful only in some cases, because high pH has been proved to be toxic in most of the cases. Moreover, in *Glycine* and *Medicago,* protoplast fusion could be promoted by Ca^{2+} ion at concentrations of 10 or 50 mM. This method has been used to bring about many intraspecific and interspecific protoplast fusions.

Electrofusion

If protoplasts are placed in a small culture vessel containing two glass capillary micro-electrodes and a mild electrical charge (5 V, electrode distance 200 μm, 10 kVm^{-1}) is applied, it impinges on the plasma membranes causing alteration, leading to fusion (Figure 17.1). This method can fuse both intra- and interspecific protoplasts and nearly 100% efficiency of protoplast fusion can be achieved without loss of viability. This method was successfully applied by Zimmermann and Scheurich in 1981. Subsequently, application of high intensity electric impulse (100 kVm^{-1}) for some microseconds results in the electric breakdown of membrane and subsequent fusion of protoplasts. Enhanced cell division and shoot regeneration are reported with electroporated protoplasts in *Prunus* and *Solanum dulcamara.*

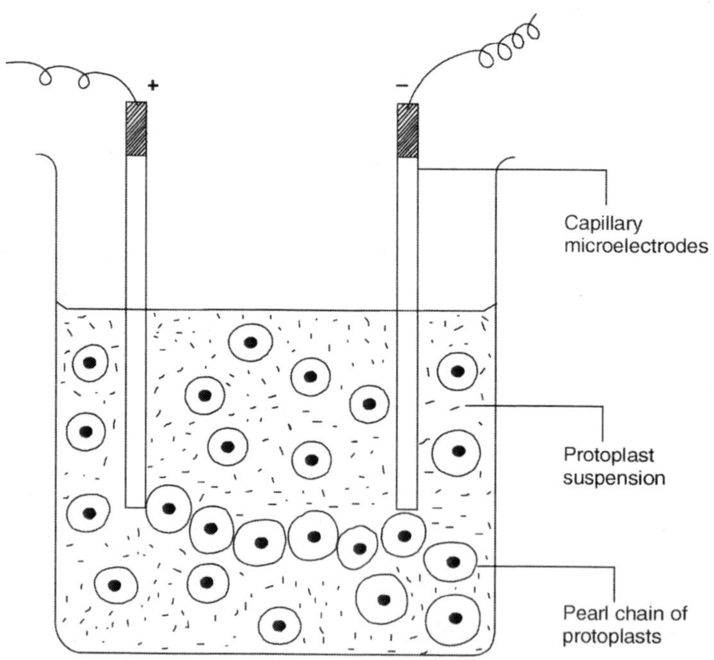

Figure 17.1 Electrofusion of protoplasts technique

Liposome

Liposomes are the artificially prepared lipid bags, which can be utilized to transfer small molecular weight compounds into protoplasts. The protoplast suspensions are treated with a vegetable oil and shaken well for a few minutes. During this time a lipid layer develops around the protoplast to form liposomes. The lipid coated protoplasts are then induced to fuse together by PEG treatment. This experiment has been successfully conducted with the protoplasts of *Catharanthus roseus,* in presence of PEG.

Other Fusogens

Some other chemicals are also used for the fusion of plant protoplasts. These are:

1. Dextran (potassium dextran sulphate) and gelatin (2.5%).
2. Polyvinyl alcohol (15%) solution in combination with 0.5% $CaCl_2$ and 0.3 M mannitol.
3. A mixture of concanavalin A, cytochalasin B, poly L-lysine and glycerol improves the yield and quality of protoplasts in fusion experiments.

Any one of the several methods discussed above could be adopted for protoplast fusion. But PEG, when used along with high Ca^{++} is found to be an efficient method for protoplast fusion and is widely accepted.

IDENTIFICATION OF HYBRID PROTOPLASTS

After the fusion of protoplasts, the protoplasts have to be identified from the mixture of parental types, homokaryons and heterokaryons. Here, true hybrids constitute about 0.5–1.0% only. The preliminary identification is based on microscopic observation; manual isolation is possible only in case where the parental protoplasts show distinguishing physical characters, like protoplasts containing chloroplasts and another showing anthocyanin-containing vacuole, but devoid of chloroplasts, or cell protoplasts that are achlorophyllous.

The fusion product could be identified on the basis of the presence of both chloroplasts and anthocyanin, and being binucleate. The protoplast of flower petal mesophyll can readily be identified. Another method is also available which is used when both types of protoplasts are similar. A hybrid cell contains two nuclei of two different parental protoplasts. Such cells can be identified using aceto-orcein or aceto-carmine nuclei staining. Carbol fuchsin stain is also utilized to identify hybrid protoplasts. Use of fluorescent compounds is also a good method to identify the heterokaryon. Fluorescein isothiocyanate or rhodamine isothiocyanate or rhodamine B is used as fluorochromes. The advantage of using fluorochromes for the identification of fusion products is that it does not depend upon the types of protoplast being used.

SELECTION OF HYBRID CELLS

In general, about 20–25% of protoplasts in the suspension undergo fusion to form heterokaryons and the rest of them remain homokaryons and unfused protoplasts. Therefore, there is a heterogeneous population of protoplasts in the suspension immediately after fusion induction. This protoplast suspension consists of the following cell types:

1. Unfused protoplasts of the two species/strains.
2. Products of fusion between two or more protoplasts of the same species (homokaryons).
3. "Hybrid" protoplasts produced by fusion between one (or more) protoplast(s) of each of the two species (heterokaryons) (Figure 17.2).

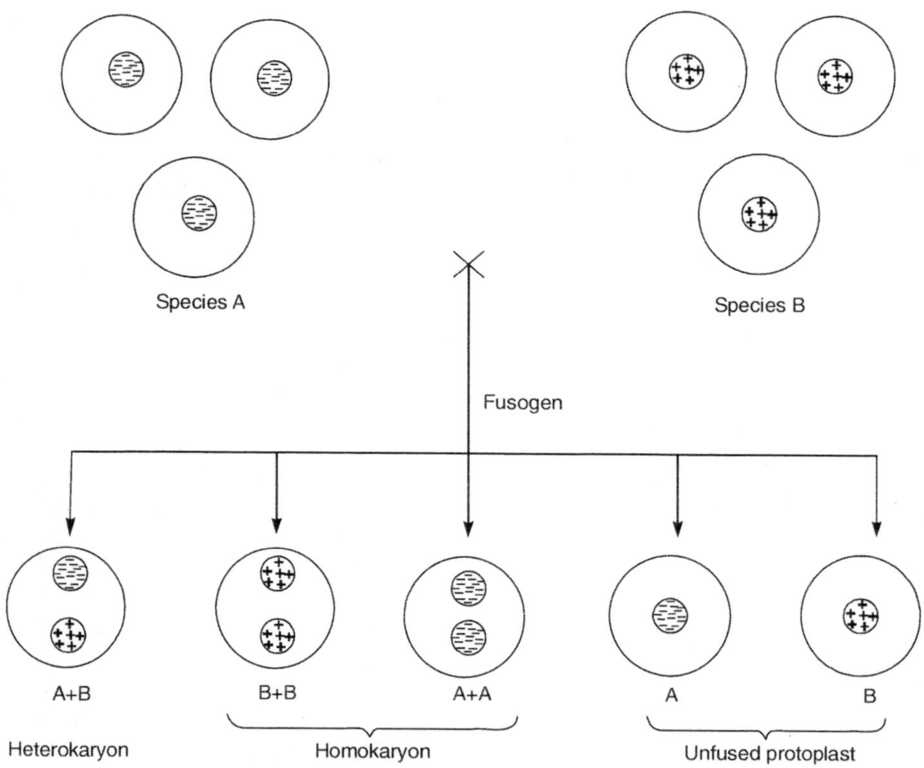

Figure 17.2 Protoplast fusion products

In somatic hybridization experiments, only the heterokaryon or hybrid protoplasts, particularly those resulting from fusion between one protoplast of each of the two species, are of main interest. However, they form only a small proportion in the population (usually 0.5–1.0%) of protoplasts.

Therefore, an effective strategy has to be employed for the identification and isolation of hybrid cells. This step is called the selection of hybrid cells and is the most critical stage of protoplast fusion technology and still it is an active area of investigation.

Methods of Hybrid Selection

The hybrid protoplasts can be screened from the heterogeneous population of protoplasts. Several screening methods have been developed. Some of the important methods are given below.

1. ***Visual selection*** Protoplasts of some plants show direct colours due to the presence of specific pigments. This method can be used to select fusion products of protoplasts that have distinct physical characters. For example, fusion between leaf mesophyll protoplasts of green colour and a cultured cell that has an anthocyanin in the protoplast, will produce hybrids that have both chloroplast and an anthocyanin vacuole together with two nuclei. These hybrid protoplasts are selected from others by visual tests.
2. ***Auxin autotrophy*** This method was developed by Carlson in 1972 in which hybrid protoplasts were selected from unfused protoplasts of *Nicotiana glauca* and *N. langsdorffii*. Protoplasts of mutant strains of these two species of tobacco cannot grow in a medium lacking growth hormones. However, the hybrid protoplasts grow well in the medium even in the absence of auxins due to genetic complementation. Hybrid protoplasts could be cultured on auxin-free culture medium in contrast to its parent protoplasts.
3. ***Selection by the use of antibiotics*** Antibiotics like actinomycin D have been used in the selection of somatic hybrids of *Petunia parodii* and *P. hybrida*. Generally, the cells of *P. hybrida* fail to grow in the presence of actinomycin D. Similarly kanamycin is used to select the hybrid of *N. sylvestris* and *N. knightiana*. Cycloheximide can also be used to select the hybrid of *Daucus* sp.
4. ***Use of herbicides*** Some plant species are sensitive to herbicides and some are resistant to these chemicals. This property can be utilized effectively for selection. Rice plants are resistant to 3,4-dichloropropionilide. Similarly, monocot tissues are relatively resistant to 2,4-dichlorophenoxyacetic acid in comparison to dicot tissues.
5. ***Use of fluorescent labels*** Non-toxic fluorescent labelled dyes can be used to select hybrid cells. Protoplasts from different strains are first incubated in octadecanoyl aminofluorescein and octadecanoyl palamine B and then allowed to fuse. After fusion of the protoplasts, fusion products are selected by their fluorescence characteristics.
6. ***Selection of hybrid callus/plants*** This is a more common and widely applicable technique. In this method, the entire population of protoplasts is initially subjected to culture on nutrient medium without applying any selection strategy for the hybrid cell. As a result, all types of protoplasts (hybrid/non-hybrid) form calli; the hybrid calli are later selected on the basis of callus morphology,

chromosome constitution, protein and enzyme banding patterns, etc. The identification/selection may be delayed till plants are regenerated and then tissues may be utilized for the selection of hybrid plants.

7. ***Use of mycotoxin*** The protoplasts of cultured soybean cells are resistant to HMT toxin (produced by *Helminthosporium maydis* T), whereas the leaf protoplasts of *Zea mays* are sensitive to this toxin. It has been observed that fusion products of soybean-*Zea mays* are efficiently able to survive on HMT toxin-containing medium.
8. ***Use of density gradient fractionation technique*** Protoplast suspension is suspended in an osmotic solution containing equal volumes of 0.35 M KCl, 0.245 M $MgCl_2$ and 0.254 M $CaCl_2$ at pH 6.0. Gradients are centrifuged at 20°C for 5 minutes at 100 g. The fused protoplasts will form a band in the intermediate density position.

REGENERATION OF SOMATIC HYBRIDS

After the callus formation from hybrid protoplasts, plants are induced to regenerate from calli cells, since this is a prerequisite for their exploitation in plant improvement. Nearly 150 plant species have been successfully regenerated from single cell or protoplast. Regeneration of plants from isolated protoplasts is possible either by organogenesis or by somatic embryogenesis. Regeneration by and large, has been achieved through organogenesis, although somatic embryogenesis has also been induced in some species.

Organogenesis starts from the callus culture under appropriate conditions of plant inoculum, medium and environmental factors. Similarly, embryogenesis can also be induced from a small group of superficial cells with dense cytoplasm and large starch grains. The developing embryos *in vitro* then pass through a sequence of growth stages that is exactly the same as seen in the development of a seed. Regeneration of plantlets through protoplast culture could be possible in the majority of the dicotyledonous plants and however, it has been proved successful to a limited extent in monocots.

A fused protoplast cell undergoes mitotic divisions and ultimately form callus tissues. Complete hybrid plant can be regenerated from such callus tissues. But plant regeneration has so far been achieved successfully only in a small number of plant species and moreover, it remains confined to some interspecific sexually compatible species only. Moreover, the well-developed callus tissue is transferred to another medium for callus proliferation. On transfer to an agarified medium with reduced concentration of glutamine (200 mg/l), rooting occurs, followed by differentiation of shoots. When BAP (4.0 mg/l) is substituted for zeatin and adenine (40 mg/l), whole plant morphogenesis occurs in tissues derived from hybrid protoplasts. On the other hand, when tissues are transferred to a medium devoid of the growth regulators, it differentiates into embryoids of different developmental patterns.

Embryos that are of globular to elongate and other similar shapes are typical of monocots. Whole plant development occurs on transfer to a semi-solid medium containing zeatin and IAA. The plantlets regenerated through hybrid protoplast culture often have thin cuticle and root hairs are poorly developed. If these plants without acclimatization are directly shifted to a situation where they are subjected to water stress they would be unable to control their adequate water balance and will gradually die. So, these plants are first kept in greenhouse for acclimatization and finally transferred to the field.

Plant regeneration from protoplast-derived callus tissues have been reported mainly from Solanaceae species. It includes 20 *Nicotiana* species, 8 *Petunia* species and 6 *Solanum* species.

Chromosome Status in Somatic Hybrids

The variability in chromosome numbers in hybrids could be due to any one of the following reasons:

1. Multiple fusions give higher chromosome numbers.
2. Asymmetric hybrids result from fusion of protoplasts isolated from actively dividing tissue of one parent and quiescent tissue of the other parent.
3. Unequal rates of DNA replication in two fusing partners.
4. Somaclonal variations in cultured cells used for protoplast isolation may also lead to variation in chromosome numbers.

Distant Somatic Hybrids and Types

In contrast to sexual hybrid cells (zygote), which contain the plasmogens from the female parent only, somatic hybrid cells obtain cytoplasmic components from both the parental species. The plasmogens appear to be distributed randomly during the mitotic cell divisions. As a result, some cells receive chloroplasts from one parental species and a small proportion of cells also retain the chloroplasts of both species. This is reflected in the plants regenerated from these cells. A somatic hybrid plant may contain chloroplasts from one parental species and mitochondria from the other fusion parent.

1. ***Symmetric or near-symmetric hybrids*** Somatic hybrid plants retain the full or nearly full somatic chromosome complement of the two parental species; these are called symmetric hybrids. Such hybrids provide unique opportunities for synthesizing novel species, which may be of theoretical and/or practical interest, for example, Pomato or Topato (Figure 17.3). Some examples of symmetric hybrids that have been produced successfully are:

   i. *Solanum tuberosum* × *Lycopersicon esculentum*

   ii. *Datura innoxia* × *Atropa belladonna*

   iii. *Arabidopsis thaliana* × *Brassica campestris*

   iv. *Atropa belladonna* × *Nicotiana chinensis*

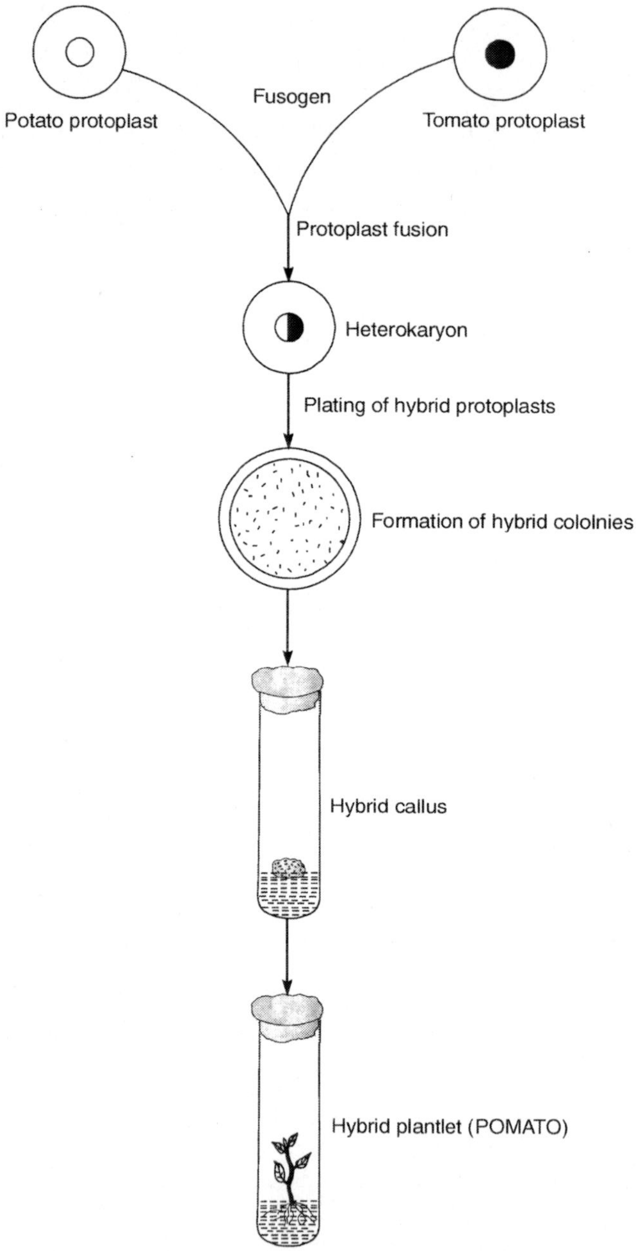

Figure 17.3 Formation of "POMATO" or "TOPATO" hybrid

2. ***Asymmetric hybrids*** Many somatic hybrids exhibit the full somatic complement of only one parental species, while all or nearly all chromosomes of the other parental species are lost during the preceding mitotic divisions; such hybrids are referred to as asymmetric hybrids. Some examples of such types of hybrids are:
    i. *Daucus carota* × *Aegopodium podagraria*
    ii. *Daucus carota* × *Petroselinum hortense*
    iii. *Hyoscyamus muticus* × *Nicotiana tabacum*
    iv. *Datura innoxia* × *Physalis minima*
    v. *Nicotiana tabacum* × *Daucus carota*

The available evidences suggest that such hybrids are likely to show introgression of chromosome segments from the eliminated genome(s). Asymmetric hybrids can be obtained even from those combinations which normally produce symmetric hybrids. In such approach, protoplasts of one of the parental species are irradiated with a suitable dose of X-rays or gamma-rays to induce extensive chromosome breakage. In such cases, chromosome segment introgression may be markedly enhanced.

APPLICATIONS OF PROTOPLAST FUSION

Protoplast fusion and somatic hybridization have opened up a new field in plant science, like:

1. Protoplast fusion provides a method of combining the different genomes of different genera and species, with the potential to overcome sexual incompatibility barrier between plants.
2. Protoplast cloning is innovative and offers novel experimental possibilities of somatic hybridization by cell fusion and regeneration, uptake of macromolecules, study of the structure of plasma membrane and probably of host-parasite relationship.

LIMITATIONS OF PROTOPLAST FUSION TECHNOLOGY

1. This technique is successfully implemented for isolation, selection of hybrids and production of entire plant only in limited families. So far, the production of somatic hybrids of agronomically important plants has not been successful.
2. The pre-fertilization barriers between intergeneric or interkingdom crosses cannot be overcome by this method.
3. In some wide crosses, elimination of chromosomes from the hybrid cell is another limitation of somatic hybridization.
4. Even now 100% fusion between two different parental protoplasts is not possible.
5. In protoplast technology there is not yet an established or well-defined protocol for isolation of protoplast and further regeneration of plants which could be applicable to all tissues, particularly to cereals and grasses.

CYBRIDIZATION

Cybrids or cytoplasmic hybrids are cells or plants containing the nucleus of one parent species only, but cytoplasm from both the parental species. The protoplast contains both nucleus and cytoplasm but the cytoplast has no nucleus at all. Therefore, cybridization brings together the nucleus of protoplast and cytoplasm of protoplast and cytoplast. Cybrids were first produced by *Palletier et al.* (1983) by fusing the protoplasts of *Brassica napus* with cytoplasts of *B. campestris.* The objective of cybrid production is to combine the plasmagenes of one species with the nuclear and cytoplasmic genes of another species. But, the mitotic segregation of plasmagenes leads to the recovery of plants having plasmagenes of one or the other species only. They are produced in variable frequencies in normal protoplast fusion experiments because of:

1. Fusion of a normal protoplast of one species with an enucleate protoplast or a protoplast having an inactivated nucleus of the other species.
2. Elimination of the nucleus of one species from a normal heterokaryon or gradual elimination of the chromosomes of one species from a hybrid cell during the subsequent mitotic divisions.

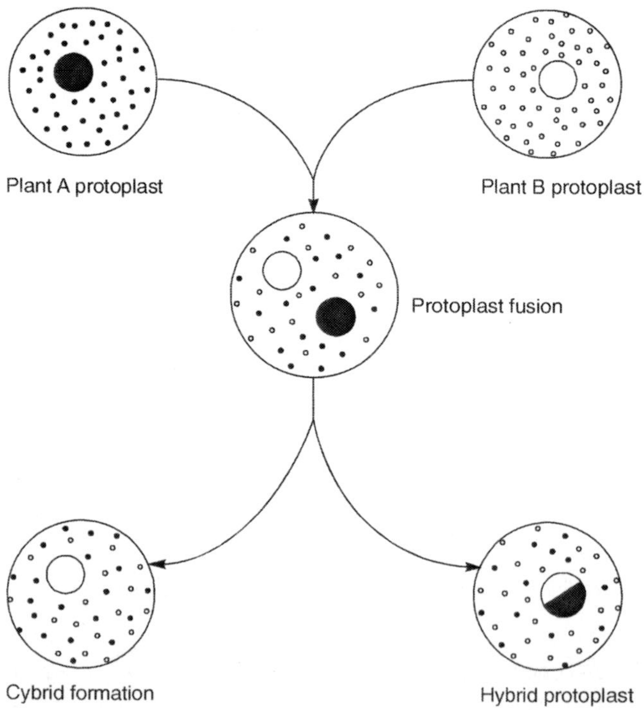

Figure 17.4 Fusion products of protoplasts (cybrid and hybrid)

During the fusion between protoplasts, it is possible that the two nuclei may first produce a true hybrid, but may remain separate in the mixed cytoplasm, thus forming a heterokaryon. If elimination of one of the nuclei occurs with retention of the other in a mixed cytoplasm (Figure 17.4), the resulting product is termed as a cybrid and the inheritance of cytoplasmic characters is strictly maternal. A cybrid is produced by the fusion of a cytoplast and protoplast. It is also known as cytoplasmic hybrid. It contains a single nucleus and cytoplasmic mixture of the cytoplasm and protoplasm.

Methods of Cybrid Production

Cybrids can be obtained by the following methods:

1. Protoplasts are isolated from two different plant species. In order to transfer herbicide resistance, protoplasts are isolated from herbicide-resistant spring oilseed rape and herbicide-sensitive winter oilseed rape and fused together. The resistant gene for the herbicide triazine is located in the chloroplast DNA of the spring oilseed rape, which gives the herbicide resistance nature of the produced cybrid.
2. Fusion of normal protoplasts from one parent with enucleated protoplasts from the other parent; enucleated protoplasts can be obtained by high-speed centrifugation (30000 g for 60–90 minutes) of protoplasts or by X-rays or gamma irradiation treatment.
3. Fusion of normal protoplast from one parent and protoplast containing non-functional nuclei from the other.
4. Selective elimination of chromosomes of one parent at a later stage after the fusion of the nuclei.

However, cybrids may be produced in relatively high frequency by the following methods:

1. Irradiating (with X-rays or gamma rays) the protoplasts of one species prior to fusion in order to inactivate their nuclei.
2. By preparing enucleate protoplasts (cytoplasts) of one species and fusing them with normal protoplasts of the other species.

Applications of Cybrids

Cybrids have several applications. Some important applications are as follows:

1. Transfer of plasmagenes of one species into the nuclear background of another species in a single generation, even in sexually incompatible combinations.
2. Recovery of recombinants between the parental mitochondrial or chloroplast DNAs (genomes).
3. Production of a wide variety of combinations of the parental and recombinant chloroplasts with the parental or recombinant mitochondria.

4. Transfer of cytoplasmic male sterility from *Nicotiana tabacum* to *N. sylvestris*, from *Petunia hybrida* to *P. axillaris*, etc.
5. Mitochondria from one parental species may be combined with the chloroplast of the other parental species.

REVIEW QUESTIONS

1. What is somatic hybridization? Describe the various methods of protoplast fusion.
2. What are fusogens? Discuss about their types and action.
3. Describe the methods of protoplast hybrid selection.
4. How are protoplast hybrids cultured?
5. What are the applications and limitations of protoplast fusion technology?
6. What are cybrids and how can they be produced?

REFERENCES

Carlson, P.S., Smith, H.H. and Dearing, R.D. (1973). Parasexual interspecific plant hybridization. *Proc. Nat. Acad. Sci.* 69: 2292–2294.

Glimelius, K., Eriksson, T. and Muller, A.J. (1991). Gene transfer via somatic hybridization in plants. *TIBTECH.* 9: 24–30.

Melchers, G., Sacristan, M.D. and Holder, A. (1978). Somatic hybrid plants of potato and tomato regenerated from fused protoplasts. *Carlsberg Res. Comm.* 43: 203–218.

Nomura, K. and Komamine, A. (1986). *In situ* hybridization on tissue sections. *Plant Tissue Cult. Lett.* 3: 92–94.

18

SOMACLONAL VARIATION: SOURCE AND SIGNIFICANCE

INTRODUCTION

Somaclonal variation is the term used to describe the variation observed in regenerated plants that have been produced by *in vitro* cell culture technique. Chromosomal rearrangements are possibly an important cause of such variations. It is the genetic variability which is expressed during cell or tissue culture.

Plants which regenerated from somatic cells during *in vitro* culture and expressed genetic variability with new heritable traits, are known as somaclonal variants. For a long time, the frequently observed variability in plant population has been ignored and normal plant regeneration from the cell culture was the ultimate objective for plant propagation. The name somaclonal variation was coined by Larkin and Scowcroft in 1981.

Somaclonal variation is not restricted to, but is particularly common in plants regenerated from callus. Somaclonal variants raised from calli are called calliclones, those raised from cells are called cellclones and those raised from protoplasts are called protoclones. If genetic variability is established in plants by culturing gametic cells, then it is said to be gametoclonal variation. The variations can be genotypic or phenotypic in nature, which in the latter case, can be of either genetic or epigenetic origin.

The mutations at the gene level are rarely reversible and these mutations may be passed from generation to generation. However, the epigenetic changes are transient in nature, and therefore, not transmissible to subsequent generations and that is why these changes have limited scope in crop improvement programme. It is also now well established that frequent genetic changes could result either from mutations and epigenetic changes or a combination of both processes. Genetic variations are irreversible and are likely to persist in the progeny of regenerated plants whereas epigenetic changes are not at all transmitted by sexual reproduction.

Genetic variability is present among cultured cells and such variability is expressed in plants derived from cultured cells and progeny of such plants are called somaclonal variants. Similarly, gametoclonal variation has been proposed for the variability present among pollen-derived cells/plants. Plant cell culture and regeneration of whole plants from cultured cells is essentially a criterion of vegetative propagation and theoretically should result in clonal uniformities.

Although conventional micropropagation has resulted to a large extent in clonal fidelity, it has become increasingly clear that under the appropriate culture conditions, a great deal of genetic variability can be recovered in regenerated plants. If cultures are established from explants that did not contain a preorganized meristem, or if cultures are maintained as callus prior to plant regeneration, the regenerated plants would be quite variable. Cultures initiated from even a single cell can show variation after repeated subculture. Distinct lines can be selected with their own particular morphology and physiology. It suggests that the *in vitro* culture contains a population of genotypes whose proportion can be altered by imposing an appropriate selection pressure. This variation can be transmitted to plants regenerated from the cell cultures. It provides an additional source of novel variation for exploitation by plant breeders.

In early reports, most of the variations were attributed to the readily detected chromosome instability of cultured plant cells. In many cases, the degree of instability was reported to be proportional to the length of time the cells remained in culture. Recognition of this spontaneous variation inherent in long-term culture led to the use of cell culture for mutagenesis and selection of genetic variants and for direct recovery of novel genotypes from cell cultures via somaclonal variation.

In theory, the regeneration of tissue explants into a mature plant should result in the production of clones of the parent plant, that is, plants which are identical in genotype and phenotype to the plant from which the explant was obtained. In practice, this is in fact the result observed in the vast majority of cases. However, it has long been recognized that occasional abnormalities occur, resulting in variant plants which are generally discussed as "artefacts" of the tissue culture process. In more recent years, it has become apparent that rather than being an unexplainable aberration in an otherwise uniform regeneration process, the appearance of variants in tissue culture may be a routine occurrence for certain types of plants and/or specific explant sources. Even more, it has been recognized that the existence of this phenomenon potentially provides a source of useful variation, which can form the basis of developing economically useful variant plant lines.

HISTORY

- S. K. Pillai in1969 observed variability in plant populations raised from tissue culture of *Geranium*.
- Krishnamurthi obtained disease-resistant variants of sugarcane from tissue culture in 1974.

- Indications of somaclonal variation in several crop plants have stimulated interest in application of this method for crop improvement. Shepard, Bidney and Shahin in 1980 developed potato variant plant lines. There are now several cases where somaclonal variations have produced agriculturally useful changes in the progeny, for example, eye-spot resistance or increased sugar yield in sugarcane and late blight resistance in potato.
- Ling *et al.* in 1985 screened a somaclonal variant of rice, resistant to *Xanthomonas oryzae* via tissue culture.
- Somaclonal variation may arise due to changes in chromosome number (aneuploidy and polyploidy) or changes in chromosome structure (deletions, duplications, translocations, inversions) or changes in DNA sequence (base mutations). It may also be due to gene amplification and gene methylation, which are considered to be epigenetic changes (Gould, 1986).
- The phenomenon of somaclonal variation has recently provided a source for achievement of variant plant lines. Evans and Sharp (2000) analysed somaclonal variations in a large number of plants regenerated from leaf explants of tomato.

EXISTENCE OF SOMACLONAL VARIATION

Somaclonal variation occurs both in monocots (*Allium, Oryza, Hordeum, Zea mays, Triticum, Saccharum,* etc.) and dicots (*Arachis, Solanum, Brassica, Lycopersicon, Nicotiana,* etc.). In general, somaclonal variation has been observed in plants regenerated from the culture of explant tissues of shoot tips, leaf base, embryos, apical meristem, flower buds, anthers, ovaries, seedlings, seeds, axillary buds, petioles, stolen tips, cotyledons, pith and stem. The regeneration source can be callus, protoplasts, adventitious shoots or androgenesis. It is known to occur both in asexual and sexual modes of transmission.

According to Gould (1986), the variations associated with the generating mechanisms are of three types:

1. *Cytogenetic variations* These changes are associated with polyploidy, aneuploidy, deletion, duplication, inversion, translocation, etc.
2. *Genetic variations* These are associated with point mutation where base sequence alteration is observed; the variations may also arise due to position effect, gene amplification, etc.
3. *Epigenetic variations* These are associated with DNA methylation due to cytokinin habituation and the variants are stress-resistant and this is due to physiological adaptations.

Somaclonal variation does provide the opportunity to generate new plant types that would be difficult or impossible to generate by conventional plant breeding.

SOMACLONAL VARIANTS VS. GM PLANTS

The concern over the safety of genetically modified organisms has forced plant geneticists to reconsider the importance of some older methods of plant improvement. Somaclonal variation, is heritable and results from *in vitro* procedures, and although the origin of somaclonal variation is not well understood, there are genetic mechanisms associated with it that have been studied in some detail. These range from gross chromosomal abnormalities to point mutations. But GM plants are created by introducing desired genes into plant cells.

Somaclonal variation during plant cell *in vitro* cultures often occurs at higher frequencies (up to 10% per cycle of regeneration) than chemical or radiation-induced mutation, making it a viable alternative to mutagenesis and a valuable tool for the plant geneticist to introduce variation into a breeding programme. Somaclonal variation has been implicated in many phenotypical changes in plants, and named cultivars have been released from somaclonal variation studies. The exploitation of somaclonal variation seems especially more applicable to older vegetatively propagated cultivars or clones which could be expected to have accumulated large numbers of mutant somatic cells. But genetic manipulation can be done in both vegetatively and sexually reproducing plants, which have some basic uniformity without expecting mutation in constituent cells.

Although recombinant DNA techniques offer promise for modification of crops, the paucity of the relative knowledge of plant genetics and biochemistry has delayed the development of recombinant DNA-based products using higher plants. However, somaclonal variation offers an opportunity to uncover the natural variability in plant cells and to use this genetic variability for producing new plant varieties. During the last few decades, even for the development of new crops or for the improvement in the characteristics of existing crops (for example, to obtain disease-free crops, etc.), the techniques of *in vitro* cell culture, in combination with genetic manipulation, have become important tools.

FACTORS CAUSING SOMACLONAL VARIATIONS

The amount of variations that can be expected *in vitro* will vary with the clone, age of the clone, use of mutagenic agents and use of selection pressure applied to single cells for stress conditions such as salt level, herbicides, microorganisms or their by-products and specific metabolites. Identification of possible somaclonal variants at an early stage of development is considered to be very useful for quality control in plant tissue culture, transgenic plant production and in the introduction of variants. Somaclonal variability often arises in tissue culture as a manifestation of epigenetic influence or changes in the genomic organization during differentiation of plants from callus culture under given *in vitro* conditions (viz. explant type, growth regulators, temperature and light conditions).

The effects of the type and concentration of plant growth regulators and frequency of subculture on somaclonal variation were studied in Cavendish banana cv. Zelig. *In vitro* grown plants at the fourth multiplication cycle were used for the investigation. Auxins (IAA, IBA and NAA) and cytokinins (BA and TDZ) were used to multiply shoots for 10 generations. DNA bands generated through RAPD-PCR were scored accordingly to analyse whether a particularly amplified DNA band was present or absent to determine the extent of somaclonal variation.

To induce mutations in cell population, the culture may be irradiated with UV rays or gamma rays or treated with chemical mutagens such as N-methyl-N-nitro-N-nitrosoguanidine/azide/EMS. These substances induce gene mutations to produce somatic variants.

The origins of somaclonal variation have been studied extensively but remain largely theoretical or unknown. Some types of somaclonal variation have been transient (epigenetic), whereas others have been either stable through repeated generations of asexual propagation or have proved to be genetically based and inherited in a Mendelian fashion. Some somaclones have been formally introduced as improved types of the original parental clone.

The types of variation which are frequently observed may differ from species to species, and it is often difficult to determine the genetic nature of the observed variation. One of the more frequent types of variation is a difference in chromosome number. Chromosomal changes are known to occur in high frequency in the early stages of callus or liquid cell cultures, and therefore, the occurrence of such abnormalities is not particularly surprising. It is possible, however, to select an appropriate culture medium which will enhance chromosomal stability in the explants used.

FREQUENCY OF SOMACLONAL VARIATIONS

Although original reports of somaclonal variation were limited to plants which were normally asexually propagated, it is now known that these techniques can be successfully applied to sexually propagated plants as well. It is easy to identify the right type of stably inherited variation, such that a tool can be extremely useful in establishing improved cultivars in a wide range of different plants. Significantly, it appears to be possible to induce variation to some extent by manipulation of the culture medium (Figure 18.1).

However, it is not possible to predict what type of somaclones, if any, would be obtained by the development of a variant of a particular type, or whether the produced somaclones will have any value. In other words, the scientists have limited option to know what randomly occurring variation might appear in culture; the possibility always exists, therefore, no agriculturally useful variants will appear.

Eukaryotic genomes are in a dynamic state of flux. This is most apparent in higher plants under *in vitro* culture where the amount of variability generated is extensive. Genetic and molecular analysis provide some understanding of the events which give rise to somaclonal variation.

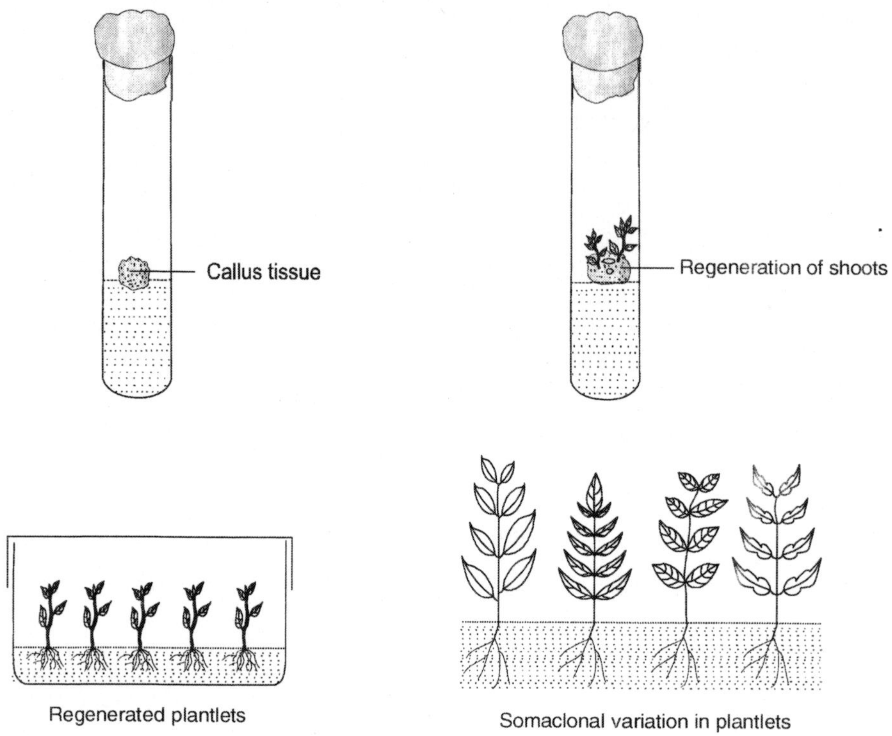

Figure 18.1 Diagramatic representation of steps in the induction of somaclonal variation in plants

SELECTION OF SOMACLONAL VARIANTS

For the selection of useful genetic somaclonal variation, two approaches have been followed:

1. Selection is exercised in cells cultured for different periods and screened for the derived traits (for example, for resistance to specific herbicides, fungal toxins, pollutants, extremes of temperatures and salinity) and from these selected cells or cultures, plants are regenerated.
2. In the second approach, selection is exercised at the phenotypic level in regenerated plants.

The former approach of selection at the cellular level has the advantage of screening millions of cells with a relatively little effort and resources. But this method cannot be used for traits like flower characters or grain yield.

APPLICATIONS OF SOMACLONAL VARIATIONS

Somaclonal variation has been suggested as a useful source in breeding programme. Some desirable traits have been raised from tissue culture and the adaptive quality of the traits may be permanently achieved through somaclonal variation mechanism. The major likely benefit of somaclonal variation is in plant improvement. It leads to the creation of additional genetic variability. Characteristics for which somaclonal mutants can be enriched during *in vitro* culture include resistance to disease, pathotoxins, herbicides and tolerance to environmental or chemical stress, as well as for increased production of secondary metabolites.

1. *Rice* The somaclonal variation showed significant improvement in the somaclone progeny relative to the parent in terms of seed weight, seed protein percentage, tiller number and time of flowering. The variety tolerant to salinity and aluminium toxicity has also been produced through this process and these have been segregated for characters such as plant height, maturity, heading date and grain yield.
2. *Wheat* Heterozygous and homozygous mutants were screened in the primary regenerants of wheat. In wheat, variants have been analysed for traits such as height and alcohol dehydrogenase synthesis, as well as complex gene loci involved in the synthesis of grain amylases.
3. *Potato* The usefulness of somaclonal variation in potato improvement programmes was first demonstrated through the recovery of disease-resistant plants in potato (resistance against late blight and early blight). The disease-resistant character may be transmitted through subsequent tuber generations.
4. *Sugarcane* Resistant varieties of sugarcane against eye-spot disease, Fiji disease and downy mildew are somaclones obtained from tissue cultures.
5. *Maize* T-toxin resistance, male fertility and mtDNA, etc. were modified in maize by somaclonal variations. Cytoplasmic male sterility and sensitivity to toxin produced by pathogens on maize may be overcome through regeneration of plantlets from the selected cell lines resistant to toxin and infection. Toxin-resistant male-fertile plants may be regenerated from cms-T maize culture grown without exposure to toxin. All this conversion is associated with mitochondria, which undergo genetic changes during cell division. In maize, a new fully functional electrophoretic variant at the alcohol dehydrogenase locus has been screened and subsequently characterized as resulting from a single nucleotide substitution.
6. *Tomato* The callus tissue of tomato, derived from leaf culture was used for somaclonal variation. Variants were screened for a number of characters such as male sterility, joint-less pedicel, fruit colour, indeterminate growth, leaf morphology, branching habit, etc.
7. *Brassica* sp. The occurrence of somaclonal variation in *Brassica* sp. has been independently reported several times. Variants were found which affected flowering time, growth habit, waxiness, glucosinolates, etc.
8. *Pelargonium* sp. A high degree of variability in culture-regenerated plants from six cultivars of *Pelargonium zonale* was observed. Changes were found in plant and organ size, leaf and flower

morphology, essential oil constituents, etc. One of the variants has been released as a new cultivar known as "Velvet Rose".

9. *Production of salt-resistant plants* Plant cell lines resistant to 4–5 times the toxic levels of the normal salt concentration have been isolated. In many cases, the plants regenerated from them were also tolerant to saline conditions. Tobacco plants regenerated from high salt-tolerant cell lines were also salt-tolerant and this feature was passed on to two successive sexual generations. This approach may permit the development of crop varieties suitable for cultivation in saline soils.
10. *Production of low-temperature-resistant varieties* Temperature is another important abiotic factor affecting survival and performance of crop plants. Cell lines resistant to chilling have been isolated in several cases, for example, chillies, *Nicotiana,* etc.
11. *Heavy-metal-resistant varieties* In some soils, an excess of heavy metal ions, such as aluminium, nickel, cobalt and molybdenum, may adversely affect the performance of crop varieties. Varieties suitable for cultivation on soil containing such heavy metal may be developed by selecting cell lines resistant to the concerned ion. The feasibility of this approach has been amply demonstrated by the selection of aluminium-resistant cell lines of N. *plumbaginifolia* and regeneration of resistant plants from these lines.
12. *Phosphate ultra-efficient plants* Somaclonal variant tomato cell lines, which are able to grow normally under phosphate-deficient condition due to a high secretion of acid phosphatase, and which greatly increases the rate of phosphate uptake were isolated.
13. *Activation of transposable elements* Somaclonal analysis in maize and alfalfa has shown that cell culture greatly enhances the activation of transposable elements, which ultimately leads to expression of somaclonal variation.

LIMITATIONS OF SOMACLONAL VARIATION

A serious disadvantage of somaclonal variation is that operations which require clonal uniformity, as in the horticulture and forestry industries where tissue culture is employed for rapid propagation of elite genotypes, are difficult to achieve. In micropropagation programme, where it is of paramount importance to produce true-to-type planting material, somaclonal variation of any kind is undesirable. Thus, it interferes with the successful performance of such industries.

STEPS TO AVOID SOMACLONAL VARIATION

As mentioned earlier, somaclonal variation is not good for true-to-type production experiments and industries. So, the following steps can be taken to avoid somaclonal variation in tissue culture:

1. It is well known that increasing numbers of subculture increases the likelihood of somaclonal variation, so the number of subcultures in micropropagation protocols should be kept to a minimal level.
2. Regular re-initiation of clones from new explants might reduce variability over time.
3. 2,4-D presence should be avoided in the culture medium for a longer period, as this hormone is known to introduce variations in cultures.

REVIEW QUESTIONS

1. What is somaclonal variation? How can it be induced?
2. Discuss the potential of somaclonal variation in crop improvement.
3. Discuss the frequency, advantages and limitations of somaclonal variation.
4. What are the applications of somaclonal variations?
5. Explain the factors that cause somaclonal variations.

REFERENCES

Burza W. and Malepszy, S. (1995). Direct plant regeneration from leaf explants in cucumber (*Cucumis sativus* L.) is free of stable genetic variation. *Plant Breeding.* 114: 341–345.

Donovan, A., Collin, H.A., Isaac, S. and Mortimer, A.M. (1994). Analysis of potential sources of variation in tissue-culture derived celery plants. *Annals of Applied Biology.* 124(2): 383–398.

Evans, D.A. and Sharp, W.R. (1986). Application of somaclonal variation. *Biotechnology.* 4: 528–532.

Gould, A.R. (1986). Factors controlling generation of variability *in vitro.* In: Vasil, I.K. (Ed.). *Cell Culture and Somatic Cell Genetics of Plants.* 3: 549–567.

Malepszy, S., Burza, W. and Smiech, S. (1996). Characterization of a cucumber (*Cucumis sativus* L.) somaclonal variant with paternal inheritance. *J. Appl. Genetics.* 37: 65–78.

19

BIODIVERSITY AND PRESERVATION OF GERMPLASM

Germplasm is the sum total of genetically variable stocks of a plant species. It is the source of the genetic potential of living organisms. Among other things, diversified germplasm allows organisms to adapt to changing environmental conditions. No single individual of any species contains all the genetic diversity of that species. This means that the total genetic potential is represented only in populations made up of many individuals. Such genetic potential is referred to as the gene pool. The potential represented in a gene pool is the foundation for our crop plants in both agriculture and forestry.

The limited number of plants that have historically fed the human population is approximately one percent of the flora of the world, and the numbers that have entered in agriculture is a small fraction of that percentage. As our human population has grown in number over the last 2000 years, and especially since the development of the science of genetics in this century, we have increasingly depended on a shorter list of the most productive and most easily stored and shipped crops. Today, only about 150 plant species with about one-quarter million local landraces are important in meeting the calorie needs of human beings. Extinction of a species or a genetic line represents the loss of a unique resource. This type of genetic and environmental impoverishment is irreversible. Throughout the world, people increasingly consume food, take medicine and employ industrial materials that owe their source to genetic resources of biological organisms.

Given the needs of the future, genetic resources can be reckoned among society's most valuable raw materials. Any reduction in the diversity of resources narrows the society's scope to respond to new problems and opportunities. If a new crop-threatening hazard (disease, insect, and so on) appears for which genetic resistance is unavailable in the breeders own germplasm, the breeder is forced to look elsewhere for the needed genes. In this event an active collection, if available, will be screened.

GERMPLASM STORAGE AND CONSERVATION

The preservation of plant biodiversity combats the risk of "genetic erosion". It refers to the the risk that the species and plant varieties may become extinct or move in the direction of extinction, producing a significant loss of the genetic variability which they contain. An alternative and more practical approach may involve the cultivation of more and more restricted number of selected cultivars. The establishment of in-field clonal collections (for vegetatively propagated species) and seed banks (for seed-propagated species) are some of the additional approaches to *ex situ* germplasm preservation.

Ex situ preservation of germplasm for higher plant species has been accomplished using either seeds or clones but storage of these under typical conditions does not provide the extreme longevity that is needed to minimize risk of loss. Germplasm is only maintained in living tissue, most often in the embryo of seeds. When the seed dies, the germplasm is lost. The best and most cost-effective way of protecting plants and animals is *in situ* (in their natural habitat), for example, in national parks or wildlife sanctuaries. However, it is often necessary to conserve many rare and endangered species *ex situ* (outside their natural habitat), for example, in institutions such as botanical gardens, zoos and aquaria, laboratories, etc.

One of the simplest ways to conserve plants *ex situ* is the seed. This is because seeds allow storage of more genetic diversity in a small space at low cost. However, some plants produce seeds that cannot be dried or stored at low temperatures. Similarly, there are also plants that cannot produce sufficient or viable seeds. These plants exist now only as male or female specimens which produce seeds either infrequently or very rarely. In these cases, the plants can be conserved in planted collections via cell culture or by cryopreservation techniques.

Conservation of plant genetic resources or germplasm by several national (National Bureau of Plant Genetic Resources (NBPGR)) and international organizations (International Board of Plant Genetic Resources (IBPGR), now renamed as International Plant Genetic Resources Institute (IPGRI)) has become a thrust area of biotechnology in recent years. For this purpose, various strategies for storage of germplasm without losing its viability have been developed and are being utilized. Germplasm can be stored in a variety of forms including seeds, buds, protoplasts, cells or tissues, well-organized plant parts like shoot tips, plantlets in culture, etc.

The techniques involving storage of tissues in culture may use the germplasm in the growing stage and their growth may be transiently suspended by the following methods:

1. Lowering the temperature
2. Adding chemical retardants or hormones
3. Reduction in oxygen concentration

Such methods, where limited growth of cultures is allowed, remain effective for about a year, so that periodic renewal of these cultures is required. However, the most popular and effective method for long-term storage of cell cultures involves storage at a very low temperature using liquid nitrogen.

GERMPLASM PRESERVATION

In India, Botanical Survey of India has released three volumes of *Red Data Book*, enlisting the threatened plants of the Indian subcontinent. There are several other agencies which are engaged in the protection and preservation of such plants. Some of the Indian organizations are Indian Council of Agricultural Research, New Delhi; National Bureau of Plant Genetic Resources, New Delhi; Council of Scientific and Industrial Research, New Delhi and Botany Departments of various Universities.

Attempts have been made to conserve the germplasm by preserving the genetic material in the following ways:

1. *In situ* preservation
2. *Ex situ* preservation

In situ Preservation

Preservation of germplasm in their natural environment by establishing biosphere reserves, national parks, gene sanctuaries, etc. is called *in situ* preservation. The need for *in situ* conservation is a relatively recent event. One hundred years ago, humans were packed less densely, communication was localized, and our demands on our surroundings were less harsh. All useful plants and animals were maintained in the open world of fields and forest. Only recently we have engaged in *ex situ* conservation of crop plant seed in cold storage, seed banks or *in situ* preserves.

In situ conservation, can take two distinct but overlapping pathways:

1. *Entire biomes* It is the entire preservation of vast tracts with the *in situ* conservation of animals and plants. This is a continuum from vast tracts of wilderness, totally devoid of human interference, to unique habitats with considerable management and protection by humans. This level of preservation will be extremely important in slowing the species extinction rate but will have little impact on regenerating degraded habitats or habitats that human beings exploit as a resource base.
2. *Specific sites* Preservation of those useful plants and animals in a dynamic interaction that does not lead necessarily to a decreased carrying capacity of the human population. The essential elements are that the plants and animals are useful to human beings and the human density does not increase to the point where the resource base is degraded. In truth, these idealized systems will not be easy to achieve by *in situ* conservation, but it will at least decelerate their disappearance and give us more time to understand and appreciate how these systems evolved. Considerable potential exists for creative institutional arrangements including tourism for *in situ* conservation, especially for historical sites.

The limitation of this type of preservation is the risk of declination of the preserved species due to environmental hazards.

Ex situ Preservation

This is the chief mode of preservation of germplasm. Providing the suitable condition in the genebank preserves the genetic materials in the form of seed or *in vitro* cultures. But for being successful in the establishment of a gene bank, considerable knowledge of genetic structure as well as elements influencing them are necessary. Usually, seeds form the most common material to conserve plant germplasm in the seed-propagated plants. But this method has certain limitations like:

1. Loss of seed viability with passage of time.
2. Seed destruction due to seed-borne pathogens, pests, etc.
3. Being confined only to seed propagating plants.

In contrast to seed propagating plants, the vegetatively propagated plants are preserved *in vitro* as shoots, meristems, embryos, etc. The advantages of *in vitro* preservation over *in situ* preservation are as follows:

1. Large amount of material can be preserved in a small area.
2. It overcomes the destruction due to environmental hazards.
3. It provides large amounts of plant material to culture under *in vitro* conditions.

GERMPLASM CONSERVATION AT LOW TEMPERATURE

The maintenance of genetic stocks in living condition for a long time is known as germplasm storage or germplasm maintenance or germplasm conservation. Germplasm provides the raw materials, which the breeder uses to develop commercial crop varieties. Conventionally, germplasm is stored in the form of seeds because they occupy a relatively small space and can be stored for many years. But there are a number of important vegetatively propagated crops like potato, sweet potato, cassava, etc. where the conventional method is not applicable. It is now possible with modern *in vitro* cell culture techniques to provide a germplasm storage procedure which uniquely combines the possibilities of disease elimination and rapid clonal propagation.

Some techniques involve storage in growing stage whereas others relate to the suspension of growth. Growth inhibition may be by temperature reduction, use of growth-retardant chemicals or hormones, reduction in oxygen concentration, etc. These methods require periodic renewal after some intervals. Another method is cryopreservation which often can store plant materials for virtually indefinite periods. This technique preserves original genetic stocks in a limited area and ensures genetic stability of cryopreserved material.

The preservation of cells, tissues and organs by cooling to sub-zero temperatures, such as (typically) 77 K or –196°C (the boiling point of liquid nitrogen) in liquid nitrogen is called cryopreservation. Cell suspensions, hybrid protoplasts, pollen grains, seeds and meristems of desired plant species are stored in this method for establishment of germplasm bank. Liquid nitrogen maintains low temperature for long-term storage. The low temperature reduces the growth rate of the stored cells and delays ageing. Therefore, the plant materials do not lose their viability for a long time. After a long period of storage, plants can be regenerated from the cryopreserved plant materials.

CRYOBIOLOGY AND CRYOPRESERVATION

Cryobiology is the branch of biology that studies the effects of low temperatures on living things. The word cryobiology is derived from the Greek words *cryo,* meaning "cold", *bios,* meaning "life", and *logos* meaning "science". It means the science of life at low temperature. In practice, this field comprises the study of any biological material or system subjected to any temperature below normal (ranging from moderately hypothermic conditions to cryogenic temperatures).

Cryopreservation is the process of freezing living material to preserve it. This is normally done at or near the temperature of liquid nitrogen, –196°C (–320° F). Greek word *cryos* means "ice", "frost" or "cold" and *preserve* means "to keep safe from harm or losses". At this temperature, physical and metabolic cellular processes are effectively stopped and the living tissue is in a state of suspended animation, free from pathogens and the risk of genetic drift.

In theory, plant tissues can be stored at low temperatures for indefinite period and whenever required, it can be recovered and grown to regenerate a whole plant. Again, cryopreservation demands the least space and reduces susceptibility to disease, mutation and environmental conditions. At least four major areas of cryobiology can be identified, which are as follows:

1. Study of cold adaptation of microorganisms, plants, invertebrates and animals.
2. Cryopreservation of cells, tissues, gametes and embryos of plants, animals and humans for purposes of long-term storage. This usually requires the addition of substances (cryoprotectants) which protect the cells during freezing and thawing.
3. Preservation of organs under hypothermic conditions for transplantation.
4. Lyophilization (freeze-drying) of some pharmaceutically important materials.

History of Cryopreservation

The history of cryobiology can be traced back to antiquity. As early as in 2500 BC, low temperatures were used in medicine in Egypt. The use of cold was recommended by Hippocrates to stop bleeding and swelling. In 1949, sperm was cryopreserved for the first time by a team of scientists, led by Christopher Polge (1926–2006). This led to a much wider use of cryopreservation; today many organs, tissues and cells are routinely stored at very low temperatures. Large organs such as hearts are usually stored and transported temporarily at cool but not freezing temperature and moreover, freezing temperature is not required for transplantation. Cell suspensions and thin tissue sections can sometimes be stored almost indefinitely at liquid nitrogen temperature (cryopreservation). Human sperms, eggs and embryos are routinely stored for fertility research and treatment.

It was not until the 1970s that it was found that plant cells could be preserved through cryopreservation techniques. Significantly, one of the earliest records on use of cryopreservation technique in plants is of Sun (1958), who attempted to preserve desiccated seedlings of *Pisum sativum* at liquid nitrogen temperature.

With the emergence of tremendous developments in the areas of biological sciences, Robert Boyle studied in detail the effects of low temperatures on animal tissues and in the early twenty first century, a baby was born from a cryopreserved egg fertilized by a cryopreserved sperm. One of the most important early workers on the theory of cryopreservation was James Lovelock who suggested that damage to red blood cells during freezing was due to osmotic stresses.

Natural Cryopreservation

Polar bears, many microscopic multicellular organisms, etc. can survive even at low freezing temperatures by replacing most of their internal water with the sugar trehalose. Sugars and other solutes that do not easily crystallize, have the effect of limiting the stresses that damage cell membranes. Trehalose is a sugar that does not readily crystallize. Mixtures of solutes can achieve similar effects. Some solutes, including salts have the disadvantage that they may be toxic at high concentrations.

Basic Concepts of Cryopreservation

1. Liquid nitrogen is made by the fractional distillation of liquefied air.
2. Scientists believe that cells can be stored in liquid nitrogen for hundreds to thousands of years.
3. Some cryopreserved cells have been successfully revived after more than half a century.
4. Plant cryopreservation is based on the principle of totipotency that majority of living cells (undifferentiated as well as fully differentiated) has the inherent capacity to divide and develop into a complete plant.
5. Cryopreservation is now the routinely used technique to store reproductive cells from animals and humans for artificial insemination.

Each individual species reacts differently to cryopreservation and so a different protocol needs to be developed for each species prior to cryostorage. Once a protocol has been developed, cryopreservation offers an effective, inexpensive and secure method, which is independent of continued electrical supply for the long-term conservation of plants.

Materials for Cryopreservation

A variety of plant material can be used for cryopreservation, including cells or tissues of *in vitro* culture, pollen, seeds or parts of seeds, embryos, tissues from the early stages of development of mosses and ferns, buds, twigs and meristematic (growing-point) tissues. The preserved germplasm can be used whenever required. Pollen required for hybridization can also be cryopreserved. For cryopreservation, cells in early exponential phase have been found to be the most highly freeze-tolerant, while those in the lag phase or stationary phase are relatively susceptible to some freezing injury. The susceptibility to freezing injury actually depends on the size of individual cells and also on the size of aggregates or clumps used for cryopreservation. Callus cultures, cell suspensions, zygotic embryos (from wide crosses) and somatic embryos, can be used as suitable materials for cryopreservation.

Cryogenic storage has been widely used for microbes, animal cell cultures, desiccation-tolerant seeds, desiccation-tolerant pollen, *in vitro* cultures including cell suspensions and other hydrated plant parts such as meristems and zygotic embryos. Recently even embryos removed from desiccation or chilling-sensitive seeds have been preserved in liquid nitrogen. The success of cryopreservation is determined by the sample's ability to survive freezing and thawing manipulations.

Cryopreservation Technique

In general, cryopreservation is easier for thin samples and small clumps of individual cells, because these can be cooled more quickly and so require lower doses of toxic cryoprotectants. Nevertheless, suitable combinations of cryoprotectants and regimes of biological materials can be prepared. Cell suspensions, hybrid protoplasts, pollen grains, seeds and meristems of desired plant species are stored by this method for establishment of germplasm bank.

The technique of cryopreservation or freeze preservation (Figure 19.1) involves four steps:

1. Freezing of tissues
2. Storage of tissues
3. Thawing
4. Regeneration of frozen material

1 ***Freezing of tissues*** Freezing may be either slow (cooling rate 0.5–4°C per minute from 0°C to –100°C and then transferring to liquid nitrogen) or rapid (–1000°C per minute), or it may be initially slow, 5°C per minute up to –30 to –50°C, held at this temperature for about 30 minutes and then can be cooled rapidly by plunging the vials into liquid nitrogen.

i. ***Slow freezing method*** It is believed that slow freezing increases the concentration of cytoplasm and increased dehydration increases the survival of cells. There are different methods of obtaining protective dehydration, including slow cooling at constant or varying cooling rates, or keeping the samples at one or more intermediate subzero temperatures. In this method, the cooling rate is slow (1°–4°C per minute) and the cooling range starts from 0°C to –100°C after which the material is transferred to liquid nitrogen.

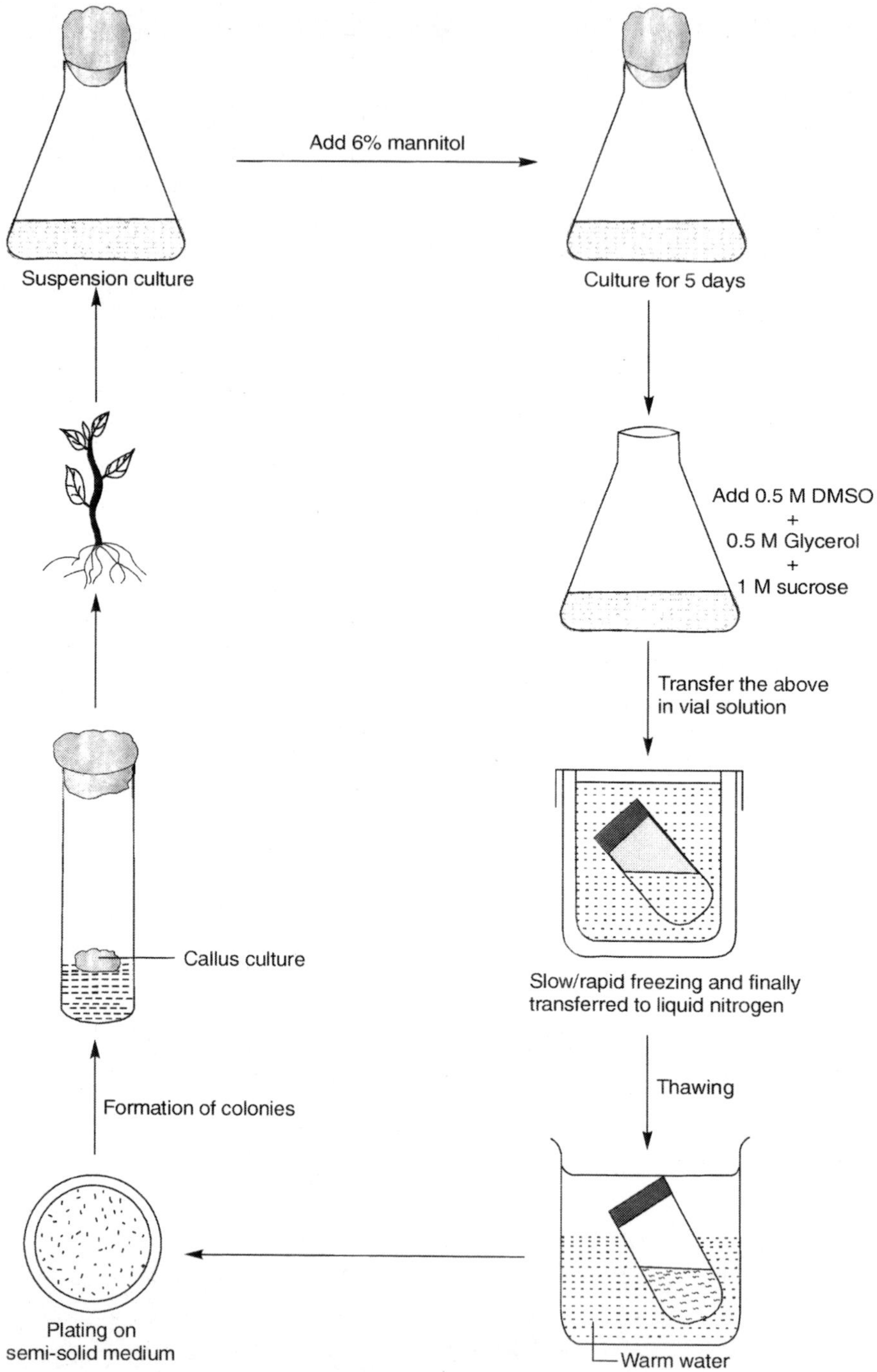

Figure 19.1 Cryopreservation and regeneration technique of cells

When plant cells are cooled progressively, ice crystal formation is usually initiated extracellularly. It is presumed that plasma membrane acts as a barrier that prevents the ice crystal formation in the cytoplasm. In the absence of ice crystals, cytoplasm remains supercooled. On further lowering the temperature, the concentration of extracellular liquid increases, as more water is converted to ice. Since the vapour pressure of the frozen solution is lower than the same concentration of the supercooled liquid, the vapour pressure of the slowly cooled cells reach an equilibrium with the external ice by efflux of water. Thus, slow freezing prevents intracellular ice formation and consequently freezing injury is prevented.

ii. ***Rapid freezing method*** Rapid freezing method is accomplished by direct immersion of the cryoprotectant-treated specimens in liquid nitrogen. The cooling rate in this method is very high (>1000°C per minute). At such high cooling rates, the intracellular fluids do not have sufficient time to equilibrate with the external ice with the possibility of intracellular ice formation, which is considered to be lethal for cells. This method is not suitable for the cryopreservation of cell cultures; it is employed to preserve shoot tips.

iii. ***Droplet freezing method*** In this method, the cryoprotectant-treated tissues are dispensed in droplets on an aluminium foil in a Petri dish. The specimens are frozen by slow cooling (1°C per minute) to a subzero temperature between –20 to –40°C prior to immersion in liquid nitrogen.

Freezing units Various types of cryostats and freezing units are available by which different rates of cooling can be easily regulated. The culture subjected to ultra cooling can be stored at –196°C in liquid nitrogen containers for various lengths of time and can be taken out and thawed when required.

2. ***Storage of tissues*** The cell culture is transferred to a small flask or beaker. The flask is insulated with a thick plastic cover to prevent damages by liquid nitrogen. The culture flask is then placed in a container containing liquid nitrogen. The container is then kept in a refrigerator. The liquid nitrogen keeps the temperature –196°C inside the culture vessel. At this temperature there is no metabolic activity in plant cells due to inactivation of cellular enzymes. This material can be stored for five years or more.

3. ***Thawing of frozen material*** Material can be kept stored in liquid nitrogen (–196°C) or in its vapour (–150°C). Rapid thawing is recommended for most cryopreservation methods. The basis of applying rapid thawing is to avoid the damaging ice recrystallization which may occur during slow warming. Thawing melts extracellular ice crystals formed during long-term storage. Immediately after melting of ice crystals, the culture flask is kept in a water bath.

   Thawing of the frozen materials is achieved by:

   i. Plunging the vials into water at 37–40°C (thawing rate of 500–750°C per minute) for 90 seconds.

ii. The material is then transferred into an ice bath till it is recultured. Thawing has to be rapid in order to avoid ice crystal formation.

Further, the post-thawing event involves the following steps:

- The liquefied cell biomass in cryovials is immediately transferred to a sterile centrifuge tube containing liquid nutrient medium with an osmotic agent (sucrose, sorbitol or mannitol) and an ethylene inhibitor or ethylene action inhibitor (silver thiosulphate).
- The contents are mixed gently and incubated on a shaker (120 rpm) for 30 minutes at room temperature.
- Then the cells are centrifuged and the pellet is transferred to two layers of sterile filter paper placed on a blotting paper (to aid in removal of excessive moisture from cells or cell biomass).
- Cells or cell biomass is incubated for two minutes.
- The upper filter paper with cells is transferred sequentially to solid nutrient medium containing osmotic agent at a decreased concentration from 0.8 M to 0.1 M sucrose (to bring about the osmotic adjustment of cells).

At each step, cells on a filter paper are incubated for 30 minutes to 1 hour and finally are transferred to a normal solid nutrient medium without the presence of any additional osmotic agent. This step of incubation should be repeated after 24 hour incubation in the dark at 25°C.

4. ***Regeneration of frozen culture material*** Reculture or regeneration means production of plants from the cryopreserved plant materials. Regeneration from cryopreserved specimens is the most reliable and accurate estimate of viability. The other technique used for vitality test of the cells such as fluorescein diacetate stain, triphenyl tetrazolium chloride test, etc. may provide a quicker method of testing cell viability. Further growth and regeneration of the cryopreserved cells after thawing are like normal cells. However, care should be taken in handling, plating and subculturing of such cells. Cryopreserved plant material may show some special requirements during reculture, like GA_3 treatment for the reculture of shoot tips of tomato, treatment with activated charcoal for the reculture of carrot plantlets, etc. Generally, meristematic cells survive better than mature differentiated cells during cryopreservation.

   Generally, the thawed material is washed several times to remove the cryoprotectants before it is recultured. But the recent trend is to reculture the material in the presence of the cryoprotectant either in diluted or undiluted state.

   i. The frozen cells or tissues may be recultured on filter paper discs kept on a suitable culture medium. The disc along with the cells is frequently transferred to fresh medium in order to dilute the cryoprotectant effects. Finally, cells are scraped off from the filtered disc and cultured directly on to the medium. At least in some species, this approach is dramatically more successful than others.

ii. Callus *de novo* growth is usually visible after one week of culture. When the new cell biomass growth is increased, callus cells are removed from the filter paper and are transferred directly to a normal solid nutrient medium.

iii. After sufficient growth (usually two to three weeks) of induced callus, a cell suspension in liquid medium is initiated from established callus.

iv. Materials subjected to cryopreservation may show some special requirements during reculture. For example, shoot tips excised from freeze preserved seedlings of tomato, require GA_3 for developing into shoots (otherwise, they form callus in the absence of GA_3), while normal shoot tips of tomato do not require GA_3.

Thus, it is necessary to determine the optimum conditions for reculture of different plant species, particularly when commonly used regimes fail. In general, meristematic cells survive better during freeze preservation than mature differentiated cells. The techniques for freeze preservation of shoot tips have been refined for germplasm storage for very long periods.

Cryoprotectants and Cryoprotection

Moist samples can survive freezing and thawing through natural processes, however, these are possible particularly in climates with low temperatures, by manipulating cellular moisture contents and freezing rates to reduce injury from desiccation or ice crystallization, or by applying chemical cryoprotectants.

A cryoprotectant is a substance that is used to protect biological tissues from freezing damage (damage due to ice formation). Arctic and Antarctic insects, fish, amphibians and reptiles create cryoprotectants in their bodies to minimize freezing damage during cold winter periods. Insects most often use sugar as cryoprotectants. Arctic frogs use glucose but Arctic salamanders store glycerol in their livers for use as cryoprotectants. Conventional cryoprotectants are glycols (alcohols containing at least two hydroxyl groups) such as ethylene glycol, propylene glycol and glycerol. Ethylene glycol is commonly used as automobile antifreeze and propylene glycol has been used to reduce ice formation in ice-cream. Dimethyl sulfoxide (DMSO) is also regarded as a conventional cryoprotectant. Glycerol is used as cryoprotectant for embryos that are cold-preserved in liquid nitrogen.

Common cryoprotectants are categorized into two groups:

1. Permeating, such as DMSO, methanol, glycerol, etc.
2. Non-permeating, such as sugars, sugar alcohols, high molecular weight dextran polymers, polyvinyl pyrrolidone, hydroxyl ethyl starch, etc.

Mixtures of cryoprotectants have less toxicity and are more effective than single cryoprotectant. These mixtures have been used for vitrification, that is, solidification without any crystal ice formation. Vitrification has important applications in preserving embryos, biological tissues and organs. Vitrification is also used in cryonics in an effort to eliminate freezing damage.

Common cryoprotectants

1. ***Dimethyl sulfoxide (DMSO)*** Dimethyl sulfoxide (DMSO) is the chemical compound with the formula $(CH_3)_2SO$. This colourless liquid is an important polar aprotic solvent that dissolves both polar and non-polar compounds and is miscible in a wide range of organic solvents as well as in water. It has a distinctive property of penetrating the skin very readily.
2. ***Ethylene glycol*** Ethylene glycol or monoethylene glycol (MEG) is an alcohol with two -OH groups (a diol), a chemical compound widely used as an automatic antifreeze. In its pure form, it is an odourless, colourless, syrupy, sweet and toxic liquid. Polyethylene glycol (PEG) is another example in this category.
3. ***Glycerol*** Glycerol is also commonly called glycerin. It is a colourless, odourless, viscous liquid that is widely used in pharmaceutical formulations. Glycerol has three hydrophilic hydroxyl groups that are responsible for its solubility in water and its hygroscopic nature. The glycerol substructure is a central component of many lipids. It is sweet-tasting and of low toxicity.
4. ***Propylene glycol*** Propylene glycol, also known by the systematic name, propane-1,2-diol, is an organic compound, usually a faintly sweet and colourless, clear viscous liquid that is hygroscopic and miscible with water, acetone and chloroform.
5. ***Trehalose*** Trehalose, also known as mycose, is a natural alpha-linked disaccharide formed by a α-1, 1-glycosidic bond between two α-glucose units. It can be synthesized by fungi, plants and invertebrate animals. It is implicated in anhydrobiosis. It has high water retention capabilities and is used in food and cosmetics.
6. ***Sucrose*** Sucrose, also called table sugar or saccharose, is a disaccharide of glucose and fructose, with the molecular formula $C_{12}H_{22}O_{11}$. Its systematic name is α-D-glucopyranosyl- (1-2)-β-D-fructofuranoside.

Functions of cryoprotectants Some cryoprotectants function by lowering a solution's or material's glass transition temperature. In this way, the cryoprotectants prevent actual freezing and the solution maintains some flexibility in a glassy phase. Many cryoprotectants also function by forming hydrogen bonds with biological molecules. Hydrogen bonding in aqueous solution is important for proper protein and DNA function. Thus, as the cryoprotectant replaces the water molecules, the biological material still retains its native physiological structure, although they are no longer immersed in an aqueous environment. This preservation is most often observed in anhydrobiosis.

Cryoprotectants are also used to preserve foods. These compounds are typically sugars that are inexpensive and do not pose any toxicity concerns. For example, many (raw) frozen chicken products contain a "solution" comprised of water, sucrose and sodium phosphates.

Removal of cryoprotectants Stored plant cells have cryoprotectants on the surface of cell walls. The cryoprotectants interfere with the growth of the cells in callus and plant regeneration. Generally, it is washed with distilled water. The cells are transferred to a nutrient medium and incubated at 25°C for callus induction.

Cryopreserved Plants

Most of the plant cells, tissues and organs which are cryopreserved and recovered successfully using this process include both gymnosperms and angiosperms (Table 19.1).

Table 19.1 Some cryopreserved plants

Gymnosperms	Angiosperms	
	Monocots	Dicots
Abies	*Avena*	*Achyrocline*
Cypressus	*Cocus*	*Atropa*
Ginkgo	*Dioscorea*	*Brassica*
Juniperus	*Hordeum*	*Berberis*
Picea	*Musa*	*Sophora*
Pinus	*Oryza*	*Lupinus*
Pseudotsuga	*Saccharum*	*Capsicum*
Sequoia	*Sorghum*	*Catharanthus*
Taxus	*Triticum*	*Conospermum*
Tsuga	*Zea*	*Datura*
Zamia		*Daucus*
Taxus brevifolia		*Digitalis*
T. canadensis		*Echinacea*
T. chinensis		*Eschscholzia*
T. cuspidata		*Glycine*
T. floridana		*Gossypium*
T. globosa		*Hyoscyamus*
T. media		*Lycopersicum*
T. nucifera		*Malus*
T. wallichiana		*Medicago*
		Nicotiana
		Panax
		Pisum
		Ruta
		Solanum
		Trichosanthes

Viability Estimation

Since ageing can be greatly slowed by cryopreservation, viability estimation generally focuses on injuries caused by freezing and thawing the samples, not on loss of viability during the storage period. Accordingly, viability in storage can be substantially tested more rapidly than can be done with most other systems. The main goal of cryopreservation research is to develop consistent protocols by manipulating natural hardening, tissue moisture content, rate of temperature change and chemical cryoprotectants to minimize injury. Germplasm viability cannot be assessed from samples while in storage; rather subsamples must be carefully thawed for testing. Then, established protocols for viability testing from conventional storage can be applied, for example, seed germination, pollen viability testing, *in vitro* tissue growth under defined conditions, etc.

Cryopreservation Equipment

Two pieces of equipment are needed to carry out cryopreservation. The first is a vitrification device and the second one is a storage unit:

1. *Vitrification device* Vitrification is a process of converting a material into a glass-like amorphous solid that is free from any crystalline structure, either by the quick removal or addition of heat or by mixing with an additive. Solidification of a vitreous solid occurs at the glass transition temperature (which is lower than melting temperature due to super-cooling). A partial dehydration of the material (in vacuum or by treating them with a specially formulated concentrated solution called plant vitrification solution, PVS) before freezing has been found to increase the survival of cells or tissues.

Vitrification is the heart of cryopreservation because it begins the freezing process and must be done properly so as not to damage or kill the cells. The devices that carry out vitrification are composed of a special low-temperature freezing unit, a control panel, and specimen holding chamber. The vitrification freezer is operated to bring the cells from their growing temperature of usually 37°C to a variety of temperature ranges spanning –20°C to –140°C. The freezing process is done in a two-step manner that has been shown through many studies to be safe for cells and whole organisms. A manual or computer-operated control panel regulates the temperature change. The control panel is hooked up to an electrical circuit that operates a pump-driven freezer. A special refrigerant liquid is pumped to the specimen-holding chamber. There are various types of freezing units that use either fluorocarbons or ethylene glycol refrigerants. Fluorocarbons are used in household freezers and ethylene glycol is found as a coolant in automobile radiators.

The vitrification unit's specimen-holding chamber comes in a variety of styles and sizes based on the types of specimens undergoing cryopreservation. Most research laboratories use small units to freeze tubes placed in long narrow tubes or miniature bottles. Larger units are used to freeze large volumes of cells or whole organisms. Companies manufacturing biotechnological products are likely to have very large cryopreservation facilities to handle litres of cells produced at commercial scale. Specimens placed in the chamber must be soaked in a special cryoprotectant before beginning the freezing process.

2. *Storage unit* The next component of the cryopreservation set-up is the storage unit. A typical storage unit is a container filled with liquid nitrogen. The storage unit is an insulated drum that traps the freezing of cold liquid nitrogen. Many cells are stored at temperatures from –78°C to –120°C. A cloud of water vapour appears when the storage container is opened because the liquid nitrogen is immediately converted to a cold gas that freezes the water in the atmosphere. Special deep freezers have been developed for holding cryopreservation specimens.

The cells must be stored under special conditions to prevent the freezers from dehydrating the specimen. Ultra-cold temperatures in the freezer dry out the air-holding area making it possible to evaporate the frozen contents from the cells. Dehydration is unlikely to happen in the wet environment of the liquid nitrogen. Many biotechnology companies have developed liquid nitrogen systems according to their facilities or needs because the storage units lose large amounts of liquid nitrogen every time they are opened. Even a closed store unit must release liquid nitrogen to prevent an explosion as the liquid nitrogen warms and expands into a gas.

LYOPHILIZATION OR FREEZE-DRYING

Lyophilization is a dehydration process typically used to preserve a perishable material or make the material more convenient for transport. Freeze-drying works by freezing the material and then reducing the surrounding pressure and adding enough heat to allow the frozen water in the material to sublime directly from the solid phase to gas.

Freeze-drying Protectants

Similar to cryoprotectants, some molecules protect freeze-dried material known as lyoprotectants. These molecules are typically polyhydroxy compounds such as sugars (mono, di and polysaccharides, polyalcohols and their derivatives). Trehalose and sucrose are natural lyoprotectants. Trehalose is produced by a variety of plants, fungi and animals that remain in a state of suspended animation during periods of drought (anhydrobiosis).

Mechanism of Freeze-Drying

There are three stages in the complete freeze-drying process:

1. *Freezing* The freezing process in a laboratory is often done by placing the material in a freeze-drying flask and rotating the flask in a bath called a shell freezer, which is cooled by mechanical refrigeration, dry ice and menthol or liquid nitrogen. On a larger scale, freezing is usually done by using a freeze-drying machine. In this step, it is important to cool the material below its eutectic point, the lowest temperature point at which the solid and liquid phase of the material can co-exist.

This ensures sublimation rather than melting. Larger crystals are easier to freeze-dry. To produce larger crystals, the product should be frozen slowly or can be cycled up and down in temperature. This cycling process is called annealing. However, in the case of food or objects with formerly living cells, larger ice crystals will break the cell walls. Usually, the freezing temperatures are between –50°C and –80°C. This freezing phase is the most critical in the whole freeze-drying process, because the product can get spoilt if it is not handled properly.

2. ***Primary drying*** During the primary drying phase, the pressure is lowered (to the range of a few millibars) and enough heat is supplied to the material for the water to sublimate. The amount of heat necessary can be calculated by using the sublimating molecules' latent heat of sublimation. In this initial drying phase about 95% of the water in the material is sublimated. This phase, should be slow, because if too much heat is added, the material's structure could be altered.

In this phase, pressure is controlled though the application of partial vacuum. The vacuum speeds sublimation, making it useful as a deliberate drying process. Furthermore, a cold condenser chamber and/or condenser plates provide a surface for the water vapour to resolidify. This condenser plays no role in keeping the material frozen; rather, it prevents water vapour from reaching the vacuum pump, which could degrade the pump's performance. Condenser temperatures are typically below –50°C.

It is important to note that in this range of pressure, heat is mainly brought by conduction or radiation and the convention effect can be considered as insignificant.

3. ***Secondary drying*** The secondary drying phase aims to sublimate the water molecules that are absorbed during the freezing process, since the mobile water molecules were sublimated in the primary drying phase. This part of the freeze-drying process is governed by the material's adsorption isotherms. In this phase, the temperature is raised higher than in the primary drying phase and can even be above 0°C, to break any physicochemical interactions that have formed between the water molecules and the frozen material. Usually, the pressure is also lowered in this stage to encourage sublimation. However, there are products that benefit from increased pressure as well.

Once the freeze-drying process is completed, the vacuum is usually broken with an inert gas, such as nitrogen, before the material is sealed. At the end of the operation, the final residual humidity in the product is around 1% to 4%, which is extremely low.

Properties of Freeze-Dried Products

1. If a freeze-dried substance is sealed to prevent the reabsorption of moisture, the substance may be stored at room temperature without refrigeration and be protected against spoilage for many years. Preservation is possible because the greatly reduced water content inhibits the action of microorganisms and enzymes that would normally spoil or degrade the substance.

2. Freeze-drying also causes less damage to the substance than other dehydration methods that use higher temperatures. Freeze-drying does not usually cause shrinkage or toughening of the material being dried.
3. Freeze-dried products can be rehydrated (reconstituted) much more quickly and easily because the process leaves microscopic pores. These pores are created by the ice crystals that sublimate, leaving gaps or pores in their place. This is especially important when it comes to pharmaceutical products.
4. Lyophilization can also be used to increase the shelf-life of some pharmaceuticals for many years.

Achievements

In vitro storage and cryopreservation are two technologies used in genebanks to conserve species with recalcitrant seeds or species that are vegetatively propagated. Cryopreservation involves storing the living tissues at ultra-low temperature to guarantee long-term preservation of germplasm in genetically unaltered state. Research works on cryopreservation have led to the development of protocols for cryopreservation of more than 200 different plant species. A wide range of species can now be routinely cryopreserved as shown in Table 19.1.

Cryopreservation is a more promising technique for the long-term conservation of plant species. A few examples are listed below:

1. Cryopreserved protoplasts of carrot have been successfully regenerated.
2. Meristems of potato, cassava, sugar cane, peanuts, etc. can be stored for several years.
3. Cell suspensions of soybean, carrot, datura, etc. are stored for two to four years.
4. Pollen grains and protoplast hybrids of peanuts, mustard, cotton, wheat, rice, etc. can be stored for several years.

Cryopreservation Facility in India

The National Bureau of Plant Genetic Resources (NBPGR), New Delhi, has this facility. The Bureau increased its storage capacity from 154,964 varieties to 600,000 varieties. The expansion project has been promoted by the realization that the country's genetic diversity is being steadily eroded by modern agriculture practices, urbanization and deforestation.

Scientists world over are availing the NBPGR seed facility through germplasm exchange programme. The export of few plant species like black pepper is banned in the national interest. Within India, anyone involved in plant or crop research can get seeds/germplasm from the NBPGR.

Risks in Cryopreservation

Phenomena which can cause damage to cells during cryopreservation are solution effects, extracellular ice formation, dehydration and intracellular ice formation.

1. ***Solution effects*** Solution effects are caused by concentration of solutes in non-frozen solution during freezing as solutes are excluded from the crystal structure of the ice. Therefore, high concentrations of solutes can be very damaging.
2. ***Extracellular ice formation*** When tissues are cooled slowly, water migrates out of cells and ice forms in the extracellular space. Too much extracellular ice can cause mechanical damage due to crushing.
3. ***Dehydration*** The migration of water causing extracellular ice formation can also cause cellular dehydration. The associated stresses on the cell can cause damage directly.
4. ***Intracellular ice formation*** Though some organisms and tissues can tolerate some extracellular ice, however, any appreciable intracellular ice is almost always fatal to cells. When vitrification solutions are not used, the cells being preserved are often damaged due to freezing during the approach to low temperatures or warming to room temperature.

Prevention of cryopreservation risks Vitrification provides the benefits of cryopreservation without the damage due to ice crystal formation. In cryopreservation, vitrification usually requires the addition of cryoprotectants prior to cooling. The cryoprotectants act like antifreeze; they lower the freezing temperature. They also increase the viscosity. Instead of crystallizing, the syrupy solution turns into amorphous ice, that is, it vitrifies. Vitrification of water is promoted by rapid cooling and can be achieved without cryoprotectants by an extremely rapid drop in temperature. The rate that is required to attain a glassy state in pure water was considered to be impossible until recently.

Two conditions usually required to allow vitrification are an increase in the viscosity and a reduction of the freezing temperature. Many solutes do both, but larger molecules generally have larger effect, particularly on viscosity. Rapid cooling also promotes vitrification.

Limitations and Drawbacks

1. In artificial cryopreservation, the solute must penetrate the cell membrane in order to achieve increased viscosity and depressed freezing temperature inside the cell. Sugars do not readily permeate through the membrane. Those solutes that do, such as dimethyl sulfoxide, a common cryoprotectant, are often toxic in high concentrations. One of the difficult compromises faced in artificial cryopreservation is limiting the damage produced by the cryoprotectant itself.
2. *Ex situ* conservation, while helpful in man's efforts to sustain and protect the environment, is rarely enough to save a species from extinction. It is to be used as a last resort or as a supplement to *in situ* conservation. It cannot recreate the habitat as a whole, the entire genetic variation of a species, its

symbiotic counterparts or those elements which might help a species adapt to its changing surroundings over time.

3. *Ex situ* conservation removes the species from its natural ecological contexts, preserving it under semi-isolated conditions whereby natural evolution and adaptation processes are either temporarily halted or altered by introducing the specimen to an unnatural habitat.
4. In the case of cryogenic storage methods, the preserved specimen's adaptation processes are frozen altogether. The downside to this is that, when re-released, the species may lack the genetic adaptations and mutations which would allow it to thrive in its ever-changing natural habitat.
5. *Ex situ* conservation techniques are often costly.
6. Seed banks are ineffective for certain plant genera with recalcitrant seeds that do not remain fertile for long periods of time.
7. Diseases and pests foreign to the species, to which the species has no natural defence, may also cripple crops of protected plants in *ex situ* plantations and in animals living in *ex situ* breeding grounds.

These factors, combined with the specific environmental needs of many species, some of which are nearly impossible to be recreated by man, make *ex situ* conservation impossible for a great number of the world's endangered flora and fauna.

REVIEW QUESTIONS

1. Define the following terms:
   i. Germplasm
   ii. Genetic erosion
   iii. Genetic wipe-out
   iv. Genetic vulnerability
   v. Threatened flora
2. What is germplasm preservation? Discuss *in situ* and *ex situ* preservation.
3. What is cryopreservation? Describe the process of cryopreservation.
4. Write notes on the following:
   i. Gene banks
   ii. Cryoprotectants
   iii. Freezable tissues

   iv. Cryopreservation equipment
   v. Reculture of cryopreserved material
5. Explain in detail about the freeze-drying process.
6. What are the achievements, risks and risk prevention measures of cryopreservation?

REFERENCES

Cohen, J.I. *et al.* (1991). *Ex situ* conservation of plant genetic resources: Global development and environmental concerns. *Science*. 253: 866–872.

Prescott- Allen, R. and Prescott-Allen, C. (1990). How many plants feed the world? *Conservation Biology*. 4, 365–374.

Towill, L.E. (1989). Biotechnology and germplasm preservation. In: Janick, J. (ed.). *Plant Breeding Reviews*. 7: 159–182.

Uma Shanker, R. and Ganeshaiah, K. N. (1997). Mapping genetic diversity of Phyllanthus emblica: Forest gene banks as a new approach for in situ conservation of genetic resources. *Current Science*. 73: 163–168.

Wilker, G. (1989). Germplasm preservation: Objective and needs. In: Knutsen, L. and Stoner, A.K. (Eds.). *Biotic Diversity and Germplasm Preservation: Global Imperatives*. Kluwer Academic Publishers, Dordrecht.

Withers, L. A. and Engelmann, F. (1998). *In vitro* conservation of plant genetic resources. In: Altman, A. (ed.). *Agricultural Biotechnology*. Marcel Dekker, Inc., New York. pp. 57–88.

Yadav, J., Choudhary, N., Shekhawat M.S., Rathore J.S., Arya V. and Shekhawat, N.S. (2002). Micropropagation of plants of arid zone: Achievements and limitations. In: Nandi *et al.* (eds.) *Role of Plant Tissue Culture in biodiversity conservation and economic development*. 11–17.

Ziv, M. (1991). Vitrification: Morphological and physiological disorders of *in vitro* plants. In: Debergh, P.C. and Zimmerman, R.H. (eds.). *Micropropagation*. Kluwer Academic Publishers, The Netherlands. pp. 45–69.

20

ARTIFICIAL (SYNTHETIC) SEED PRODUCTION TECHNOLOGY

INTRODUCTION

Plants are traditionally propagated either by seeds or by vegetative propagules. The natural seed is a ripened ovule consisting of diploid zygotic embryo and its coats. The zygotic embryo is a very important structure. It grows into seedling by seed germination. The seed coats are derived from the integuments. The concept of artificial seed or synthetic seed has been developed from somatic embryoids, which are formed adventitiously from *in vitro* cultured somatic tissues.

Artificial seeds are living seed-like structures, which are made experimentally by a technique where somatic embryoids derived from plant cell *in vitro* culture are encapsulated by a hydrogel. These encapsulated embryoids behave like true seeds if grown in soil and can be used as substitutes for natural seeds.

A mature plant seed consists of an embryo and its food store (endosperm), surrounded by a seed coat (testa). The seed ensures that the next generation of plant exists within it. The embryo is made up of one or two cotyledons attached to a central axis. The upper part of the axis, called epicotyl contains a plumule at its tip, which on germination grows into the shoot system. Similarly, the lower part of the axis, hypocotyl, consists of a radicle that later grows into the root system.

The endosperm is the food reserve that the embryo uses as a source of nutrition during the early stages of germination. Before the germinating embryo is able to produce its own food through photosynthesis, the endosperm provides vital nutrients to the embryo. The testa protects the embryo from injury and drying out. It also makes sure that the embryo remains viable before germination.

As germination occurs, water is absorbed and the seed coat breaks, allowing the radicle to emerge from the seed.

ARTIFICIAL OR SYNTHETIC SEED

Synthetic seeds or "synseeds" are defined as artificially encapsulated somatic embryos, shoot buds, cell aggregates or any other tissue that can be used for sowing as a seed. It also possesses the ability to convert into a plant under *in vitro* or *ex vitro* conditions and these synseeds retain such potential even after storage. Earlier, synthetic seeds referred only to the somatic embryos that were of economic use in crop production and plant delivery to the field or greenhouse. The term "artificial seed" was coined by Murashige in 1977. This technique was later perfected by Redenbaugh and his coworkers. In India, the technique of encapsulation of somatic embryos is commonly used in many modern research laboratories.

Artificial seeds have been proposed as a new low-cost and a high-efficiency propagation system. They have opened new avenues in storage and delivery of new plant lines produced through biotechnological approaches. The main advantage of the system is the combination of high-volume production and low-cost propagation. Encapsulation of cells, embryos, somatic tissues and somatic embryos has been attempted in several crops. It has become increasingly popular as a simple way of handling cells, tissues and embryos and protecting them against external gradients and also as an efficient storage and delivery system. The acceptance of the technology, however, depends on the value of the propagated crop and the cost of competing products.

CONCEPT OF ARTIFICIAL SEED TECHNOLOGY

The basic concept of somatic embryo encapsulation to produce an analogue to true seeds is based on the similarity of somatic embryos (SE) with zygotic embryos in terms of morphology, physiology and biochemistry (Figure 20.1). The gross morphology of somatic embryos is very similar in appearance to zygotic embryos. An important difference between somatic and zygotic embryos is that the latter typically ceases growth, becoming quiescent or dormant as water is lost, followed by storage tissue maturation and seed coat hardening. This arrested growth phase is a major factor accounting for efficient storage and handling qualities of natural seed. A similar arrested growth phase induced during somatic embryo development is essential to match the efficiency of seed propagation and is a pivotal step in the development of artificial seed technology. The concept of artificial seed is still in its infancy phase. There is an immense potential for the use of this technique for the propagation and preservation of elite germplasm.

The concept of "synthetic seed" or "artificial seed" (synseed) has evolved from a futuristic idea into a real field of experimentation and research by Murashige (1977). Gray (1987) has traced the the early work on the subject. Since then, synseed technology has evolved because of increasing research in the recent years. The original definition of a synthetic seed given by Murashige (1978) was as follows:

"a synthetic seed is an encapsulated single somatic embryo", that is, a clonal product that could be handled and used as a real seed for transport, storage and sowing, and which would eventually grow, either *in vitro* or *ex vitro*, into a plantlet. This definition limited synseed manufacture to the use of a somatic embryo, that is, a bipolar propagule regenerated through somatic embryogenesis that could be contained in a capsule, which would allow manipulation and seeding.

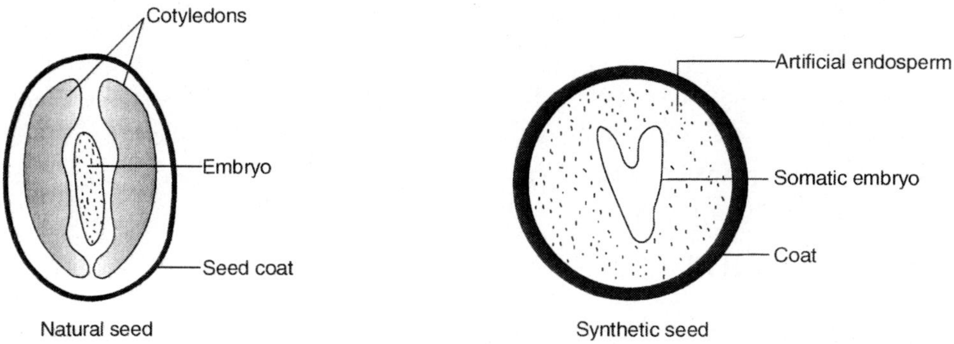

Figure 20.1 Natural and artificial seed

Bapar, *et al.* (1987) subsequently proposed the making of synseeds through the encapsulation of *in vitro* derived propagules, different from somatic embryos, especially in non-embryogenic species; in *Morus indica*, for instance, they proposed the use of encapsulated axillary buds. Despite the increasing number of studies that have considered the use of non-embryogenic *in vitro* derived plant propagules for the constitution of synseeds, the limitation of Murashige's initial definition has been maintained till recent years. Currently, a broader definition is accepted by the international scientific community. A more current definition is the one given by Aitken-Christie *et al.* (1995), which considers synseeds as "artificially encapsulated somatic embryos, shoots or other tissues which can be used for sowing under *in vitro* or *ex vitro* conditions." This definition extends the concept of the synseed to any type of vegetative propagule, characterizing it only by its use as a seed-like structure. To complete this definition, it should be emphasized that the propagule must be able to grow into a plantlet after sowing. Also, the *in vitro* origin of the explants could be discretionary, since examples are also available of buds from mature trees used as synseeds.

OBJECTIVES OF SYNTHETIC SEED TECHNOLOGY

The basic objective in developing synthetic seed technology is to achieve and maintain a clonal propagation system that will enable the vegetative propagule to be stored for long periods of time and at the same time enable multiplication of the plant (for example, *Medicago sativa*). There are economic objectives in developing this technology in alfalfa but the technology that is being developed with this experimental system can also be applied to many other plants.

Artificial seeds were first introduced in the 1970s as a novel analogue to the plant seeds. The production of artificial seeds is especially useful for plants which do not produce viable seeds. It represents a method to propagate these plants. Artificial seeds are small sized and provide further advantages in storage, handling, shipping and planting. Various forms of synthetic seeds have been envisioned over time. The first were simply hydrated somatic embryos produced from vegetative cells in plant tissue culture. These had the particular advantage of enabling rapid clonal multiplication of some plants, but the labour cost was high and the propagules were very delicate. This was partially overcome with the development of alginate capsules that encapsulated a single embryo in a protective coating enabling mechanized handling. These hydrated, encapsulated embryos could only be stored in low temperatures for a few weeks. The capability of prolonged storage was achieved when the somatic embryos could be dried so their moisture content is less than 20%. This is the current state of the technology. In the future, coatings for these embryos would be developed that will provide protection and improve the ease of handling these propagules.

PRODUCTION OF ARTIFICIAL SEEDS

The production techniques of synthetic seeds are discussed below:

Source of Artificial Seed Synthesis

Several sources of plant cell and tissue culture have been employed in the production of synthetic seeds. These can be somatic embryos, shoot buds, and shoot tips, organogenic or embryogenic callus.

Thus, the concept of synthetic seeds has been set free from its bonds to somatic embryogenesis and the term has been linked not only to its use (storage and sowing) and product (plantlet) but also to other techniques of micropropagation like organogenesis and enhanced axillary bud proliferation system. Artificial seeds can be produced by encapsulating a plant propagule in a matrix which will allow it to grow into a plant. Plant propagules may consist of shoot buds or somatic embryos that have been grown aseptically in tissue culture. In culture, these plant propagules can easily grow into individual plants. Its growth can also be controlled using chemicals provided in the culture media.

Mature somatic embryos are collected from the callus or suspension cultures. A 2.0% solution of sodium alginate is filled in a burette and allowed to drip in drops into a 100 mM $CaCl_2$ solution. As the sodium alginate bead or drop forms at the tip of the burette, a somatic embryo is inserted into it with the help of a spatula before the drop falls into the $CaCl_2$ solution (Figure 20.2). The beads become hardened as calcium alginate is formed. After about 20 to 30 minutes the artificial seeds are removed, washed with water and used for planting. Successful results have already been obtained in the production of artificial seeds in some crops such as *Apium graveolens, Daucus carota, Zea mays, Lactuca sativa, Medicago sativa, Brassica* sp., *Gossypium hirsutum, Valeriana* sp., *Santalum* sp., etc. In the production of artificial seeds, an artificial endosperm can be synthesized within the encapsulation matrix. The encapsulation matrix is a hydrogel made of natural extracts from various sources like:

- Seaweed (agar, carrageenan or alginate)
- Plants (gum arabic or tragacanth)
- *Cyamopsis tetragonolobus*, locust bean gum (Seed gums).
- Microorganisms (dextran, gellan or xanthan gum).

These compounds will gel when mixed with or dropped into an appropriate electrolyte (copper sulphate, calcium chloride or ammonium chloride). Ionic bonds are formed to produce stable complexes. Useful adjuvants such as nutrients, plant growth regulators, pesticide and fungicide can be supplied to the plant propagule within the encapsulated matrix. In most cases, a second coat covering the artificial endosperm is required to simulate the seed coat.

Types of Artificial Seeds

Artificial seeds are basically of two types: Hydrated seeds and Desiccated seeds.

1. *Hydrated artificial seeds* The system consists of somatic propagules (embryo or shoot buds) encapsulated in a hydrogel like sodium alginate, gelrite, agar, carageenan, etc. The most common gelling agent used is sodium alginate, along with calcium salts as complexing agents. Somatic embryos are sieved from the suspension cultures that are mixed with sodium alginate and dropped in calcium chloride solution. When sodium alginate drops come in contact with calcium chloride solution, surface complexion begins and firm round beads are formed, each containing 1–2 propagules. These beads are allowed to stand for 60–90 minutes and washed with water followed by air-drying, and stored at 4°C. Axillary, apical/ adventitious buds can also be encapsulated by this method. The buds are trimmed to the smallest possible size and mixed with sodium alginate. In another method, the propagules are mixed in temperature-dependent gels such as agar, or gelrite and are gelled to form capsules by lowering the temperature.

2. *Desiccated artificial seeds* In this method, the artificial seeds are produced by coating a mixture of somatic embryos or shoot buds with polyethylene glycol (PEG). The coated mixture is then allowed to dry for several hours on Teflon surface under sterile conditions. The dried mixture in the form of wafers is then placed *in vitro* in the culture medium, allowed to re-hydrate and stored for embryo survival and conservation.

Capsule hardness can be controlled by the concentration of the complexing agent solution and the complexation time. Size of the capsule is determined by the size of the shoot bud or somatic embryo and the inner diameter of the pipette used. The capsules or artificial seeds are collected by decanting off the complexation solution and rinsed in water. The artificial seeds should be pliable enough to cushion and protect the embryo, so that they can allow germination and growth of the shoot bud or somatic embryo. It should be rigid enough to withstand rough handling during manufacture, transportation and planting. For the artificial seeds to remain dormant until planting, a thin layer of water-soluble resin is used to coat the encapsulation matrix.

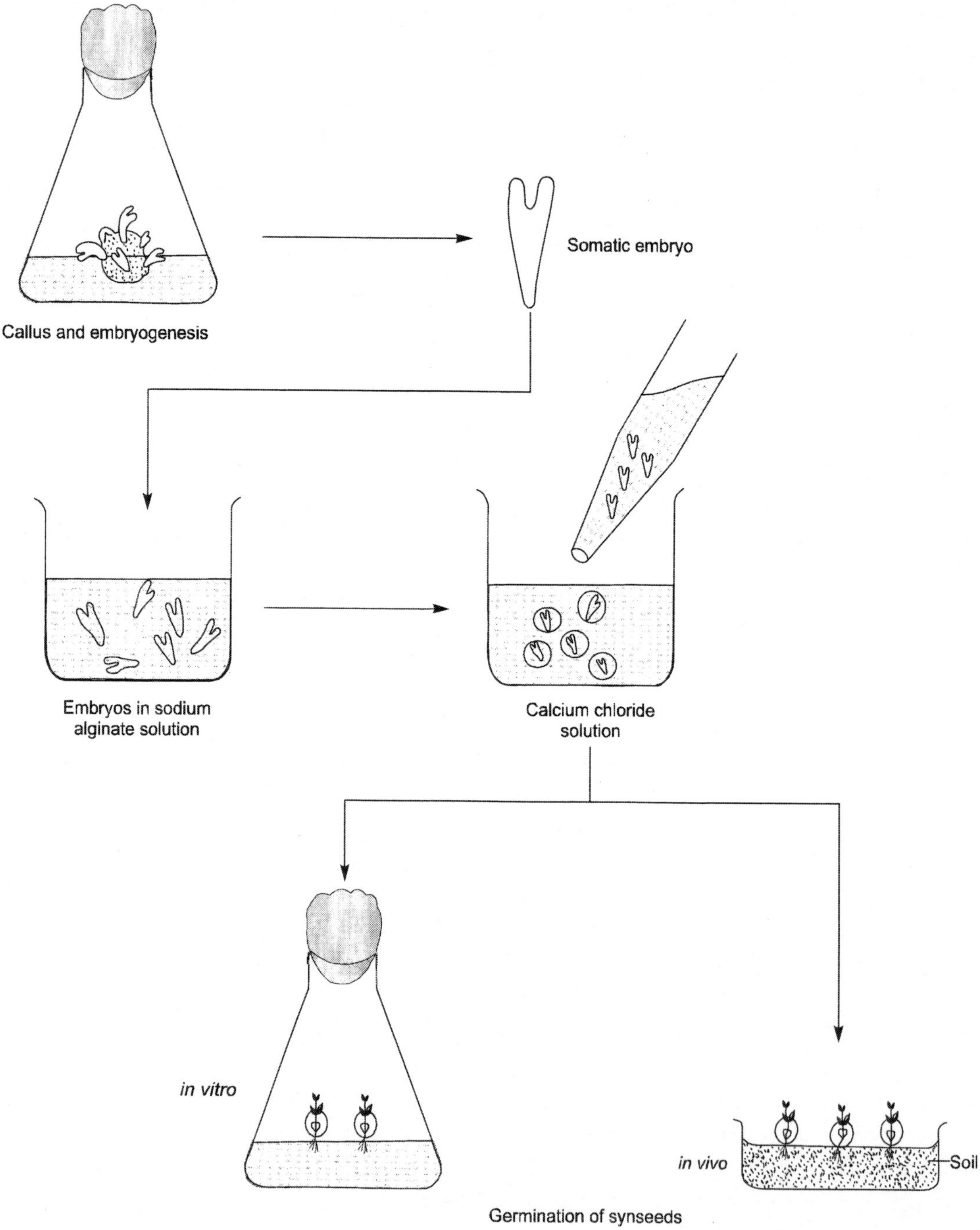

Figure 20.2 Artificial seed production technique

Hydrophobic Coating

New compounds have been tried for hydrophobic coating of artificial seeds. A recent compound Elvax 4260, manufactured by DuPont has shown significant impediment to capsule drying. This compound also reduces the stickiness and beads can be planted using a seed planter. Conventionally, the beads can be rolled in talcum powder to make them flowable and stored for a short period in culture room.

Artificial seeds have been successfully produced in the following crops for commercial use:

Alfalfa, Apium graveolens, Cymbidium (orchid), *Daucus carota, Dendrobium* (orchid), *Dioscorea floribunda, Gossypium hirsutum, Medicago sativa, Morus indica, Picrorhiza kurroa, Pogostemon patchouli, Phalaenopsis* (orchid), *Rheum emodi, Santalum album, Spathoglottis plicata* (orchid), *Azadirachta* etc.

Somatic Embryos as Artificial Seeds

The use of somatic embryos as artificial seeds for large-scale clonal propagation of plants is close to becoming a reality. The quality of the artificial seed depends on the temporal, quantitative and qualitative supply of growth regulator and nutrients along with an optimal physical environment. Desiccation of somatic embryos provides a quiescent phase analogous to true seeds, facilitating the convenience of year round production, storage and distribution. Somatic embryos possess the ability to express desiccation tolerance in response to an external chemical or physical stimulus. The mechanisms of desiccation tolerance involve stabilization of membranes in dry state and prevention of oxidative degradation of biomolecules. Encapsulation of embryo may control the water uptake and release of nutrients. It also provides mechanical protection required for field planting.

Recent advancements in the area have revealed that besides somatic embryos, encapsulation of cells and somatic tissues obtained following tissue culture techniques has become popular as a simple way of handling cell and tissue, protecting them against strong external gradients, and as an efficient delivery system.

PRINCIPLE OF SYNTHETIC SEED TECHNOLOGY

In order to produce a synthetic version of a seed, the somatic embryo must be forced to follow the same developmental programme as the zygotic embryo in the seed. After pollination, a zygotic embryo of a dicotyledonous species develops through a series of morphological stages termed globular, heart and torpedo. Cotyledons develop and expand, as the storage reserves of protein, starch and/or oil are deposited. At some time before the embryo achieves its maximum weight, it acquires the ability to tolerate drying. Then, the seed's vascular connections to the maternal plant are severed, it stops importing nutrients and begins to lose water. Seeds of most crop plants can survive drying and can be stored for several years. Once they are hydrated, germination commences culminating in the emergence of a radicle and then the mobilization of the storage reserves by the seedling.

Intensive research efforts have been made on synthetic seed production techniques in a number of plant species for the last two decades. Despite these achievements in synthetic seed technology, practical implementation of the technology is yet to be fully realized due to limitations encountered with the production, development, maturation and subsequent conversion of the micropropagules into plantlets under *in vitro* or *ex vitro* conditions. Development of artificial seeds requires sufficient control of somatic embryogeny from the explants to embryo production, embryo development and their maturation as well. The mature somatic embryos must be capable of germinating out of the capsule or coating to form vigourous normal plants.

A number of *in vitro* technologists have tried to improve the quality and quantity of somatic embryos via modification of culture conditions such as medium composition, growth regulators (types and concentrations), physical state of the medium as well as incubation conditions like temperature, illumination, etc.

Induction Somatic Embryo Formation

Induction of somatic embryogenesis requires a change in the fate of a vegetative (somatic) cell. In most cases, an inductive treatment is required to initiate cell division and establish a new polarity in the somatic cell. In alfalfa, the inductive treatment most commonly requires 2,4-D only, but other auxins such as 2,4,5-T are also effective. The auxin response is quite complex. Some auxins such as IAA and IBA are ineffective, and still others will stimulate the formation of embryos and callus, but not somatic embryos. Inorganic components in the medium such as potassium, and organic components such as proline can modulate the embryogenesis or callus response, but they cannot replace auxin. In order to induce somatic embryogenesis, in general, explants are required to be inoculated on 2,4-D supplemented nutrient media irrespective of monocot or dicot tissues. However, significantly in case of monocot plants, explant starts callusing at one end of the leaf-base segment (Figure 20.3) on callus induction medium, followed by differentiation of somatic embryos (Figure 20.4) on 2,4-D-free nutrient medium. Furthermore these somatic embryos can be easily germinated into plantlets on the same hormone-free medium.

In simple terms, the 2,4-D differentially activates the cell cycle of many cells in the explants (such as petiole); cells in the vascular cambium develop directly into a callus, whereas some subepidermal cells develop into a somatic embryo instead of callusing. The initial somatic embryos, which are only small dense cell clusters at this stage, are embedded in a callus mass of non-differentiated cells. To liberate these proembryonic structures, and to stimulate the formation of more embryos, the callus is dispensed in a liquid medium to form a suspension culture.

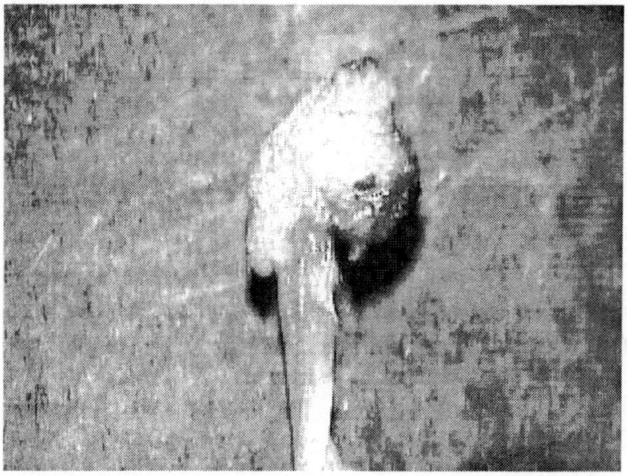

Figure 20.3 Callus formation at one end of leaf-base segment (Plate 3.5)

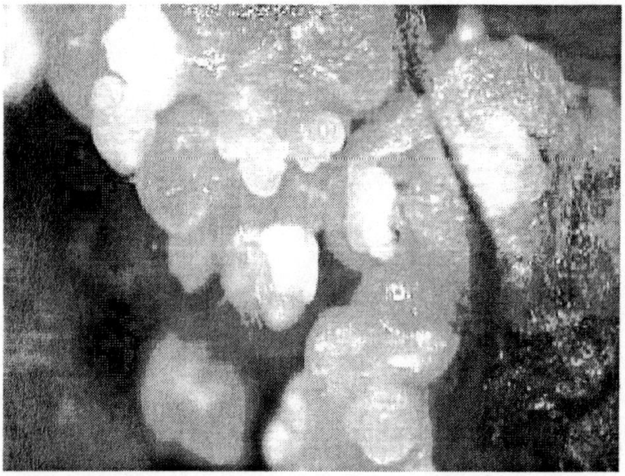

Figure 20.4 Embryogenic callus showing formation of numerous somatic embryos in a millet crop (Plate 3.6)

Somatic Embryo Formation by Suspension Culture

In suspension, proembryogenic cell clusters form and these can be separated from the single cells and the larger clumps of callus by sequentially sieving through nylon membranes of 224–500 mm pore size. The suspension, however, cannot be maintained in this embryogenic state. If the 224–500 mm embryogenic fraction is placed back into suspension, the single cell population increases, the embryogenic cell clusters disperse and they become a progressively smaller proportion of the culture. Identifying techniques to stabilize these embryogenic cultures is critical for the scale-up of this technology. The organic and inorganic components of the liquid suspension culture medium have been precisely defined. Also, physical factors including inoculant density (weight of callus added to 40 ml of liquid suspension) are also critical and show precise optimal values.

Embryo Maturation

To allow development of somatic embryos from a sieved suspension culture, the fraction collected on the 224 mm screen is spread in a thin layer on somatic embryo development medium. At approximately 4 days after sieving, green globular somatic embryos appear (Figure 20.5), which later on enlarge and develop through the heart stage into the torpedo-shaped embryos in 7–10 days. Generally, at the torpedo stage, somatic embryos in culture typically germinate precociously into a seedling.

Figure 20.5 Globular somatic embryos on medium from layer of cells (Plate 2.3)

In callus induction medium, the embryos develop through the morphological stages such as globular, heart or torpedo-shaped (Figure 20.6). Once the majority of embryos reach the torpedo stage (7–10 days after sieving), they are transferred to an enriched medium containing a high level of sucrose, nitrogen and sulphur to prevent precocious germination and to enable deposition of storage reserves. Subsequently, the embryos accumulate storage proteins and carbohydrates. Storage protein accumulation is facilitated by the inclusion of an organic nitrogen source, glutamine and an inorganic sulphur source, potassium sulphate. Glutamine plays a regulatory as well as a nutritive role in somatic embryo maturation. When included in the medium, glutamine is converted to 5-oxoproline by autoclaving. Other amino acids cannot substitute for glutamine, but 5-oxoproline can mimic the glutamine effect. Whether 5-oxoproline is simply a convenient form of nitrogen for transport or whether it has a regulating role in metabolism is unknown.

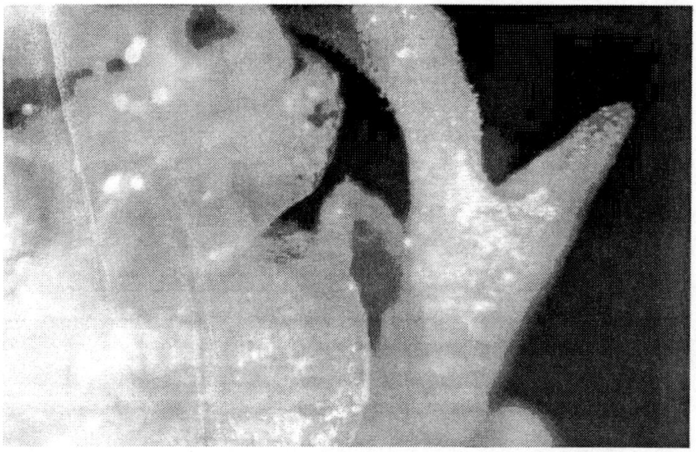

Figure 20.6 Torpedo shaped somatic embryo (Plate 2.4)

The final stage of maturation is achieved by transferring the embryos to a medium containing ABA. This growth regulator apparently triggers a process leading to the expression of desiccation tolerance. Other physical treatments including cold, heat, osmotic or nutrient stress can elicit a similar response, presumably because they stimulate the endogenous synthesis of ABA. If the embryos germinate precociously, there is not sufficient time to accumulate storage reserves or acquire desiccation tolerance and thus the embryo does not achieve the quiescence stage. Imposing an osmotic stress on the embryos by incorporating 5.0–6.0% sucrose in the nutrient medium instead of the normal 3.0% that prevents precocious germination, maintains complete embryogenic development.

Because nutrition plays such a critical role in embryo maturation, physical factors that influence nutrition such as plating density (the number of somatic embryos per Petri plate) must also be considered. Light intensity (optimal at 75 mmol $m^{-2}s^{-1}$) is also critical at this stage of the culture process. Therefore, practices such as stacking Petri plates, which create variable physical environments, should be avoided. Somatic embryos are then removed from the medium, washed to remove sugar and other nutrients, and dried. The standard method of drying is to place the somatic embryos in a sealed chamber over a saturated salt solution designed to give specific relative humidities (Senaratna *et al.*, 1990). The embryos are transferred daily for 1 week to a progressively lower relative humidity chamber and finally dried at ambient conditions. At this stage, the embryos would have reached approximately 15% moisture and can be stored for a year or more with good viability. Subsequently, the embryos are germinated on moist filter paper.

Desiccation Tolerance

Desiccation tolerance is a characteristic of somatic embryos that must be induced and therefore, requires a pretreatment with ABA or stress, to elicit the desired response. The type of pretreatment used, the duration for which it is applied and the stage of embryo that is treated are critical factors that must be considered. Three days on a medium containing 20 mM ABA is sufficient to induce tolerance, but chilling requires almost 3 weeks. The rate of drying has secondary importance. If the embryos are immature, slow drying over one week is optimal, but if there are large numbers of fully mature embryos, rapid drying in an LAF cabinet is preferable.

Desiccation tolerance is a quantitative characteristic, not a qualitative one and embryos can have varying degrees of tolerance. This is especially true among different species, but is also evident with different inductive treatments. Drying the somatic embryos promotes metabolic changes, particularly in sugar metabolism. In alfalfa, somatic embryos contain high amounts of sucrose, glucose and fructose before drying. However, after drying, the amounts of glucose and fructose become very low, and stachyose had increased proportionately. Raffinose is also present. Given the importance of these sugars in desiccation tolerance in seeds, these changes might be quite significant.

Germination and Seedling Vigour

Dry somatic embryos lack the vigour normally associated with seedlings from normal seeds. The reason for this is not obvious, although there are several possibilities. The dry somatic embryos may lack storage proteins or some other critical component required after germination by the seedling. Storage protein levels have been enhanced, leading to some improvement in embryo vigour. Somatic embryos store starch and sucrose, whereas seeds store a hemicellulose in the cell walls of the endosperm called galactomannan.

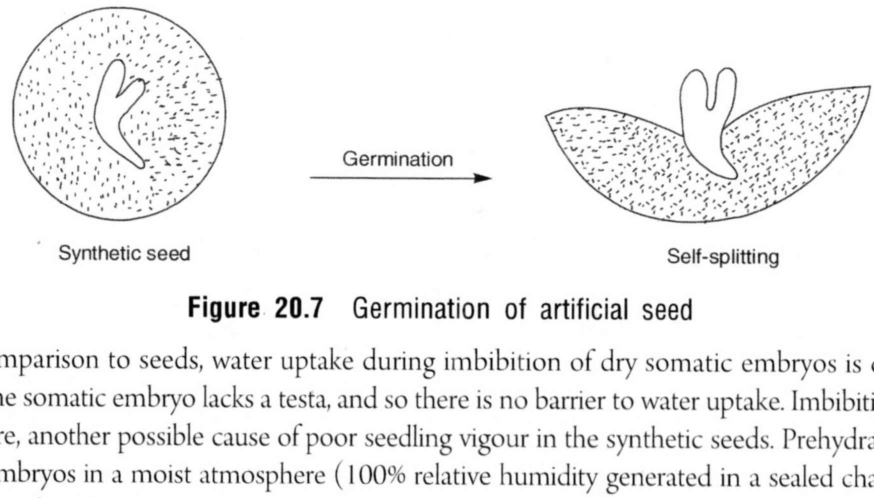

Figure 20.7 Germination of artificial seed

In comparison to seeds, water uptake during imbibition of dry somatic embryos is quite rapid because the somatic embryo lacks a testa, and so there is no barrier to water uptake. Imbibitional injury is, therefore, another possible cause of poor seedling vigour in the synthetic seeds. Prehydration of the somatic embryos in a moist atmosphere (100% relative humidity generated in a sealed chamber over water) for 24 hours improves the vigour of somatic embryos. The most probable cause of poor seedling vigour is abnormal apical or plumule development. In germinating somatic embryos, there are two separate processes: 1) germination, and 2) conversion. Germination is defined as the process culminating in the emergence of the radicle, whereas conversion is defined as the emergence of the radicle and the shoot. Rarely does a somatic embryo form a shoot without roots, but root emergence without plumule emergence is common (Figure 20.7). Therefore, although germination rates of seeds and zygotic embryos are quite similar, conversion rates are much lower in dry somatic embryos.

Artificial Endosperm

Seed coat or testa and endosperm are absent in somatic embryos which provide nutrition and protection for zygotic embryos in developing seeds. To fulfil these deficiencies, addition of nutrients and growth regulators to the encapsulation matrix results in increase in efficiency of germination and viability of encapsulated somatic embryos. Again, this is advantageous in that synthetic seeds can be stored (at 4°C) for a longer time (up to six months) without losing viability.

ADVANTAGES OF SYNSEED TECHNOLOGY

1. Synthetic seed technology offers many useful advantages on a commercial scale for the propagation of a variety of crop plants, especially crops for which true seeds are not used or not readily available for multiplication (for example, potato) or the true seeds are expensive (for example, cucumber and geranium), for hybrid plants (for example, hybrid rice) and many vegetatively propagated plants which are more prone to infections (for example, day lily, garlic, potato, sugarcane, sweet potato, grape and mango). This technology would also be useful for multiplying genetically engineered plants (transgenic plants), somatic and cytoplasmic hybrids (obtained through protoplast fusion techniques), sterile and unstable genotypes.
2. Besides, artificial seeds are useful for the preservation of desirable elite genotypes (cryopreservation). They are also valuable tools in experimental research to study the process of zygotic embryogenesis and understanding the role of endosperm in normal embryo development and germination.
3. Dormancy is the common feature of natural seeds, but by means of artificial seeds the dormancy period can be reduced to a great extent, thereby shortening the life cycle of a plant. They are useful in preserving the germplasm. Synthetic seeds are applicable for large-scale monoculture as well as for mixed genotype plantation. The viability of artificial seeds is good and these can be made in large numbers when required.
4. The storage and transportation are easier than natural seeds. The germination of artificial seeds is uniform and all the clones are identical and these can be tested initially in the greenhouse prior to sowing in the field.
5. Production of artificial seed is not dependent on the reproductive phase of the plant. It can be produced at any time and season of the year. Artificial seed coating has the capacity to hold growth-promoting agents. It gives protection from meiotically unstable genotypes. Seed-borne diseases are totally eliminated by the use of artificial seeds for plantation.

6. At the Biotechnology Division of BARC, research on the development of protocols for synthetic seeds using somatic embryos, axillary buds and shoot tips is in progress in five economically important plants. These are sandalwood, rice, mulberry, banana and cardamom.
7. Artificial seeds have the potential for providing an inexpensive plant delivery system. The process of planting can be easily mechanized and this allows the direct delivery of tissue-cultured plant propagules to the field. It also provides rapid bulking up for the production of genetically engineered individual plants.
8. Artificial seeds make a promising technique for propagation of transgenic plants, plants not producing seeds, polyploids with elite traits and plant lines with problems in seed propagation. Being clonal in nature, the technique cuts short the laborious selection procedure of the conventional recombination breeding and can bring the advancements of biotechnology to the doorsteps of the farmer in a cost-effective manner.

APPLICATIONS OF SYNSEED TECHNOLOGY

By combining the benefits of a vegetative propagation system with the capability of long-term storage and clonal multiplication, synthetic seeds have many diverse applications in agriculture. A similar system could be developed in other crops, although the specific commercial applications will undoubtedly differ. In plant breeding programmes of cross-pollinating species, this technology might provide an alternative way to store germplasm instead of a field or greenhouse nursery. Synthetic seeds also offer the opportunity to store genetically heterozygous plants or other plants with unique gene combination that cannot be maintained by conventional seed production due to genetic recombination that occurs at each generation of seed increase.

Many plants are sterile and do not set seed. Somatic embryogenesis would, in many instances, be preferable to make cuttings as a means to propagate these plants. Other plants, particularly some tropical species, produce recalcitrant seeds that cannot be dried. Therefore, long-term storage of these species in seed banks is not currently possible. Somatic embryos might be an alternative that would enable long-term storage of these species, but the success of the somatic embryo approach will depend on the cause of the seed's recalcitrance. If its inability to tolerate drying is due to a genetic trait that makes the seed unable to respond to the inductive signals that trigger the expression of desiccation tolerance (for example, lacking ABA receptors), then the synthetic seed approach will not work. However, if the inability is due to a genetic trait that prevents the plant from producing the required signal (for example, block in ABA synthesis), then the synthetic seed approach might work.

Synthetic seed technology may also find application in the short-term storage of propagules to enable the synchronized planting of commodities in a greenhouse. One of the limitations of current micropropagation procedures are that the tissue culture facility and the greenhouse production facility must be physically linked, as well as their production must be synchronized to meet peak market demands.

Synthetic seeds could be a means to uncouple this linkage, enabling the micropropagation to be done year round at a site distance from the production facility.

As stated earlier, the exact application of synthetic seeds will vary from species to species. In self-pollinated crops that currently have good seed production systems, synthetic seeds will not have any practical applications, but in cross-pollinating species, especially those where seed production is difficult and expensive, synthetic seeds offer many advantages and opportunities.

PROBLEMS IN SYNTHETIC SEED TECHNOLOGY

Artificial seed technology starts with an embryo. A somatic embryo is generated by utilizing somatic embryogenesis technology. Somatic embryos are similar in size, shape and composition to zygotic embryos, but most somatic embryos lack certain key components that are required for storage, processing or germination. Specifically, they lack the nutrients and the protective seed coats. This is where the real challenge in artificial seed technology lies. To overcome this challenge, two general approaches have been developed to germinate somatic embryos:

1. Utilization of aseptic techniques to germinate somatic embryos on sterilized growth media in sterile germination chambers.
2. Encapsulation of the somatic embryos in various types of gels, which are then sown into growing mixes.

Both methods have shown great promise; however, neither approach has been proven practical in commercial environments.

1. *The problems with in vitro germination* The first approach involves sowing naked somatic embryos onto a sterilized, semi-solid or liquid media that is contained within a solid support such as a Petri dish. These types of proprietary *in vitro* processes work, but they have serious limitations. Each embryo must be manually handled throughout the germination and transplanting steps. The risk of contamination is high and the variability in results is wide. Though a number of automated systems have been explored, none have been found to be commercially viable. Indeed, there seems to be no system that can reliably and cost-effectively produce germinated *in vitro* embryos at large-scale volumes for transplanting into conventional propagation systems.

2. *Gel encapsulation* The second approach for germinating somatic embryos involves the encapsulation of somatic embryos with gels. The end products of these proprietary techniques are various forms of artificial seeds, which can then be sown into a conventional growing system using standard seeding technology. Though promising, this method also has serious problems:

i. The process of encapsulation requires sterile conditions in order to reduce the risk of microbial contamination.

ii. The semi-solid physical characteristics of encapsulating gels are not compatible with commercial equipment currently used for plant propagation. The embryos tend to clump together during handling, creating difficulty in dispensing and generally clogging up the equipment.

iii. A final difficulty with encapsulated somatic embryos is their inability to take in nutrients by hydrolysing large molecules, like a zygotic embryo does inside the protective coating of the seed. As a result, the encapsulation technology for somatic embryos includes nutrients such as sugars, fertilizers and oxygen within the encapsulation medium (or gel).

Unfortunately, when these encapsulated embryos are germinated in non-sterile environments, they tend to be highly susceptible to microbial invasion during the manufacturing, storage and germination steps. The most significant problem with artificial seed technologies utilizing encapsulation or other coatings is that germination vigour is substantially reduced, which produces less vigorous plants.

LIMITATIONS IN SYNTHETIC SEED TECHNOLOGY

The initial cost of production is high and the storage of artificial seeds is also a great limitation. The limited storage time for artificial seeds is probably due to the anaerobic environment in the capsule. This is a problem for somatic embryos because these are not developmentally arrested and therefore, continue very active respiration. To overcome this problem, two possible solutions are available:

1. By having a smaller ratio of capsule volume to embryo volume so that gas diffusion can readily occur.
2. By inducing an arrested state in the embryo, using a growth control agent in the encapsulation.

Synthetic or artificial seeds are single encapsulated somatic embryos. Frequently, the term "ambling" is used for the plants originated from synthetic seeds. Generally, Na-alginate is used for the encapsulation of the embryo. Also new self-breaking gel beads have been developed. Until now the quality and fidelity of somatic embryos are the limiting factors for the development and scaling up of artificial seed production. However, because of certain inherent problems, the rate of production of uniform and high-quality embryos is much lower as a result of which the preparation of efficient and quality seeds has been successful in only a few crop plants like carrot and alfalfa. Although results of intensive studies in the field of synthetic seed technology seem promising for propagating a number of plant species, practical implementation of the technology is constrained due to the following reasons:

1. Limited production of viable micropropagules useful in synthetic seed production.
2. Anomalous and asynchronous development of somatic embryos.
3. Improper maturation of the somatic embryos that makes them inefficient for germination and conversion into normal plants.

4. Lack of dormancy and stress tolerance in somatic embryos that limit the storage of synthetic seeds.
5. Poor conversion of even apparently normal matured somatic embryos and other micropropagules into plantlets that limit the value of the synthetic seeds and ultimately the technology itself.

Once the technique is perfected it will be a boon to the seed production industry. Use of artificial seeds will become popular for greenhouse production. Additional studies, however, will be needed to understand the areas including the hardening treatments, coating and duration of storage, embryo conversion frequency and the economics of artificial seeds as compared to natural seeds.

REVIEW QUESTIONS

1. Define artificial or synthetic seeds with the objectives of this technology.
2. How can somatic embryos be induced?
3. How can synthetic seeds be formed by the use of cell suspension culture?
4. Write notes on the following:
   i. Embryo maturation
   ii. Desiccation tolerance
   iii. Seedling vigour
   iv. Artificial endosperm
5. Write in detail about the process of artificial seed production.
6. What are the advantages and applications of synthetic seed technology?
7. Write down the limitations of synthetic seed technology.

REFERENCES

Awal, A., Taha, R.M and Hasbullah, N.A. (2007). *In vitro* formation of synthetic seed of *Begonia hiemalis Fotch. Int. J. Environ. Sci.* 2: 189–192.

Bapat, V.A. and Rao, P.S. (1987). Sandalwood plantlets from synthetic seeds. *Plant Cell Rep.* 7: 434–436.

Ghosh, B. and Sen, S. (1994). Plant regeneration from alginate encapsulated somatic embryos of *Asparagus cooperi* Baker. *Plant Cell Rep.* 13: 381–385.

Kitto, S.L. and Janick, J. (1985). Production of synthetic seeds by encapsulating asexual embryo of carrot. *J.Am. Soc. Hortic. Sci.* 110: 277–282.

Redenbaugh, K., Fujii, J.A., Slade, D. and Kossler, M. (1987). Artificial seeds-Encapsulated Somatic Embryos. In: Bajaj, Y.P.S. (ed.). *Biotechnology in Agriculture and Forestry* 17 (High Tech and Micropropagation 1). Springer-Verlag, Berlin. pp. 395–416.

Sarkar, D. and Naik, P.S. (1997). Synseeds in potato: An investigation using nutrient-encapsulated *in vitro* nodal cutting segments. *Sci. Hortic.* 73: 179–184.

Senaratna, T., Mckersie, B.D. and Bowley, S.R. (1990). Artificial seeds of alfalfa (*Medicago sativa* L.). Induction of desiccation tolerance in somatic embryos. *In vitro Cell Dev.* Biol. 26: 85–90.

Sharma, T.R., Singh, B.M. and Chauhan, R.S. (1994). Production of disease-free encapsulated buds of *Zingiber officinale* Rosc. *Plant Cell Rep.* 13: 300–302.

Soneji, J.R., Rao, P.S. and Mhatre, M. (2002). Germination of synthetic seeds of pineapple (*Ananas comosus* L. Merr.). *Plant Cell Rep.* 20: 891–894.

Vikrant and Rashid, A. (2003). Somatic embryogenesis or shoot formation following high 2,4-D pulse-treatment of mature embryos of *Paspalum scrobiculatum* L. *Biologia Plantarum.* 46: 297–300.

Vikrant and Rashid, A. (2003). Somatic embryogenesis and plant regeneration from mesocotyl and leaf-base segments of *Paspalum scrobiculatum* L. a minor millet. *In vitro Cellular and Developmental Biology.* 39: 485–489.

21

SECONDARY METABOLITE PRODUCTION-I

INTRODUCTION

Higher plants synthesize a wide variety of low molecular weight compounds in addition to the essential primary metabolites (for example, carbohydrates, lipids, proteins, vitamins and minerals). These are called secondary metabolites. These secondary metabolites offer protection against pests, they act as attractants and as the plant's own hormones. Secondary metabolites accidentally appear as a result of stress and disease; such metabolites may be synthesized *de novo* (i.e., gene activation) and are referred to as *phytoalexins*. Other metabolites produced in response to stress are referred to as *phytoanticipins*.

In contrast to the primary metabolites, secondary metabolites do not have nutritional importance for human beings. Although, they are usually found in very small amounts (less than 1.0% dry weight), they have an effect on human metabolism. Secondary metabolites carry out a number of protective functions in the human body. A diet which is rich in plant foods contains a variety of secondary metabolites and contributes to protecting the human body against cancer and cardiovascular illnesses. Moreover, these secondary metabolites can boost the immune system, protect the body from free radicals and kill pathogenic germs and provide much more uses.

Most herbivores and plant pathogens possess mechanisms that ameliorate the impacts of plant metabolites, leading to evolutionary associations between particular groups of pests and plants. Some herbivores can store (sequester) plant toxins to protect themselves against enemies. These may also inhibit the growth of competitor plants (allelopathy). Significantly, pigments (such as terpenoid carotenes, phenolics and flavonoids) of coloured flowers along with terpene and phenolic odours, attract pollinators.

Most pharmaceutical industries are based on plant's extracted chemical products. Moreover, secondary metabolites (the alkaloids, nicotine and cocaine; the terpene, cannabinol) are also widely used for recreation and stimulation of physical activity along with mental relaxation. The study of such plants used for human consumption is called ethnopharmacology.

Higher plants are the rich source of bioactive constituents used in pharmaceutical industry. Some of the plant-derived natural products include: 1) drugs, such as morphine, codeine, cocaine, quinine, etc., anti-cancer catharanthus alkaloids, belladonna alkaloids, colchicines, phytostigminine, pilocarpine, reserpine, etc. 2) steroids, like diosgenin, digoxin and digitoxin. Technological development for the production of secondary metabolites has been directed primarily at cell culture. Plant cell culture offers the advantages of year-round availability of plant material, process and understanding of chemical reactions. The production of secondary metabolites in plants is possible as a result of plant development, which includes differentiation in metabolism as well as morphogenesis. Moreover, the production of secondary metabolites is the expression of a plant genome under developmental control.

Plant cell culture is viewed as a potential means of producing useful plant products such that conventional agriculture with all its concerned problems can be circumvented. These problems may be environmental factors (drought, floods, etc.), disease and uncontrollable variations in the crop quality. Thus, the production of useful and valuable secondary metabolites in large bioreactors is an attractive proposal. There are a number of examples of plant cells producing metabolites during *in vitro* culture, however, in contrast, production of these metabolites are not observed in the plant *in vivo*, for example, *Lithospermum erythrorhizon* cultures have been observed to synthesize rosmarinic acid during culture. It is apparent that the choice of original plant tissues having high yields of the desired phytochemical is important in establishing high-yielding cultures.

HISTORY

- Since the dawn of history, the human society for health and well-being has collected plants for drugs, colours, spices and fragrances. The 21,000 plus organic compounds listed as secondary metabolites show an enormous diversity and thus allowed chemical description to complement conventional classification of plants by morphological features. However, the large-scale cultivation of tobacco and a variety of vegetable cells was examined from the late 1950s to the early 1960s by Tulecke and Nickell. Their results stimulated more recent studies on the industrial application of plant cell culture in many countries.
- A Japanese company, Meiji Seika, has elucidated the fundamentals of production of *Panax ginseng* cells in large volumes. Furthermore, these research groups have claimed that cultured ginseng cells stimulated physiological activities in animals in a similar fashion as elicited by native ginseng roots. These cells are used as health foods in Japan. Several research workers also have conducted extensive

pharmacological screening of numerous cell cultures and found various novel products of great interest, including plasmin inhibitory proteins in *Scopolia japonica* cells and plant virus inhibitors in cultured cells of *Phytolacca americana* and other species. The virus inhibitors of *P. americana* are now being studied by many research groups in the world because of their activity against AIDS and other animal viruses.

- Zenk and his colleagues, in 1976, successfully selected high-alkaloid producing cell lines of *Catharanthus roseus* using a method similar to the microbial mono-colony isolation technique. A number of laboratories in industries and universities followed the same approach, and in fact, some researchers were successful in increasing significantly the level of secondary metabolites such as ubiquinone-10, biotin and various plant pigments produced by cell cultures.
- A combination of a plant cell culture process and a simple chemical coupling reaction was invented by a Canadian company, Allelix, to manufacture vinblastine, as a commercially feasible process. Similarly, sanguinarine production was also studied by Kurz *et al.*, since this alkaloid has a tremendous market value in dental care.
- The recent biotechnology boom has triggered interest in plant cell cultures. Currently, a number of companies and academic institutions in the US, Japan, Canada, India and Europe are investigating intensively the production of a very promising anti-tumour compound, taxol, using *in vitro* technology.

CLASSIFICATION OF SECONDARY METABOLITES

Plant secondary metabolites can be defined as compounds that have no recognized role in the maintenance of fundamental life processes in the plants that synthesize them, but they do have an important role in the interaction of the plant with its environment. Secondary metabolites can be classified on the basis of the following characters: 1) chemical structure (for example, having rings, containing a sugar) 2) chemical composition (presence or absence of nitrogen), 3) solubility in various solvents, 4) pathways by which they are synthesized (for example, phenylpropanoid, which produces tannins).

A simple classification of secondary metabolites includes three main groups:

1. The terpenes (made from mevalonic acid, composed almost entirely of carbon and hydrogen).
2. Phenolics (made from simple sugars, containing benzene rings, hydrogen and oxygen).
3. Nitrogen-containing compounds (extremely diverse, may also contain sulphur).

SOURCES AND EFFECTS OF MAJOR SECONDARY METABOLITES

1. *Carotenoids* Carotenoids are organic pigments occurring in plants and are mostly found in red, orange and yellow fruits and vegetables. Other vegetables such as broccoli, spinach or curly kale also contain carotenoids. Carotenoids have antioxidative effects and prevent cancer. In addition to this, they boost the immune system and reduce the risk of heart attacks.

2. *Phytosterols* Phytosterols are found in plant foods such as sunflower seeds, sesame, nuts and soybeans. Phytosterols protect against colon cancer, and lower cholesterol levels. Phytosterols are chemically similar to cholesterol and, therefore, they compete against each other for absorption in the body.

3. *Saponins* Saponins are flavour additives, which are found in legumes and spinach. Saponins boost the immune system, lower the cholesterol levels in the blood and reduce the risk of intestinal cancer.

4. *Glucosinolates* Glucosinolates are also flavour additives, which are found in all types of cabbages, mustard, radish and cress. Glucosinolates prevent infections and inhibit the development of cancer.

5. *Flavonoids* Flavonoids are organic pigments occurring in plants that give plants a red, violet or blue colour. Flavonoids have a particularly broad spectrum of efficacy. Flavonoids inhibit the growth of bacteria and viruses, protect the cells against the damages of free radicals, protect against cancers and heart attacks, have a repressive effect against inflammations and influence blood coagulation.

6. *Protease-inhibitors* Protease-inhibitors are found in plants that are rich in protein such as legumes, potatoes, wheat, etc. and they inhibit the decomposition of protein. Protease inhibitors protect the body against cancers and regulate the blood sugar levels.

7. *Terpenes* Terpenes are plant flavours (for example the menthol in peppermint oil or the essential oils in herbs and spices). Terpenes decrease the risk of cancer.

8. *Phytoestrogens* Phytoestrogens are natural plant hormones which are similar to the sexual hormones. Phytoestrogens are mostly found in wheat, legumes and wheat products. Phytoestrogens protect the body against hormone-dependant cancers such as breast, uterine and prostate cancer.

9. *Sulphides* Sulphides are compounds containing sulphur which are mostly found in plants that belong to the lily family such as onions, leeks, asparagus and garlic. Sulphides inhibit the growth of bacteria, lower cholesterol levels, protect the body from free radicals and have preventive effects against cancer.

10. *Phytic acid* Phytic acid is found in wheat, legumes and flax seeds. Phytic acid was considered undesirable for a long time because it binds trace elements such as iron and zinc and it also affects various digestive enzymes. However, new studies have proved that phytic acid has an antioxidant effect in the large intestine.

Table 21.1 shows some important secondary metabolites and their sources.

Table 21.1 Important secondary metabolites obtained from plants

Secondary metabolite	Plant
Rosmarinic acid	*Coleus blumei, Ocimum basilicum*
Capsidiol	*Capsicum annuum*
Shikonin	*Lithospermum erythrorhizon*
Taxol	*Taxus chinensis*
Taxol, baccatin III	*Taxus brevifolia, T. cuspidata*
5'-phosphodiesterase (PDase)	*Catharanthus roseus*
Acridone expoide	*Ruta graveolens*
Rishitin	*Jimson* sp.
Ajmalicine	*Catharanthus roseus*
Tropane	*Datura stramonium*
Isoflavonoids	*Lotus corniculatus*
Anthraquinones	*Morinda citrifolia*
Azadirachtin	*Azadirachta indica*
Berberine	*Thalictrum rugosum*
Capsidiol, debneyol, scopoletin, nicotine	*Nicotiana tabacum*
Thiophene	*Tagetes patula*
Codeine, morphine	*Papaver somniferum*
Diterpenoid tanshinones	*Salvia miltiorrhiza*
Glyceollins, apigenin, genistein, luteolin	*Glycine max*
Saponin	*Panax ginseng*
Scopoletin	*Lycopersicon esculentum*
Hyoscyamine, scopolamine	*Hyoscyamus niger, H. muticus*
Methoxymellein, 4-hydroxybenzoic acid	*Daucus carota*
Raucaffrincine	*Rauwolfia canescens*
Salidroside	*Rhodiola sachalinensis*
Thujaplicin	*Cupressus lusitanica*
Sanguinarine	*Eschscholtzia californica, Sanguinaria canadensis, Papaver bracteatum*

(Contd.)

Table 21.1 (Continued)

Secondary metabolite	Plant
Indole glucosinolates, camalexin	*Arabidopsis thaliana*
Kinobeon A	*Carthamus tinctorius*
Scopoletin	*Ammi majus*
Sesquiterpenes	*Hyoscyamus muticus*
Soyasaponin, resveratrol	*Vitis vinifera*
Tanshinone	*Salvia miltiorrhiza*
Taxuyunnamine C	*Taxus chinensis*
Tcibulin 1, tcibulin 2	*Allium cepa*
Tropane alkaloids	*Glycine max*
Diosgenin	*Panax ginseng*

STEPS IN THE PRODUCTION OF SECONDARY METABOLITES IN CULTURES/BIOREACTORS

1. Choice of Explants

In theory, any part obtained from any plant species can be employed to induce callus tissue, however, the successful production of callus depends upon plant species and their inherent potential of regeneration. Dicotyledons are rather amenable for callus tissue induction, as compared to monocotyledons; similarly, the callus of woody plants generally grows slowly. Stems, leaves, roots, flowers, seeds and other parts of plants are used as explants, but younger and fresh tissues, as explant source, are preferable in comparison to mature tissues.

Procured explants must be sterilized prior to culture by using ethanol, sodium hypochlorite and/or mercuric chloride in order to remove the entire possible surface living microorganisms. The period of explant treatment in these solutions depends upon the plant species, their parts and age. The explants must then be rinsed thoroughly with sterilized water.

2. Nutrient Culture Media

To induce callus from an explant and to cultivate the callus and cells in suspension, various kinds of media have been designed. Agar or its substitutes are added into the media to prepare solid medium for callus induction. One of the most commonly used media for plant cultures is MS medium, which was developed by Murashige and Skoog (1962) for tobacco tissue culture. The significant feature of the MS medium is its very high concentration of nitrate, potassium and ammonia. Simultaneously, B_5 medium, established by Gamborg *et al.* (1968), is also being used by many research groups. The levels of inorganic nutrients in the B_5 medium are lower than in MS medium.

Many other nutrient media have been developed and slightly modified. Nutrient compositions of some typical media are described in Table 21.2. However, it is not always necessary to test many kinds of basal media when a callus is induced. It would be better to use only one or two kinds of basal media in combination with different kinds and concentrations of phytohormones. The most suitable medium composition should be optimized afterwards in order to obtain higher level of products as well as higher growth rate.

Table 21.2 Various nutrient media for plant *in vitro* studies (mg/l)

Components	Murashige–Skoog (1962)	White (1963	Gamborg (1968)	Nitsch (1951)	Heller (1953)	Schenk–Hildebrandt (1972)	Nitsch–Nitsch (1967)	Kohlenbach–Schmidt (1975)	Knop (1865)
$(NH_4)_2SO_4$	-	-	134	-	-	-	-	-	-
$MgSO_4 \cdot 7H_2O$	370	720	500	250	250	400	125	185	250
Na_2SO_4	-	200	-	-	-	-	-	-	-
KCl	-	65	-	1,500	750	-	-	-	-
$CaCl_2 \cdot 2H_2O$	440	-	150	25	75	200	-	166	-
$NaNO_3$	-	-	-	-	600	-	-	-	-
KNO_3	1,900	80	3,000	2,000	-	2,500	125	950	250
$Ca(NO_3)_2 \cdot 4H_2O$	-	300	-	-	-	-	500	-	1,000
NH_4NO_3	1,650	-	-	-	-	-	-	720	-
$NaH_2PO_4 \cdot H_2O$	-	16.5	150	250	125	-	-	-	-
$NH_4H_2PO_4$	-	-	-	-	-	300	-	-	-
KH_2PO_4	170	-	-	-	-	-	125	68	250
$FeSO_4 \cdot 7H_2O)$	27.8	-	27.8	-	-	15	27.85	27.85	-
Na_2EDTA	37.3	-	37.3	-	-	20	37.25	37.25	-
$MnSO_4 \cdot 4H_2O$	22.3	7	10 $(1H_2O)$	3	0.1	10	25	25	-

(Contd.)

Table 21.2 (Continued)

Components	Murashige–Skoog (1962)	White (1963	Gamborg (1968)	Nitsch (1951)	Heller (1953)	Schenk–Hildebrandt (1972)	Nitsch–Nitsch (1967)	Kohlenbach–Schmidt (1975)	Knop (1865)
$ZnSO_4 \cdot 7H_2O$	8.6	3	2	0.5	1	0.1	10	10	-
$CuSO_4 \cdot 5H_2O$	0.025	-	0.025	0.025	0.03	0.2	0.025	0.025	-
H_2SO_4	-	-	-	0.5	-	-	-	-	-
$Fe_2(SO_4)_3$	-	2.5	-	-	-	-	-	-	-
$NiCl_2 \cdot 6H_2O$	-	-	-	-	0.03	-	-	-	-
$CoCl_2 \cdot 6H_2O$	0.025	-	0.025	-	-	0.1	0.025	-	-
$AlCl_3$	-	-	-	-	0.03	-	-	-	-
$FeCl_3 \cdot 6H_2O$	-	-	-	-	1	-	-	-	-
$FeC_6O_5H_4 \cdot 5H_2O$	-	-	-	10	-	-	-	-	-
KI	0.83	0.75	0.75	0.5	0.01	1.0	-	-	-
H_3BO_3	6.2	1.5	3	0.5	1	5	10	10	-
$Na_2M_0O_4 \cdot 2H_2O$	0.25	-	0.25	0.25	-	0.1	0.25	0.25	-
Sucrose	30,000–	20,000–	20,000–	50,000 or 36,000	20,000–	30,000–	20,000–30,000	10,000–	-
Glucose									-
Myo-Inositol	100	-	100	-	-	1,000	100	100	-
Nicotinic Acid	0.5	0.5	1.0	-	-	0.5	5	5	-
Pyridoxine HCl	0.5	0.1	1.0	-	-	0.5	0.5	0.5	-
Thiamine HCl	0.1-1	0.1	10	1	1	5	0.5	0.5	-
Ca-Pantothenate	-	1	-	-	-	-	-	-	-
Biotin	-	-	-	-	-	-	0.05	0.05	-
Glycine	2	3	-	-	-	-	2	2	-
Cysteine HCl	-	1	-	10	-	-	-	-	-
Folic Acid	-	-	-	-	-	-	0.5	0.5	-
Glutamine	-	-	-	-	-	-	-	14.7	-

3. Callus Induction

The sterilized explants are cut into approximately 0.5–1.0 cm long segments using a sterilized scalpel and each segment is transferred with tweezers to a solid agar medium in a flask or a Petri dish. The explant tissues inoculated on the nutrient medium is further incubated aseptically at around 25 ± 2°C for several weeks or more, which leads to the induction and development of callus tissue as a morphogenic response. Moreover, these calluses are further subcultured by transferring a small piece of callus to fresh solid medium. After several subsequent transfers, the callus becomes soft and fragile.

4. Suspension Culture

The growth rate of the suspension-cultured cells is generally faster than that of the solid culture. The former is more desirable, particularly in the production of useful metabolites on a large scale. A piece of the callus is transferred to a liquid medium in a vessel such as an Erlenmeyer flask and these vessels are kept on a rotary shaker. The culture conditions depend upon the plant species and other factors, but, in general, the cells are cultivated at 100 rpm on a rotary shaker at 25 ± 2°C. By repeated subculture for several generations, a fine cell suspension culture containing small cell aggregates and single cells are established. The time required to establish the cell suspension culture varies greatly and depends on the tissue of the plant species and the medium composition. The cells in suspension are also used for a large-scale culture with jar-fermentors and bioreactors.

For commercialization, it is necessary to progress through several stages of suspension culture and increase the volume at each stage until the requisite bioreactor size is attained. In principle, it is anticipated that such large-scale suspension cultures will be suitable for industrial production of useful plant chemicals such as pharmaceuticals and food additives similar to that of microbial fermentation. Generally, the culture period in plant cell cultures is longer than that of microbial cultures and it is crucial to protect plant cultures against microbial contamination.

Table 21.3 indicates some of the problems that are encountered with plant cell cultures. The sensitivity to shear is due to the large size of the cells as well as the relatively inflexible cellulose cell wall. Thus, with normal blade impellers, the cells may twist, which will inhibit mitosis and, for this reason, air-lift fermentors are recommended. The large size of the plant cell contributes to its comparatively high doubling time (12 hours to several days), which thus prolongs the time required to run a successful fermentation.

The vacuole of the cell is the major site of product accumulation, and since the product secretion is uncommon, the high metabolite yields as that seen in microorganisms that secrete product cannot be expected.

Cultured plant cells often produce reduced quantities and different profiles of secondary metabolites when compared to the intact plant and these quantitative and qualitative features may change with time.

The poor product expression is often attributed to a lack of differentiation in cultures. On the other hand, there are cases of cultures that over-produce metabolites compared to the whole plant.

Table 21.3 Comparison between plant cell culture and microbial cultures

Characteristics	Plant cell cultures	Microbial cultures
Size	>10 μ	2 μ
Water content	>90%	75%
Time for duplication	Days	<1 hour
Stress sensitivity	Sensitive	Insensitive
Fermentation time	Weeks	Days
Accumulation of secondary metabolites	Vacuoles	Medium

5. Fermentors or Bioreactors

In order to cultivate plant cells on a large scale, fermentors with different sizes are useful. Since the beginning of the 1960s, various types of fermentors have been designed. The simplest vessel is a carboy system described by Tulecke and Nickell in 1959, which consists of a rubber-stoppered 20 litre carboy fitted with four tubes (air-in, air-out, medium-in and sample-out). Filtered compressed air is employed for oxygen supply, aeration and agitation of the medium.

However, the most common types of systems on the bench are a stirred-jar fermentor which is used for microbial cultivation, although some minor alterations are made (Figure 21.1). It is suggested that an agitation speed of 50 to 100 rpm is most appropriate for the growth of tobacco cells in stirred-jar fermentors. It is true that cultured plant cells are more fragile than bacterial cells, however, in case, if cell lines differ in their resistance to shear effects, a single optimum agitation speed cannot be employed for all lines.

Fermentors installed with agitators, which are similar to those of microbial cultures have been employed for commercial production of shikonin and ginseng cells, although, some modification in equipment was made to optimize physical conditions. A German company has equipped five sophisticated fermentors of up to 75,000 litres capacity for plant cell cultures. This company has also cultivated *Echinacea purpurea* cells for the production of immuno-biologically active polysaccharides.

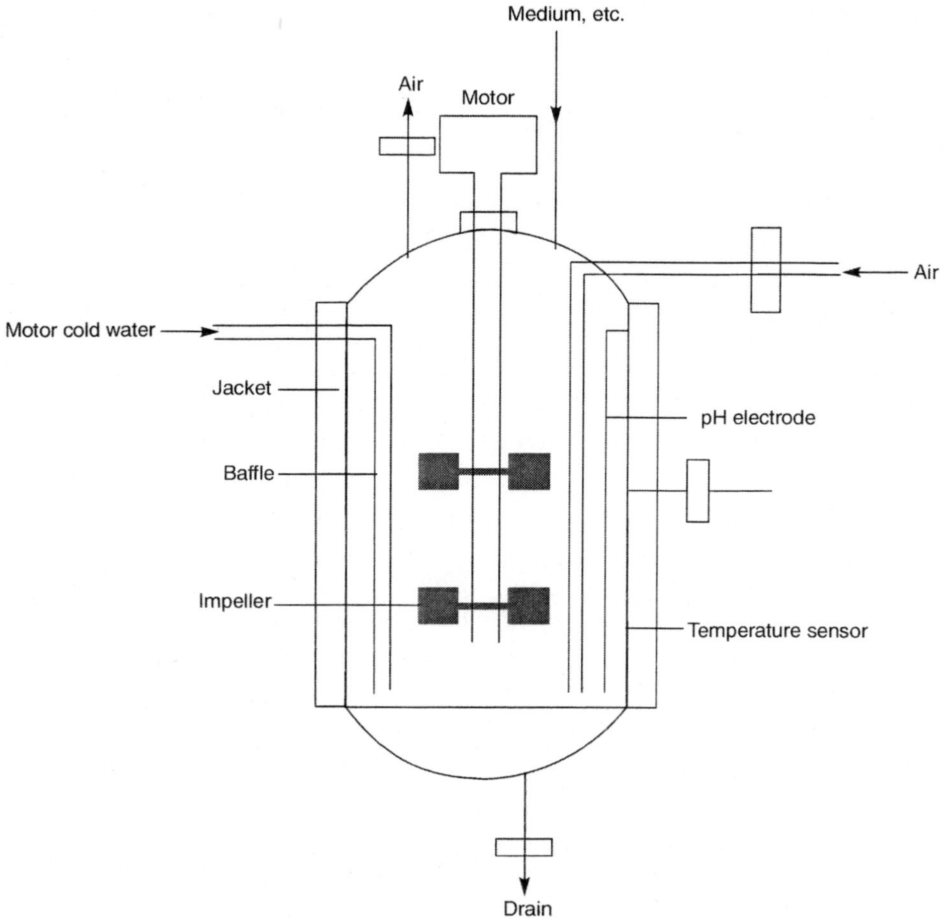

Figure 21.1 Stirred tank bioreactor

ENHANCEMENT IN SECONDARY METABOLITE PRODUCTIVITY

Several products have been found to be accumulated in cultured cells at a higher level than those in native plants through optimization of culture conditions. Generally, some of the products like ginsenosides in *Panax ginseng,* rosmarinic acid in *Coleus blumei,* shikonin in *Lithospermum erythrorhizon,* diosgen in in *Dioscorea,* ubiquinone-10 in *Nicotiana tabacum* accumulate in much higher levels in cultured cells than in the intact plants. However, in many cases, it has been observed that quantities of desired products were very low or sometimes not detectable in dedifferentiated cells such as callus tissues or suspension-cultured cells. In order to obtain products in concentrations high enough for commercial manufacturing, efforts have been made to stimulate or restore biosynthetic activities of cultured cells using various methods. The following are the possible approaches that probably may increase productivity of cultured plant cells.

Standardization of Protocols

1. *Nutrient medium composition* A number of chemical and physical factors that affect cultivation have been tested extensively with various plant cells. These factors include media components, phytohormones (growth regulators), pH, temperature, aeration, agitation, light, etc. This is the most fundamental approach in plant cell culture technology. Sucrose and glucose are the preferred carbon source for plant cell or tissue cultures. In many cases, the concentration of the carbon source affects cell growth and also the quantity of secondary metabolites. The maximum yield of rosmarinic acid 3.5 g/l is produced by cell suspension cultures of *Salvia officinalis* when 5% of sucrose is being used, but remarkably, it is just 0.7 g/l, if the medium is supplemented with 3% sucrose.

Among a number of other components in the nutrient medium, growth hormones such as auxins and cytokinins have shown the most remarkable effects on growth and productivity of plant metabolites. In general, an increase of auxin levels, such as 2,4-D, in the medium, stimulates dedifferentiation of the cells and consequently reduces the level of secondary metabolites. Therefore, it is suggested that auxins should be added to the medium for callus induction, but at a low concentration and furthermore, these are to be removed for the production of metabolites.

Moreover, it is also seen that in *C. roseus*, cytokinins stimulate alkaloid synthesis, which is induced by removing auxin from the medium. However, in contrast, production of L-DOPA by *Mucuna pruriens*, ubiquinone-10 by *N. tabacum* and diosgenin by *Dioscorea deltoidea* have been stimulated by high levels of 2,4-D. Significantly, kinetin is one of the most popular cytokinins, however, 4-chloro-2-diphenylurea is also reported to stimulate production of an antitumour compound, tripdiolide, by *Triptorygium wilfordii* cell cultures. Gibberellic acid is also effective on plant cell cultures. It is recently reported that the growth of callus of a taxol-producing plant, *Taxus cuspidata*, was significantly promoted by the addition of gibberellic acid to the medium.

2. *Temperature, light, pH and oxygen* Temperature, pH, light and oxygen are the basic parameters, the effects of which must be properly analysed in the study of secondary metabolite production. A temperature of 17–25°C is normally used for the induction of callus tissues and growth of cultured cells. But, each plant species may prefer a different temperature range. The pH of the nutrient medium is usually adjusted between 5 and 6 before autoclaving . The optimum pH is determined and controlled using a small-scale bioreactor or a jar-fermentor with pH control equipment.

Presently, the use of stainless steel tanks for growth of plant cells on an industrial scale, in general, eliminates the use of light. However, since there are cases of light-stimulated secondary metabolite production, light factor must be investigated in detail, as it could help elucidate regulatory factors. The effect of quality of light on production of different secondary metabolites in cultures is shown in Table 21.4. Modification of fermentors with lighting facilities has also been carried out. These include modification of impellers or agitators for microbial fermentors, improvement of air-lift fermentors and use of new type reactors such as rotary-drum fermentors. Each plant species has different optimized conditions, both for growth of the cells and for production of useful products, so it is necessary to optimize the conditions in each case separately.

Table 21.4 Increased production of secondary metabolites by the use of light in cultures

Quality of light	Secondary metabolites and source
Red	Podophyllotoxins from *Podophyllum peltatum*
	Chlorogenic acid from *Haplopappus gracilis*
White	Flavonoids from *Citrus aurantium*
	Anthocyanins from *Daucus carota* and *Rosa* sp.
	Cardiac glycosides from *Digitalis lanata*
	Alkaloids from *Ruta graveolens, Ephedra gerardiana,* etc.
Blue	Anthocyanins from *Haplopappus gracilis*
Green	Anthocyanins from *Populus* sp.
Ultraviolet	Glyceollin from *Glycine max*

3. *High cell density* To increase the productivity of secondary metabolites, high cell density cultures have been investigated. Using a newly designed fermentor and optimized culture medium, *Coptis japonica* cells are grown at higher cell density, up to 75 g/l of cell mass. However, the highest yield of berberine 3.5 g/l is produced intracellularly at relatively lower cell density, 55 g/l of the cell mass.

4. *Absorption of products* Most products are generally accumulated intracellularly by cultured plant cells, but some compounds are reported to be secreted into the media where they may be degraded or their synthesis may be regulated by feedback inhibition. In such cases, it is helpful to remove the metabolites from the medium just as they are released from cells. This is achieved by the addition of a compound into the medium, which either adsorbs (for example, activated charcoal, polyethylene glycol, XAD-2, XAD-4, XAD-7, etc.), dissolves (for example, Miglyol) or encapsulates (for example, B-cyclodextrin) these metabolites. *Cinchona ledgeriana* cells excrete anthraquinones in the liquid medium. Robins *et al.* reported that addition of a resin, XAD-7, into its suspension culture stimulates the production of anthraquinones up to 539 mg/l which is approximately 15 times increase compared to the medium without resin. The pigments are mostly found to be absorbed by the resin.

5. *Selection of high-producing cell strains* The physiological characteristics of individual plant cells are not always uniform. Biochemical content of cell cultures depends on the plant species, the genotype of the species, the specific plant of a strain, and even the explants from which they are derived. For example, pigment-producing cell aggregates typically consist of producing and non-producing cells. Cell cultures are highly heterogeneous for biochemical production in that different cells show different levels of production. This variation is used to advantage by screening a large number of clones for isolation of high-producing clones. It has been observed in cell lines of *Catharanthus roseus,* where higher levels of ajmalicine and serpentine were accumulated as determined by radioimmunoassay. A strain of *Euphorbia milii* was also recognized to accumulate about seven times higher amounts of anthocyanins than that of the parent strain after 24 selections.

However, the cell cloning is especially compatible with selection for high pigment production, since selection can be achieved visually or with the use of simple spectrophotometric analysis. Cell cloning is undoubtedly a very useful technique to increase the level of secondary metabolites and it should be applied as widely as possible. However, it is not clear exactly why cultures contain both high- and low-yielding cells.

6. *Addition of precursors* Addition of appropriate precursors or related compounds to the culture media sometimes stimulates secondary metabolite production, but this approach is advantageous if the precursors are inexpensive. For example, amino acids have been added to cell suspension culture media for production of tropane alkaloids, indole alkaloids and ephedrin and some stimulative effects have also been observed. Phenylalanine is one of the biosynthetic precursors of rosmarinic acid. Addition of this amino acid to *Salvia officinalis* suspension cultures stimulates the production of rosmarinic acid and reduces the production time as well. Similarly, addition of phenylalanine into the agar medium of *Taxus cuspidata* cells stimulates the biosynthesis of taxol.

However, biotransformation using intact cells or immobilized cells is an alternative way of producing a product by adding precursors into the culture media. *Rauwolfia serpentina* forms a new indole alkaloid, 6-hydroxytaumacline, in significant amounts when the cells are cultivated in the presence of ajmalicine. Arbutin, a skin depigmentation agent, is produced by biotransformation of hydroquinone using *C. roseus* cells. Addition of the precursor into the liquid culture medium of this cell line also produces arbutin efficiently.

7. *Biotransformation* Modification of an organic compound by simple chemically defined reactions, catalysed by enzymes present in cells, into a final product that can be recovered is called biotransformation. Instead of the addition of a particular compound as a precursor into the culture medium of plant cells, a suitable substrate compound may be biotransformed to a desired product using plant cells. This approach has been extensively applied in the fermentation industry using microorganisms and their enzymes. The interest in bioconversion is mainly because the product of the process is more valuable than the precursor used. For example, L-aspartic acid and L-malic acid are being manufactured commercially from fumaric acid, using microorganisms. Biotransformation of β-methyldigitoxin to β-methyldigoxin using *D. lanata* cells has been extensively investigated (Figure 21.2) because digoxin has a large market value as a cardiac glycoside.

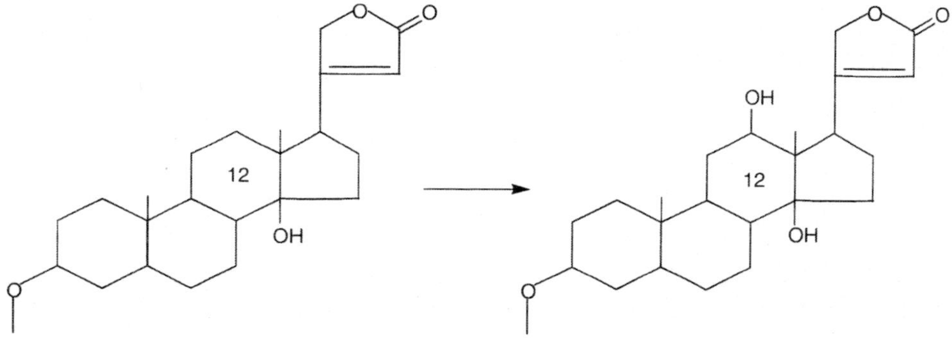

Digitoxin Digoxin

Figure 21.2 Structure of digitoxin and digoxin

Biotransformation process, including addition of substrates into the cultures, is one of the most commercially practical approaches in plant cell culture because of economic reasons. However, the availability of inexpensive precursors is a key problem.

8. *Elicitation* An elicitor may be defined as a substance which, when introduced in small concentrations to a living cell system, initiates or improves the biosynthesis of specific compounds. Elicitation is the process of induced or enhanced biosynthesis of metabolites due to addition of trace amounts of elicitors.

Elicitors can be classified on the basis of their 1) nature (abiotic elicitors or biotic elicitors) or 2) origin (exogenous elicitors and endogenous elicitors). Abiotic elicitors are substances of non-biological origin, predominantly inorganic salts and physical factors acting as elicitors like, Cu and Cd ions, Ca^{+2} and high pH. However, biotic elicitors are substances of biological origin. They include polysaccharides derived from plant cell walls (pectin or cellulose) and microorganisms (chitin or glucans) and glycol proteins or intracellular proteins whose functions are coupled to receptors and act by activating or inactivating a number of enzymes or ion channels. Exogenous elicitors are substances originating outside the cell, like polysaccharides, polyamines and fatty acids, whereas endogenous elicitors are substances originating inside the cell, like galacturonide or hepta-glucosides, etc.

Microbial infections of intact plants often elicit the synthesis of specific secondary metabolites. In case of fungal pathogens, the regulatory molecules have been identified as glucan polymers, glycoproteins and low-molecular weight organic acids. There is a strong correlation between stress and secondary metabolism in cultured cells. The response to elicitor polysaccharides seems to be similar to the defence response to microorganism attack. Recent developments in phytochemical elicitation have shown that simple inorganic and organic molecules can induce product accumulation. Sodium orthovanadate and vanadyl sulphate induce the accumulation of isoflavone glucosides in *Vigna angularis* cultures and indole alkaloid accumulation in *Catharanthus roseus* cultures, respectively. Other substances that have been found to stimulate alkaloid accumulation in *C. roseus* include sodium chloride, potassium chloride, sorbitol and abscisic acid.

9. *Immobilization of cells* In this method, cells are entrapped or immobilized either in gels, such as agarose, agar, polyacrylamide, chitosan, gelatin, gellan and calcium alginate, to produce beads, or in a membrane or metal screen cylinder. The cells entrapped in the membrane are kept in a chamber through which the nutrient medium is circulated from a medium recycle chamber. The medium flows parallel to the membrane cylinder and diffuses across the screen into the cell population. Simultaneously, the secondary metabolites diffuse into the medium from cells. A simple immobilized cell bioreactor is shown in Figure 21.3.

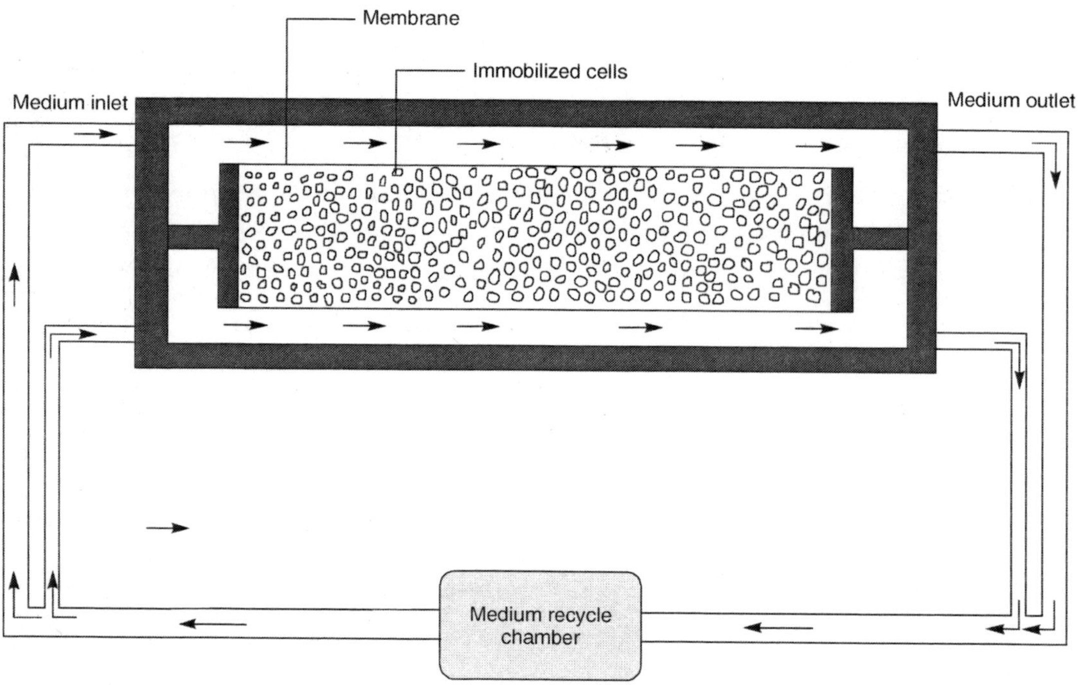

Figure 21.3 A schematic representation of a simple immobilized cell bioreactor

Immobilization of plant cells is considered to be of importance in growth and development in plant cell culture technology, because of the potential benefits that include:

1. Extended viability of cells in the stationary stage, enabling maintenance of biomass over a prolonged time period.
2. Simplified downstream processing (if products are secreted).

3. The (putative) promotion of differentiation that could be linked with enhancing secondary metabolism.
4. Higher cell density enabling a reduced bioreactor size, thereby reducing costs and the risk of contamination.
5. Reduced shear sensitivity.
6. Promotion of secondary metabolite secretion, in some cases.
7. Greater flow rates when flow-through reactors are used.

An immobilization system which could maintain viable cells over an extended period of time and release the bulk of the product into the extracellular medium in a stable form could dramatically reduce the costs of phytochemical production in plant cell culture. However, an immobilized system also has some limitations as described below:

1. Immobilization is normally limited to cases where production is decoupled from cell growth.
2. The initial biomass must be grown in suspension cultures.
3. Secretion of product into the extracellular medium is imperative.
4. At the site where secretion occurs, there may be chances of extracellular degradation of the products.
5. When gel entrapment is used, the gel matrix introduces an additional diffusion barrier.

Due to these constraints, a system with commercial potential has not yet been developed in plant cell cultures, however, various immobilization methods have been developed (for example, entrapment, adsorption and covalent coupling). Some preliminary results with *C. roseus* have been obtained with immobilized cells and indicated that agar, agarose and carageenan are all suitable immobilization matrices suitable for maintenance of cell viability; but alginate is superior in terms of ajmalicine production, however, the possibility that alginate acts as an elicitor of secondary metabolism cannot be ruled out. Moreover, agar has been shown to stimulate shikonin accumulation in *L. erythrorhizon* cultures.

Adsorption immobilization has been successfully used with a number of plant species, like:

- ***Capsicum frutescens*** Cells immobilized on polyurethane foam produced 50 times as much capsaicin as suspension cells.
- ***Solanum nigrum*** Cells accumulated glycoalkaloids to levels exceeding those found in suspensions.

In general, it appears that mild immobilization either through gel entrapment or surface adsorption enhances productivity and prolongs the viability of cultured cells. Immobilized cells can also be used as biocatalysts for biotransformations. Such a system compares favourably with the use of freely suspended cells, since, in the case of immobilization, the catalyst is theoretically reusable and the product is easily separated from the biomass.

10. *Product secretion (feedback inhibition)* Many secondary metabolites produced by cell cultures have been reported to be accumulated intracellularly. However, it may be possible to produce much higher levels of product, if it is secreted into the medium. The product intracellularly accumulated sometimes inhibits its own synthesis by regulation mechanisms such as product inhibition and repression. For many immobilized plant cell systems to work, it is essential that a significant amount of product is released into the medium. Two types of immobilization system with *C. roseus,* that is, gel entrapment in polysaccharide beads and in polyacrylamide sheets, both exhibit alkaloid release by mechanisms that do not appear to be associated with losses in viability. *Capsicum frutescens* cells immobilized on polyurethane releases capsaicin entirely into the medium, although other species immobilized by the same method retains the product intracellularly.

Permeabilizing agents can increase the permeability of the cell but it will affect cell viability. DMSO, electroporation, ultrasonication, chitosan, Triton X-100, etc. are some of the examples of permeabilizing agents.

HAIRY ROOT CULTURE

The use of *Agrobacterium rhizogenes* has been receiving attention recently in secondary metabolism research. It inserts the *Ri* plasmid into wounded tissue, causing the growth of very fine and profuse adventitious roots, called "hairy-roots". These roots can be cultured on hormone-free medium and there are several examples of enhanced accumulation of secondary metabolites, relative to non-transformed tissue:

1. All hairy root clones of *Hyoscyamus* plants grow faster than ordinary root cultures and produce the similar level of tropane alkaloids to that accumulated in the intact plants.
2. Hairy roots of *Beta vulgaris* have twice as much betacyanin and three times as much betaxanthin as seedling roots. On mg/g dry weight basis, the concentrations of these betalains were equal to or greater than those reported for storage roots.
3. Hairy root cultures of *Lupinus polyphyllus* and *L. hartweigii* produce higher amounts of isoflavine glucosides than those of hormone-dependent cells.

To cultivate hairy roots in large-scale fermentors, a fermentor comprising a vessel defining a fermentation chamber and a space-filling wire lattice of inoculation points designed to be located within the chamber is required. At the start of the fermentation process, the fermentation vessel is inoculated with suitable small lengths of hairy roots which then grow substantially to fill the vessel. It has been found to be highly advantageous if the lengths of roots which form the inoculants are distributed throughout the vessel. To achieve this, a space-filling lattice of inoculation points is provided in the fermentation vessel. The inoculating lengths are suspended in a suitable medium which is then passed into the vessel.

The inoculants' lengths lodge at the inoculation points and thereafter grow in a conventional fashion. The biotechnological application of hairy root cultures is promising for a number of reasons like:

1. Stable and high-level production.
2. Fast and regulator-independent growth.
3. Suitability for adaptation to fermentor/bioreactor systems.

It is, therefore, one of the most feasible techniques from the industrial point of view.

IN VITRO-SYNTHESIZED PRODUCTS—A COST FACTOR ASSESSMENT

In spite of remarkable advances in plant cell culture technology, the production cost of metabolites is still high. Significantly, industrial plant cell culture techniques would be introduced only if the plant product under consideration is produced at a price equal to or preferably lower than the field-produced product. The only factor determining the industrial realization of plant cell culture is the price at which a given product can be produced.

The assessment of product cost makes the future of commercial plant cell culture technology quite viable, but the crucial problems to be overcome are enhancing product levels, recycling of biomass (for example, through immobilization) and raising the biomass concentration. In comparison with microbial cultures, doubling times are long and product yields are low. Furthermore, since production is often not growth-linked, there are many cases where two-stage productions are necessary, thus increasing the costs markedly.

Therefore, it is more imperative to choose products that are particularly expensive to produce in the field. Shikonin is a good example, in that the long growing period (3–5 years) and strict climatic requirements mean that the cost of the plant raw material is high. It has been suggested that a quick way to assess the attractiveness of a cell culture method versus conventional agriculture, is to calculate a specific biosynthesis rate "based on the final dry weight and the total time of fermentation or land occupation". With shikonin, there is 830-fold increase in plant cell culture, while with ajmalicine from *C. roseus,* there is only a 24-fold improvement.

REVIEW QUESTIONS

1. What are secondary metabolites? How is the production of these metabolites in culture better than *in vivo* production?
2. Explain the process of secondary metabolite production in culture.
3. Compare plant cell culture with microbial cell cultures.

4. What are fermentors or bioreactors? Describe their use in secondary metabolite production.
5. How can we enhance the production of secondary metabolites in culture?
6. Explain about the optimization of culture conditions.
7. Write short notes on the following:
   i. Biotransformation
   ii. Precursors
8. What are elicitors and their role in enhancement of production of secondary metabolites?
9. How are immobilized cells useful in the production of secondary metabolites?
10. Write notes on hairy root cultures.

REFERENCES

Akkerwar, D. M. (1999). Natural colors of commercial importance. In: *Book of Papers.* Convention on Natural Dyes. Dept. of Textile Technology, IIT, Delhi. pp. 311–312.

Constabel, F. (1988). Principles underlying the use of plant cell fermentation for secondary metabolite production. *Biochem. Cell Biol.* 66: 658–664.

Constabel, F. and Tyler, R.T. (1994). Cell culture for production of secondary metabolites. In: Vasil, I. K. and Thorpe, T.A. (eds.). *Plant Cell and Tissue Culture.* Kluwer Academic Publishers, The Netherlands. pp. 271–292.

Deno, H., Suga, C., Morimoto, T. and Fujita, Y. (1987). Production of shikonin derivatives by cell suspension cultures of *Lithospermum erythrorhizon* VI. Production of shikonin derivatives by two-layer culture containing on organic solvent. *Plant Cell Reports.* 6: 197–199.

Fujita, Y., Hara, Y., Ogino, T. and Suga, C. (1981a). Production of shikonin derivatives by cell suspension cultures of *Lithospermum erythrorhizon.* I. Effects of nitrogen sources on the production of shikonin derivatives. *Plant Cell Rep.* 1: 59–60.

Fujita,Y., Hara. Y., Suga, C. and Morimoto, T. (1981b). Production of shikonin derivatives by cell suspension cultures of *Lithospermum erythrorhizon.* II. A new medium for the production of shikonin derivatives. *Plant Cell Rep.* 1: 61–63.

Fujita,Y., Tabata, M., Nishi, A. and Yamada, Y. (1982). New medium and production of secondary compounds with the two-stage culture method. In: Fujiwara, A. (ed.). *Plant tissue culture.* Maruzen, Tokyo. pp. 399–400.

Fujita, Y., Takahashi, S. and Yamada, Y. (1985). Selection of cell lines with high productivity of shikonin derivatives by Protoplast culture of *Lithospermum erythrorhizon* cells. *Agric. Biol. Chem.* 49: 1755–1759.

Fujita, Y. and Hara, Y. (1985). The effective production of shikonin by cultures with an increased cell population. *Agric. Biol. Chem.* 49: 2071–2075.

Fujita, Y. (1988). Shikonin: production by plant (*Lithospermum erythrorhizon*) cell cultures. In: Bajaj, Y. P. S. (ed.). *Biotechnology in Agriculture and Forestry*. Vol. 4. *Medicinal and Aromatic plants I*. Springer-Verlag, Berlin. pp. 225–236.

Kamboj, V. P. (2000). Herbal medicine. *Current Science*. 78: 35–39.

Kumar, S., Singh, J., Shah, N. C. and Ranjan, V. (1997). Indian medicinal and aromatic plants facing genetic erosion. CIMAP. Lucknow. pp. 49–50.

Mizukami, H., Konoshima, M. and Tabata, M. (1977). Effect of nutritional factors on shikonin derivative formation in *Lithospermum* callus cultures. *Phytochemistry*. 16: 1183–1186.

Sim, S. J., and Chang, H.N. (1997). Shikonin production by hairy roots of *Lithospermum erythrorhizon* in bioreactors with in situ separation. In: Doran, P.M. (ed.). *Hairy Roots Culture and Applications*. Harwood Academic Publishers. pp. 219–225.

Sukh Dev. (1997). Ethanotherapeutics and modern drug development: The potential of Ayurveda. *Current Science*. 73: 909–928.

Tyler, V.E. (1999). Photomedicines: Back to the future (Invited review). *J. Nat. Prod.* 62: 1589–1592.

Verpoorte, R., Van der Heijden, R. ten Hoopen, H. J. G. and Memelink, J. (1998). Metabolic engineering for the improvement of plant secondary metabolite production. *Plant Tissue Culture and Biotechnology*. 4: 3–20.

22

SECONDARY METABOLITE PRODUCTION-II

COMMERCIAL PRODUCTION OF SECONDARY METABOLITES

Higher plants contain a variety of metabolic products which are useful as medicines, food additives, perfumes, etc. However, non-compromising reduction in plant resources, increase in labour cost and other problems in order to obtain these high value-added substances from natural plants, have forced to develop an alternative approach towards the use of plant cell culture for production of these metabolic products. Since, plant cell culture is not affected by changes in environmental conditions such as climate or natural depredation, improved production is possible anywhere or at any time as discussed in Chapter 21.

Although studies on the production of useful metabolite by plant cell culture have been carried out on an increasing scale since the early 1950s, the large-scale *in vitro* cultivation of tobacco and various vegetable cells was examined only in the late 1950s and early 1960s. Furthermore, their results stimulated more recent studies on the industrial application of *in vitro* cell technology.

PHARMACEUTICALLY IMPORTANT CHEMICALS

Alkaloids

A variety of alkaloids have been used as pharmaceuticals and most of them are plant metabolites. The typical tropane alkaloids, atropine, hyoscyamine, scopolamine and cocaine are widely used as blockers of the parasympathetic nervous system such as anodyne and antispasmodic. Study on production of these useful alkaloids by plant cell cultures has been carried out for more than 25 years; however, industrial production has not yet succeeded because of the low production ability of the cultured cells. Plants used for these studies are mainly *Atropa belladonna, Hyoscyamus niger, Datura meteloides,* etc.

Various approaches to increase their productivity have been tried by many research groups and significantly at present, vinblastine, an antitumour alkaloid, is most likely to be produced commercially using a combination process of plant cell culture and chemical synthesis.

1. *Morphine alkaloids* Codeine is an analgesic and cough-suppressing drug and *Papaver somniferum* is a traditional commercial source of codeine. Also, morphine can be converted into codeine. Mature capsules of *P. bracteatum* accumulate up to 3.5% of thebaine which also can be converted to codeine. Many workers have tried to produce codeine by undifferentiated cells of these plants and a significant amount of success has also been achieved.

2. *Berberine* Berberine is an isoquinoline alkaloid which is produced and distributed in roots of *Coptis japonica* and cortex of *Phellondendron amurense.* Moreover, berberine chloride is used in the treatment of intestinal disorders in the Orient and generally, it takes 5 to 6 years to produce coptis roots as the raw material.

Significantly, the Mitsui Petrochemical, Japan, has improved the berberine productivity by the addition of 10^{-8} M gibberellic acid into the medium. This company is able to produce berberine in a large scale at a level of 1.4 g/l of their optimized medium within 2 weeks. Furthermore, they established a "high-density cell culture" process to produce berberine much more efficiently. In order to achieve a cell mass of 70 g/l on a dry weight basis, stirring without damaging the cells, supply of sufficient amounts of oxygen and that of appropriate nutrients should be optimized. In *Thalictrum minus* cell suspension cultures, addition of a polyamine spermidine has been found to stimulate the production of berberine.

3. *Tropane alkaloids* Scopolamine and hyoscyamine are being used commercially as anesthetic and antispasmodic drugs. These alkaloids occur in leaves of members of the family Solanaceae including *Duboisia myoporoides* and *Datura leichhardtii, Scopolia, Atropa* and *Hyoscyamus.*

The concentrations of scopolamine and hyoscyamine in cultured cells of *Atropa belladonna* are generally very low in spite of many efforts to increase the yield using various approaches. Therefore, the plant cell culture has not yet been employed to manufacture these alkaloids at commercial level. For example, additional tropic acid into *Scopolia japonica* suspension cultures as a precursor can increase the level of alkaloids up to 15 times.

4. *Cardinolides* Cardiac glycosides or cardinolides are products of *Digitalis* species. Some of these compounds have been employed in the treatment of heart diseases. For commercial production, *Digitalis* plants are being cultivated in the fields of several countries. The studies of plant cell cultures for the production of cardiac glycosides began more than 30 years ago.

Nutritional factors including growth regulators, sugars, nitrogen sources and vitamins in the medium have significant effect on differentiation of shoots and other organs and simultaneously, the secondary metabolites such as cardiac glycosides are often synthesized. The *in vitro* differentiated tissue culture is a prerequisite for secondary metabolite production in some cases, however, in general, the method requires much longer culture time than the suspension cell culture and consequently it is not efficient. Various secondary metabolites other than digitoxin and digoxin have been found in callus tissues of *D. lanata* and *D. purpurea*. These include cholesterol, campesterol, stigmasterol, β-sitosterol, 4-hydroxy-digitolutein and others.

In contrast to the *de novo* synthesis, the biotransformation process with *Digitalis* plant cells seems to be more promising from a commercial point of view. Leaf and root cultures of *D. lanata* and shoot-forming callus tissues of *D. purpurea* accumulate increased levels of digoxin and/or digitoxin when progesterone is added to the cultures. Among many studies, the biotransformation from digitoxin to digoxin using *Digitalis lanata* cells is the most interesting approach in terms of commercial application since digoxin has a higher demand as a drug for heart diseases, than digitoxin. It is advantageous that *Digitalis* leaves contain a larger amount of digitoxin which can be used as a substrate.

5. *L-DOPA* L-DOPA, L-3,4-dihydroxyphenylalanine, is an important intermediate of secondary metabolism in higher plants and is known as a precursor of alkaloids such as, betalain, melanine and others. It is also a precursor of catecholamines in animals and is being used as a potent drug for Parkinson's disease. The callus tissue of *Mucuna pruriens* accumulates 25 mg/l L-DOPA in the medium containing very high concentration of 2,4-D. Immobilized cells of *M. pruriens* within alginate can synthesize DOPA from tyrosine up to 2% of dry cell weight. The synthesized DOPA secretes mostly into the medium.

6. *Valepotriates* Members of Valerianaceae, such as *Nardostachys jatamansi*, *Valeriana wallichii* and *V. officinalis* var. *angustifolia* have been used as folk medicines in India. Although the active principles in these plants have not been identified, a group of compounds having biological activities such as sedative, tranquilization, cytotoxicity and antitumour activities were named as "valepotriates". *Fedia cornucopiae* and *V. locusta* cells during culture produced higher levels of the compounds than the intact plants. Isolation of cell lines resistant to trifluoroleucine and to nystatin, treatment with colchicine, cultivation of the cells in two-phase culture media and addition of several bioregulators have been carried out intensively. As valepotriates have monoterpene skeletons, L-leucine is considered as a precursor. The isolated cell lines of *V. wallichii* resistant to trifluoroleucine, a leucine analogue have been isolated but the yield of valepotriates in the cells doesn't increase although the intracellular level of leucine increases.

Antitumour Compounds

The plant kingdom is one of the attractive sources of novel antitumour compounds (Figure 22.1). These antitumour compounds include maytansine, tripdiolide, homoharringtonine, bruceantin, ellipticine, thalicarpine, indicine-N-oxide, and baccharin. In addition to these compounds, some of the higher plant products such as vinblastine, vincristine, podophyllotoxin derivatives including etoposide and camptothecin and its derivatives have already been marketed as very important anticancer drugs.

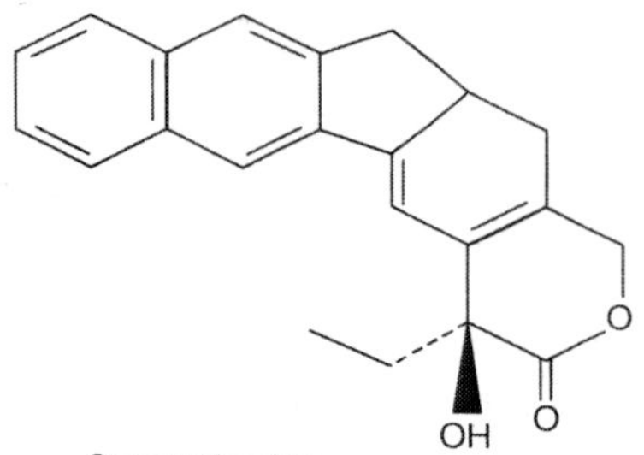

Camptothecine

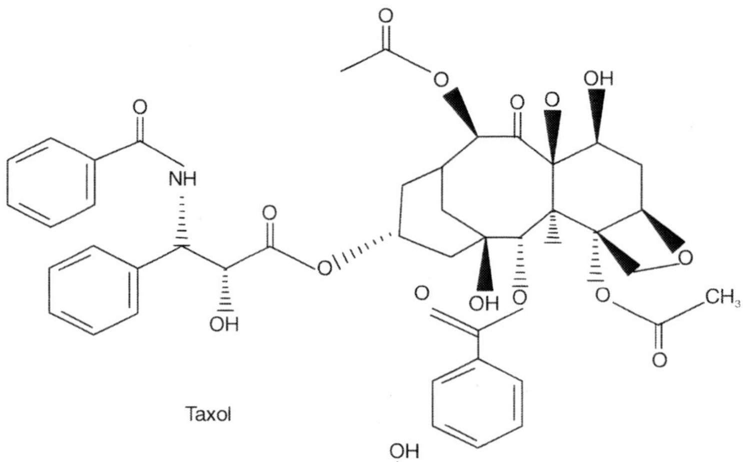

Taxol

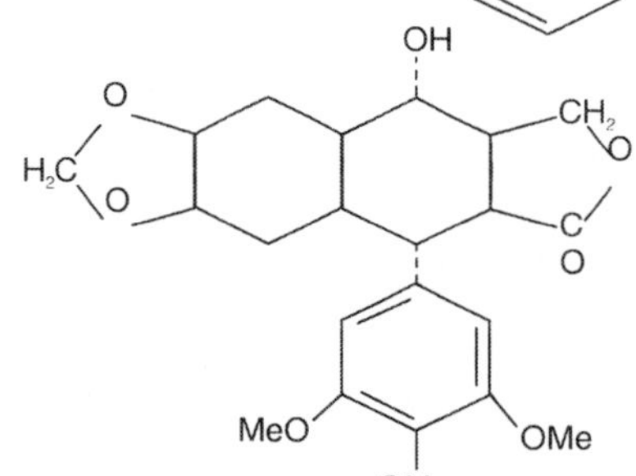

Podophyllotoxin

Figure 22.1 Secondary metabolites as anticancer agents

However, the concentration of these active compounds in plants is generally low, the growth rate of the plants is slow and the accumulation pattern of these compounds is highly susceptible to geographical or environmental conditions. Therefore, it is not an easy task to produce these compounds economically by extracting from the intact plants. Furthermore, as indicated for taxol production, the over exploitation of native plants is becoming a serious problem in terms of environmental preservation. Thus, plant *in vitro* culture technology is undoubtedly one of the appropriate approaches to solving these problems. Also many interesting antitumour compounds isolated from higher plants have very complicated chemical structures. Therefore, the research interest has grown for over a decade and a couple of processes based on plant cell culture technology are likely to be applied for commercial production of antitumour drugs. The major antitumour compounds produced by this technology are as follows:

1. *Vinca alkaloids* The dimeric indole alkaloids, vinblastine, and vincristine have become highly valued drugs in cancer chemotherapy due to their potent antitumour activity against various leukemias, Hodgkin's disease and solid tumours. They are currently produced commercially by extraction from *Catharanthus roseus* plants, but the process is not efficient because of very low concentrations of the alkaloids in the plant. It is reported that the concentration of both vinblastine and vincristine is only 0.0005% on a dry weight basis.

However, production of both alkaloids by *de novo* synthesis using the callus or the suspension-cultured cell of *C. roseus* is so far not promising because the productivity of the cultured cells has been proved to be very low. An alternative way was developed for the production of vinblastine and established an economically feasible process consisting of production of catharanthine by plant cell fermentation and a simple chemical or an enzymatic coupling.

The results show that the MS medium is the most favourable for catharanthine production but the optimal levels of phytohormones for growth and production are varied in different cell lines. For example, one line requires no phytohormones but another line requires 0.1 mg/l NAA and 0.1 mg/l kinetin. Addition of various chemically defined compounds to the medium as "inducers" is found to stimulate the production efficiently. Among them, effects of vanadyl sulphate, abscisic acid and NaCl, on the production of catharanthine are significant.

2. *Taxol* It is a diterpene amide and has shown activity against mouse melanoma tumour, human mammary xenograft and colon xenografts. The mode of action of taxol is rather unique because it stabilizes microtubules and inhibits depolymerization. Taxol is now being manufactured by extraction from the bark of wild-grown *Taxus brevifolia* trees. The demand for taxol will be undoubtedly increasing since it will be applied for other cancers including breast and lung cancers in the near future, but its supply is limited. Other related plants such as *T. canadensis* and *T. cuspidata* also contain taxol and other related compounds. Generally, the concentration of taxol in *Taxus* plants is very low. Therefore, harvesting *Taxus* trees for production of taxol commercially causes a serious problem from the environmental point of view.

As alternative ways, plant tissue and cell cultures as well as chemical transformation processes from baccatin extracted from needles of the plant and the total chemical syntheses have been investigated.

A cell line of *T. brevifolia* in suspension culture produces 3.9 mg/l taxol in the medium after 26 days of culture. The level of taxol was determined by reverse-phase HPLC. It would be interesting to note that all the produced taxol is secreted into the medium, which was very unusual for plant cell cultures. Taxol is one of the most suitable and desirable plant products to plant cell culture research because of its shortage of supply and high-value.

3. *Ginseng* The roots of *Panax ginseng*, have been widely used as a tonic and precious medicine since ancient times, particularly in the oriental countries, including Korea and China. It is effective for gastro-enteric disorders, diabetes and weak circulation, and has been used as an adjuvant to prevent various disorders, rather than a medicine to cure disorders. Thus, ginseng has been recognized as a miraculous medicine in preserving health and longevity. It has been known that the root contains various saponins and sapogenins. Among them, ginsenoside-Rb has a sedative activity, while ginsenoside-Rg has stimulative activities. Although there are several species of ginseng, the commercially important species, *P. ginseng* grows in the area of 30–48° north latitude such as Korea. The cultivation of ginseng in the field requires four to seven years, and it is impossible to plant consecutively for 20 to 50 years but the demand for the plant has increased dramatically in the world, and its price has soared. These are reasons why many research groups have tried to produce ginseng through plant tissue cell cultures.

Furthermore, callus tissues and redifferentiated roots of *P. ginseng* are able to accumulate saponins and sapogenins that are known from the intact plant. The callus tissues and roots are cultivated on both MS solid and liquid media containing vitamins, sucrose, 2,4-D and suitable natural nutrients such as soybean powder or beef extract for several weeks at 25–28°C. The concentrations of crude saponins in the callus (21.1%), in the crown gall (19.3%) and in the redifferentiated root (27.4%) are much higher than those in the natural root (4.1%). The saponins are found to contain ginsenoside-Rb and ginsenoside-Rg.

4. *Rosmarinic acid* Rosmarinic acid has been found in the members of the families Laminaceae and Boraginaceae. Rosmarinic acid and some related compounds are reported to have physiological or pharmaceutical activities. Oxidized rosmarinic acid is also reported to show antithyrotropic activity and rosmarinic acid itself has been shown to effectively suppress the complement-dependent components of endotoxin shock in rabbits. However these compounds have not yet been used as commercial drugs.

Cultured cells of several plant species such as *Coleus blumei, Anchusa officinalis* and *Lithospermum erythrorhizon* are found to accumulate rosmarinic acid. It is of interest that production of the acid is stable and that high levels are produced. It is described that the production of rosmarinic acid appears to be constitutively expressed in *C. blumei* cells, some of which had been in continuous culture for 10 years with no reduction in metabolite yield. Dedifferentiated cell cultures of *C. blumei* and *A. officinalis* accumulates almost exclusively, rosmarinic acid, with levels higher than in the intact plants, when both types of cells are cultivated in a Gamborg and Eveleigh's B_5 medium. The yield of the acid, before optimization, was about 1.4 g/l for *C. blumei* and 0.7 g/l for *A. officinalis*. Both species grow very well and reach up to 16 g/l within 30 to 40 hours of the culture period.

5. *Camptothecin* *Camptotheca acuminata*, a native of China, has been found to produce a potent antitumour alkaloid, camptothecin. The callus is induced from the stem of *C. acuminata* on MS medium containing 0.2 mg/l 2,4-D and 1 mg/l kinetin and furthermore, this induced callus can be transferred to the liquid medium for the formation of cell suspensions. Gibberelline, L-tryptophan and conditioned medium stimulate the growth of the cells. After 15 days of cultivation in suspension, the concentration of camptothecin in the cells is 0.0025% on a dry weight basis, which is about 1/20 of the level in the intact plant.

6. *Arbutin* Arbutin is widely distributed in various plants of the family Ericaceae such as *Arctostaphylos uva-ursi* and *Vaccinium vitisidaea*. The level of arbutin in the bark of *Pyrus communis* reaches up to 28%. *Arctostaphylos uva-ursi* has been used as an urethral disinfectant and its major active principle, arbutin, is shown to suppress the synthesis of melanin in human skin. A Japanese cosmetic company, has developed arbutin as an additive for the company's product lines because of its preventive activity towards pigmentation of skin.

Although arbutin is commercially available by a chemical process, researchers have investigated an alternative process using plant cell cultures including biotransformation. They showed that cultured cells of *Datura innoxia* have a remarkably high capability for glucosylation of hydroquinone to form arbutin, and hydroquinone is totally converted to arbutin within 10 hours after administration. The glucosylation is catalysed by an enzyme, uridine diphosphate glucose (UDPG)-hydroquinone glucosyltransferase. Moreover, they selected *C. roseus* cells as a producer of the enzymes since arbutin is formed efficiently when hydroquinone is added into the suspension culture.

7. *Homoharringtonine* Homoharringtonine, together with harringtonine and isoharringtonine, has been isolated from *Cephalotaxus harringtonia*. These alkaloids are complex esters of the inactive alcohol, cephalotaxine. These compounds inhibit the growth of murine leukemia. Homoharringtonine also shows activity against colon tumours, melanoma and leukemia in mice. Since the chemical synthesis of homoharringtonine is not efficient for commercialization, studies on *in vitro* cell culture have been conducted. MS medium containing 1 mg/l kinetin and 3 mg/l NAA is favourable to induce the callus from *C. harringtonia*, while the medium without growth regulators strongly promotes organogenesis. The radioimmunoassay established by the same group shows the levels of cephalotaxine and its esters are only 1/300 of the intact plant.

8. *Podophyllotoxin* *Podophyllum peltatum*, contains an antitumour lignan, podophyllotoxin. It is used against certain viral diseases and skin cancer. A semi-synthetic derivative of podophyllotoxin, etoposide (V-16), has been found to be active against brain tumour, lymphosarcoma and Hodgkins' disease.

Production of podophyllotoxin by *P. peltatum* cell cultures was first attempted by Kadkade and he found that a combination of 2,4-D and kinetin in the medium supports the high production. Moreover, red light stimulates production. Embryogenic root cultures are induced from a callus of the plant in a liquid MS medium supplemented with 1 mg/l NAA, 0.2 mg/l kinetin and 500 mg/l casein hydrolysates. The roots are then transferred to the medium without growth regulators. Furthermore, the accumulation of 1.6% of podophyllotoxin is observed in the dried tissues, which is six times higher than in the mother plant. To increase the yield of podophyllotoxin, a complex of the precursor, coniferyl alcohol, and β-cyclodextrin can be added to *Podophyllum hexandrum* cell suspension cultures.

AGRICULTURALLY IMPORTANT CHEMICALS

Virus Inhibitors

A large number of chemically synthesized compounds and natural molecules have been examined for their inhibitory effects on plant viruses and some of them have potent activity as protectors against virus infection.

In order to screen cultured cells that are high producers of plant viral inhibitors, a variety of callus extracts have been analysed regarding their inhibitory activity towards tobacco mosaic virus (TMV) infection, using tobacco disc method. Among these extracts, *Phytolacca americana* callus was selected as the most potent producer of the inhibitor. The inhibitor accumulates in the suspension-cultured cells and reaches a maximum level on the ninth day of culture using MS medium containing 1 mg/l 2,4-D. Furthermore, the cell suspension of *P. americana* is again homogenized and the supernatant is diluted up to 100 times with water. Significantly, the diluted solution appears to inhibit TMV infection on tobacco and tomato plants.

Another plant *Agrostemma githago* has been used as a more potent producer of the plant virus inhibitor. Remarkably, the growth of suspension-cultured cells of *A. githago* appears somewhat faster than *P. americana*. Active principles of *P. americana* and *A. githago* have been isolated with column-lite chromatography and electrofocusing. At least four basic proteins are obtained from *P. americana* cells whose molecular weights vary from 1.10×10^4 to 3.15×10^4. Among them, the highest molecular weight component contains sugars in the molecule. On the other hand, only one basic protein is isolated as the principle from *A. githago* culture and its molecular weight is 2.5×10^4. The proteins obtained from *P. americana* have been widely investigated because of their activities against HIV.

Simultaneously, various plants also have been tested and *Mirabilis jalapa* was found to be a producer of an anti-plant-virus protein. The callus is induced from the leaves of *M. jalapa* and its suspension-cultured cells established are found to accumulate the protein intracellularly. Optimization of production and cell growth, as well as selection of high-producing cell lines was conducted extensively. One of the lines produced 95 mg/l of the protein in the optimized medium, based on MS medium, on the seventh day of cultivation.

FOOD COLOURING AGENTS

1. *Saffron* Saffron is the name of a spice, which is made from the stamens of *Crocus sativus* and is prized for their use as a flavouring agent and colourant. The stigma of the plant contains crocin (yellow pigment), safranal (a fragrance) and picocrocin (bitter substance). The plant is grown mainly in India and Spain. It requires about 30,000–35,000 hand-picked blooms to produce 1 lb of dry saffron. Crocin, being a glycoside, is water-soluble and is not soluble in oils and fats. Saffron is sensitive to pH changes and is unstable towards light and oxidative conditions, but it is moderately resistant to heat. It is used in baked goods, soups, meat and curry products, cheese, and confectionary and as a condiment for rice in Indian foods. Saffron also has medicinal value for stomach ailments.

The very high value of the product is mainly due to the fact that the life of the flowers is very short, making harvesting difficult. Thus, this is an ideal target for plant tissue cultures. People have approached this problem through the propagation of saffron stigma-like structures *in vitro*. Further studies show that crocin and picrocrocin are present and, after heat treatment (as done with field-grown stigmas), safranal is produced. The composition of these phytochemicals corresponded with that of similarly treated young, intact stigmas.

2. *Shikonin* Shikonin and its derivatives such as acetyl shikonin and isobutyl shikonin that are accumulated in roots of *Lithospermum erythrorhizon* and *Arnebia hispidissima* are reddish purple pigments, and these have been used in traditional dyeing. The plant has also been used as an herbal medicine. Because of a shortage of this plant, mass cultivation of *L. erythrorhizon* cells to produce shikonin compounds has been realized. Therefore, its culture conditions are optimized extensively to increase the level of the products.

L. erythrorhizon produces shikonins in White's medium but the cell growth is poor in the same medium. On the other hand, Linsmair–Skoog's (LS) medium was recognized to support growth but not shikonin production. Therefore, there is a need for a two-stage culture for mass production of shikonin compounds. To proliferate the cells, LS medium was used at the first stage and then the cells are transferred into White's medium for production of shikonin compounds. In order to improve the yields further, optimization of components in both media is carried out extensively, and MG-5 and MG-9 media are established.

Since most cultured cells in liquid and solid media occur as aggregates, selection of high-producing cell lines from the aggregated cells is not effective and it is also quite labour-intensive. The Mitsui group of Japan prepared protoplasts from the cultured cells with appropriate enzymes and selected protoplasts producing large amounts of shikonin compounds using a cell sorter. The selected protoplasts are generated to cell lines and cultivated in suspension. From 48 cell lines, they obtained a cell line having 1.8-fold the productivity of the parent line. The cell line showed stable production of shikonin compounds.

Shikonin and its derivatives are being manufactured commercially by the company. A major application of the pigment is in bio-cosmetic industries (biolipstic, bionailpolish, etc.).

Significantly, a hairy root culture of *L. erythrorhizon* with *Agrobacterium rhizogenes* does not produce shikonin on solid MS medium but produces the pigment in the root culture medium and also secretes it into the medium. Addition of absorbents, XAD-2, XAD-4, charcoal and so on, increases the concentration of shikonin produced. The roots are cultivated in a 2.0 litre air-lift type fermentor connected to a XAD-2 column (25 g) through a peristaltic pump and 5 mg/day of shikonin is continuously produced during a period of more than 220 days.

3. *Anthocyanins* Anthocyanins are a large group of water-soluble pigments responsible for many of the bright colours seen in flowers and fruits. They are composed of an anthocyanidin and more than one sugar moiety and normally change colour over the pH range due to the existence of four pH-dependent forms. Thus, at low pH they are red and at higher pH value (over 6) they turn blue. They are commonly used in acidic solutions in order to impart a red colour to soft drinks, sugar confectionary, jams and bakery toppings. The major source of anthocyanins for commercial purposes is grape pomaces and waste from juice and wine industries, but other potential sources have also been investigated.

Crude preparations of anthocyanins are used extensively in the food industry and it has been claimed that pure anthocyanins are priced high ($1,250–2,000/kg), but crude materials are rather inexpensive. Hence, production of anthocyanins through *in vitro* culture technology is highly imperative. Callus induction could be achieved from *Euphorbia milii* leaves on MS medium containing 2,4-D, NAA, natural sources such as malt extract and yeast extract. As a major component in the callus, cyanidin 3-arabinoside has been analysed. The callus consisting of cell aggregates is cut to small pieces and cultivated on solid agar media. High-producing cell aggregates are selected visually and they are transferred to fresh agar media. This procedure is repeated 28 times and one of the cells was determined to produce 1.32% anthocyanins in the cells. However, in contrast, the levels of the pigments in flowers and leaves are 0.28% and less than 0.01%, respectively.

Significantly, the accumulation of anthocyanins could be enhanced by a high osmotic potential in *Vitis vinifera* cell suspension cultures. On addition of sucrose or mannitol in the medium, in order to increase the osmotic pressure, the level of anthocyanins accumulated, increases to 1.5 times, 550 μg/10 cells.

4. *Safflower yellow* This yellow pigment is obtained from the floret of the safflower plant, *Carthamus tinctorius*. The major pigment is carthamin, which exists at levels of up to 30% in the flowers, and there is also a red pigment in concentrations of about 0.5%. Carthamin is the quinoid form of isocarthamin, the glucoside of 2',3',4',6'-tetrahydro-chalcone. It is stable to heat and light and is used in baked goods and beverages.

The callus is obtained from flower bud explants and could also be put into suspension. Medium optimization has been performed. The production of alpha-tocopherol (the tocopherol with the highest vitamin E activity) has been described for safflower cultures. Selection with various media components and precursor feeding experiments has enhanced the production. It is found that addition of cellulose, chitin or chitosan increases production of the red pigment. These polysaccharides appear to show an eliciting activity. Addition of 1 mM D-phenylalanine and removal of Mg alone or both Mg and Ca from the culture medium also increases the production.

CHEWING-GUMS (CHICLE)

Chicle is the most important raw material in chewing-gums and is made from the latex of *Achras sapota*. Chicle contains approximately 60% of resin and 15% of rubber. The resin consists of lupeol, α-amyrin and β-amyrin, and the rubber fraction contains *cis*- and *trans*-1,4-polyisoprene; however, the biosynthetic pathway of chicle and its regulation mechanisms have not been elucidated, although components of chicle are suggested to be synthesized through the mevalonic acid pathway.

In *A. sapota* tissue cultures, the callus induced from young shoots on a LS medium has been shown to produce lupeol acetate, palmitate and stearate, but neither α and β-amyrin nor rubber. On the other hand, cells cultivated in suspension produced triterpenes as well as phytosterols such as campesterol, cholesterol, β-sitosterol and stigmasterol. In spite of many efforts for optimization of the culture conditions in suspension, the yield of triterpenes was not high enough for commercial application of the cell culture; therefore the researchers induced callus tissue of *Dyera costulata* and *Couma macrocarpa*, both of which are known to produce latexes for chicle as well. The callus tissues of both plants produce triterpenes at higher levels than those in the intact plants, but not rubber.

REVIEW QUESTIONS

1. What are alkaloids? Describe the pharmaceutically important alkaloids which are produced by *in vitro* methods.
2. Write an assay on antitumour compounds produced by cell cultures.
3. What are plant virus inhibitors?
4. Explain the secondary metabolites used as food additives.
5. Write notes on the following:
   i. Chicle
   ii. Mucilage

REFERENCES

Balandrin, M.F. and Klocke, J.A. (1988). Medicinal, aromatic and industrial materials from plants. In: Bajaj Y.P.S. (ed.). *Biotechnology in Agriculture and Forestry*; Vol. 4. *Medicinal and Aromatic plants I.* Springer-Verlag, Berlin. pp.3–36.

Curtis, M.E. (1983). Harvesting profitable products from plant tissue culture. *Biotechnology.* 1: 649–657.

Fujita, Y. (1990). The production of industrial compounds. In: Bhojwani, S.S. (ed.). *Developments in Crop Science, 19. Plant tissue culture: Applications and limitations.* Elsevier Science Publishers, B.V. The Netherlands. pp.259–275.

Huffman, G.A. et al. (1984). Hairy-root inducing plasmid: Physical map and homology to tumour-inducing plasmids. *J. Bacteriol.* 157: 269–76.

Kurtz, W.G.W. and Constabel, F. (1998). Production of secondary metabolites. In: Altman, A. (ed.). *Agricultural Biotechnology,* Marcel Dekker, Inc., New York. pp.183–224.

23

TRANSGENIC PRODUCTION-I

INTRODUCTION

One of the most important events in the human history, perhaps took place when man gradually stopped gathering food and became a farmer. That was probably the beginning of genetic engineering. Through classical breeding, countless plants were domesticated by man either by creating genetic diversity by crossing individuals or by selecting offsprings that showed some useful trait/s. Classical plant breeding, however, has its limitations as it depends on sexual compatibility and often takes 10–15 years to release a new variety. Due to the unsuccessful crosses and narrow gene pool available within a species, genetic engineering, as an additional tool, is being used to keep pace with the burgeoning population in the developing countries. Genetic engineering is defined as the manipulation of plant genomes via the introduction of a characterized DNA segment.

Genetic engineering offers many opportunities for improving agriculture and public health. An elite variety could be modified for a single trait with the gene/s coding for insect resistance, disease (viral and fungal) resistance or herbicide tolerance from entirely different organisms. Other quality traits such as protein and carbohydrate content, modified plant oil and fatty acid composition for health reasons, enhanced flavour and texture and longer shelf life could also be introduced. Furthermore, the potential benefits include higher yields and enhanced nutritional value for crops and livestock, reduction in pesticide and fertilizer use, development of disease resistance and the use of plants to produce valuable heterologous molecules, which yield many new products.

Transgenic plants possess a gene or genes that have been transferred from a different species. Although DNA of another species can be transferred and integrated in a plant genome by natural processes, the term "transgenic plants" refers to plants created in a laboratory using recombinant DNA technology. The aim is to design plants with specific characteristics by artificial insertion of genes from other species or sometimes entirely different kingdoms. Transgenic plants are unique because they develop from only one plant cell, whereas in normal sexual reproduction; young plants are created when a pollen and an ovule fuse. In a similar laboratory procedure, two plant cells that have had their cell walls removed can be fused to create an offspring.

Significantly, a transgenic crop plant contains a gene or genes which have been artificially inserted instead of the plant acquiring them through pollination. The inserted gene sequence known as the "transgene" may come from another unrelated plant, or from a completely different species; such as transgenic Bt corn, which produces its own insecticide, contains a gene from a bacterium. Plants containing transgenes are often called genetically modified or GM crops, although in reality, all crops have been genetically modified from their original wild state by domestication, selection and controlled breeding over long periods of time.

HISTORY OF PLANT TRANSGENESIS

- The production of transgenic plants in wide crosses by plant breeders has been a vital aspect of conventional plant breeding for about a century. The first historically recorded interspecies transgenic cereal hybrid was actually developed between wheat and rye.
- In the 20th century, the introduction of alien germplasm into common foods was repeatedly achieved by traditional crop breeders by artificially overcoming fertility barriers.
- Novel genetic rearrangements of plant chromosomes, such as insertion of large blocks of rye (*Secale*) genes into wheat chromosomes have also been exploited widely for many decades.
- By the late 1930s with the introduction of colchicine, perennial grasses were being hybridized with wheat with the aim of transferring disease resistance and perenniality into annual crops, and large-scale practical use of hybrids was well-established, leading to the development of "*Triticosecale*" and other new transgenic cereal crops.
- In 1983, the era of plant transformation was initiated when *Agrobacterium*-mediated gene delivery was used for producing transgenic plants. Following years of unsuccessful experiments with variations in feeding isolated DNA to plant tissues and organs, gene transfer became a reality soon after it was discovered that the soil bacterium *Agrobacterium tumefaciens* and *A. rhizogenes* are natural genetic engineers due to their ability to transfer and integrate DNA into plant genomes through a unique integrative gene transfer mechanism.
- Historically, in 1985, Plant Genetic Systems (Ghent, Belgium), was the first company to develop genetically engineered (tobacco) plants with insect tolerance by expressing genes encoding for insecticidal proteins from *Bacillus thuringiensis* (Bt). These transgenic plants were developed using a disarmed version of the Ti plasmid of *Agrobacterium tumefaciens*, a pathogenic bacterium, which can

transfer part of its "T-DNA" into the plant genome. The genes causing crown gall disease were removed from the Ti plasmid, while leaving the DNA transfer mechanism intact. By replacing the tumour-causing genes with foreign genes and the subsequent conversion of plant cells to kanamycin antibiotic resistance by the transfer of a bacterial selectable marker gene (*nptII*-neomycin phosphotransferase) into the plant cells, the expression of foreign genes into plants was achieved.

- The production of transgenic plants depends on the stable introduction of foreign DNA into the plant genome, followed by regeneration of host cells into intact plants, and the subsequent expression of the introduced gene(s). Initial successes in transformation studies were limited to solanaceous crops, in general and tobacco (*Nicotiana tabacum* L.), in particular. This dramatically changed in the 1980s and the 1990s, when it was possible to transform a wide range of plants, many of which are of agronomic importance using genetically engineered avirulent strains of *Agrobacterium* as vectors. However, host-range limitations of *Agrobacterium*-mediated gene transfer prompted the search for alternate gene transfer systems, leading soon to the development of "direct gene transfer" to protoplasts. Further limitations in these gene transfer systems led to the exploration of various other approaches such as pollen.

STRATEGIES OF TRANSGENIC PRODUCTION

The basic strategy for transforming dicot or monocot plants is to cut leaf discs (causing cell injury) and then incubate with either *Agrobacterium* carrying recombinant ← Ti-vectors or directly treat the recombinant plasmids with target tissues by any physical transferring approaches. These discs can be transferred to shoot-inducing medium and simultaneously selected for markers carried on the T-DNA and plasmid DNA. After a few weeks, the discs with the regenerated shoots are transferred to root-inducing medium and after another few weeks plantlets can be transferred to soil. However general strategies of transgenic production is discussed below.

1. Basic concepts of gene/DNA and traits
2. Identification and isolation of gene for specific traits
3. Designing of gene constructs for transformation
4. Gene transformation techniques
5. Selection and regeneration of putative transgenics

1. Basic Concepts of Gene/DNA and Traits

The construction of transgenic plants is possible due to the universal presence of DNA (deoxyribonucleic acid) in the cells of all living organisms. DNA stores the organism's genetic information and controls the metabolic processes of life. Genes are specific segments of DNA that encode the information necessary for assembly of a specific protein. The proteins then function as enzymes to catalyse biochemical reactions, or as structural or storage units of a cell, to contribute to the expression of specific traits in an organism.

The general sequence of events by which the information encoded in DNA is expressed in the form of proteins via an mRNA, an intermediate product, is shown below (Figure 23.1).

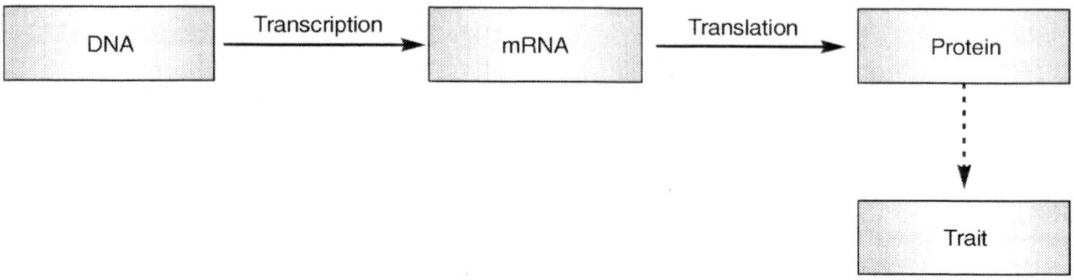

Figure 23.1 A general pathway of flow of genetic information

The transcription and translation processes are controlled by a complex set of regulatory mechanisms, so that a particular protein is produced only when it is needed. Even species that are very different from each other have similar mechanisms for converting the information in DNA into proteins; thus, a DNA segment from bacteria can be manipulated and translated into a functional protein when inserted into a plant.

2. Identification and Isolation of Genes for Specific Traits

Identification and location of genes for agriculturally important traits is currently the most limiting step in the transgenic process. There is still very little information available about the specific genes required to enhance yield potential, improve stress tolerance, modify chemical properties of the harvested product, or otherwise affect plant characters. Usually, identifying a single gene involved with a trait is not sufficient; it is also needed to understand how the gene is regulated, what other effects it might have on the plant, and how it interacts with other genes active in the same biochemical pathway. There are research programmes all over the world that mainly concentrate on new technologies to rapidly sequence and determine functions of genes of the most important crop species. These efforts should result in identification and isolation of a large number of genes potentially useful for producing transgenic varieties.

3. Designing of Gene Constructs for Transformation

Once a gene has been isolated and cloned (amplified in a bacterial vector), it must undergo several modifications or manipulations before it can be effectively inserted into a plant (Figure 23.2). These steps are explained below:

i. A promoter sequence must be added for the gene to be correctly expressed. The promoter is the on/off switch that controls when and where in the plant the gene will be expressed. Currently, most promoters in transgenic crop varieties have been constitutive, that is, causing gene expression throughout the life cycle of the plant in most tissues. The most commonly used constitutive promoter is CaMV35S, from the cauliflower mosaic virus, which generally results in a high degree of expression in plants. However, some of the other promoters are more specific and respond to cues in the plant's internal or external environment such as light-inducible promoter, the promoter from the *cab* gene encoding the major chlorophyll a/b binding protein.

ii. Sometimes, the cloned gene is modified to achieve over-expression in a plant. For example, the Bt gene for insect resistance is of bacterial origin and has a higher percentage of A–T nucleotide pairs compared to plants, which prefer G–C nucleotide pairs. A minor modification by substitution of A–T nucleotides with G–C nucleotides in the *Bt* gene is possible without significantly changing the amino acid sequence. The result indicates that there is enhanced production of the gene product in plant cells.

iii. The termination sequence that signals to the cellular machinery about the end of the gene sequence has also been a limiting factor and somehow terminator sequence influences the rate of transgene gene of interest (GOI) expression in transgenic plants.

iv. A selectable marker gene is added to the gene "construct" in order to identify plant cells or tissues that have successfully integrated the transgene. This is necessary because achieving incorporation and expression of transgenes in plant cells is a rare event, occurring in just a few percent of the targeted tissues or cells.

Selectable marker genes encode proteins that provide resistance to agents that are normally toxic to plants, such as antibiotics or herbicides. Only plant cells that have integrated the selectable marker gene will survive when grown on a medium containing the appropriate antibiotic or herbicide. As for other inserted genes, marker genes also require promoter and termination sequences for proper function.

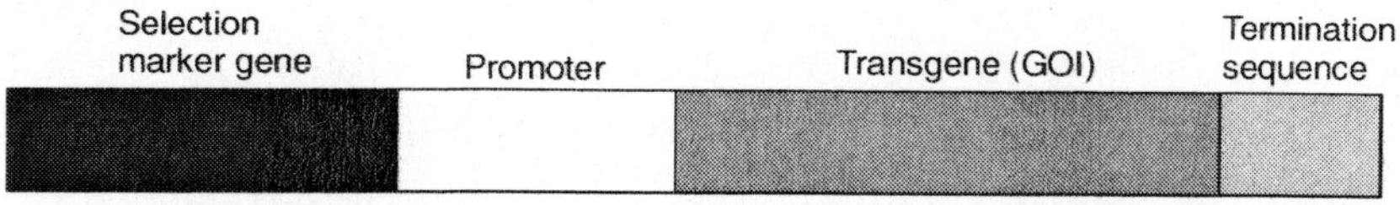

Figure 23.2 Schematic representation of a transgene construct containing necessary components

4. Gene Transformation Techniques

Transformation is the heritable change in a cell or organism brought about by the uptake and establishment of introduced DNA. There are two main methods of transforming plant cells and tissues. Foreign manipulated gene can be transferred either directly to the target tissue via biolistics, electroporation and others, or via *Agrobacterium*. In this chapter, direct method of gene delivery has been discussed in detail and the next chapter deals with the biology of *Agrobacterium* and its role as a transformation vector in the production of plant transgenics.

1. *Direct gene transformation technique* DNA can be introduced directly into plant protoplasts by techniques that are similar to those used for animal and yeast cells. The freely accessible plasma membrane allows easy entry of DNA. The advantage of this procedure is that it can be used on any plant from which protoplasts can be obtained. It was first reported by Paszkowski *et al.* (1984) in tobacco. The simple application of naked DNA under defined conditions to plant protoplasts has been shown to result in the uptake and integration of DNA into the plant. It was developed as an alternative to *Agrobacterium*-mediated gene transfer because of host-range limitations. Direct uptake of DNA into protoplasts can be promoted by chemical and physical treatment, for example, with polyethylene glycol (PEG), and application of electric pulses (electroporation) respectively, or a combination of both. The basic procedure involves the preparation of protoplasts from plant tissue by enzymatic digestion, the addition of DNA to the protoplast suspension, the uptake of DNA stimulated by various methods, the selection for expression of a transformed gene usually applied at some point in the regeneration process from the treated protoplast to plants.

A barrier to a more widespread use of genetic manipulation techniques is the difficulty of transforming some of the major crop plant species. Some cereal and grass species are recalcitrant to genetic manipulation. Nevertheless, because of the agronomic importance of monocots as staple food plants, there is great interest in transforming these species. Cereal transformation has been achieved by several different techniques of direct gene transfer. The first successful method was protoplast transformation and subsequently particle bombardment, cell electroporation and vortexing tissues with silicon carbide whiskers have all resulted in transgenic cereal plants.

i. ***Biolistics method*** The "gene gun" method (also known as microprojectile bombardment or biolistics) for the transformation of regenerable tissue (Figure 23.3) offers an easier way to transform recalcitrant higher plants. As an alternative to protoplast transformation, particle bombardment was first

Figure 23.3 A set of microprojectile system, arranged inside a laminar flow chamber for gene transformation experiment.

described as a method for the production of transgenic plants in 1987, especially for transformation of more recalcitrant cereals. It is the introduction of substances into intact cells and tissues through the use ofhigh-velocity microprojectile via a mechanism that breaches cell walls and membranes. A microprojectile should be small enough to enter a cell or a tissue in a non-lethal manner, and should be capable of carrying DNA on its surface or in its interior. Typically these are made of high density metals such gold or tungsten, which are more or less spherical, and approximately 1.5–3.0 μm fold particles are used.

In plant research, the major applications of biolistics include transient gene expression studies, production of transgenic plants and inoculation of plants with viral pathogens. Because of the low cost, more heterogeneous tungsten particles are also widely consumed as a physical factor but transformation efficiency is highest with gold particles in the range of 0.7–1.0 μm mean diameter. This method is capable of circumventing the host-range restrictions of *Agrobacterium tumefaciens*, and the regeneration problems of protoplast transformation.

Microprojectile-mediated transformation is a mechanical method of introducing DNA into almost any plant species and genotype. This mechanical method may be highly advantageous when major biological barriers exist to either *Agrobacterium*- or protoplast-mediated transformation. Heavy microprojectiles (tungsten/gold) coated with DNA are accelerated into cells and tissues. The cells can survive the intrusion of particles, thus this technique facilitates the transport of genes into any type of intact cells or tissues and it was demonstrated for the first time in maize cells by Klein *et al.* in 1987.

Gene constructs for biolistics can be in the form of circular or linear plasmids or a linear expression cassette. Typically DNA is loaded onto 1.5–3.0 μm gold bead at a rate of up to 40 μg DNA/mg of gold, using $CaCl_2$ to precipitate the DNA to the gold, and particle delivery system PDS-1000/He uses a shock wave generated by the sudden release of compressed helium to accelerate a thin plastic sheet into a metal followed by ethanol treatment (Figure 23.4). Embryogenic cell cultures are the best choice of explant for biolistic transformation because they can be spread out as uniform targets of cells and also because they have a high recovery capacity. Application of mannitol or sorbitol as osmotica in bombardment medium has been effective and causes higher rates of stable transformants for all suspension-cultured cells.

Particle-mediated plant transformation technology has been used to transform several plant species, including relatively recalcitrant species such as maize, soybean and wheat. Stable transformation using biolistics method has been achieved in tobacco and many other crops.

Hypersensitive responses to *Agrobacterium* that leads to plant cell death are eliminated via biolistics as well as the need to kill *Agrobacterium* after transformation. The mode of the biolistic device operation is easy. Biolistics is the method of choice for the study of transient gene expression, 24–48 hours after bombardment as well as for plastid transformation. Co-integration of this system with viral vectors enhances expression through increased copy number by self-replication and integration. *In situ* fertilization using *in vitro* biolistic-based targeted microspores is useful for transgenic plant production without any tissue culture phase.

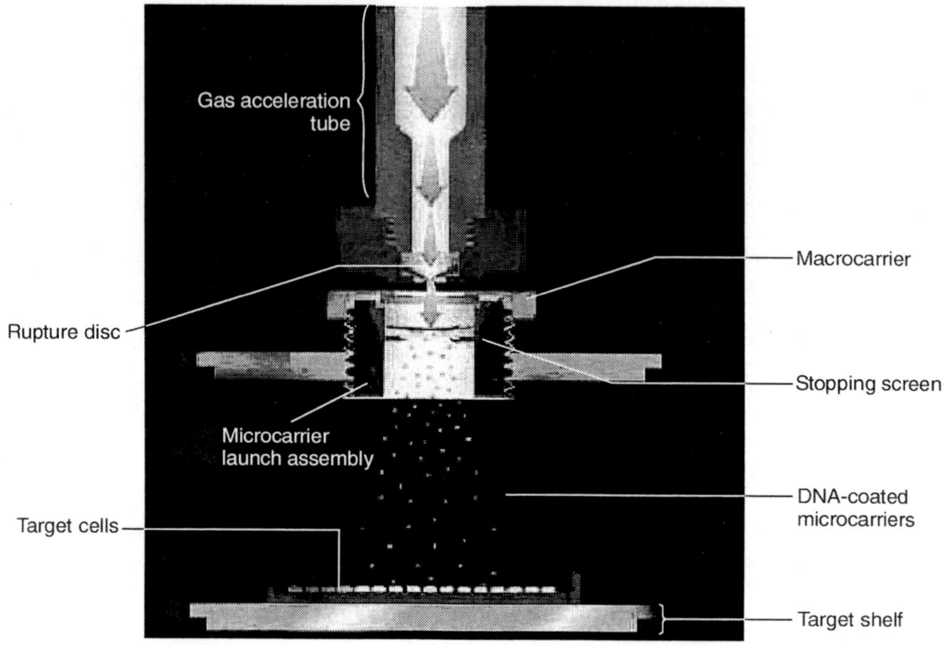

Figure 23.4 Design of a microprojectile experiment

Disadvantages of biolistics The transformation efficiency might be lower than with *Agrobacterium* and it is more costly as well. Intracellular targets are random and DNA is not protected from damage. Many researchers have not used biolistics because of the high frequency for complex integration patterns and multiple copy insertions that could cause gene silencing and variation of transgene expression. But some laboratories have overcome this problem by reducing the quantity of DNA loaded onto the microcarriers and/or by use of linear cassettes. The ability of successful integration of Yeast Artificial Chromosomes (YACs) with insert up to 150 kb and of 106 kb BAC plasmid into the plant genome by particle bombardment is an attractive proposition, especially for functional complementation analysis. But it is limited to the copy number, and rearrangement of the introduced DNA is prone to gene silencing and genomic changes. However, to date it is still considered as a very efficient tool for production of phenotypically normal, fertile transgenic japonica and indica rice cultivars.

ii. ***Agrolistics method*** Agrolistics is a refinement to microprojectile bombardment technology where *vir* genes are co-bombarded. Efforts to reduce the incidence of recombination and reduction in the amount of DNA bombarded into each cell have led to the development of "agrolistics" methods, which was designed to increase the efficiency of biolistics and to yield simpler integration patterns. Use of agrolistics has been shown to increase the number of transgenic plants that have the clear integration of transgene as well as to reduce the frequency of degraded transgene integrations. This technology requires further development but has the potential as a refinement to alternative of biolistics.

iii. ***Electroporation-mediated gene transfer*** Electroporation is the application of strong electric field pulses to cells and tissues. It is known to cause some type of structural rearrangement of the cell membrane resulting in a temporary increase in porosity and providing a local driving force for ionic and molecular transport through the pores. *In vitro* introduction of DNA into cells is the most common application of electroporation. Several physical factors such as created transmembrane potential by the imposing pulse electric field, extent of membrane permeation, duration of the permeated state, mode and duration of molecular flow, global and local (surface) concentrations of DNA, form of DNA, tolerance of cells to membrane permeation and the heterogeneity of the cell population may affect the electro-transfection efficiency.

In order to efficiently apply electric field, especially to expel the tissue culture step, electrophoresis-mediated gene transfer was developed to force negatively charged foreign DNA into the host cells. It was proposed as an alternative method to biolistics, but is not nearly as efficient. A voltage of 25 mV and amperage of 0.5 mA for 15 minutes are the most often used parameters. On an average, this technique gives about 55% survival rate in which up to 57% of the survivors expressed a marker gene.

Transient expression from electroporated plant cells has been used to study the various purposes of molecular biology and plant biotechnology. Electroporation coupled with a transient expression assay is rapid, allowing for the reproducible detection of gene products within hours of introduction of DNA, in contrast to a stable transformation.

a. Electroporation of protoplasts A widely used physical treatment for enhancing the frequency of transformation of protoplasts is to subject them to electric pulses. Electroporation causes the uptake of DNA into protoplasts by temporary permeabilization of the plasma membrane to macromolecules. This is achieved by application of a high-intensity electric field to protoplasts contained in a buffer between two electrodes. DNA diffusion occurs immediately after the electric field is applied and until the pores in the membrane re-seal.

Electroporation treatment is believed to create transient pores in the cell membranes through which the DNA present in the external solution gains entry under appropriate conditions (pH, concentration). This leads to either transient expression of introduced DNA, or stable integration of the DNA into a small proportion of the protoplasts, and the transgenic tissues and plants can be regenerated. Suitable conditions must be identified so that the pore formation is reversible and the protoplasts can be recovered from the treatment.

The efficiency of gene transfer and viability of protoplasts is influenced by the amplitude and duration of the electric pulse and composition of the electroporation medium. Electroporation has been used successfully for transient and stable transformation of protoplasts from a wide range of species and tissue sources. Stable transformation of maize protoplasts and carrot protoplasts has been known to us.

b. Electroporation of entire tissues Comparison of electroporation technique with *Agrobacterium* and particle gun delivery approaches has introduced the *in vivo* electrophoresis transformation as an alternative method by using entire tissues like intact meristems. However, electroporation-mediated gene transfer into intact meristems was reported by Chowrira *et al.* in 1995. Moreover, gene transformation could be possible via electroporation of intact leaf tissue of rice. The applicability of this technique to other monocots including wheat, maize and barley has also been seen.

Electroporation has several advantages over biolistics in that it does not require the expensive particle gun apparatus, associated consumable supplies and licensing and has worked well for stable transformation experiments. Also, the efficiency of electroporation-mediated transformation has been proved to be relatively high (6.25% for embryos and 54.6% for callus clusters), which is fully comparable with the best results obtained for this species after microparticle bombardment techniques. However, when using protoplasts as explants, regeneration of transformants, limits its use for stable transformation. *Zea mays* embryos are suitable explants which enable the recovery of fertile transgenic plants. It is mentionable that use of pollen electroporation is also able to escape regeneration phase. Despite the apparent advantages of electroporation, it is not used very often because sometimes it produces irreproducible results and non-Mendelian segregation of progeny.

iv. *Polymer-based transfection (polyfection)-mediated gene transformation* PEG is the most widely used chemical treatment for facilitating DNA uptake into plant protoplasts. PEG acts to increase the permeability of cell membranes and has been used as an efficient protoplast fusion agent. PEG-mediated transformation involves mixing freshly isolated protoplasts with DNA and immediately adding a given concentration of PEG dissolved in a buffer. PEG concentration can affect protoplast viability and gene transfer efficiency. Production of transgenic plants is dependent on the regeneration competency of the resultant transformed callus.

The most common method of delivering foreign DNA into plant protoplasts involves treatment with PEG as a hydrophilic long-chain polymer.

Protoplast transformation has several advantages. Many cells can be handled with ease, many independent transformants can be produced, the selection of transformants is relatively simple and no specialized equipment is needed. However, regenerating whole plants from transgenic protoplasts has proven difficult. The system is only rarely employed due to the low frequency of transformants and the inability of many species to be regenerated into whole plants from protoplasts. However, it has been reported that the efficiency of PEG-based method is 10–20% for *Physcomitrella* transformation and 65 to 28% for rice protoplasts, according to plasmid size (4.5 to 12 kb).

Recently, a fast PEG-mediated transient expression system was reported which was suitable for the production of biopharmaceuticals. In addition, long-chain cations such as poly-L-ornithine, poly-L-lysine, dextran sulphate and high concentrations of divalent ions such as Zn^{2+} and Cu^{2+} have been applied to achieve high levels of DNA uptake. Among the polycations, combined polybrene-spermidine-based callus treatment was introduced as an efficient method for cotton transformation, which is less toxic than the other polycations, and spermidine protects DNA from shearing because of its condensation effect.

2. *Other gene transformation methods* Due to host-range limitations of *Agrobacterium* and the absence of regeneration protocols from protoplasts for direct gene transfer, alternate approaches of gene transfer have been explored. Some of the alternative gene transformation approaches are pollen tube pathway, electrophoresis, microlaser treatment, sonication, incubation of dry embryos in DNA, silicon fibre-mediated gene transfer, microinjection, etc.

i. ***Viral based transformation*** A number of biological alternative strategies have been developed over the last few years for gene delivery in higher plants, but none have led to stable plant transformation. Viral-based transformation systems provide very high expression temporarily in plant cells. Potato virus X, barley stripe mosaic virus and geminivirus vectors are some of plant viral vectors that are commonly used. Substrates of viral-mediated transformation could be viral RNA or more conveniently, viral cDNA which can be used via direct inoculation, and it has proven to be the most efficient way of introducing cDNA-derived viral RNA into plants. Modified viruses in which an antibiotic-resistance gene replaces the viral coat protein genes, can be used to enhance integration. Rapidity and high-level expression of transgene are the benefits of virus-based transient RNA and DNA replicons versus non-viral systems.

However, some of the disadvantages of this method include integration of large number of copies of transgenes, chromosomal rearrangement, appearance of disease symptoms and low frequency for stable transformation. Viral vectors can also serve as efficient initiators of post-transcriptional gene silencing (PTGS) or virus induced gene silencing (VIGS) in natural virus infections. Also, inducible expression is applicable by viral-based transformation methods, which is a useful tool for marker gene elimination.

ii. ***Silicon carbide-mediated transformation (SCMT) method*** One of the recently developed methods for delivery of DNA to plants is silicon carbide-mediated transformation (SCMT). Physical and chemical characteristics of silicon carbide fibres make them capable of puncturing the cells without killing them. The advantages of this system are that they are rapid, inexpensive and easy to set-up. Also, this method is effective on a variety of cell types. Some of its disadvantages include low transformation efficiency, cell damage and health hazard due to inhalation of the fibres. The efficiency of SCMT depends on the fibre size, parameters of vortexing, shape of the vessels used, plant species and nature of explant and also characteristics of the plant cells, especially the thickness of the cell wall. There are several known examples of SCMT success including maize, rice, wheat, tobacco, *Lolium*, *Festuca arundinacea* and *Agrostis stolonifera*. SCMT is comparable in efficiency to biolistics in some instances and therefore, it represents as alternative option for embryogenic tissue transformation. Also, silicon carbide fibres can be used for wounding, to improve frequency of *Agrobacterium*-based transformation.

iii. ***Liposome-mediated transfection (Lipofection)*** A derivation of PEG-mediated transformation is the liposome-mediated transformation technique. Cationic liposomes are positively charged lipids and are increasingly used for DNA uptake due to their favourable interactions with negatively charged DNA and cell membranes. In this approach, foreign DNA must be encapsulated in a spherical lipid bilayer termed as "liposome" to prepare lipoplexes. After endocytosis, the DNA is free to recombine and integrate into the host genome.

As with other transformation systems, a variety of vectors including viral vectors can be applied to this system. Successful transformation based on this system was reported for tobacco, wheat and potato. This method is relatively non-toxic, simple to perform with readily available chemical reagents, is highly reproducible and efficient and requires no sophisticated equipment. Transformation of intact YACs into plant cell was achieved via lipofection-like particle bombardment. A lipofection–PEG combination method has been proved to be more efficient than each one of them separately. It is suggested that lipoplex size is a major determining factor in terms of lipofection efficiency where large lipoplex particles show, in general, higher lipofection efficiency than small particles.

iv. ***Injection-based methods*** **Microinjection** is one of the most effective techniques of transforming animal cells and has been used for the transformation of plant cells as well. An important feature of the method is the use of low melting point (LMP) agarose, both for holding protoplasts during microinjection and for their subsequent culture. This direct and precise delivery of DNA into the plant cell or its nucleus is achievable by means of a glass microcapillary-injection pipette. The procedure is very slow and requires an expensive micromanipulator. However, it allows the introduction of not only plasmids but also whole chromosomes into plant cells. Several plant species such as tobacco, petunia, oilseed rape and barley have been transformed using microinjection. Also, the technique has been used to study intercellular communication and macromolecular trafficking in plants. This technique is very precise from the delivery point of view, is capable of transformation of any protoplast or single-cell system and has extremely high transformation efficiencies, but requires expensive equipment and tedious procedures.

The injection of inheritable materials using a hypodermic syringe is called **Macroinjection**. Macroinjection is an approach to the delivery of exogenous DNA that could circumvent the expense, time and genome stress associated with the transformation of cells in culture and subsequent plant regeneration. Also, plant regeneration from transformed protoplasts, still remains a problem. Therefore, cultured tissues, which facilitate the development of immature structures, provide an alternate cellular target for transformation. These immature structures include immature embryos, meristems, immature pollen, germinating pollen, etc. The main disadvantage of this technique is the probability for the production of chimeric plants with only a part of the plant transformed. However, from this chimeric plant, transformed plants of single cell origin can be subsequently obtained. To escape these problems pollen-tube pathway (PTP), utilizing the normal fertilization cycle is developed to eliminate the regeneration phase.

PTP-based transformation is an injection/delivery of naked DNA into ovaries to produce transformed progeny. In this approach, the stigma is cut off and a drop of DNA solution is applied to the cut end of the style of recently pollinated plant florets. This procedure was used for the first time in rice by Luo and Wa in 1988. Afterward, other species such as wheat, soybean, *Petunia hybrida* and watermelon have been subjected. In rice, the transformation efficiency was higher after excising the palea than after cutting the top floret. A variation is the injection of a bacterial inoculum or plasmid DNA into inflorescences with pollen mother cells in the premeiotic stage without removing the stigma. Such an approach has been employed for rye. Although the results of the described experiments are encouraging, the final transformation efficiency was about 10-fold lower than with biolistics. However, this approach has recently been introduced as a potential method for stable plant transformation.

v. ***Wave and beam-mediated transformation*** Ultrasound treatment increases membrane permeability, thereby facilitating the entrance of macromolecules into cells for transformation. Ultrasound for transfection has been widely applied in animal cells or tissues. However, there has been little research using ultrasound in plant cells or tissues. Ultrasound has been reported to mediate gene uptake in plant protoplast, suspension cells and intact pieces of tissues.

In this technique, explants are suspended in a few millilitres of sonication medium in a microcentrifuge tube. Plasmid DNA (and possibly carrier DNA) is then added and after rapid mixing, the samples are ready for sonication. The cells are finally transferred to fresh growth medium. In this technique, sound frequency and exposure time determine the uptake efficiency. Transient expression of chloramphenicol acetyltransferase (*cat*) gene was reported in sugar beet (*Beta vulgaris* L.) and tobacco (*Nicotiana tabacum* L.) via a brief exposure of the protoplasts to 20 kHz ultrasound in the presence of plasmid DNA. Also, ultrasound treatment can be effective for transfection of virus particles.

Stable transformation of tobacco by leaf pieces sonication was achieved via 1500–2000-fold longer time ultrasound treatment compared to protoplast sonication. Another example for intact tissue sonication-based transformation is potato tuber disc ultrasonication. Much of the ultrasound technique is aimed at sonication-assisted *Agrobacterium*-mediated transformation (SAAT) in plant cells or tissues.

SAAT is a new technology and this method involves subjecting the plant tissue to brief periods of ultrasound in the presence of *Agrobacterium*. Experiments have been demonstrated that SAAT tremendously improves the efficiency of *Agrobacterium* infection by introducing large numbers of microwounds into the target plant cells or tissues. It is reported that in all tissues tested, the SAAT treatment greatly enhanced the levels of transient expression. In addition, 2.2-fold, 2.5-fold and 4.1-fold increase in transformation frequency was consequently achieved by sonication, sonication with $CaCl_2$ treatment and sonication with acetosyringone treatment, in contrast to manual wounding-based method. Also, SAAT could be efficiently useful for transformation of woody trees, particularly *Eucalyptus* sp.

Recently, laser beam has been applied to introduce genetic materials. Laser-mediated transformation works by a focused laser microbeam to puncture momentarily-made self-healing holes (~ 0.5 ?m) in the cell wall and membrane. Therefore, exogenous DNA could simply be taken up by cells. This newly developed method requires further assessment for both the different experimental conditions and plant species.

vi. ***Desiccation-based transformation*** Since dried embryos have the unique ability to take up DNA during rehydration, it was hypothesized that they could be mixed with a nutrient solution containing the foreign DNA for direct transformation. DNA would be taken up as the embryo rehydrates and seedlings would germinate in the presence of a selection medium, which is used to assess the incorporation of the foreign DNA. The old and immature approach is potent at least theoretically to expel many problems associated with biological or chemical and/or physical methods but it needs further research.

5. Selection and Regeneration

Selection of putative transgenic tissues Following the gene insertion process, plant tissues are transferred to a selective medium containing an antibiotic or herbicide, depending upon which suitable selectable markers are used. Only plants expressing the selectable marker gene will survive and it is assumed that these plants will also possess the transgene of interest. Thus, subsequent steps in the process (Figure 23.5) will only use these surviving plants.

The genes coding for hygromycin phophotransferase (*hpt*) and neomycin phosphotransferase (*nptII*) have been widely used as selectable marker genes in *Agrobacterium* transformation experiments. Several strategies for production of marker-free transgenic plants include co-transformation, a site-specific recombination system, and intragenomic relocation of transgenes via transposable elements.

Regeneration of transgenic plant To obtain whole plants from transgenic tissues such as immature embryos, they are grown under controlled environmental conditions in a series of media containing nutrients and hormones, a process known as tissue culture. Once whole plants are generated and seed is produced, evaluation of the progeny begins. This regeneration step has been a stumbling block in producing transgenic plants in many species, but specific varieties of most crops can now be transformed and regenerated.

Field testing of transgenics In the field of transgenic technology, production of transgenic plants is an extensive evaluation process to verify whether the inserted gene has been stably incorporated without detrimental effects to other plant functions, product quality, or the intended agroecosystem. Initial evaluation includes:

i. Activity of the introduced gene.

ii. Stable inheritance of the gene.

iii. Unintended effects on plant growth, yield, and quality.

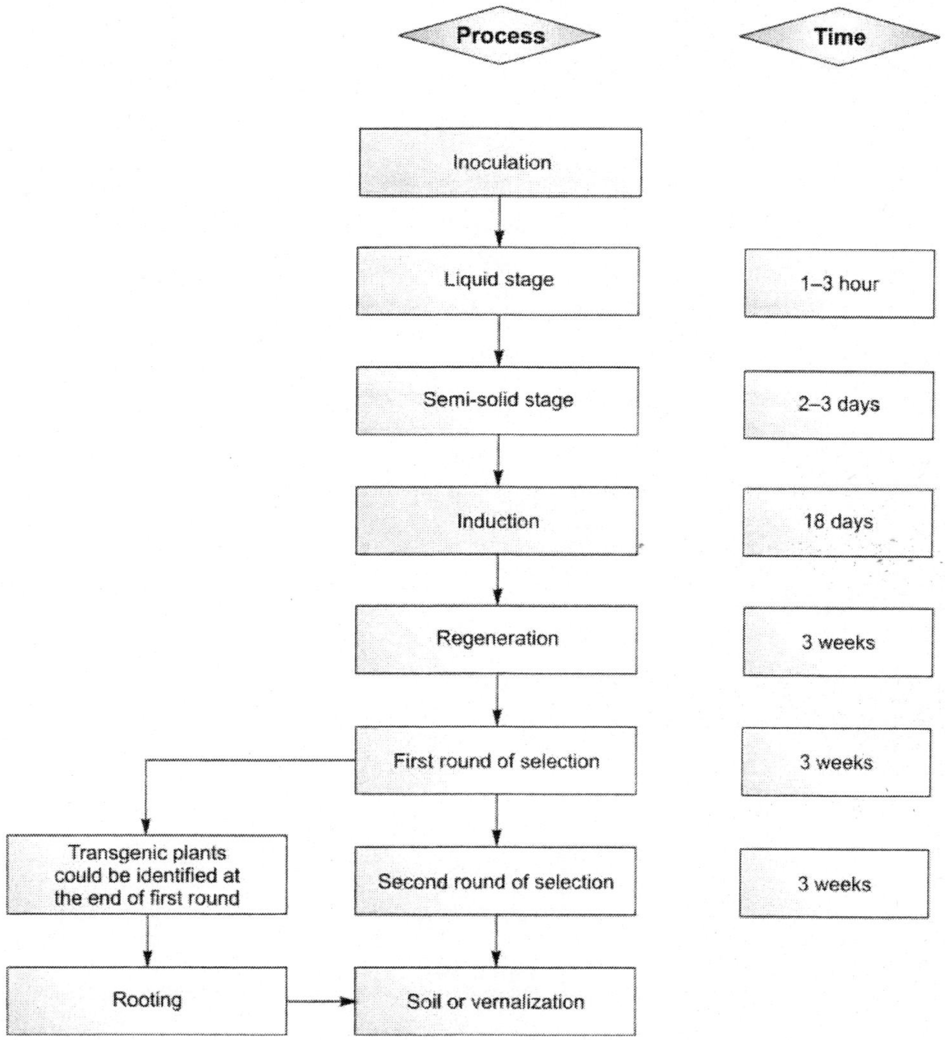

Figure 23.5 A detailed representation of selection steps of putative transgenics

If a plant passes these tests, it is most likely that it will not be used directly for crop production, but will be crossed with improved varieties of the crop. This is because only a few varieties of a given crop can be efficiently transformed, and these generally do not possess all the producer and consumer qualities required of modern cultivars. The initial cross to the improved variety must be followed by several cycles of repeated crosses to the improved parent, a process known as backcrossing. The goal is to recover as much of the improved parent's genome as possible with the addition of the transgene from the transformed parent.

The next step in the process is multi-location and multi-year evaluation trials in greenhouse and field environments to test the effects of the transgene and overall performance. This phase also includes evaluation of environmental effects and food safety.

FUTURE OF TRANSGENIC TECHNOLOGY

New techniques for producing transgenic plants will improve the efficiency of the process and will help resolve some of the environmental and health concerns. Among the expected changes are the following:

1. More efficient transformation, that is, a higher percentage of plant cells will successfully incorporate the transgene.
2. Better marker genes to replace the use of antibiotic resistance genes.
3. Better control of gene expression through more specific promoters, so that the inserted gene will be active only when needed.
4. Transfer of multigene DNA fragments to modify more complex traits.

REVIEW QUESTIONS

1. Write the history and scope of transgenic production.
2. What is transgene? Write the main strategies of gene transformation.
3. Discuss the method of gene delivery by particle projectile technology.
4. What are the possible alternative techniques of transgenic production?
5. Write about the mode of selection and regeneration of putative transgenic.

REFERENCES

Aziz, N. and Machray, G.C. (2003). Efficient male germ line transformation for transgenic tobacco production without selection. *Plant Mol., Biol.* 51: 203–211.

Banks, M.S. and Evans, P.K. (1976). A comparison of the isolation and culture of mesophyll protoplasts from several Nicotiana species and their hybrids. *Plant Sci. Lett.* 4: 409–416.

Bart, R., Chern, M., Park, C.J., Bartley, L. and Ronald, P.C. (2006). A novel system for gene silencing using siRNAs in rice leaf and stem-derived protoplasts. *Plant Methods.* 2: 13.

Chowrira, G.M., Akella, V and Lurquin, P.F. (1995). Electroporation-mediated gene transfer into intact nodal meristems in plants: Generating transgenic plants without *In vitro* tissue culture. *Mol. Biotechnol.* 3: 17–23.

Darbani, B., Safar, F., Mahmoud, T., Saeed, Z., Shahin, N. and Neal Stewart Jr., C. (2008). DNA-delivery methods to produce transgenic plants. *Biotechnology.* 7: 385–402.

Darbani, B., Farajnia, S., Noeparvar, S.H., Stewart, C.N. and Mohammadi, S.A. (2008). Plant Transformation: Needs and futurity of the transgenes. *Biotechnology.* 7.

D' Halluin, K., Bonne, E., Bossut, M., Beuckeleer, M. De and Leemans, L. (1992). Transgenic maize plants by tissue electroporation. *The Plant Cell.* 4: 1495–1505.

Ercolano, M.R., Ballvora, A., Paal, J., Steinbiss, H.H., Salamini, F. and Gebhardt, C. (2004). Functional complementation analysis in potato via biolistic transformation with BAC large DNA fragments. *Mol. Breed.* 13: 15–22.

Gordon-Kamm, W.J., Spencer, T.M., Mangano, M.L., Adams, T.R. and Daines, R.J. (1990). Transformation of maize cells and regeneration of fertile transgenic plants. *Plant Cell.* 2: 603–618.

Jia, H., Pang, Y. Chen, X. and Fang, R. (2006). Removal of the selectable marker gene from transgenic tobacco plants by expression of Cre recombinase from a *Tobacco mosaic virus* vector through agroinfection. *Transgenic Res.* 15: 375–384.

Luo, Z.X. and Wa, R. (1988). A simple method for the transformation of rice via pollen-tube pathway. *Plant Mol. Biol. Rep.* 6: 165–174.

Marillonnet, S., Thoeringer, C., Kandzia, R., Klimyuk, V. and Gleba, Y. (2005). Systemic *Agrobacterium tumefaciens*-mediated transfection of viral replicons for efficient transient expression in plants. *Nature Biotechnol.* 23: 718–723.

Paszkowski, J., Shillito, R.D., Saul, M., MandÃ¡k, V. and Hohn, T. (1984). Direct gene transfer to plants. *EMBO J.* 3: 2717–2722.

Potrykus, I. (1991). Gene transfer to plants: Assessment of published approaches and results. *Annu. Rev. Plant Physiol. Plant Mol. Biol.* 42: 205–225.

Rhodes, C.A., Pierce, D.A., Mettler, I.J., Mascarenhas, D. and Detmer, J.J. (1988). *Genetically transformed maize plants from protoplasts. Sci.* 240: 204–207.

Schaefer, D.J. (2002). A new moss genetic: Targeted mutagenesis in *Physcomitrella patens. Annu. Rev. Plant Biol.* 53: 477–501.

Songstad, D.D., Somers, D.A. and Griesbach, R.J. (1995). Advances in alternative DNA delivery techniques. *Plant Cell Tissue Organ Culture.* 40: 1–15.

Storm, M.M.H., Van der Schoot, C., Prins, M., Kormeling, R. and Van Lent, J.W.M. (1998). A comparison of two methods of microinjection for assessing altered plasmodesmal gating in tissue expressing viral movement proteins. *Plant J.* 13: 131–140.

Vik, N.I., Hvoslef-Eide, A.K., Gjerde, H. and Bakke, K. (2001). Stable transformation of poinsettia via electrophoresis. *Acta Horticult.* 560: 101–103.

Wang, K., Drayton, P., Frame, B., Dunwell, J. and Thompson, J.A. (1995). Whisker mediated plant transformation: An alternative technology. *In vitro Cell. Dev. Biol. Plant,* 31: 101–104.

Zhang, L.J., Chen, L.M., Xu, N., Zhao, N.M. and Li, C.G. (1991). Efficient transformation of tobacco by ultrasonication. *Biotechnol.* 9: 996–997.

24

TRANSGENIC PRODUCTION-II

INTRODUCTION

Agrobacterium-mediated gene transfer, the first widely adopted means of creating transgenic plants, remains the most popular technique. Transformation via *Agrobacterium* has been successfully practised in dicots (broad-leafed plants like soybeans and tomatoes) for many years, but only recently has it been effective in monocots (grasses and their relatives). In general, the *Agrobacterium* method is considered preferable to other techniques like gene gun, because of the greater frequency of single-site insertions of the foreign DNA, making it easier to monitor.

Probably the greatest advantage of the system is that it offers the potential to generate transgenic cells at relatively high frequency without significant reduction in plant regeneration rates. Plants are usually transformed with relatively simple constructs, in which the gene of interest is coupled to the promoter of plant, viral or bacterial origin. Some promoters confer constitutive expression, while others may be selected to permit tissue-specific expression, or environmentally inducible expression.

AGROBACTERIUM AS A NATURAL GENETIC ENGINEER

Agrobacterium tumefaciens is a gram-negative soil bacterium causing crown gall tumours at wound sites of infected dicotyledonous plants (Figure 24.1). It attaches to plant cells and then transfers part of its tumour-inducing (Ti) plasmid called, T (transferred) DNA, to some of these plant cells. The T-DNA becomes integrated into one of the chromosomes of the plant cell and expression of the gene located on the T-DNA leads to the formation of proteins involved in the production of an auxin and a cytokinin, which causes the tumourous phenotype. *Agrobacterium tumefaciens* can genetically transform plant cells by transferring a defined piece of DNA (known as T-DNA) from its tumour-inducing (Ti) plasmid into the genome of infected plants resulting in crown gall tumours. The tumour provides substances for *Agrobacterium tumefaciens* to grow on the plant with a set of 25 expressible *vir* genes into seven operons within the Ti-plasmid (Figures 24.2 and 24.3). Other genetic elements include the *Agrobacterium* chromosomal gene (*Chv*). The right and left border of T-DNA are necessary to constitute the T-DNA transfer machinery.

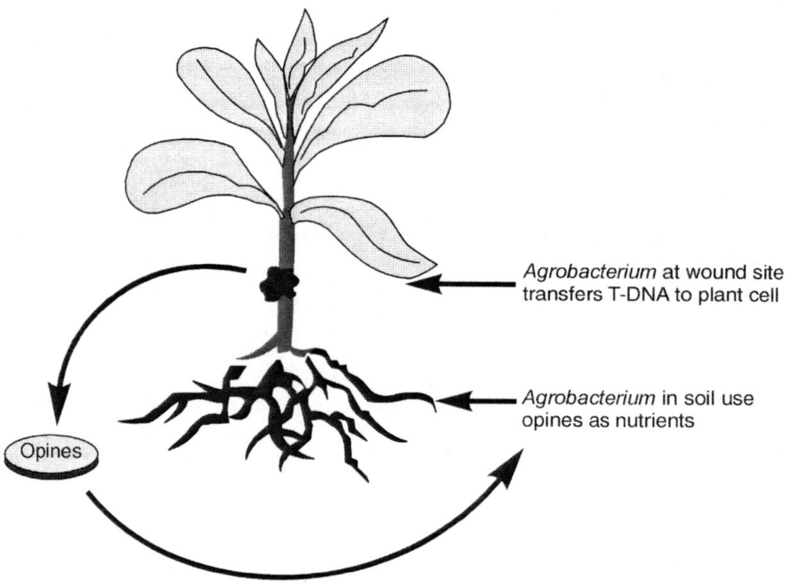

Figure 24.1 *Agrobacterium* causes the formation of crown gall disease in host plant

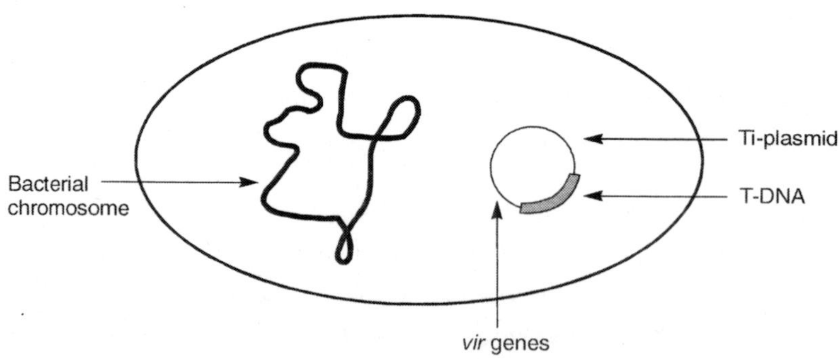

Figure 24.2 *Agrobacterium* cell carrying Ti-plasmid along with chromosomal DNA

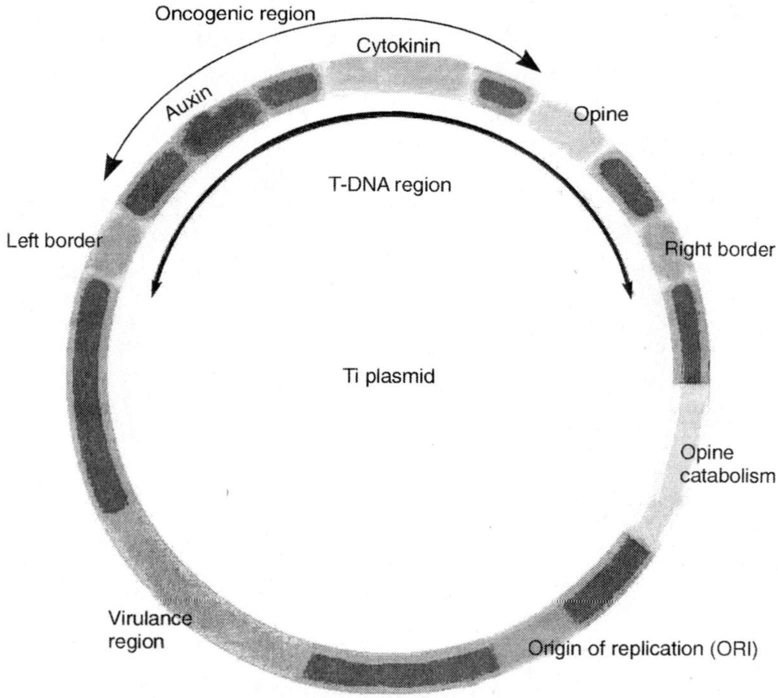

Figure 24.3 Ti-plasmid structure associated with T-DNA and other essential regions

Several Ti-plasmid-based vector systems are available for plant transformation. Typically, binary vector comprising octopine type (Figure 24.4) *vir* helper strain such as LBA 4404 that harbours the disarmed *Ach* Ti-plasmid and a binary vector such as pBin 19 are very commonly used for transformation.

The cauliflower mosaic virus (CaMV) 35S RNA promoter is often used because it directs high levels of expression in most tissues. Genetic transformation that is effected by *Agrobacterium* is dependent on the use of disarmed Ti-plasmid. The *Agrobacterium* genes that are responsible for tumour formation can be replaced with foreign genes that are expressed following infection of plant cells. The inserted genes could be:

1. A selectable marker, such as the gene for neomycin phosphotransferase (*npt II*), which confers resistance to antibiotic kanamycin;
2. A reporter gene such as the gene for β-glucuronidase (*GUS*) that encodes a scorable marker β-glucuronidase (GUS) which undergoes a useful histochemical colour reaction.
3. A sequence that encodes a promoter, such as 35S from cauliflower mosaic virus, for expressing the different genes that have been introduced into the plasmid;
4. Genes that have agricultural interest.

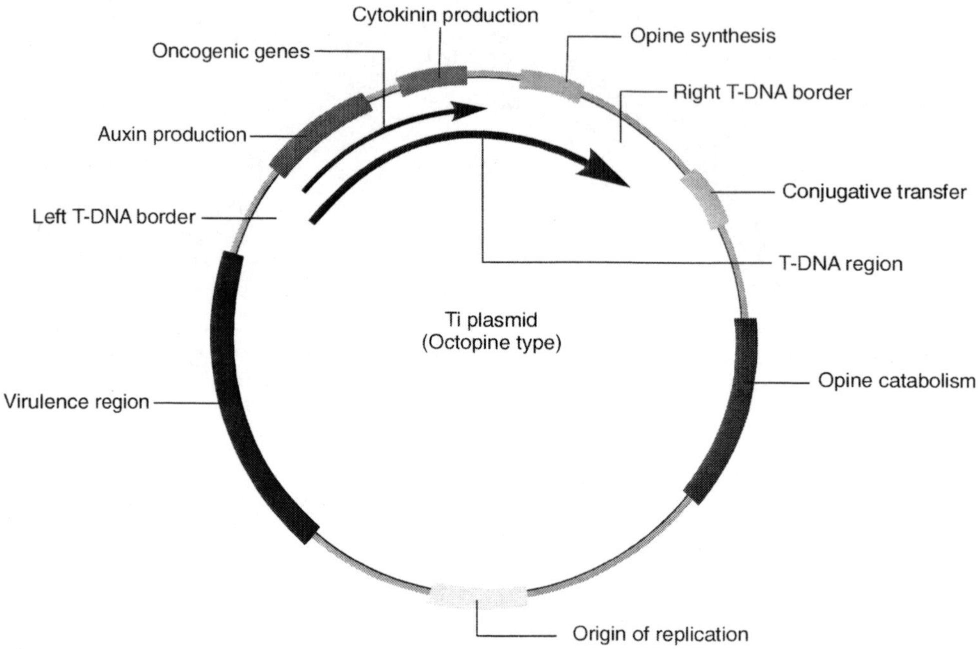

Figure 24.4 A typical map of a octopine type of Ti-plasmid along with other associated regions

Agrobacterium-mediated transformation involves incubation of cells or tissues with the bacterium, followed by regeneration of plants from the transformed cells directly and/or via the callus stage. Transformation mediated by *Agrobacterium* has provided a reliable means of creating transgenics in a wide variety of species that are amenable to tissue culture and regeneration. *Agrobacterium*-mediated gene transfer remains the most common method of transforming dicotyledonous plants, including legumes. Legumes are natural hosts to *Agrobacterium* and many grain legumes produce stably transformed callus and plants.

MECHANISM OF TRANSGENESIS

The process of gene transfer from *Agrobacterium tumefaciens* to plant cells involves several essential steps:

1. Colonization of bacterial cells to plant cell surface
2. Induction of virulence in bacterial system
3. Formation of T-DNA transfer complex
4. Mechanism of T-DNA complex transfer
5. Integration of T-DNA into plant genome.

A hypothetical model depicting the most important stages including the involvement of various gene products during the process of transgenesis is presented here.

1. Colonization of Bacterial Cells to Plant Cell Surface

Bacterial colonization is an essential and the earliest step in tumour induction and it takes place when *A. tumefaciens* is attached to the plant cell surface (Figure 24.5). Mutagenesis studies show that non-attaching mutants lose the tumour-inducing capacity. The polysaccharides of the *A. tumefaciens* cell surface play an important role in the colonizing process. There are some evidences indicating that capsular polysaccharides may play a specific role during the interaction with the host plant. In a particular case of *A. tumefaciens*, it was observed that the attachment of wild-type bacterium to plant cells was directly correlated with the production of an acidic polysaccharide.

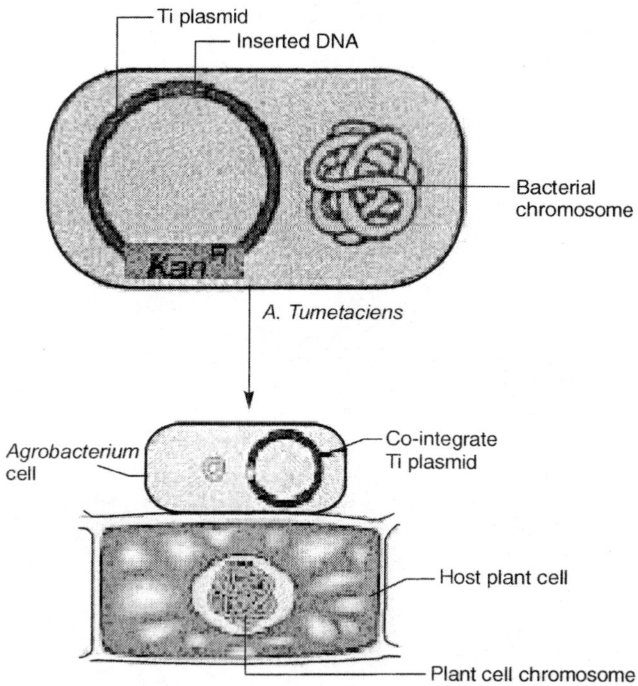

Figure 24.5 The process of *agrobacterium* colonization and attachment with host plant cell

Moreover, the chromosomal 20 kb *att* locus contains the genes required for successful bacterium attachment to the plant cell. It is suggested that genes placed at the left side of *att* are involved in molecular signalling events, while the right side genes are likely to be responsible for the synthesis of fundamental components. Furthermore, the *att* left side has an operon composed of nine open reading frames (ORF). Moreover, four of these ORFs show homology to the genes involved in the so-called periplasmic binding protein-dependent (or ABC) transport system. The mutant analysis indicates a failure in the production and accumulation of specific compounds essential for bacterial attachment. The ABC transporter encoding genes, therefore, may be involved in the secretion of this substance or in the introduction into bacteria of some plant-originated activators, specific for attachment.

2. Induction of Virulence in Bacterial System

The T-DNA transfer is mediated by products encoded by the 30–40 kb *vir* region of the Ti plasmid (Figures 24.6 and 24.7). This region is composed of at least six essential operons (*virA, virB, virC, virD, virE, virG*) and two non-essential (*virF, virH*) operons. The number of genes per operon differs, *virA, virG* and *virF* have only one gene; *virE, virC, virH* have two genes, while *virD* and *virB* have four and eleven genes, respectively. The only constitutively expressed operons are *virA* and *virG,* which code for a two-component (VirA–VirG) system and activate the transcription of the other *vir* genes.

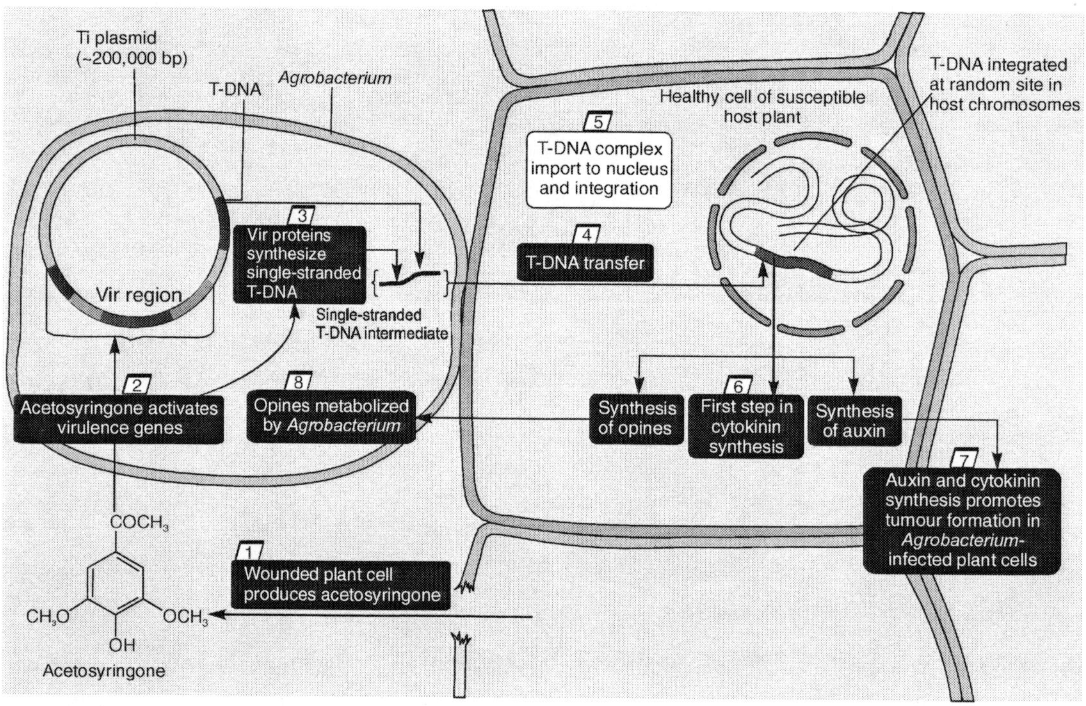

Figure 24.6 Important steps in the infection and T-DNA transfer at the molecular level

VirA is a transmembrane dimeric sensor protein that detects signal molecules, mainly small phenolic compounds, released from wounded plants. The signals for VirA activation include acidic pH, phenolic compounds such as acetosyringone, and certain class of monosaccharides, which acts synergistically with phenolic compounds.

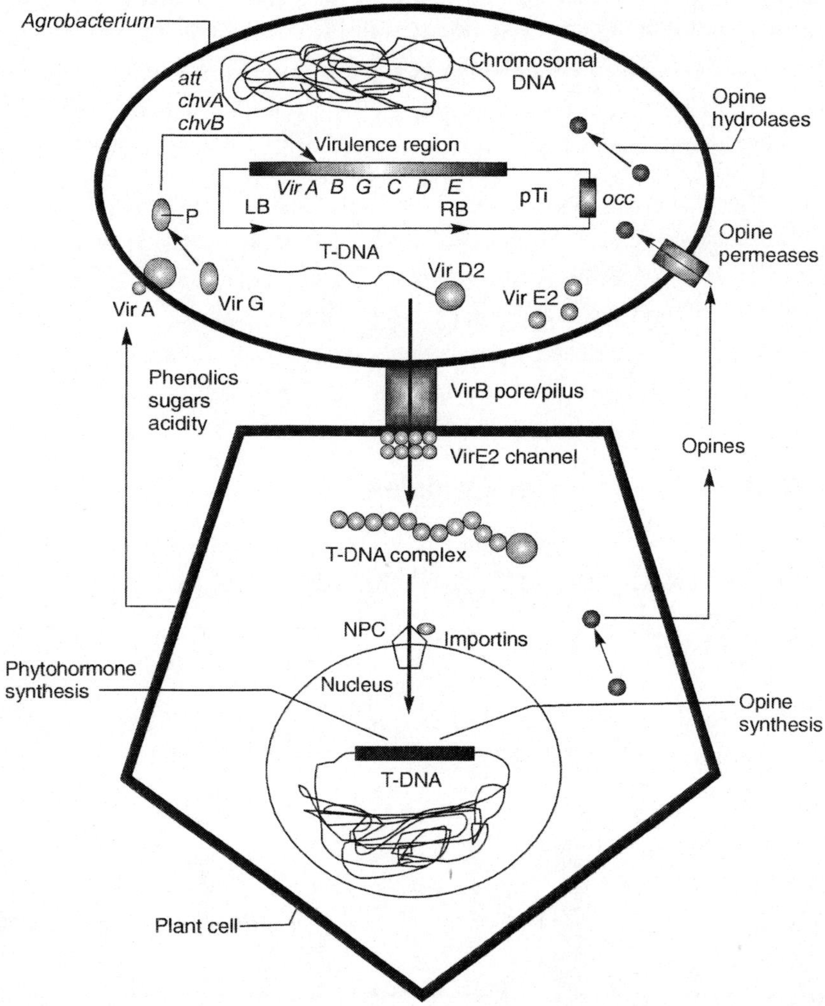

Figure 24.7 Molecular mechanism of induction of virulence and T-DNA transfer

Significantly, VirA protein can be structurally defined into three domains: the periplasmic domain and two transmembrane domains (TM1 and TM2). The TM1 and TM2 domains act as transmitter (signalling) and receiver (sensor), whereas the periplasmic domain is important for monosaccharide detection. The TM2 is the kinase domain and plays a crucial role in the activation of VirA protein, by phosphorylating itself on a conserved His-474 residue in response to signalling molecules from wounded plant sites. Monosaccharide detection by VirA protein is an important amplification system and responds to low levels of phenolic compounds. Significantly, the induction of this system is only possible through the periplasmic sugar (glucose/galactose) binding protein ChvE, which interacts with VirA protein.

Furthermore, activated VirA protein has the capacity to transfer its phosphate to a conserved aspartate residue of the cytoplasmic DNA binding protein VirG and now VirG protein functions as a transcriptional factor regulating the expression of other *vir* genes when it is phosphorylated by VirA protein. The C-terminal region of VirG protein is responsible for the DNA binding activity, while the N-terminal is the phosphorylation domain and shows homology with the VirA receiver (sensor) domain.

The activation of *vir* system also depends on external factors like temperature and pH. At temperatures greater than 32°C, the *vir* genes are not expressed because of a conformational change in the folding of VirA protein and induce the inactivation of its properties. The effect of temperature on VirA is suppressed by a mutant form of VirG ($VirG^c$), which activates the constitutive expression of the *vir* genes. However, this mutant cannot confer virulence capacity at that temperature to *Agrobacterium*, probably because the folding of other proteins that actively participate in the T-DNA transfer process are also affected at high temperature.

3. Formation of T-DNA Transfer Complex

The activation of *vir* genes generates single-stranded (ss) molecules representing the copy of the bottom T-DNA strand. Any DNA placed between T-DNA borders will be transferred to the plant cell as single-stranded DNA, and integrated into the plant genome (Figures 24.7 and 24.8). These are the only *cis*

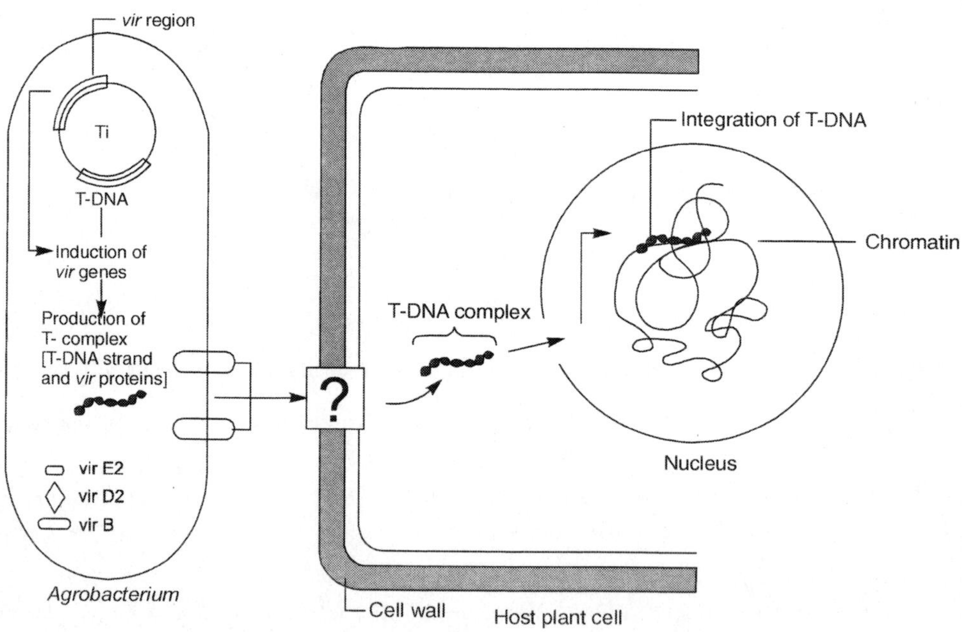

Figure 24.8 The complete process of T-DNA transfer from *agrobacterium* to host plant cell

acting elements of the T-DNA transfer system. The proteins VirD1 and VirD2 play a key role in this step and recognize the T-DNA border sequences and nick (endonuclease activity) the bottom strand at each border. The nick sites are assumed as the initiation and termination sites for T-strand recovery. After endonucleotidic cleavage, VirD2 remains covalently attached to the 5'-end of the ssT-strand. This association prevents the exonucleolytic attack at the 5'-end of the ssT-strand and distinguishes the 5'-end as the leading end of the T-DNA transfer complex.

Extensive mutation or deletion of the right T-DNA border indicates that there is almost complete loss of T-DNA transfer capacity, while mutation at the left border results in lower transfer efficiency. This fact indicates that T-strand synthesis is initiated at the right border and it proceeds in the 5' to 3' direction. Moreover, the termination process takes place even when the left border is mutated or completely absent, although with lower efficiency. Left border may act as a starting site for ssT-strand synthesis but the efficiency is much lower. This difference may be because of the presence of an enhancer or "overdrive" sequence next to the right border. This enhancer has been found to be specifically recognized by Vir C1 protein. It has been observed that if a deletion is made on *vir C* operon, it causes the attenuation of virulence of the *Agrobacterium* strains.

4. Mechanism of T-DNA Complex Transfer

The transferring vehicle to the plant nucleus is a ssT-DNA–protein complex. It must be translocated to the plant nucleus by passing through cell membranes, the plant cell wall and cellular spaces (Figures 24.8 and 24.9). According to the most accepted model, the ssT-DNA–Vir D2 complex is coated by the 69 kDa Vir E2 protein, a single-strand DNA-binding protein. This cooperative association prevents the attack of nucleases and also extends the ssT-DNA strand by reducing the complex diameter to approximately 2 nm, and supports the translocation through membrane channels easier. However, that association does not stabilize T-DNA complex inside the *Agrobacterium* cell. Furthermore, VirE2 contains two plant nuclear location signals, (NLS) and VirD2 one. This fact indicates that both proteins presumably play an important role, once the complex is in the plant cell and they mediate the complex uptake to the nucleus. The deletion of NLS in one of these proteins reduces but does not totally inhibit the ssT-DNA transfer and its integration into plant genome. It indicates that the other partner can at least partially assume the function of the absent protein.

It is known that VirE1 is essential for the export of VirE2 to the plant cell, however, other specific functions are still uncharacterized. Bacterial strains mutated in *vir*E1, cannot export VirE2, which is accumulated inside the bacterium. Such mutants can be complemented if co-infected with a strain that can export VirE2, indicating that this protein can be exported independently and that the transfer of VirE2 as part of the ssT-DNA complex is not necessary for the transmission event.

An alternative model proposes that the transfer complex is a single-stranded DNA covalently bound at its 5'-end with VirD2, but uncoated by VirE2. The independent export of VirE2 to plant cell is presented as a natural process, and once the naked ssT-DNA–VirD2 complex is inside the plant cell, it is coated by VirE2. It is also possible that the process can be performed by one of the proposed alternative ways according to the infection conditions.

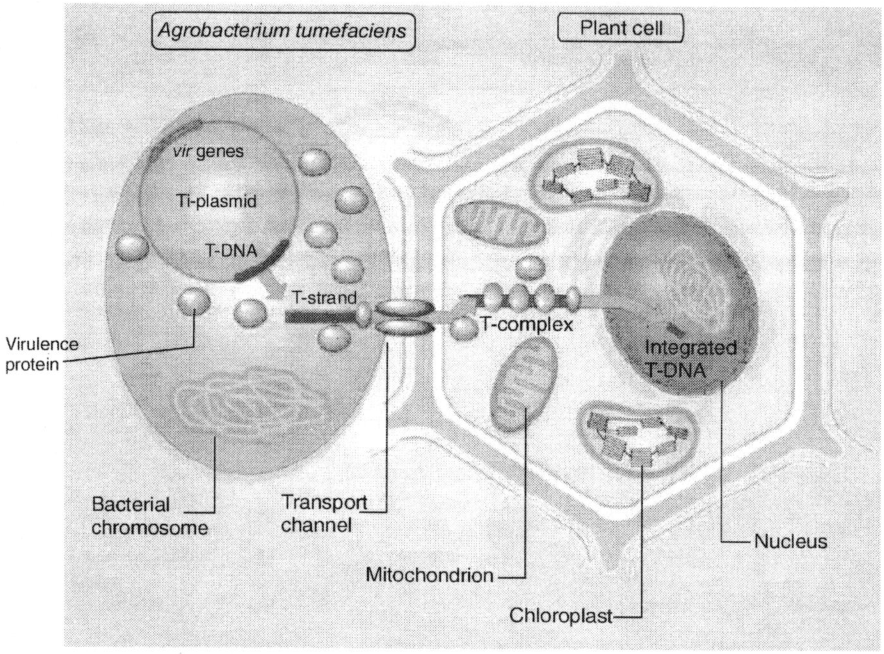

Figure 24.9 Mechanism of T-DNA complex transfer from bacterial cell to plant cell

Previously, it was described that only 9.5 kb *virB* operon is involved in the generation of a suitable cell surface structure for the ssT-DNA complex transfer from bacterium to plant. However, later on, it was suggested that the VirD4 protein is also required for the ssT-DNA transport. VirD4 is a transmembrane protein but predominantly located at the cytoplasmic side of the cytoplasmic membrane. The function of VirD4 is the ATP-dependent linkage of protein complex necessary for T-DNA translocation.

Some of the initial steps of biogenesis of ssT-DNA complex apparatus have been recently identified. First, VirB7 and VirB9 monomers are exported to the membrane and processed. They interact with each other to form covalently cross-linked homo- and heterodimers. Although the role of both types of dimers in the biogenesis of the transfer apparatus is widely known, it is likely that only heterodimers are essential. Subsequently, the VirB7–VirB9 heterodimer is sorted to the outer membrane. The next step involves the interaction with the other Vir proteins for assembling the transfer channel with the contribution of the transglycosidase VirB1. It is known that VirB2 through VirB11 are essential for

DNA transfer, suggesting that these proteins are fundamental components of the transfer apparatus while VirB1 has a lesser contribution in this process.

Two additional *vir* operons, present in the octopine Ti-plasmid, are *virF* and *virH*. The *virF* operon encodes for a 23 kD a protein that functions once the T-DNA complex is inside the plant cells. The role of VirF seems to be related with the nuclear targeting of the ssT-DNA complex but its contribution is less important. Moreover, the *virH* operon consists of two genes that code for VirH1 and VirH2 proteins. These Vir proteins are not essential but could enhance the transfer efficiency, detoxifying certain plant compounds that can affect the bacterial growth. If that is the function of VirH proteins, then they play a role in the host-range specificity of bacterial strain for different plant species.

5. Integration of T-DNA into Plant Genome

Inside the plant cell, the ssT-DNA complex is targeted to the nucleus, crossing the nuclear membrane. Two Vir proteins have been found to be important in this step: VirD2 and VirE2, and probably VirF, which has a minor contribution to this process. The nuclear location signals (NLS) of VirD2 and VirE2 play an important role in nuclear targeting of the delivered ssT-DNA complex.

VirD2 has one functional NLS. The ssT-DNA complex is a large (up to 20 kb) nucleoprotein complex containing only one 5'-end covalently attached VirD2 protein per complex. But the complex is coated by a large number of VirE2 molecules (approximately 600 molecules for a single 20 kb T-DNA), and each of them has two NLS. The two NLS of VirE2 have been considered important for the continuous nuclear import of ssT-DNA complex, probably by keeping both sides of nuclear pores simultaneously open. The nuclear import is probably also mediated by specific NLS-binding proteins, which are present in plant cytoplasm.

The final step of T-DNA transfer is its integration into the plant genome and expression of transgenes in the host system (Figure 24.10). The exact mechanism involved in the T-DNA integration has not been characterized but it is considered that the integration occurs by illegitimate recombination.

According to the illegitimate recombination model, paring of a few bases, known as micro-homologies, are required for a pre-annealing step between T-DNA strand coupled with VirD2 and plant DNA. These homologies are very low and provide just a minimum specificity for the recombination process by positioning VirD2 for the ligation. The 3'-end or adjacent sequences of T-DNA find some low homologies with plant DNA resulting in the first contact (synapses) between the T-strand and plant DNA and forming a gap in 3'–5' end strand of plant DNA. Displaced plant DNA is subsequently cut at the 3'-end position of the gap by endonucleases, and the first nucleotide of the 5'-end attaches to VirD2 pairs with a nucleotide in the top (5'–3') plant DNA strand. The 3'-end overhanging part of T-DNA together with displaced plant DNA are digested away, either by endonucleases or by 3'–5' exonucleases. Then, the 5'-end attached to VirD2 end and the other 3'-end of T-strand (paired with plant DNA during the first step of integration process) joins the nicks in the bottom plant DNA strand.

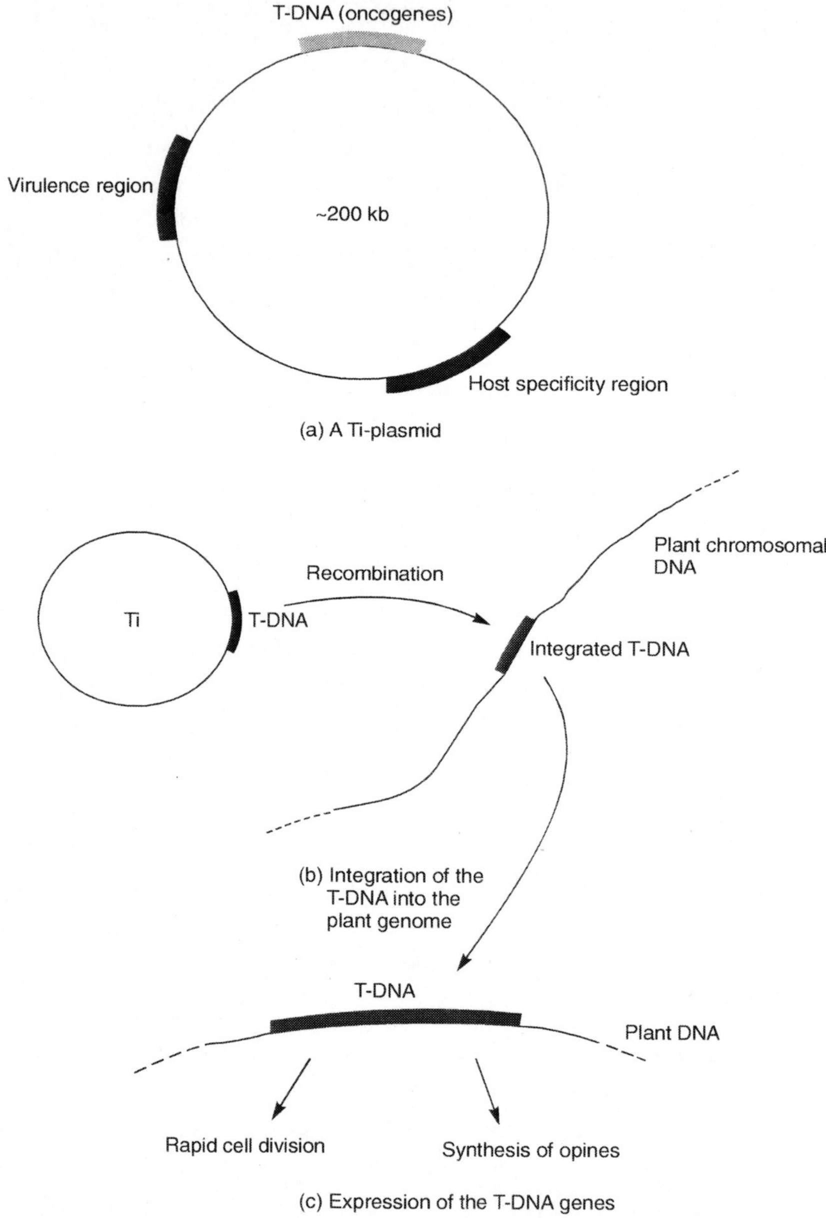

Figure 24.10 Structure of Ti-plasmid with mode of T-DNA integration and expression in plant cell

VirD2 has an active role in the precise integration of T-strand in the plant chromosome. The release of VirD2 protein may provide the energy containing in its phosphodiester bond, at the Tyr-29 residue, with the first nucleotide of T-strand, providing the 5'-end of the T-strand for ligation to the plant DNA. This phosphodiester bond can serve as electrophilic substrate for nucleophilic 3'-OH from nicked plant DNA. When the mutant VirD2 protein is transferred attached to the T-strand, the integration process takes place with the loss of nucleotides at the 5'-end of the T-strand.

Once the introduction of T-strand in the 3'–5' strand of the plant DNA is completed, a torsion followed by a nick into the opposite plant DNA strand is produced. This situation activates the repair mechanism of the plant cell and the complementary strand is synthesized using the early inserted T-DNA strand as a template.

SILENCING OF TRANSGENES

Analysis of the large collection of genetically modified plants generated in recent years has expanded our knowledge of physiological processes and gene regulation mechanisms tremendously. However, transgenes do not always behave as expected. This has revealed the existence of hitherto unknown cellular processes. There is considerable variation in the expression of transgenes in individual transformants, which is not due to differences in copy number. Thus, gene activity is not exclusively determined by the strength of the promoter which controls transcription; epigenetic effects also influence expression levels. This sometimes leads to gene inactivation either by blocking transcription or by inhibiting mRNA accumulation. The mechanisms by which silencing is achieved are still poorly understood. Despite the different silencing systems being examined, several cases of gene silencing have features in common, which gives us insight into the factors involved.

Gene silencing also occurs in untransformed plants where it reduces expression of endogenous genes. A number of mutations in various plant species appear to result from epigenetic gene silencing. For example, paramutation in maize and tomato probably involve gene-to-gene interactions. Although paramutation was known long before the discovery of transgene-mediated silencing, it is only recently that the underlying molecular mechanisms have become apparent. Several features resemble those associated with transgene-mediated silencing.

Various possibilities have been raised to explain variation in transgene expression levels among transformants that is independent of copy number. All these imply that integrated transgenes cannot be regarded as independent transcription units. Transgenes, often as part of the *Agrobacterium tumefaciens* T-DNA, integrate at different chromosomal locations. If they become inserted into euchromatin, in a transcriptionally active region, expression may be influenced by regulatory sequences of nearby host genes. If they insert in or near repetitive DNA or heterochromatin, they can be inactivated.

Another important factor associated with gene silencing is the number of transgenes per integration site. The T-DNA transfer system can insert two or more T-DNAs at the same chromosomal site.

These T-DNAs can be arranged "head-to-tail" as a direct repeat (DR), and "head-to-head" or "tail-to-tail" as an inverted repeat (IR). Transgenes of T-DNAs that are organized as IRs often show low expression indicating that the genes are silenced to some degree.

There are two kinds of gene silencing. First, transcriptional gene silencing (TGS), which results from promoter inactivation; and second, post-transcriptional gene silencing (PTGS) which occurs when the promoter is active but the mRNAs fail to accumulate. Even though this clear difference suggests two distinct silencing mechanisms, the two seem related, in particular, when one invokes interactions between homologous DNA sequences. This notion is inspired by observations that DNA methylation, which is often associated with TGS, is sometimes also found associated with PTGS.

CONCLUSION

Plant genetic transformation is of particular benefit to molecular genetic studies, crop improvement and production of pharmaceutical materials. These applications and the desire for higher transformation efficiency have stimulated work on inventing and improving various methods. Apart from the elimination of tissue culture for *Arabidopsis* and a few other species, no great breakthroughs on basic transformation techniques have taken place since *Agrobacterium* and biolistics. *Agrobacterium*-based methods are usually superior for many species including dicots and monocots. For some angiosperms and non-angiosperms, biolistics is usually used. The others are typically not done on a routine basis. Biolistics is by far the most widely used direct transformation procedure both experimentally, in research, and commercially. In addition, it has proven to be the most efficient. Chloroplast targeting is advantageous for direct transformation methods in sharp contrast to *Agrobacterium*-based methods. High-level recombinant protein expression is one advantage for chloroplast targeting but it has some drawbacks, such as low transformation frequency, prolonged selection procedures under high selection pressure and appearance of green tissue dependency for transgene products. However, maternal inheritance of transgenes is another advantage for transplastomes, which has benefits relative to biosafety. In plant, transformation methods are immensely attractive; however, this has proven to be effective and efficient in *Arabidopsis*.

REVIEW QUESTIONS

1. "*Agrobacterium* is a natural genetic engineer" explain it.
2. Write in detail about the biochemical steps of *Agrobacterium* infection during transgenesis.
3. What is T-DNA? Write its significance for transgenic production.
4. Write an essay on molecular mechanism of T-DNA transfer by *Agrobacterium*.
5. Discuss briefly the silencing of transgenes in transgenic plant.

REFERENCES

Berger, B.R. and Christie, P.J. (1993). The *Agrobacterium tumefaciens virB4* gene product is an essential virulence protein requiring an intact nucleoside triphosphate-binding domain. *Journal of Bacteriology,* 175: 1723–1734.

Berger, B.R. and Christie, P.J. (1994). Genetic complementation analysis of the *Agrobacterium tumefaciens virB* operon: *virB2* through *virB11* are essential virulence genes. *Journal of Bacteriology*. 176: 3646–3660.

Binns, A.N., Beaupre, C.E. and Dale, M. (1995). Inhibition of VirB-mediated transfer of diverse substrate from *Agrobacterium tumefaciens* by the IncQ plasmid RSF1010. *Journal of Bacteriology*. 177: 4890–4899.

Bradley, L.R., Kim, J.S. and Matthysse, A.G. (1997). Attachment of *Agrobacterium tumefaciens* to Carrot Cells and Arabidopsis wound sites is correlated with the presence of a cell-associated, acidic polysaccharide. *Journal of Bacteriology*. 179: 5372–5379.

Brink, R.A. (1993). Paramutation. *Annual Review of Genetics*. 7: 129–152.

Chang, C.H. and Winans, S.C. (1992). Functional roles assigned to the periplasmic, linker and receiver domains of the Agrobacterium tumefaciens VirA protein. *Journal of Bacteriology*. 174: 7033–7039.

Dürrenberger, F., Crameri, A., Hohn, B. and Koukolikova-Nicola, Z. (1989). Covalently bound VirD2 protein of *Agrobacterium tumefaciens* protects the T-DNA from exonucleolytic degradation. Proceedings of the National Academy of Sciences. USA. 86: 9154–9158.

English, J.J., Mueller, E. and Baulcombe, D.C. (1996). Suppression of virus accumulation in transgenic plants exhibiting silencing of nuclear genes. *Plant Cell.* 8: 179–188.

Firth, N., Ippen-Ihler, K. and Skurray, R.A. (1996). Structure and function of the F factor and mechanism of conjugation, In: Neidhardt, F.C., Curtis, R., Ingraham, J.L., Lin, E.C.C., Low, K.B., Magasanik, B., Reznikoff, W.S., Riley, M., Schaechter, M. and Umbarger, H.C. (eds.). *Escherichia coli and Salmonella: Cellular and Molecular Biology,* 2nd edition. American Society for Microbiology. Washington DC. pp. 2377–2401.

Fullner, K.J., Lara, J.C. and Nester, E.W. (1996). Pilus assembly by Agrobacterium T-DNA transfer genes. *Science.* 273: 1107–1109.

Gustavo, A. de la Riva, Joel González-Cabrera, Roberto Vázquez-Padrón and Camilo Ayra-Pardo. (2008). *Agrobacterium tumefaciens*: a natural tool for plant transformation. *Molecular Biology and Genetics.* Vol. 1. Issue. 3.

Hageman, R. (1993). Studies towards a genetic and molecular analysis of paramutation at the sulfurea locus of *Lycopersicon esculentum* Mill. In: Yoder, J.I. (ed.). *Molecular Biology of Tomato.* Technomic Publishers Co., Inc., Lancaster, Basel. pp. 75–82.

Hille, J., Wullems, G. and Schilperoort, R.A. (1983). Non-oncogenic T-region mutants of *Agrobacterium tumefaciens* do transfer T-DNA into plant cells. *Plant Molecular Biology.* 2: 155–163.

Hooykaas, P.J.J. and Shilperoort, R.A. (1992). *Agrobacterium* and plant genetic engineering. *Plant Molecular Biology.* 19: 15–38.

Jones J.D.G., Gilbert, D.E., Grady, K.L. and Jorgensen, R.A. (1987). T-DNA structure and gene expression in petunia plants transformed by *Agrobacterium* C58 derivatives. *Molecular and General Genetics.* 207: 478–485.

Kanemoto, R.H., Powell, A.T., Akiyoshi, D.E., Regier, D.A., Kerstetter, R.A., Nester, E.W., Hawes, M.C. and Gordon, M.P. (1989). Nucleotide sequence and analysis of the plant-inducible locus *pinF* from *Agrobacterium tumefaciens. Journal of Bacteriology.* 171: 2506–2512.

Kertbundit, S., Degreve, H., Deboeck, F., Van Montagu, M. and Hernalsteens, J.P. (1991). In vivo random beta-glucuronidase gene fusions in *Arabidopsis thaliana.* Proceedings of the National Academy of Sciences of the United States of America. 88: 5212–5216.

Matthysse, A.G. (1986). Initial interactions of *Agrobacterium tumefaciens* with plant host cells. *Critical Reviews in Microbiology.* 13: 281–307.

Pan, S.Q., Charles, T., Jin, S., Wu, Z.L. and Nester, E.W. (1993). Preformed dimeric state of the sensor protein VirA is involved in plant-Agrobacterium signal transduction. Proceedings of the National Academy of Sciences. USA. 90: 9939–9943.

Puchta, H. (1998). Repair of genomic double-strand breaks in somatic cells by one-side invasion of homologous sequences. *Plant Journal.* 13: 331–339.

Rogowsky, P.M., Powell, B.S., Shirasu, K., Lim, T.S., Morel, P., Zyprian, E.M., Steck, T.R. and Kado, C.I. (1990). Molecular characterization of the *vir* regulon of *Agrobacterium tumefaciens*: complete nucleotide sequence and gene organization of the 28.63 kbp regulon cloned as single unit. *Plasmid.* 23: 85–106.

Tinland, B., Schoumacher, F., Gloeckler, V., Bravo, A.M., Angel, M. and Hohn, B. (1995). The Agrobacterium tumefaciens virulence D2 protein is responsible for precise integration of T-DNA into the plant genome. *EMBO Journal.* 14: 3585–3595.

Zupan, J.R. and Zambryski, P.C. (1995). Transfer of T-DNA from *Agrobacterium* to the plant cell. *Plant Physiology.* 107: 1041–1047.

25

TRANSGENIC PRODUCTION-III

INTRODUCTION

Transgenic technology enables plant breeders to bring together in one plant useful genes from a wide range of living sources, not just from within the same crop species but also from closely related plants. A plant breeder tries to assemble a combination of genes in a crop plant which will make it as useful and productive as possible. Depending on where and for what purpose the plant is grown, desirable genes may provide features such as higher yield or improved quality, pest or disease resistance, or tolerance to heat, cold and drought.

This technology provides the means for identifying and isolating genes controlling specific characteristics in one kind of organism, and for moving copies of those genes into another quite different organism, which will then also have those characteristics. This powerful tool enables plant breeders to do what they have always done to generate more useful and productive crop varieties containing new combinations of genes and also expands the possibilities beyond the limitations imposed by traditional cross-pollination and selection techniques.

Combining the best genes in one plant is a long and difficult process, especially as traditional plant breeding has been limited to artificially crossing plants within the same species or with closely related species to bring different genes together. For example, a gene for a protein in soybean could not be transferred to a completely different crop such as corn using traditional techniques. Varieties containing genes of two distinct plant species are frequently created by classical breeders who deliberately force hybridization between distinct plant species when carrying out interspecific or intergeneric wide crosses with the intention of developing disease-resistant crop varieties.

Classical plant breeders use a number of *in vitro* techniques such as protoplast fusion, embryo rescue or mutagenesis to generate diversity and produce plants that did not exist in nature. Since 1930, such traditional techniques have allowed crop breeders to develop varieties of basic food crops such as wheat, which resist devastating plant diseases such as rusts. "Hope" is one such wheat variety with a gene from a wild grass. "Hope" saved American wheat growers from devastating stem rust outbreaks in the 1930s.

Methods used in traditional breeding, which generate plants with DNA from two species by non-recombinant methods are widely familiar to professional plant scientists, and serve important roles in securing a sustainable future for agriculture by protecting crops from pests and helping more efficient use of land and water.

HISTORY OF TRANSGENIC INDUSTRY

- Transgenic plants have been engineered to produce a variety of products using their high capacity and self-assembling fermentors, which operate in non-sterile conditions. The first target for research was herbicide-resistant transgenic plants. Herbicide-tolerant plants have a number of commercially and environmentally desirable properties and provide more cost-effective and eco-friendly weed control. Disease resistance has also been targeted as an advantageous trait in the major crops and a number of strategies have been employed to achieve this goal. Virus resistance is important for good crop yields and it reduces the amount of chemicals required for controlling the insect vectors that transmit the virus.
- Storage lipid modification via engineering of the fatty acid biosynthesis enzymes has also been achieved, which contribute to the production of healthier foods as well as production of chemical feed stocks. Moreover, crop losses occur due to cold and freeze damage. Cold tolerance could be induced in cold-susceptible plants by a process called "acclimatization" and one of the changes that occur is the production of organic compounds and antifreeze proteins, which lower the freezing temperature of tissue liquids. Significantly, Georges *et al.* (1990) expressed a fusion between chloramphenicol acetyl transferase and a synthetic gene encoding for the antifreeze protein in maize tissues. They observed that tissue extracts, which showed the presence of the protein, displayed a reduction in ice crystal formation.
- Introduction of bacterial insecticide genes from *Bacillus thuringiensis* into plants to confer pest resistance by inhibition of insect vectors has also been achieved. Cotton, which is transgenically insect-resistant, is the first insect-resistant plant to be commercialized. Significantly, in transgenic peas, gene expression was observed in the seed and the protein was accumulated at sufficient levels to prevent infestation with the bruchid beetle weevils. Moreover, the 2S storage albumin of Brazil nut (*Bertholeita excelsa*) has been a popular target for increasing the level of sulphur amino acids in legumes, because it contains multiple methionine and cystine residues and therefore, in the direction of transgenic

efforts. Saalbach *et al.* (1994) have constructed a chimeric 2S albumin cDNA that is constitutively expressed throughout the seeds and reveals that the Brazil nut protein is synthesized in this organ.

- The ability to create transgenic organisms offers the opportunity of programming plants to accumulate foreign proteins in seeds, roots or leaves, to enhance their quality, value or even provide a new source of valuable pharmaceuticals, including antibiotics. The "Flavr Savr" tomato is the first genetically engineered plant product to be released, which has a synthetic gene that inhibits the expression of polygalacturonase, which normally accelerates fruit softening and contributes to the over-ripening of tomatoes. Transgenic plants with improved seed protein and oil content in corn, cotton plants resistant to boll worm, plants with the ability to synthesize antibodies and vaccines, plants which synthesize biodegradable plastics are some of the important examples of genetically engineered plants.
- Transgenic plants have been deliberately developed for a variety of reasons: longer shelf-life, disease resistance, herbicide resistance, pest resistance, non-biological stress resistances, such as to drought or nitrogen starvation, and nutritional improvement. The first modern recombinant crop approved for sale in the US, in 1994, was the Flavr Savr tomato, which had a longer shelf-life. The first conventional transgenic cereal created by scientific breeders was actually a hybrid between wheat and rye in 1876. The first transgenic cereal may have been wheat, which itself is a natural transgenic plant derived from at least three different parental species.

"Ex vitro" TRANSGENESIS

Natural movement of genes between species, often called horizontal gene transfer or lateral gene transfer, can occur because of gene transfer mediated by natural processes. This natural gene movement between species has been widely detected during genetic investigation of various natural mobile genetic elements, such as transposons, and retrotransposons that naturally translocate to new sites in a genome, and often move to new species over an evolutionary time scale. There are many types of natural mobile DNAs, and they have been detected abundantly in food crops such as rice. These various mobile genes play a major role in causing dynamic changes to chromosomes during evolution.

Genetically mobile DNA constitutes a major fraction of the DNA of many plants, and the natural dynamic changes to crop plant chromosomes caused by this natural transgenic DNA mimics many of the features of plant genetic engineering currently pursued in the laboratory, such as using transposons as a genetic tool, and molecular cloning. There is new scientific literature about natural transgenic events in plants, through movement of natural mobile DNAs called MULEs between rice and *Setaria,* a millet crop. It is becoming clear that natural rearrangements of DNA and horizontal gene transfer play a pervasive role in natural evolution. Importantly many, if not most, flowering plants evolved by transgenesis, that is, the creation of natural interspecies hybrids in which chromosome sets from different plant species are added together. There is also a long and rich history of interspecies cross-breeding with traditional methods.

TRANSGENIC CROPS IN INDIA

Several transgenic crops have been developed in India or imported from abroad either for direct field testing or for involving them in plant breeding programmes to develop varieties that are adapted to Indian conditions, but carried transgenes. Transgenic crops are being developed in several laboratories in India, and many transgenic crops have been successfully produced so far, however, in other cases, transgenic work is still in progress and transgenic plants will soon become available.

Transgenic crops that were field tested in India during 1996–2001 are cotton, brinjal, tomato and rice. Bt-cotton, which was grown without approval in large areas (> 10,000 hectares) in Gujarat in the year 2001, was later field tested (Mahyco-Monsanto) and approved for commercial cultivation in the year 2002. Several other transgenic crops have been approved for large-scale field trials, for instance, the first large-scale Bt-rice (IR64 and Basmati, carrying Bt-gene) field trials were conducted in the year 2001. Similar trials have also been conducted in brinjal and tomato. In the private sector, Proagro also has conducted large-scale field trials of hybrid *Brassica juncea*, developed from transgenic male-sterile (due to barnase) and restorer lines (due to barstar).

A list of higher plants, where transgenic plants have been produced is given in Table 25.1.

Resistance Against Abiotic Stresses

Serious crop losses are also caused due to a variety of abiotic stresses, including drought, cold (frost), salinity, metal toxicity, etc. Transgenic plants resistant to these stresses have been produced, utilizing transgenes that alter the desired biosynthetic pathways, although even in the year 2008–09, these transgenic plants were only in a preliminary stage of development and perhaps will be commercialized very soon after subsequent field trials.

Herbicide-resistant Transgenic

Due to increasing concern about contamination of environment by herbicides, new herbicides are being developed that are safer and biodegradable. This has necessitated the development of resistance in crop plants against these new and safer herbicides. Herbicides, normally affect processes like photosynthesis or biosynthesis of essential amino acids. Two approaches have, therefore, been used for the development of herbicide resistant plants:

1. In the first approach, either the target molecules is made insensitive to herbicide or the target protein is made to overproduce.
2. In the second approach, a pathway is introduced that will detoxify the herbicide.

Modification of the target molecules The target molecule has been modified for developing resistance against at least three herbicides (glyphosate, sulphonylureas and imidazolinones).

The transgenic petunia plants, resistant to glyphosate (active ingredient of Roundup herbicide) have been developed by transfer of a gene for EPSPS (5-enol-pyruvylshikimate 3-phosphate synthase), that overproduces this enzyme. This overexpressing gene was isolated from plants selected for herbicide resistance. In some other cases, a gene *aroA* was isolated from the bacteria *Salmonella typhimurium* or *E. coli* and was transferred to tomato and/or tobacco. Mechanism of action of different herbicides and their basis of resistance is listed in Table 25.2.

Table 25.2 Mechanisms of action of different herbicides

Active principle of herbicide	Inhibited pathway	Target product	Use	Basis of resistance
Amino acid biosynthesis inhibitors				
Glyphosate (Roundup)	Aromatic amino acid biosynthesis	EPSPS	Broad spectrum	Overexpression of *EPSPS* gene; bacterial *aroA* gene
Sulphonylurea and Imidazolinones	Branched chain amino acids	*ALS*	Selected crops	Mutant ALS gene
Phosphinothricin (Basta)	Glutamine biosynthesis GS	Broad spectrum	Gene amplification	*Bar* gene detoxification
Photosynthesis inhibitors				
Atrazine (Lasso)	Photosystem II	QB (32kDa protein)	Selected crops	Mutant *PsbA gene. GST* gene: detoxification
Bromoxynil (Buctril)	Photosynthesis	–	Selected crops	bxn gene: detoxification

Another class of herbicides includes sulphonylurea compounds (active ingredients of Glean and Qust herbicide) and imidazolinones that inhibit the enzyme acetolactate synthase (ALS), which is involved in the biosynthesis of branched chain amino acids like leucine, isoleucine and valine. Transgenic tobacco plants expressing a mutant ALS gene from tobacco or *Arabidopsis* could be produced that are tolerant to sulphonylurea herbicides.

Similarly, transgenic tobacco plants have also been produced by incorporation of genes for resistance against

1. L-phosphinothricin (PPT), which is the active ingredient of herbicide "Basta" and inhibits glutamine synthase (GS), and
2. Atrazine, which inhibits photosynthesis. These genes were isolated from *Medicago sativa* and *Amaranthus* hybrids respectively. The transgenic tobacco plants expressed low level of resistance

against the corresponding herbicides. Scientists of Plant Genetic Systems (PGS) at the German Company, Hoechst, isolated a gene from *Streptomyces hygroscopicus,* which encodes an enzyme, capable of inactivating the herbicide "Basta".

Transgenic plants with this particular gene have been produced and field tested, demonstrating effectiveness of this gene for protection against the herbicide "Basta".

Detoxification or degradation of herbicide

1. Detoxification or degradation of herbicides is the basis of selective effect of herbicide, so that the latter kills the weeds and not the crop. A number of detoxifying enzymes have been identified in plants as well as in microbes. Some of these enzymes are
   i. Glutathione-S transferase or GST (in maize and other plants), which detoxifies the herbicide atrazine;
   ii. Nitrilase (coded by gene *bxn* in *Klebsiella pneumoniae*), which detoxifies the herbicide bromoxynil, and
   iii. Phosphinothricin acetyl transferase or PAT (coded by *bar* gene in *Streptomyces* spp.), which detoxifies the herbicide PPT (L-phosphinothricin).
2. Transgenic tomato plants, using both *bxn* gene from *Klebsiella* and *bar* gene from *Streptomyces,* and transgenic plants in potato, oilseed rape (*Brassica napus*) and sugarbeet, using only *bar* gene from *Streptomyces,* have been obtained and were found to be herbicide resistant. Other target crops for engineered herbicide tolerance include soybean, cotton and corn. Herbicide resistant transgenic crops are already being commercially grown in large areas in several countries.

Insect-resistant Transgenic Plants

Serious losses in crop yields are caused due to insect pests. There are about 67,000 pest species that damage crops. Of these, 9,000 species are insects and mites, which are responsible for major yield losses in several of our important crops, particularly the tropical crops. Therefore, transfer of genes(s) providing resistance against insects in crop plants leading to the production of insect-resistant transgenic crops is a major application of biotechnology in the agricultural sector. A list of complementary systems of insect resistance in crops is given in Table 25.3. The insect-resistant transgenic crops carrying Bt-toxin genes have been shown to be very effective in controlling insect damage and are actually better than the currently used methods of insect control through insecticidal sprays.

This not only allows saving in cost and time, but also reduces health risks and provides ecological benefits, since Bt-toxins are highly specific against certain insects without affecting other specific insects (insecticides instead kill a broad spectrum of insects).

Table 25.3 Complementary systems of insect resistance in crops

System	Gene transfer	Patented	Comments
Substances occurring in plants			
Sugar (glucosinolates polysaccharides)	No	No	The relative concentration of different sugars determines whether or not certain insects prefer a given plant or species. (saccharose is preferred; not used in breeding)
Terpenoids	No	No	Large group of substances including pyrethrin I, sesquiterpenoid and phytoecdysteroids. Some terpenoids are common in cotton.
Alkaloids and glycoalkaloids	No	No	Includes nicotine, which has long been used as biological pesticide produced from tobacco extracts. Also used in classical breeding such as in potato (demissin and α-tomatin). Several *Brassica* species also contain sulphides (glycoside), which are toxic to some insects and attract others.
Flavanoids	No	No	—
Phenols	No	No	Overexpression of polyphenoloxidase leading to increased production of quinonin, which is toxic for insects. Unlikely to be used in genetic engineering due to complex chemistry and possible toxicity to mammals.
Protein antimetabolites			All are secondary metabolites.
i. Amino acids and primary storage proteins	No	No	For example, in wheat, gluten proteins are not digestible by some insects (*Eurygaster integriceps*) and thus confer partial resistance against insects. The presence of certain amino acids generally increases resistance in rice.
ii. Lectin inhibitors (also referred to as plant peptide hormones)	Yes	Yes	Lectins are common in the grains of cereals, particularly during germination. Certain lectins also have antifungal properties. Much work is going on with the snowdrop lectin (GNA); particularly promising for sucking insects (Homoptera), which cannot be controlled with Bt.

(Contd.)

Table 25.3 (Continued)

System	Gene transfer	Patented	Comments
iii. Protease inhibitors (includes trypsin and chymotrysin inhibitors)	Yes	Yes	Act on exogenous proteolytic enzymes. Trypsin inhibitor has particularly broad insect spectrum and has been demonstrated to be synergistic with Bt; much work is being done with the cowpea protease inhibitors (CpTI).
iv. α-amylase inhibitors	Yes	Yes	Widely occurring in seeds, particularly dicots. For gene transfer, the most effective so far has been the α-amylase inhibitor genes from the common bean.
Substances not occurring in plants			
Other toxins	Yes	Yes	Spider and wasp toxins have been transferred to plants for experimental purposes. Unlikely to be applied in crops due to the effect of some toxins on mammals.
Smart proteins	Yes	Yes	Computer-aided design of novel proteins.

In classical plant breeding, following three systems for insect resistance are utilized:

1. Morphological barriers to insects (e.g., hairy leaves);
2. Insect repellant or toxic substances released constitutively by the plant, or induced by insect damage, and
3. Toxins that have either a repellant affect (e.g., quinonin) or deadly effect (e.g., proteolytic enzymes, such as trypsin inhibitor, found in peas).

A new insect resistance technology involving the use of smart proteins (computer designed) and Vasointestinal peptide (VIP) family has been in use since 1997. This technology supports an approach of pyramiding of different genes active against the same insect. It is predicted that as many as five or more foreign genes against the same insect will be available in most insect-resistant transgenic crops.

Some of the novel systems include the following:

1. Control of insect growth through toxicity of their purine metabolic pathway;
2. Use of *Beauveria,* virulent with an active avaricide, against Lepidoptera;
3. Use of 3-hydroxy-steroid oxidase against Lepidoptera and boll weevil;
4. Use of a gene for carbonarin anti-insectant metabolites isolated from *Aspergillus carbonarius.*

However, the use of smart proteins, VIPs, engineered proteins (computer-aided) and manipulation of metabolic pathways (e.g., production of azadirachtin, a toxic compound from neem tree) are believed to provide long-term insect resistant transgenic plants.

Bt and biopesticides *Bacillus thuringiensis*, commonly known as Bt, is a gram positive soil bacterium. It is known for several decades that some strains of Bt kill certain insects (Lepidoptera, Coleoptera and Diptera), due to an insecticidal protein called δ-endotoxin, which disrupts the function of digestive system of these insects.

It is also known that Bt protein is not harmful to mammals and is degraded within 20 seconds in the digestive tract of a mammal, thus permitting safe handling and use of Bt formulations. Different Bt strains are effective against different insects thus making it necessary to make a choice while using them as biopesticides. Although Bt was registered in USA in 1961, even today only <1% of all pesticides are based on Bt.

Most of these formulations contain spores and crystalline inclusion that are released on lysis of Bt during growth. The potency of these formulations is 300 times that of synthetic pyrethroids commonly used and the toxin breaks down quickly in sunlight thus making it environment friendly.

Bt-based biopesticides have the following disadvantages.

1. They are relatively extensive.
2. Agricultural machinery is needed for their application.
3. Application need to be repeated several times in every crop season.
4. Sunlight causes breakdown of the active ingredient.
5. Water washes the protein from the plant.

Most of these difficulties, however, have been overcome in the production of transgenic crops.

Bt Endotoxins and their genes Bt toxins, symbolized as "Cry", meaning "crystalline" (reflecting the crystalline appearance of δ-endotoxin), were initially classified into four distinct classes, based on their host range. They are the following classes of Bt toxins.

1. Cry I (active against Lepidoptera),
2. Cry II (active against Lepidoptera and Diptera),
3. Cry III (active against Coleoptera) and
4. Cry IV (active against Diptera).

Many more classes have been added to these classes (e.g., CryV, CryVI and Cry IX) and new strains have been shown to be active against nematodes and other pests.

So far, 50 genes for 58 Cry proteins are known, of which 28 genes were isolated from 14 different Bt subspecies that have been shown to be active against insects. Additional 30 Cry proteins described in the literature do not have any insecticidal properties, although they may be effective against nematodes and mites.

These *Cry* genes are being variously modified to acquire new activities, thus extending the range of their insecticidal/nematicidal properties. The different Bt endotoxins (Cry) and their activity against specific insects are listed in Table 25.4.

Insect-resistant plants based on Bt The toxin genes (*bt2*) from *B. thuringiensis,* mentioned above, have been isolated and used for *Agrobacterium* Ti- plasmid-mediated transformation of a number of crop plants including tobacco, cotton and tomato plants. The transgenic plants were resistant to *Manducta sexta,* a pest of tobacco.

Furthermore, experiments on feeding the leaves of these plants to larvae of *Manducta sexta,* showed 75–100% mortality of the larvae, while the control plants carrying no transgenes were severely damaged. The presence of the gene *bt2* as well as that of the toxin protein synthesized under its control was also demonstrated by appropriate experiments. When inheritance of insect resistance was studied using crosses with normal control plants, F_1 showed resistance and F_2 generation exhibited expected segregation.

Since, a native gene often does not express the gene at high level, truncated versions of genes are synthesized with altered codon usage and elimination of certain sequences (codon usage means that for same amino acids, codons that are efficient in prokaryotes but not in eukaryotes, are replaced by other degenerate codons known to be efficient in eukaryotes).

BT-Crops as substitutes for insecticides The world crop loss without the use of pesticides and other control measures is estimated to be 70% of the expected production. It is also estimated that despite all efforts and measures, pests still destroy 50% of world production amounting to about US$ 300 billion. In view of this, insect-resistant crops offer great promise. Estimates of potential substitution value of Bt-transgenic plants for insecticides have been made and it is believed that a total cost spent on insecticides, can be replaced by Bt-transgenic crops.

The Bt-transgenic crops, while provide the above benefit, will also add expenses due to the cost of technology. However, in case of rice, International Rice Research Institute (IRRI) in the Philippines developed Bt-rice with a gene donated by Ciba of Switzerland. The germplasm will be made freely available to select national programmes and other interested parties. Similarly, at CIMMYT in Mexico, CryIA(b) and CryIA(c) are used in corn. The production of transgenic crops against insect resistance by using different *Cry* genes is listed in Table 25.5.

Table 25.4 Bt endotoxin (Cry) and their activity against specific insect species

Cry protein	Origin (Bt subspecies)	Major target insects	
		Order	Common names
CryIA(a)	*kurstaki*	L	Silkworm, tobacco horn worm, European corn borer
CryIA(b)	*berlineri*	L and D	Tobacco horn worm, cabbage worm, mosquito
CryIA(c)	*kurstaki*	L	Tobacco budworm, cabbage hopper, cotton bolloworm
CryIA(d)	*aizawai*	L	Several Lepidoptera
CryIA(e)	*alesti*	L	Tobacco budworm
CryIB	*thuringiensis*	L	Cabbage worm
CryIB(c)	*morrisoni*	L	Several Lepidoptera
CryIC	*entomocidus*	L and D	Cotton leaf form, mosquito
CryIC(b)	*galleriae*	L	Beet army worm
CryID	*aiawai*	L	Beet army worm, tobacco horn worm
CryIE	*keayae*	L	Cotton leaf worm
CryIE(b)	*aizawai*	L	Several Lepidoptera
CryIF	*aizawai*	L	European corn borer, beet army worm
CryIG	*galleriae*	L	Greater wax moth
CryIIA	*kurstaki*	L and D	Gypsy moth, mosquito
CryIIB	*kurstaki*	L	Gypsy moth, cabbage hopper, tobacco horn worm
CryIIC	*shanghai*	L	Tobacco horn worm, gypsy moth
CryIIIA	*San diego*	C	Colorado potato beetle
CryIIIA(a)	*tenebrionis*	C	N/A
CryIIIB	N/A	C	Colorado potato beetle
CryIIIC	*tolworthi*	C	Colorado potato beetle
CryIIID	*kurstaki*	C	(*Aedes* and *Culex*)
CryIVA	*israelensis*	D	Mosquito (*Aedes* and *Culex*)
CryIVB	*israelensis*	D	Mosquito (*Aedes*)
CryIVC	N/A	D	Mosquito (*Culex*)
CryIVD	N/A	D	Mosquito (*Aedes* and *Culex*)
CryV	N/A	L and C	European corn borer, spotted cucumber beetle
CryIX	*galleriae*	L	Greater wax moth

N/A = not available; L = Lepidoptera; D = Diptera; C = Coleoptera

Table 25.5 Information about Cry genes, used for production of insect-resistant transgenic crops

Transgenic crop	Transgene(s)
Potato	CryIA(b), Cry IIIA(a)
Corn	CryIA(b) CryIA (c), CBI-Bt
Cotton	Cry 1A(a) CryIA (b), CryIA (c)
Tobacco	CryIA(b), CryIA(c)
Tomato	CryIA(b)
Canola (rapeseed)	CryIA(b)

While International Agricultural Research Centre (IARCs) may develop and make available the transgenic crops in the developing world, for several other crops being developed by private sector, the farmers will have to pay for the investments made, so that the seed of transgenic crops will have to be purchased at a premium. However, it is believed that the seeds of transgenic varieties may be sold at prices identical to those of non-transformed variety as a means of boosting the market for transgenic crop varieties. Moreover, efficacy of CpT_I as insecticides on different insects is given in Table 25.6.

Table 25.6 Insecticidal efficacy of CpT_I in artificial diets and in transgenic plants

Insects killed by CpT_1 in artificial diets	Insects, against whom resistance is noticed in transgenic tobacco plants (with CpT_1)
I. Lepidoptera	
Heliothis virescens	*Heliothis virescens*
H. zea	*Spodoptera littoralis*
Spodoptera littoralis	*Spodoptera littoralis*
Chilo partellus	*Autographa gamma*
II. Coleoptera	
Callosobruchus maculatus	These insects do not attack tobacco plants used as control
Anthonomus grandis	
Diabrotica	
Undecimpunctata	
Tribolium confusum	
Costelytra zealandica	
III. Orthoptera	
Locusta migratoria	

Gene for alpha-amylase inhibitor (α-AI-Pv) The insect pest resistant transgenic plants, produced using Bt-gene, protect themselves from insects mainly in the field, but not during storage. The first transgenic plants, whose seeds, during storage, were resistant to pests (bruchid beetles) were produced in the year 1994 in pea. This was achieved by inserting a gene for α-amylase inhibitor (α-AI-Pv), isolated from common bean (*Phaseolus vulgaris*), and driven by a strong seed-specific promoter (from phytohemagglutinin gene of bean; phytohemagglutinin is a lectin). These transgenic plants produced seeds that contained 1.0–1.2% α-AI-Pv and were resistant to cowpea weevils and Adzuki bean weevils.

Genes for phytocystatins, (cysteine proteinase inhibitors) Phytocystatins, which are cysteine proteinase inhibitors occur in a variety of higher plants, such as rice and corn. They probably have dual function: 1) They regulate the action of endogenous cysteine proteinases in seeds during their ripening; for instance, in rice, oryzacystatin (OC) inhibits oryzains and in corn, corncystatin (CC) inhibits corn proteinases. 2) They recognize and inhibit exogenous proteinases, such as digestive enzymes in the gut of insect pests (cysteine proteinases occur in the guts of many insects and are necessary for the dietary life processes of these insects).

Therefore, the development of these insect pests can be inhibited by these cysteine proteinase inhibitors, which are also called phytocystatins. Oryzacystatin (OC) occurs in rice seeds in concentrations (0.001–0.002%), which are insufficient for effective protection against insect pests. Moreover, it is quite deficient in the leaves, which are generally eaten by the insects. Therefore, for imparting resistance against insects, transgenic plants need to be developed, which have higher level of expression of these phytocystatins (e.g., OC or CC).

Attempts to produce rice plants with overexpression of OC were not successful. However, in rice plants, high level of expression (both in seed and leaves) of corn cystatin or CC (~2% of total soluble proteins) could be successfully achieved. They offer additional advantages, since CC has a wide inhibitory spectrum against various cysteine proteinases.

Genes for other insecticidal secondary metabolites Secondary metabolites produced by plants have also been implicated in resistance to insect attack. However, biosynthesis of each of these metabolites involves a series of steps (sometimes even more than one biosynthetic pathway may be involved), each controlled by a separate gene. Furthermore, these genes are tissue specific in expression. These features make the production of transgenic plants difficult in this case. Proteinase inhibitors discussed in the previous section are also secondary metabolites, but their transfer is easier, since each is single gene controlled.

The production of transgenic plants by transfer of genes for entire multienzyme biosynthetic pathway (or for its augmentation) has now been possible. For this purpose, not only the transfer of genes is required, but the gene expression needs to be regulated, otherwise it leads to "yield penalty" and also to toxic effect when consumed by humans and livestock.

Defined promoter sequences are now available, which in the transgenic plants, allow a control over temporal and spatial expression of genes for biosynthetic pathways of secondary metabolites. Utilizing these facilities, transgenic plants with insect resistance due to secondary metabolites will be available in the future.

Simultaneously some of the genes have been identified in plants which are known to provide resistance against pathogen infections and recently a number of transgenic plants have been obtained against different pathogens as listed in Table 25.7.

Table 25.7 Some pathogens for which resistance has been transferred in some crop plants

Pathogen	Disease	Resistance gene	Source of gene	Transgenic crop
Pseudomonas syringae	Wild fire	Acetyl transferase gene	–	Tobacco
Alternaria longipes	Brownspot	Chitinase gene	*Serratia marcescens*	Tobacco
Rhizoctonia solani	Damping off	Chitinase gene	Bean	Tobacco and *Brassica napus*
Phytophthora infestans	Late blight	Osmotin gene	Potato	Potato
Cercospora nicotianae	Frogeye	Chitinase and glucanase genes	Rice and alfalfa	

Use of satellite RNA Satellite RNAs are species of RNA associated with specific strains of some plant RNA viruses, although it is not necessary for their replication. Replication of this satellite RNA depends on the virus, so that it gets packaged with it to cause infection elsewhere. Therefore, satellite RNA depends on virus for its replication and transmission, even though it is unrelated to viral genome. Presence of sat-RNA leads to reduction in severity of disease symptoms, and therefore has been used for developing resistance against specific viruses.

Using the above principle, transgenic tobacco plants were produced which carried a DNA fragment, which when transcribed, gives a species of satellite RNA which can reduce the severity of disease. Following two such attempts were made:

1. In one case, transgenic tobacco plants using DNA for RNA satellite of cucumber mosaic virus (CMV) were produced and when satellite-free aggressive CMV strains were used for infection, the presence of satellite RNA (transcribed from DNA) in transgenic plants led to reduction in disease symptoms. These transgenic plants are also protected against other related viruses, such as, tomato aspermy virus.

2. In the other example involving tobacco ringspot virus (TobRV), transgenic tobacco plants could be produced, which carried DNA version of satellite RNA of tobacco ringspot virus (STobRV). These transgenic plants are protected against the deleterious effects of subsequent TobRV infection. The advantage of using satellite RNA strategy over coat protein is that satellite RNA is not constitutively expressed like coat protein, but is expressed only after virus infection.

Resistance Against Stress (Drought, Salt, Heat, Freezing, etc.)

A number of genes are now known to be responsible for providing resistance against stresses such as heat, cold, salt and heavy metals. Stress tolerance in plants can be achieved in two ways:

1. Improving protection from stress as in oxidative stress, which is protected by super-oxide dismutases and other systems;
2. Reducing sensitivity to stress, as in improving salt tolerance by an osmolyte and chilling tolerance by increasing the level of polyunsaturated fatty acids. Studies are also being conducted on metabolites like proline and betaines that are implicated in stress tolerance.

The products of genes, whose expression is induced by stresses like drought, salt and freezing can be broadly classified in two groups:

1. Proteins that protect cells from dehydration; these include enzymes that are involved in the production of osmoprotectants, late embryogenesis abundant (LEA) proteins, antifreeze proteins, chaperones and detoxification enzymes.
2. Proteins that are involved in inducing transcription of stress responsive genes; these include transcription factors, protein kinases, and the enzymes involved in phosphoinositide metabolism.

Genes encoding proteins belonging to both these groups have been utilized and will be utilized in future for the production of stress-resistant transgenic plants.

For instance, resistance against chilling was introduced into tobacco plants, by introducing a gene for "glycerol phosphate acyl transferase" enzyme from *Arabidopsis* (*Arabidopsis* is resistant to chilling). This enzyme, encoded by nuclear genome and later transported to chloroplast, determines the level of unsaturation of fatty acids in the phosphatidyl-glycerol of chloroplast membranes.

The plants with high proportion of *cis*-unsaturated fatty acids (e.g., spinach, *Arabidopsis*) are resistant to chilling, and so are the transgenic tobacco plants carrying the gene for the above enzyme.

It has also been shown that a *cis*-acting promoter element, called dehydration response element (DRE) plays an important role in regulating gene expression in response to stresses like drought, salt and freezing. The transcription factors designated as DREB IA and DRE 2B (DREB = DRE binding protein) specifically bind to DRE and activate transcription of genes containing the DRE sequence in *Arabidopsis*. The cDNAs for both these proteins have been isolated. Each of these cDNAs, along with CaMV35S promoter, was used for production of transgenic plants which gave strong constitutive expression of stress-inducible genes (*rd29A, kinl, cor6.6/kin2, cor47/rd17, cor15a* and *erd10*) and were tolerant to stresses like drought, salt and freezing.

Transgenic plants with chilling resistance Many of our crop plants (e.g., rice, maize and soybean) are injured or killed by exposure to low non-freezing temperatures in the range of 0–15°C. They are described as chilling-sensitive plants and carry lipids with only saturated fatty acids. It has been shown that presence of even one double bond in one of the 16- or 18- carbon saturated fatty acids found in the lipids, may lead to resistance against injury by low temperature.

Some successful efforts have been made by utilizing genes from higher plants. These efforts include the following:

1. Saturation of the level of PG (a minor component of membrane lipids), was slightly reduced to 94%, in transgenic tobacco plants carrying a gene for glycerol 3-phosphate acyltransferase from *Arabidopsis thaliana,* which itself is chilling resistant. This led to an increase in chilling resistance in transgenic plants in comparison to wild plants.
2. Expression of plastidic w-3 desaturase from *A. thaliana* in the plastids of transgenic tobacco also led to a 10% increase of trienoic fatty acids over wild plant. This also imparted resistance against short-term exposure to chilling (1°C). In the above examples of transgenic chilling-resistant tobacco plants, genes from a chilling resistant higher plant (*A. thaliana*) were utilized. But the scope of using these desaturases from higher plants is limited.

Therefore, according to a report published in 1996, a broad-specificity of desaturase gene from a cyanobacterium (*Anacyctis nidulans*) was used, that was successfully introduced and its expression driven by CaMV 35S promoter was made. The enzyme was targeted into the plastids by the transit peptide of pea Rubisco small subunit. The transgenic tobacco plants thus produced had a highly reduced level of saturated fatty acid content in most membrane lipids and exhibited a significant increase in chilling resistance.

Transgenic Plants for Easy Transport and Long Shelf-life

A number of examples are available, where transgenic plants suitable for food processing have been developed.

1. Bruise resistant tomatoes were developed which expresses antisense RNA against polygalacturonase (PG) that attacks pectin in the cell walls of ripening fruit and thus softens the skin.
2. Tomatoes and cantaloupe melon fruits exhibiting delayed ripening were developed, either by using antisense RNA against enzymes involved in ethylene production (e.g., ACC synthase or ACC oxidase), or using gene for ACC deaminase, which degrades 1-aminocyclopropane 1 carboxylic acid (ACC), an immediate precursor to ethylene.

This increases the shelf-life of tomato. The tomato and cantaloupe melons expressing antisense ACC synthase or ACC oxidase gene exhibit delayed ripening, of fruits that are still attached to the plant. However, in tomatoes expressing ACC deaminase gene, fruits displayed delayed ripening when detached from the plant and allowed to ripen on the shelf. When fruits are allowed to stay on the plant longer, they accumulate higher amounts of soluble sugars, thus achieving better flavour. Tomatoes obtained thus (using ACC synthase) were named "Flavr Savr".

Figure 25.1 depict the production of transgenic "Flavr Savr" tomato using an antisense RNA gene:

i. A gene of tomato codes for polygalacturonase (PG) that causes the fruit to soften.

ii. An antisense gene for PG is isolated and sequenced.

iii. An antisense PG gene (the new "Flavr Savr" gene) is constructed, cloned in bacteria and then transferred to tomato.

iv. Transgenic Flavr Savr tomato with PG gene and antisense gene produced.

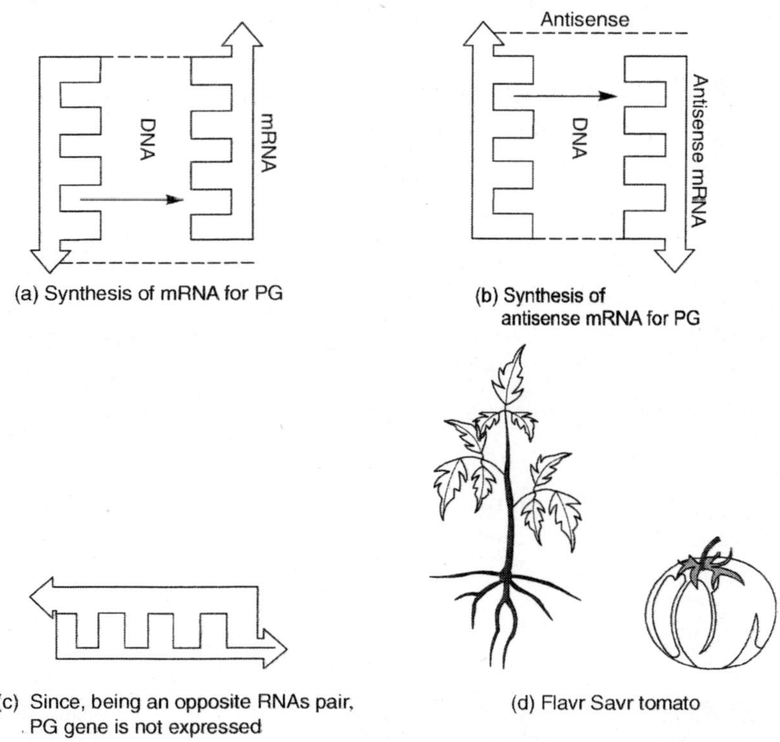

Figure 25.1 Production of transgenic Flavr Savr tomato

3. Tomatoes with elevated sucrose and reduced starch could also be produced using sucrose phosphate synthase gene.
4. Tomatoes with high level of total solids and soluble solids were produced using isopentenyl transferase (*ipt*) gene from *Agrobacterium tumefaciens*. This is a key enzyme in the biosynthesis of plant growth regulator, cytokinin. Expression of the gene was modulated using a promoter permitting ovary-specific expression. This led to increase in cytokinin level in tomato ovaries giving higher level of solids.
5. Starch content in potatoes could be increased by 20–40% by using a bacterial ADP glucose pyrophosphorylase gene (ADP GPPase).

Transgenic Crops, Rich in Vitamin A ('Golden Rice' and 'Golden Mustard')

World wide, it is estimated that 124 million children suffer from vitamin A deficiency, which annually causes irreversible blindness in 500,000 children and results in large-scale deaths of these children. It is estimated that vitamin A nutrition may prevent approximately 1 to 2 million deaths every year among children aged between 1 and 4 and an additional 0.25 to 0.5 million among higher age group of children.

In view of these concerns, transgenic crops like rice and mustard, described as "golden rice" and "golden mustard" that are rich in vitamin A have either been developed or are being developed (the term "gold" refers, to the golden colour of the rice grain or mustard oil due to vitamin A). Golden rice was developed by using two genes from *Narcissus* (phytoene synthase = *psy*; lycopene β-cyclase = *lcy*) and one gene from *Erwinia* (phytoene desaturase = *crtl*).

Extensive research during 1996–2000, by Ingo Potrykus in Switzerland and Peter Beyer in Germany, led to the development of transgenic "Golden Rice", containing beta-carotene and carotenoids that are precursors of vitamin A. It is believed to be one potential solution to vitamin A deficiency, although improved diet including green vegetables, animal products and vitamin pills are other possible alternative solutions. In the year 2000, efforts were made to donate golden rice to alleviate poverty and poor health in the third world.

In this connection, several private companies having proprietary rights (intellectual property rights) on technologies used in this research licensed these technologies free of charge for humanitarian uses of golden rice in developing countries. During the year 2001, sample seed of "Golden Rice" was also provided to International Rice Research Institute (IRRI) in Philippines, to investigate its safety and utility.

In December 2000, a collaborative project to develop "Golden Mustard" was launched between the Tata Energy Research Institute (TERI) in India, and Monsanto and Michigan State University (MSU) in USA. "Golden Mustard" yields cooking oil with high content of beta-carotene (pro-vitamin A).

The technology thus provides rice and mustard with high level of vitamin A and in future, the technology will certainly be extended to other crops like maize.

Steps involved, sources of genes and important components of gene constructs, used for the development of pro-vitamin A rice are given in Figure 25.2.

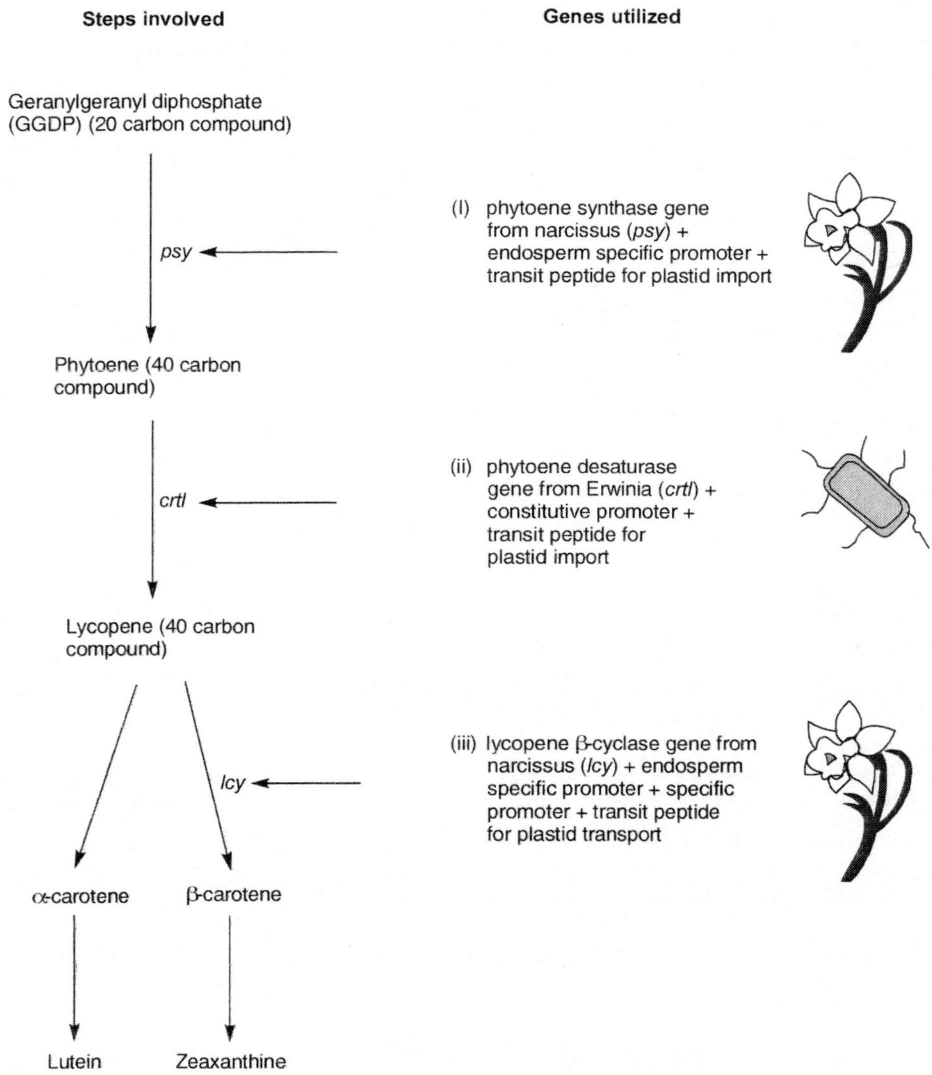

Figure 25.2 Development of pro-vitamin A rice

FUTURE TRANSGENIC PRODUCTS

Tomato

Because tomatoes are one of the world's most popular vegetables, they have benefitted from a long history of genetic improvement that continues in the transgenic age. Lycopene, a naturally occurring constituent of tomato, is a nutritional factor related to vitamin A. Tomato varieties with transgenically enhanced lycopene content are under investigation. Another trait of interest is delayed ripening. Tomatoes that ripen slower can remain on the vine longer and develop improved flavour, compared to commercial varieties that are picked at the green stage. The Flavr Savr® tomato, one of the earliest approved transgenic crop varieties, was a delayed ripening variety. Because the trait was incorporated into a variety that performed poorly otherwise, it was not a commercial success.

Salty soils are an increasing problem in many parts of the world. Many crop plants, including tomatoes, are killed by high salt levels in soil and irrigation water. The development of a salt-tolerant tomato offers the possibility that tomatoes could be grown on land that was previously unavailable for agriculture. Scientists at the University of California and the University of Toronto have developed a tomato plant that is able to tolerate high levels of salt and that holds the salt in its leaves, so that the fruit will not taste salty.

Golden Rice

Millions of people in the world suffer from vitamin A deficiency, which leads to vision impairment and increased susceptibility to diarrhea, respiratory diseases and measles. Rice is a staple food in many countries, particularly in Asia, but does not contain vitamin A or its immediate precursors. By inserting two genes from daffodil and one gene from a bacterial species into rice plants, Swiss researchers have produced rice capable of synthesizing beta-carotene, the precursor of vitamin A.

Canola

Canola is a major oilseed crop. Transgenic research has focused on improving the nutritional quality of canola oil by enhancing the Vitamin E content or by modifying the balance of fatty acids.

Modified Turfgrass for Lawns and Recreational Areas

There are a number of environmental concerns associated with the current use of turfgrass. These include:

1. The amount of chemicals applied (fertilizers, herbicides, fungicides, and even green dyes);
2. The large quantity of water required to keep lawns healthy, especially in the West;
3. The energy required to mow lawns.

New transgenic turfgrass varieties will address some of these concerns by incorporating genes for herbicide tolerance, disease and insect resistance, reduced growth rates (meaning less mowing), and tolerance to drought, heat, and cold. The first such product on the market will most likely be Roundup Ready® varieties of Kentucky bluegrass, creeping bentgrass, and buffalograss. Weed control in lawns of these varieties can be accomplished with the herbicide Roundup, which is more environment friendly than currently used herbicides like 2,4-D.

Edible Vaccines

Food crops, engineered to produce edible vaccines against infectious diseases, make vaccination more readily available to children around the world. Because of their palatability and adaptation to tropical and subtropical environments, bananas have received considerable research attention as a vehicle for vaccine delivery. Transgenic bananas and tomatoes containing inactivated viruses that cause cholera, hepatitis B, and diarrhea have been produced and are currently undergoing evaluation.

Sunflower

A disease-resistance trait, an anti-pest trait, and a herbicide-resistance trait are all being pursued, but no commercial varieties are now on the market. White mould (*Sclerotinia*) is a serious problem for sunflower producers in some areas. Resistance to this disease would expand the area in which sunflowers can be grown and might improve yields in areas of current cultivation. Resistance to the Argentina looper, an insect that eats sunflower leaves, is also being investigated. Some research has been done on developing sunflowers that can tolerate being sprayed with the weed killer, Roundup. This trait would allow a farmer to spray his field for weeds while not killing the crop.

Coffee and Tea

Decaffeinated coffee is now made by treating coffee beans to remove the caffeine. One method uses organic solvents to extract the caffeine, which causes some consumers to be concerned that residues from the solvents will remain in the coffee they drink. Other methods are criticized for removing some of the desirable, flavour-producing components along with the undesirable caffeine. Two scientists in Hawaii and Scotland have identified different genes that lead to the production of caffeine in coffee beans and tea leaves. If these genes can be "turned off" in some plants, coffee and tea trees could be developed that would produce naturally decaffeinated products with full flavour and aroma.

The harvesting of coffee beans now requires many passes through the orchards because the beans ripen at different times. There are efforts to develop a method for making all the beans ripen at the same time so that harvesters could pick all the beans during one pass through the orchard.

Grapes Ripening

Grape vines (*Vitis vinifera*) are susceptible to several diseases that reduce the amount and the quality of wine grapes and table grapes or even kill the vine. Genes that confer resistance to particular diseases would reduce the cost of battling diseases in the vineyard. Researchers at the University of Florida have patented a method for producing grape vines that carry a silkworm gene to provide protection from Pierce's disease, a fatal bacterial disease that affects grapes and several other plants.

Tobacco

Nicotine-free tobacco is now being grown for a projected introduction of nicotine-free cigarettes. Previous attempts to make low-nicotine products removed some of the flavours along with the nicotine. Genetically engineered nicotine-free tobacco doesn't synthesize nicotine in the leaf.

Forest Trees

Forest trees such as poplar, aspen, and spruce have been transformed with various genes to provide resistance to insects, tolerance to herbicides, and higher levels of the commercial product. For example, reducing the lignin content of a tree can make it easier to recover wood pulp.

REVIEW QUESTIONS

1. Write an essay on history and significance of transgenic industry.
2. Discuss briefly about the main areas where transgenic crops have been developed.
3. What is BT-endotoxin gene? How it can be significant as insecticides?
4. Write an outline on production of transgenic crops against abiotic stresses.
5. What is antisense RNA? Write briefly the production of transgenic "Flavr Savr" tomato using an antisense RNA technology.
6. Write short notes on
   i. Golden rice
   ii. Golden mustard.
7. Write an essay on some transgenic products.

REFERENCES

Arencibia, A., Vázquez, R., Prieto, D., Téllez, P., Carmona, E., Coego, A., Hernández, L., de la Riva, G. and Selman-Housein, G. (1997). Transgenic sugarcane plants resistant to stem-borer attack. *Molecular Breeding.* 3: 247–255.

Cavan, G., Biss, P. and Moss, S.R. (1998). Herbicide resistance and gene flow in wild-oats *(Avena fatua* and *Avena sterilis* ssp. *ludoviciana). Annals of Applied Biology.* 133: 207–217.

Cheng, X.Y., Sardana, R., Kaplan, H. and Altosaar, I. (1998). *Agrobacterium-transformed* rice expressing synthetic cry1Ab and cry1Ac genes are higly toxic to striped stem borer and yellow stem borer. Proceedings of the National Academy of Sciences. USA 95: 2767–2772.

Enríquez-Obregón, G.A., Vázquez-padrón, R.I., Prieto-sansonov, D.L., de la Riva, G.A. and Selman-Housein, G. (1998). Herbicide resistant sugarcane (*Saccharum officinarum* L.) plants by *Agrobacterium*-mediated transformation. *Planta.* 206: 20–27.

Forster, B.P., Lee, M.A., Lundqvist, U., Millam, S., Vamling, K. and Wilson, T.M.A. (1997). Genetic engineering of crop plants: from genome to gene. *Experimental Agriculture.* 33: 15–33.

Gallo-Meagher, M. and Irvine, J.M. (1996). Herbicide resistant transgenic sugarcane plants containing the bar gene. *Crop Science.* 36: 1367–1374.

Hubbell, B.J. and Welsh, R. (1998). Transgenic crops: engineering a more sustainable agriculture? *Agriculture and Human Values.* 15: 43–56.

Mannion, A.M. (1998). Future trends in agriculture: the role of biotechnology. *Outlook on Agriculture.* 27: 219–224.

Mogilner, N., Zutra, D., Gafny, R. and Bar-Joseph, M. (1993). The persistence of engineered *Agrobacterium tumefaciens* in agroinfected plants. *Molecular Plant-Microbe Interactions.* 6:673–675.

Ortiz, R. (1998). Critical role of plant biotechnology for the genetic improvement of food crops: perspectives for the next millennium. *Electronic Journal of Biotechnology.* Vol.1, No. 3.

Rubio, T., Borja, M., Scholthof, H.B. and Jackson, A.O. (1999). Recombination with host transgenes and effects on virus evolution: an overview and opinion. *Molecular Plant-Microbe Interactions.* 12:87–92.

Scott, S.E. and Wilkinson, M.J. (1999). Low probability of chloroplast movement from oilseed rape (*Brassica napus*) into wild *Brassica rapa. Nature Biotechnology.* 17:390–392.

Ye, X., Al-Babili, S., Klöti, A., Zhang, J., Lucca, P., Beyer, P. and Potrykus, I. (2000). Engineering the provitamin A (-carotene) biosynthetic pathway into (carotenoid-free) rice endosperm. *Science.* 287:303–305.

Zhang, H.X. and Blumwald, E. (2001). Transgenic salt-tolerant tomato plants accumulate salt in foliage but not in fruit. *Nature Biotechnology.* 19(8):765–768.

26

GM CROPS AND THEIR IMPACTS

INTRODUCTION

The intentional production of transgenic plants by recombinant DNA techniques has started recently (in mid-70s) and it has been a controversial development in the field of biotechnology. However, this technology is being opposed vigorously by many NGOs, and several governments, particularly in Europe and even in India. Significantly, the US has adopted this technology most widely whereas Europe has almost no genetically modified (GM) crops, in practice.

Transgenic plants are identified as a class of genetically modified organisms (GMO). Although commercial factors, especially high regulatory and research costs, have so far restricted the use of transgenic crop varieties as major traded commodity crops, however, recently R&D projects to enhance crops that are locally important in developing countries are being pursued, such as insect protected cowpea for Africa and insect protected Brinjal eggplant for India.

Transgenic plants have been extensively used for bioremediation of contaminated soils. Moreover, the toxic effects of mercury, selenium and organic pollutants such as polychlorinated biphenyls (PCBs) have been reduced from soils by growing transgenic plants carrying genes for bacterial enzymes.

The term "cisgenic" is being used by some plant breeders and scientists who deal with artificial gene transformation technique that could theoretically have been produced by conventional plant breeding methods. Breeders and scientists explains that "cisgenically" produced organisms do not have the same degree of novelty as transgenic organisms. Furthermore, it does not involve environmental issues that are not already present in conventional crossbreeding.

Environmental activists, religious organizations, public interest groups, professional associations and other scientists and government officials have all raised concern about GM foods and criticized agribusiness for pursuing profit without concern for potential hazards, and failure of adequate regulatory exercise at government level.

It is suggested that cisgenic modification is useful for plants that are difficult to crossbreed particularly by conventional means (such as potatoes). Moreover pure plants in the cisgenic category do not require the same level of legal regulation as other genetically modified organisms. The advent of GM crops has been introduced in the list of table discussion at the global level in general and in India, in particular. Lots of views have been proposed by the people belonging to various walks of life in favour and against the GM crops use and their possible impacts, on the whole in general and on environment in particular.

AGRICULTURAL IMPACTS

Outcrossing of transgenic plants not only causes the potential environmental risks, it also proves to be challenging for farmers and food producers to bring them into practice. Many countries have adopted different legislations for the cultivation of transgenic and conventional plants as well as for derived food and feed, however, consumers want the freedom to buy GM derived products or conventional products. Therefore, it is more emphatic for farmers and producers to separate both GM products or conventional products. This requires coexistence measures at the field level as well as traceability measures throughout the whole food and feed processing chain. Significantly research scientists have been investigating to develop the techniques for farmers to avoid outcrossing and mixing of transgenic and non-transgenic crops, and also for processors to ensure and verify the separation of both production units.

The potential ecological risks to apply genetic engineering in agriculture involve the apprehensions that some transgenic crops could become noxious weeds, and others could become a channel through which new genes may move to wild plants, which themselves in course of time could also become weeds. Moreover, the resulted new weeds could adversely affect farm crops as well as wild crops and its ecosystem.

Wild plants are presently being genetically modified to contain parts of a viral gene in order to become virus-resistant. There are arguments that widespread use of transgenic virus-resistant plants in agriculture may lead to emergence of new strains of viruses which may infect a new host. There are concerns that the creation of new viral strains and the expansion of the virus's hosts number may increase the risks of new viral diseases that adversely affect crops and other plants. Furthermore, mechanisms have been identified that genetically modified (GM) plants could possibly give rise to new plant diseases.

Another ecological impact is the possibility that field or forest plants, targeted to express toxic substances like pesticides and pharmaceutical drugs, may poison certain non-target organisms. Transgenes for insecticidal or fungicidal compounds that are introduced into crops to inhibit pests may unintentionally kill non-target and beneficial insects and fungi. Therefore, transgenic crops used to manufacture drugs or industrial oils and chemicals could ultimately harm animals, insects and soil microorganisms.

In addition, this transgenic technology involves the fear factor that the rapid spread of transgenic crops may come out as a threat to traditional crop varieties and wild plants that are the major sources of crop genetic diversity. Some traits of organisms may take decades or even longer to manifest themselves. An organism that may appear to be "safe" in the short term could eventually prove to be dangerous, in the long term.

Herbicide Tolerance

Weed control is one of the farmer's biggest challenges in crop production, because poorly controlled weeds drastically reduce crop yield and quality. There are many herbicides in the market that control only certain types of weeds, and are approved for use only on certain crops at specific growth stages. However, residues of some herbicides remain in the soil for a year or more. Thus, farmers must pay close attention to the herbicide history of a field when planning what to plant there.

Herbicide tolerant crops resolve many of those problems because they include transgenes providing tolerance to the herbicides, glyphosate or glufosinate. These herbicides are broad-spectrum, meaning that they kill nearly all kinds of plants except those that have the tolerance gene. Thus, a farmer can apply a single herbicide to his field of herbicide tolerant crops, and he can use glyphosate or glufosinate effectively at most crop growth stages as needed. Another important benefit is that this class of herbicides breaks down quickly in the soil, eliminating residue carry-over problems and reducing environmental impact. Herbicide tolerant varieties are popular with farmers because they enable less complicated, more flexible weed control. These varieties are commonly marketed as Roundup Ready® or Liberty Link® varieties.

Bt Insect-Resistant Crops

"Bt" is the short form for *Bacillus thuringiensis,* a soil bacterium whose spores contain a crystalline (Cry) protein. When it infects insects, the protein breaks down to release a toxin, known as a delta-endotoxin in the insect gut. This toxin binds to and creates pores in the intestinal lining, resulting in ion imbalance, paralysis of the digestive system, and after a few days, insect death. Different versions of the *Cry* genes, also known as "Bt genes", have been identified. They are effective against different orders of insects, or affect the insect gut in slightly different ways. A few examples are shown in Table 26.1.

The use of Bt to control insect pests is not new. Insecticides containing Bt and its toxins (for example, Dipel, Thuricide, Vectobac) have been sold for many years. Bt-based insecticides are considered safe for mammals and birds, and safer for non-target insects than conventional products. Recently, a modified version of the bacterial *Cry* gene has been incorporated into the plant's own DNA, so that the plant's cellular machinery produces the toxin. When the insect chomps on a leaf or bores into a stem of a Bt-containing plant, it ingests the toxin and will die within a few days.

Table 26.1 Cry genes and their toxicity

Cry gene	Insect orders
CryIA(a), CryIA(b), CryIA(c)	Lepidoptera
Cry1B, Cry1C, Cry1D	Lepidoptera
CryII	Lepidoptera, Diptera
CryIII	Coleoptera
CryIV	Diptera
CryV	Lepidoptera, Coleoptera

Some of the Bt insect-resistant crops are:

1. *Corn* Primarily for control of European corn borer, but also corn earworm and Southwestern corn borer.
2. *Cotton* For control of tobacco budworm and cotton bollworm.
3. *Potato* For control of Colorado potato beetle. Bt potato has been discontinued as a commercial product.

Corn hybrids resistant to corn rootworm Corn rootworm (*Diabrotica* spp.) is a serious pest of corn in many corn growing areas. It damages roots of young corn seedlings, resulting in reduced growth and poor standability of the plant. This insect is responsible for the application of heavy amount of insecticide to corn fields. Moreover, to control this pest, the insecticide must be applied directly to the soil, where it may leave residues or leach into the ground water. By replacing these chemical insecticides, corn rootworm resistant hybrids may provide major benefits to environmental quality. Although rootworm-protected hybrids apparently offer pest management and environmental benefits, there are serious concerns about development of resistance to Bt in this adaptable insect.

Bt crops vs chemical pesticides The use of Bt varieties has dramatically reduced the amount of chemical pesticides applied to cotton. Yields and profits also improved in Bt-cotton fields. The benefits from Bt-corn, however, were not as clear-cut. Due to the difficulty of effectively controlling corn borers with insecticides, most farmers do not apply chemical controls to their conventional corn fields. Thus, Bt hybrids for example, substituted for chemical pesticides on only about 20% of the total US Bt-corn area. Profitability of Bt-corn is not as certain as for cotton; it will vary over years and locations, depending on the intensity of the corn borer population.

Bt toxins vs insect pests resistance Although Bt genes have proven to be quite effective in the short term for protecting against crop insect damage, as well as in reducing fungal contamination of corn, there are serious apprehensions that widespread use of Bt varieties will accelerate development of resistance to Bt in the target pests. This could mean the loss of Bt as an effective, environment friendly insecticide. In response to these concerns, the Environmental Protection Agency at global level has mandated measures to reduce the risk of resistance development. These measures depend on a combination of high dose of the Bt toxin and planting of refuges. A refuge refers to an area planted to a non-Bt variety that is physically close to a field planted with a Bt variety.

The EPA has suggested that farmers growing Bt-corn must plant at least 20% of their total corn acreage to a non-Bt variety. The reason is that the few Bt-resistant insects surviving in the Bt field would likely mate with susceptible individuals that have matured in the non-Bt refuge. Thus, the insect genes (alleles) for resistance to Bt would be swamped by the susceptible alleles. Whether this strategy will work or not remains unclear. Some of the potential problems with the refuge strategy are:

- The frequency of Bt-resistant alleles in insect populations may be greater than assumed, in refuge models.
- Resistance to Bt in European corn borer may be semi-dominant rather than recessive.
- Resistant insects surviving in the Bt field may mature several days later than susceptible insects in the refuge, thus preventing their mating.

Crop-to-Crop Gene Flow

Hybridization of transgenic crops with nearby conventional crops raises concerns about separation distances to ensure purity of crops and about who must pay if unwanted genes move into a neighbour's crop. As "Identity Preservation" and segregation of GM from non-GM crops become factors in marketing products, it will be important to ensure that hybridization is not occurring in the field.

Many factors influence the potential for gene flow from crop to crop. Some crops are highly outcrossing, with pollen carried to other fields by wind and by insects. Other species are highly self-pollinating, with little potential for pollen transfer to neighbouring plants. Because of the differences among crops species, every case must be evaluated individually for potential to contribute to gene flow from transgenic to conventional crops.

If GM pollen pollinates plants in a neighbouring field, then the issue of genetic trespass may arise. What level of GM presence, if any, should be allowed in products that are sold as organic or conventional? Should GM farmers and companies bear responsibility for preventing gene flow, or should conventional and organic farmers pay to protect their products from gene flow? Should GM versions of outcrossing plants be banned as too risky, while GM versions of self-pollinating plants are permitted? These issues have already prompted several lawsuits and they will continue to be a factor in the development and use of transgenic plants for years to come.

HUMAN HEALTH IMPACTS

The introduction of transgenic crops and foods into the existing food production system has generated a number of questions about possible negative consequences. People with concerns about this technology have reacted in many ways, from participating in letter-writing campaigns to demonstrating in the streets to vandalizing institutions where transgenic research is being conducted.

Some of the important aspects where GM crops might influence human health are:

Influence on Allergenicity

The possibility that there might be an increase in the number of allergic reactions to food as a result of genetic engineering has a powerful emotional appeal. There is a fair chance that introducing a gene into a plant may create a new allergen and cause an allergic reaction in susceptible individuals. However, there is no evidence so far that genetically engineered foods are more likely to cause allergic reactions than conventional foods. Tests of several dozen transgenic foods for allergenicity have uncovered only a soybean that was never marketed and the now-famous StarLink corn. Although the preliminary finding is that StarLink corn is probably not allergic, the scientific debate continues. Every year some people discover that they have developed an allergy to a common food such as wheat or eggs, and some people may develop allergies to transgenic foods in the future (Figure 26.1), but there is no evidence that transgenic foods pose more of a risk than conventional foods do.

Horizontal Transfer and Antibiotic Resistance

The use of antibiotic resistance markers in the development of transgenic crops has raised concerns about whether transgenic foods will play a part in our loss of ability to treat illnesses with antibiotic drugs. At several stages of the laboratory process, developers of transgenic crops use DNA that codes for resistance to certain antibiotics, and this DNA becomes a permanent feature of the final product although it serves no purpose beyond the laboratory stage.

Fate of Antibiotic Pills

One aspect of this topic is the risk of horizontal gene transfer, that is, transfer of DNA from one organism to another, outside of the parent-to-offspring channel. Transfer of a resistance gene from transgenic food to microorganisms that normally inhabit our stomach and intestines, or to bacteria that we ingest along with food, could help those microorganisms to survive an oral dose of antibiotic medicine. Although horizontal transfer of DNA does occur under natural circumstances and under laboratory conditions, it is probably quite rare in the acid environment of the human stomach.

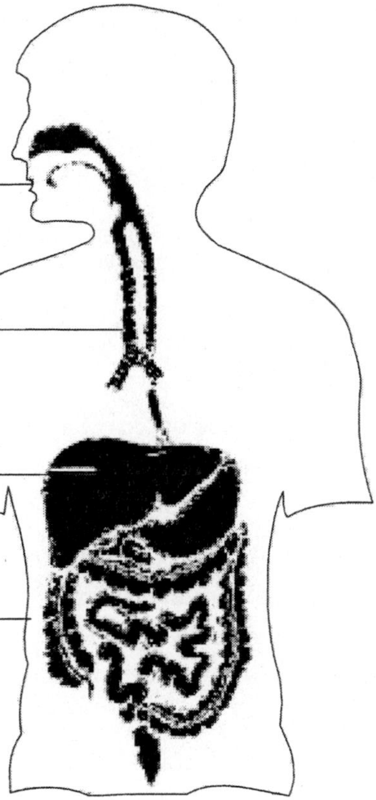

Figure 26.1 Common sites for allergic reactions

Another concern is that the enzyme product of the DNA might be produced at low levels in transgenic plant cells. While high processing temperatures would inactivate the enzyme in processed foods, ingestion of fresh or raw transgenic foods could result in the stomach containing a small amount of an enzyme that inactivates an orally administered dose of the antibiotic. This issue was raised during the approval processes for Calgene's Flavr Savr tomato and Ciba-Geigy's Bt corn 176. In both cases, tests showed that orally administered antibiotics would remain effective. While the risks from antibiotic resistance genes in transgenic plants appear to be low, steps are being taken to reduce the risk and to phase out their use.

Ingestion of Foreign DNA

When scientists make a transgenic plant, they insert pieces of DNA that did not originally occur in that plant. Often these pieces of DNA come from entirely different species, such as viruses and bacteria.

DNA is the blueprint for life and all living things contain DNA in many of their cells. What is the fate of this ingested DNA? Most of it is broken down into more basic molecules when we digest a meal. A small amount is not broken down and is either absorbed into the blood stream or excreted in the feces. We suspect that the body's normal defense system eventually destroys this DNA. Further research in this area would help to determine exactly how humans have managed to eat DNA for thousands of years without noticing any effects from the tiny bits that sneak into the bloodstream.

So far there is no evidence that DNA from transgenic crops is more dangerous to us than DNA from the conventional crops, animals, and their attendant microorganisms that we have been eating all our lives.

Action of Promoter DNA

When scientists use transgenic technology to put a new gene into a plant, they put in additional pieces of DNA to direct the activity of that gene. One of these pieces is the "promoter" that turns the concerned gene on.

The most widely used promoter is the cauliflower mosaic virus 35S promoter, often abbreviated as the CaMV promoter or the 35S promoter. This promoter was obtained from the virus that causes cauliflower mosaic disease in several vegetables, such as cauliflower, broccoli, cabbage and canola. There are concerns that the CaMV promoter might be harmful if it were to invade our cells and turn on our indigenous genes.

A multi-step chain of events occur for the CaMV promoter to escape the normal digestive breakdown process, penetrate a cell of the body and insert itself into a human chromosome. While there have been no tests to determine whether the CaMV promoter has invaded human tissues, experiments with mice indicate that normal body defenses eliminate stray fragments of foreign DNA that sneak into the blood stream from the digestive tract.

There is some indication that the CaMV promoter poses little threat to human health. People have been eating it in small quantities for hundreds of years in the form of vegetables that are infected with the disease. However, vegetables heavily infected with CaMV are unappetizing, and there have been no documented negative effects on health from eating the virus or its promoter.

Variations at Nutrient Levels

How do genetically engineered foods compare with conventional foods in nutritional quality? This is an important issue, and one for which there will probably be much research in the future, as crops that are engineered specifically for improved nutritional quality are marketed. However, there have been only a few studies till date comparing the nutritional quality of genetically modified foods to their unmodified counterparts.

The central question for GE crops that are currently available is whether plant breeders have accidentally changed the nutritional components that we associate with conventional cultivars of a crop. Since isoflavones are thought to play a role in preventing heart disease, breast cancer and osteoporosis, the isoflavone content of Roundup Ready soybeans has been investigated by several research teams.

The studies completed so far do not resolve the issue of whether Roundup Ready soybeans have isoflavone levels comparable to conventional varieties, but the differences found in experiments appear to be small or moderate in comparison with natural variation in isoflavone levels. Additional evidence may clarify the arguments for and against Roundup applications as a risk factor in soybean cultivation.

Industry studies submitted in support of applications for permission to sell transgenic crops indicate that the nutritional components that are commonly tested are similar in transgenic foods and conventional foods.

ECOLOGICAL IMPACTS

The potential impact of genetically modified (GM) crops on nearby ecosystem is one of the greatest concern associated with the transgenic technology. It is possible that transgenes may cause a significant impact if transgenic plants increase in frequency in an uncontrolled manner in the natural habitat. These ecological effects are more apparent when the transgenic crops are surrounded by conventionally bred crops. Some of the risk factors related to GM crops at the ecological level must be taken into consideration.

- Is the transgenic plant capable of growth outside a cultivated area?
- Whether the transgenic plant can pass its genes to a local wild species, or not and also offspring are fertile or sterile?
- Does the introduction of the transgene confer a selective advantage to the plant or to hybrids in the wild?

Many domesticated plants can mate and hybridize with wild relatives when they are grown in proximity, and it is quite possible that whatever genes the cultivated plant carried can then be passed to the hybrid. This mechanism of gene flow applies equally to transgenic plants and conventionally bred plants. In any case there are advantageous genes that may have negative impact on an ecosystem upon release. Although there may be fears over "mutant superweeds" overgrowing local wildlife. This may not be normally a significant concern. Although hybrid plants are very common and in most cases these hybrids are not fertile due to polyploidy, and will not multiply or survive long after the original domestic plant is removed from the population. However, this does not mean that the possibility of a negative impact of GM crops on conventional crop is out of question.

In some cases, the pollen from a domestic plant may travel many miles through wind before fertilizing another plant. This can make it difficult to assess the potential harm of crossbreeding; many of the relevant hybrids are far away from the test site. Therefore, the possible solutions for this concern are to develop systems that are designed to prevent transfer of transgenes, such as Terminator Technology and the genetic transformation of the chloroplast only, so that only the seed of the transgenic plant would bear the transgene. However, terminator technology involves some controversy and is not being preferred. But poor farmers might be forced to depend upon producers for valid seed supply whereas chloroplast genetic engineering technique has not much concern but certainly involves technical constraints that still need to be standardized.

There are at least three possible ways of hybridization leading to escape of a transgene:

1. Hybridization with non-transgenic crop plants of the same species and variety.
2. Hybridization with wild plants of the same species.
3. Hybridization with wild plants of closely related species, usually of the same genus.

However, there are a number of factors which must be taken into consideration prior to hybrids production:

1. The transgenic plants must be close enough to the wild species for the pollen to reach the wild plants for pollination and fertilization.
2. The wild and transgenic plants must flower at the same time.
3. The wild and transgenic plants must be genetically compatible.
4. In order to persist, these hybrid offsprings must be viable and fertile and also should carry the transgene. It is suggested that hybridization with wild plants of related species could be a possible escape route for transgenic plants.
5. It is known that some crop plants have been found to hybridize with wild counterparts.
6. As a basic part of population genetics it is known that the spread of a transgene in a wild population will be directly related to the fitness effects of the gene in addition to the rate of influx of the gene to the population. Advantageous genes will spread rapidly, neutral genes will spread with genetic drift and disadvantageous genes will only spread if there is a constant influx.
7. The ecological effects of transgenes are not known, but it is generally accepted that only genes which improve resistance/tolerance tendencies in relation to abiotic factors would provide hybrid plants sufficient advantages to become weedy or invasive. Abiotic factors are parts of the ecosystem which are not alive, such as climate, salt and mineral content, and temperature.
8. Genes that are responsible to provide fitness in relation to biotic factors could disturb the (sometimes fragile) balance of an ecosystem. For instance, a wild plant receiving a pest resistance gene from a transgenic plant might become resistant to one of its natural pests, such as a beetle. This could allow the plant to increase in frequency, while at the same time animals higher up in

the food chain, which are at least partly dependent on that beetle as food source, might decrease. However, the exact consequences of a transgene with a selective advantage in the natural environment are almost impossible to predict.

Damage to the Environment

The possible chemical contamination of surface water and ground water by microorganisms or plants with unusual or accelerated metabolic processes is a special concern because of the crucial importance of water for all life. It may be impossible to recall and difficult to control harmful GEOs, especially those that may contaminate ground water.

Monarch Butterfly

There is common suggestion that Bt corn pollen might kill Monarch butterfly larvae and it has galvanized public interest in the effect of transgenic crops on the environment. It is stated that pollen from Bt corn causes high mortality rates in monarch butterfly caterpillars. Monarch caterpillars consume milkweed plants, not corn but the fear is that if pollen from Bt corn is blown by the wind on to milkweed plants in neighbouring fields the caterpillars could eat the pollen and perish. Unfortunately Bt toxins kill many species of insect larvae indiscriminately, and it is not possible to design Bt toxin that would only kill crop-damaging pests and remain harmless to all other insects.

Crop-to-Weed Gene Flow

Hybridization of crops with nearby weeds may enable weeds to acquire traits they didn't have, such as resistance to herbicides. Research results indicate that crop traits may escape from cultivation and persist for many years in wild populations. Genes that provide a competitive edge, such as resistance to viral disease, could benefit weed populations around a crop field.

Many cultivated crops have sexually compatible wild relatives with which they hybridize under favourable circumstances. The likelihood that transgenes will spread can be different for each crop in each area of the world. For example, there are no wild relatives of corn in the United States or in Europe for transgenic corn to pollinate, but such wild relatives exist in Mexico. Soybeans and wheat are self-pollinating crops, so the risk of transgenic pollen moving to nearby weeds is small. However, small risk must balance against the fact that there are wild relatives of wheat and soybean. Thus, each crop must be evaluated individually for the risk of gene flow in the area where it will be grown.

Antibiotic Resistance to Soil Microbes

There is also concern that transgenic plants growing in the field will transfer their antibiotic resistance genes to soil microorganisms, thus causing a general increase in the level of antibiotic resistance in the environment. However, many soil organisms have naturally occurring resistance as a defense against other organisms that generate antibiotics, so genes contributed occasionally by transgenic plants are unlikely to cause a change in the existing level of antibiotic resistance in the environment.

Leakage of GM Proteins into Soil

Many plants leak chemical compounds into the soil through their roots (Figure 26.2). There are concerns that transgenic plants may leak different compounds than conventional plants do, as an unintended consequence of their changed DNA. There is speculation that this may lead to concern about whether the communities of microorganisms living near transgenic plants may be affected.

The interaction between plants and soil microorganisms is very complex, with the microorganisms that live around plant roots also leaking chemical compounds into the soil. Much more research must be done in order to understand the relationships that occur between microorganisms and conventional crops. Moreover, there should be sincere attempts to discover whether transgenic plants are really changing the soil environment, and whether they are changing it in good ways or bad ways.

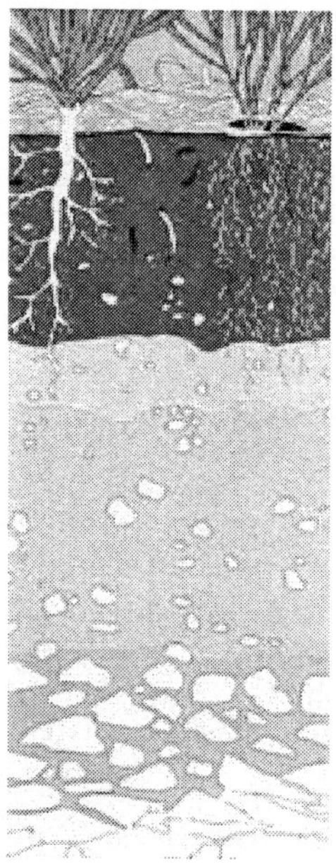

Figure 26.2 Leakage of chemical compounds into the soil through plant roots

Source: Rural Life Center, Kenyon College, Gambier, Ohio.

REVIEW QUESTIONS

1. Discuss about the social opinions regarding use of GM crops.
2. What are GM crops? Discuss its ecological impacts.
3. Discuss briefly implications of GM crops on human health.
4. Write a brief account of impacts of transgenic technology in agriculture.
5. Write short notes on (a) GM crop (b) Bt-gene and endotoxin (c) Bt-toxin and insect resistance.

REFERENCES

Barrett, C., Cobb, E., McNicol, R. and Lyon, G. (1997). A risk assessment study of plant genetic transformation using *Agrobacterium* and implications for analysis of transgenic plants. *Plant Cell, Tissue and Organ Culture.* 47:135–144.

Bengtsson, B.O. (1997). Pros and cons of foreign genes in crops. *Nature.* 385:290.

Bergelson, J., Purrington, C.B. and Wichmann, G. (1998). Promiscuity in transgenic plants. *Nature.* 395:25.

Borja, M., Rubio, T., Scholthof, H.B. and Jackson, A.O. (1999). Restoration of wild-type virus by double recombination of tombusvirus mutants with a host transgene. *Molecular Plant-Microbe Interactions.* 12: 53–162.

Borlaug, N. (1997). Feeding a world of 10 billion people: the miracle ahead. *Plant Tissue Culture and Biotechnology.* 3:119–127.

Franck-Oberaspach, S.L. and Keller, B. (1997). Consequences of classical and biotechnological resistance breeding for food toxicology and allergenicity. *Plant Breeding.* 116:1–17.

Harding, K. and Harris, P.S. (1997). Risk assessment of the release of genetically modified plants: a review. *Agro-Food-Industry Hi-Tech,* November/December 1997. pp. 8–13.

Straughan, R. (1991). Social and ethical issues surrounding biotechnological advance. *Outlook on Agriculture.* 20:89–94.

27

PLASTID ENGINEERING

INTRODUCTION

The public debate on the acceptability of genetically modified (GM) plants revolves around a mixture of issues. Besides the respectable emotional and ethical concerns on the introduction of GM plants, several rational arguments plead for the careful application of transgenic techniques before the release of plants into the environment.

The genetic information of plants is distributed among three cellular compartments: the nucleus, the mitochondria, and the plastids. Each of these compartments carries its own genome and consequently, expresses heritable traits. Plastids and mitochondria do not obey the Mendelian rules and usually exhibit uniparental transmission to the next generation.

Significantly, the insertion of foreign DNA sequences into the nuclear plant genome is associated with several potential risks. *Agrobacterium*-mediated transformation and direct DNA transfer into plant cells often result in the insertion of vector sequences associated with the gene of interest. These additional sequences may affect the endogenous and transgenic expression in an unpredictable fashion. Furthermore, co-transferred plasmid DNA, bacterial antibiotic genes and plant genome DNA may spread between and within species co-existing in the environment.

Therefore, an alternative approach of nuclear manipulation is necessary. Plastid transformation is an environment-friendly technology in plant genetic engineering that minimizes out-crossing of transgenes to related weeds or crops and reduces the potential toxicity of transgenic pollen to non-target insects. Because the plastid genome is highly polyploid, transformation of chloroplasts permits the introduction of thousands of copies of foreign genes per plant cell, and generates extraordinarily high levels of foreign protein.

SIGNIFICANCE OF PLASTID ENGINEERING

The cytoplasm and chloroplast contain different proteases, and a protein might survive better in one compartment than in the other. A specific type of plastid might be a good place to accumulate certain proteins or their biosynthetic products that would be harmful if they were present in large amounts in the cytoplasm or in a plastid of a different type. The various forms of plastid (amyloplasts, chromoplasts, etc.) have desirable properties for conducting reactions and for accumulating proteins or products of enzymes, and they can be exploited using known nuclear gene transformation methods.

There are several types of plastids:

1. Chloroplasts (chlorophyll-containing plastids)
2. Chromoplasts (yellow, orange or red carotenoid-containing plastids)
3. Amyloplasts (starch-storing plastids)
4. Elaioplasts (oil-containing plastids)
5. Proplastids (plastid precursors found in most plant cells plastids)
6. Etioplasts (partially developed chloroplasts that form in dark-grown seedlings plastids).

The conversion of photosynthetic chloroplasts into yellow carotenoid-rich chromoplasts is seen in the ripening of bananas; the conversion of chloroplasts to lycopene-containing red chromoplasts is seen in the ripening of tomatoes. Each compartment of the eukaryotic cell is unique. A particular biochemistry will be favoured in one compartment (for example, chloroplasts or chromoplasts) while the environment in another compartment (for example, the cytoplasm) will be unfavourable.

The photosynthetic apparatus of the chloroplast captures light energy and converts it into biologically useful energy that is, available for 1) manufacturing sugars and amino acids 2) reducing nitrate to ammonium 3) making starch 4) synthesizing complex organic compounds and 5) assembling chloroplast proteins that are required for the chloroplasts themselves to function. However, non-green plastids use energy produced elsewhere in the cell for metabolism. It is emphatic to modify the existing biosynthetic chains, which are already linked to the energy-converting machinery of the chloroplast, to produce and accumulate desirable new products.

The development of technologies to engineer the chloroplast genome of the green alga *Chlamydomonas reinhardtii* and the higher plant *Nicotiana tabacum* has opened up the possibility to target transgenes to the plastid genome by chloroplast transformation. These technologies offer a great potential for the biotechnology of the future and a number of most attractive advantages over conventional transgenic plants (generated by transformation of the nuclear genome) such as

1. High levels of transgene expression and foreign protein accumulation of 40% or more of the total soluble cellular protein (presumably resulting from the polyploidy of the plastid genetic system and/or the high stability of foreign proteins).
2. The possibility of expressing multiple transgenes as operons ("transgene stacking") due to efficient translation of polycistronic messenger RNAs (mRNAs) in plastids.
3. Absence of position effects in plastids due to lack of a compact chromatin structure and efficient transgene integration by homologous recombination.
4. Absence of epigenetic effects (gene silencing).
5. Transgene containment due to uniparental maternal inheritance of chloroplasts in higher plants (that is, absence of pollen transmission of transgenes).

Plastid transformation has several additional advantages when compared with nuclear transformation:

1. The high ploidy level of the plastome—5–80 (*Chlamydomonas*) or 500–10,000 (*Nicotiana*) DNA copies per cell—enables a high degree of expression. The large number of copies is calculated from the number of DNA copies per chloroplast multiplied by the number of chloroplasts per cell. Chloroplast transformation therefore has the potential to enhance the yield of transgenic products. Moreover, it appears that gene silencing is absent in plastids and expression is therefore more likely to be stable.
2. Chloroplast transformation vectors use two targeting sequences that flank the foreign genes and insert them, through homologous recombination, at a precise, predetermined location in the organelle genome. This results in uniform transgene expression among transgenic lines and eliminates the "position effect" often observed in nuclear transgenic plants. Gene silencing, frequently observed in nuclear transgenic plants, has not been observed in genetically engineered chloroplasts.
3. The ability to express foreign proteins at high levels in chloroplasts and chromoplasts, and to engineer foreign genes without the use of antibiotic resistant genes, make this compartment ideal for the development of edible vaccines. Moreover, the ability of chloroplasts to form disulphide bonds and to fold human proteins has opened the door to high-level production of biopharmaceuticals in plants. Furthermore, foreign proteins, observed to be toxic in the cytosol, are non-toxic when accumulated within transgenic chloroplasts.

HISTORY

- When the concept of chloroplast genetic engineering was developed in the 1980s, it was possible to introduce isolated intact chloroplasts into protoplasts and regenerate transgenic plants. Therefore, early investigations of chloroplast transformation in vascular plants focused on the development of chloroplast systems capable of efficient, prolonged protein synthesis and expression of foreign genes.

- The development of the gene gun as a transformation device by John Sanford enabled plant chloroplasts to be transformed without the use of isolated plastids.
- Chloroplast genetic engineering was accomplished through autonomously replicating chloroplast vectors in dicot plastids and transient expression in monocot plastids, and by stable integration of selectable marker genes into the chloroplast genomes of *Chlamydomonas reinhardtii* and tobacco using the gene gun. However, it is only recently that genes conferring agronomically valuable traits have been introduced via chloroplast genetic engineering. For example, integrating the *cry* genes into the chloroplast genome generated plants that were insecticidal to *Bacillus thuringiensis* (Bt)-sensitive and highly resistant insects.
- Chloroplasts have also been engineered recently to generate plants tolerant to bacterial and fungal diseases, drought and herbicides. Exceptionally high accumulation of foreign proteins (up to 46% of total soluble protein) has been reported recently for chloroplast transgenes. This feature makes the compartment ideal for low-cost production of biopharmaceuticals.
- Stable expression of a pharmaceutical protein in tobacco chloroplasts was first reported for GVGVP, a protein-based polymer with varied medical applications, and subsequently for human somatotropin.
- Chloroplast genomes have been engineered to express pharmaceutical peptides, human serum albumin (HSA), monoclonals and antigens composed of oligomeric proteins with stable disulphide bridges.
- Several recent advances have made chloroplasts even more attractive as biopharmaceutical reactors. These include engineering of multiple foreign genes as operons; transformation without the use of antibiotic markers; elimination of resistance genes subsequent to transformation; and the transformation of edible crops such as potato and tomato.

CHLOROPLAST GENETIC MANIPULATIONS

Recent success in genetic engineering, the chloroplast genome for resistance to herbicides, insects, disease and drought, and also for the production of biopharmaceuticals, has opened the door to a new era in biotechnology. Tomato chromoplasts have been successfully engineered for high-level transgene expression in fruits, in addition to hyper-expression of vaccine antigens, and the use of plant-derived antibiotic-free selectable markers for oral delivery of edible vaccines and for biopharmaceuticals.

The plastid genome of higher plants is a circular double-stranded molecule of 120 to 160 kilobases. Identical copies of this genome are present in all cells and in all plastid types (for example, undifferentiated proplastids, photosynthesis performing chloroplasts, carotenoid-accumulating chromoplasts, and starch-storing amyloplasts). A remarkable feature of the plastid genome is its extremely high ploidy level, for example, a single tobacco leaf cell may contain as many as 100 chloroplasts, and each chloroplast contains 100 identical copies of the plastid genome, resulting in an extraordinarily high ploidy degree of up to 10,000 plastid genomes per cell.

Moreover, the chloroplast genomes of flowering plants encode perhaps 150 genes; the remainder of the hundreds or thousands of plastid proteins are the products of nuclear genes. The compositions of chloroplasts and other plastids can be modified by inserting genes into the nuclear genome; a chimeric gene can be constructed with an N-terminal sequence that targets its protein product to the chloroplast.

In higher plants, chloroplast transformation is routinely done only in tobacco, *N. tabacum*. The main obstacle to extending the technology to other species and, most importantly, to major crops is probably because of their limitations such as unavailability of stable and reproducible regeneration protocols for transplastomic plants. Although some progress has been made recently with *Arabidopsis* and potato chloroplast transformation, the production of fertile transplastomic plants in any species is still limited.

Of late, the development of a stable plastid transformation system for tomato has been achieved in the direction of establishment of stable and fertile transplastomic lines. These transplastomic tomato plants produced fruits and viable seeds, which transmitted the transgene in a uniparentally maternal fashion as expected for a plastid-encoded trait. Moreover, there was efficient transgene expression not only in chloroplasts of green leaves but also in chromoplasts of tomato fruits, demonstrating the potential of the system for biotechnological applications.

METHODS OF PLASTIDS TRANSFORMATION

In 1988, it was reported that a deletion in the chloroplast gene *atpB* that renders the green alga *Chlamydomonas reinhardtii* incapable of carrying out photosynthesis could be corrected by introducing the wild-type gene. The DNA was deposited on tungsten microprojectiles ~1 mm in size and propelled into cells spread on an agar plate using a gunpowder charge. Later, compressed gases were used for biolistic transformations. The DNA became incorporated into the *Chlamydomonas* chloroplast chromosome by homologous recombination. Thus, unlike nuclear transformation, DNA can be directed to a specific site within the plastid chromosome. Blowers *et al.* showed that foreign DNA, bordered by chloroplast DNA sequences can be incorporated and stably maintained in the *Chlamydomonas chloroplast* chromosome.

In general, two methods for delivering DNA into plastids have been developed recently: 1) A polyethylene glycol-mediated method and 2) Direct injection of DNA *in situ*. *Nicotiana plumaginifolia* and *Nicotiana tabacum* with transformed chloroplasts have been produced by exposing protoplasts to polyethylene glycol and reporter DNA bordered by chloroplast sequences to enable homologous recombination to occur. The polyethylene glycol method holds the capacity to generate more cells with transformed plastids more readily than by the biolistic procedure. However, the limiting step in this case is the regeneration of plants from protoplasts. Moreover, Knoblauch *et al.* have shown that DNA is expressed after it is injected directly into individual chloroplasts in photosynthetic leaf cells of tobacco.

CONVENTIONAL METHODS OF PLASTOME (PLASTID GENOME) TRANSFORMATION

The genetic transformation of plastids follows the principles of homologous recombination. For successful transformation, the transgene has to be flanked by plastomic sequences, ensuring the integration of the transgene at its defined position in the plastome (Figure 27.1). The principal methods to introduce DNA into chloroplasts are biolistic bombardment and the use of polyethylene glycol. In biolistic bombardment, a particle gun propels gold or tungsten particles coated with the appropriate DNA at target leaves. The chance that a chloroplast is hit by a particle in a nondestructive manner is small; nevertheless, the enormous amounts of particles make the method relatively efficient. Following protoplast isolation, a simple protocol leads to a polyethylene glycol induced pore formation at the cell membranes for the entry of genetic material. However, PEG-mediated plastid transformation technique has relatively less success rate than biolistic bombardment.

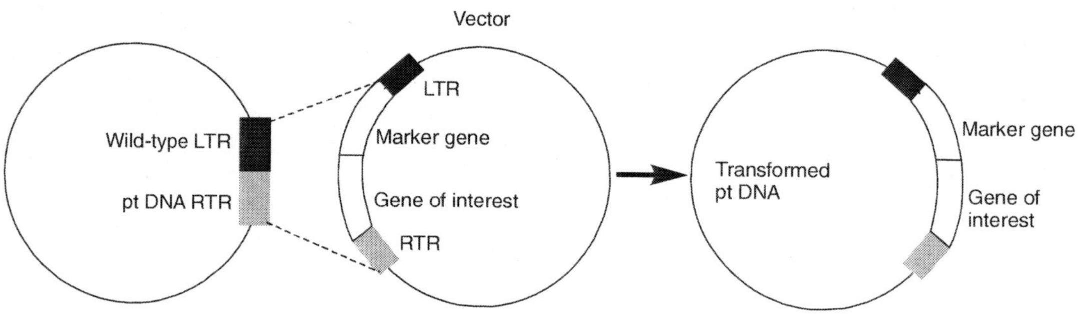

Figure 27.1 A Representative model of vector for plastid transformation

INJECTION OF DNA INTO CHLOROPLASTS

A novel development is the injection of DNA material into chloroplasts using syringes with extremely narrow tips. In animal cells, the nuclear injection of DNA through capillaries with tip diameters of one micron is a common practice, while the microinjection of DNA in plant cells has been less successful. The rigidity of the cell walls demands rigorous electrode penetration which inflicts considerable damage because of the turgor of the plant cell. Electrodes with smaller tips would cause less damage, but require extremely high pressures for fluid injection.

A novel microinjection principle combines a reduced tip size with the required injection force. Micropipettes with a tip diameter of 0.1 micron are drawn from capillaries of borosilicate or quartz glass, backfilled with a mixture of silicone oil and galinstan (an alloy of gallium, indium and tin) and sealed with a miniature glass cap. The fluid in the tip is expelled by the pressure exerted by warming the mixture in the capillary. This galinstan expansion femtosyringe (GEF) allows the injection of volumes in the femtolitre range into nuclei. The GEF tip is also suitable for insertion into chloroplasts. The injection of constructs containing the *bla* gene into the filamentous cyanobacterium *Phormidium laminosum* resulted in a stable transformation conferring ampicillin resistance.

Likewise, the injection of plasmid DNA containing the *gfp* gene under the control of a chloroplast ribosomal RNA promoter into chloroplasts of marginal mesophyll cells of intact tobacco plants led to the clearly visible expression of GFP in the chloroplasts. This procedure has a number of advantages. 1) The cells survive the injection, 2) The transformed cell can be spotted easily, and 3) the cellular content remains intact. It appears that GFP moves to other chloroplasts in the injected cell. Because cell division in the leaf tissues is thought to be fully completed at the time of injection, the division of chloroplasts is therefore unlikely and there is no evidence that cytosolic GFP is taken up by chloroplasts. Thus, GFP traffic between chloroplasts must have a different explanation.

Furthermore, the usefulness of plastid transformation for biotechnology has been shown strikingly by McBride *et al.*, who introduced the gene encoding the *B. thuringiensis* lepidopteron protoxin into the chloroplast chromosomes of tobacco plants using the microprojectile method. In the regenerated transgenic plants, this protoxin constitutes 2–3% of the soluble leaf protein. This very high level of transgene expression is attributed to the presence of ~50 chloroplasts per leaf cell and 60–100 chromosomes per plastid. Each leaf cell contains ~3000 copies of each gene, present once in each plastid chromosome. However, some chloroplast genes encode proteins that are amongst the most abundant in the cell, but others are much less highly expressed. Also, the mRNAs for some nuclear genes that are present in <10 copies per cell are as abundant as the most abundant plastid-encoded mRNAs.

REPORTER GENE FOR CONSTRUCT DESIGNING

The insertion of reporter genes in the DNA construct that is integrated into the plastome, such as that coding for the green fluorescent protein (GFP) or resistance genes against lethal agents (e.g., spectinomycin and streptomycin), is mandatory for tracing the transformed cells. Several new reporter gene strategies for the identification of plastome transformation have been developed over the past few years. Vectors carrying the bacterial gene *aphA-6*, coding for an aminoglycoside phosphotransferase that detoxifies kanamycin or amikacin, were successfully inserted into the plastome of *Chlamydomonas*. Transformants were identified by their survival on a selection medium.

In a second novel strategy, the FLARE-S system, the aminoglycoside 3" adenyl transferase (*aadA* gene), which confers resistance against spectomycin and streptomycin, is translationally fused to the *gfp* gene of *Aequorea victoria* (Figure 27.2). The drugs used in the selection procedure suppress chlorophyll production and inhibit shoot formation on plant regeneration media. The transformed lines can thus be distinguished by their ability to form green shoots on bleached wild-type leaf sections. During cultivation on the selection medium, however, chimeric plants can develop in which visual distinction of the transgenic tissues is impossible. The aforementioned fusion strategy lends visual support to the identification of the transplastomic leaf sectors and helps isolate sections of leaf that can be used for a second cycle of plant regeneration.

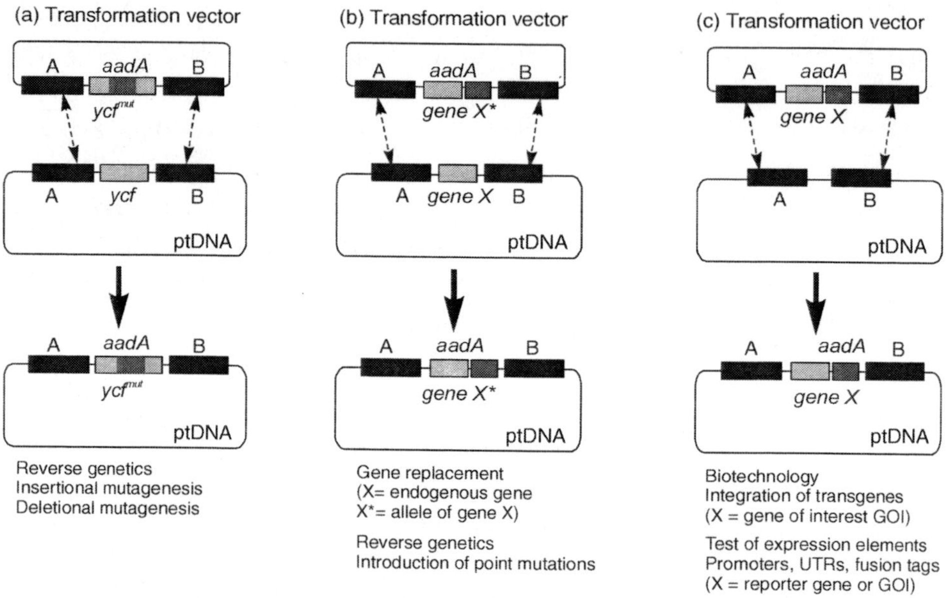

Figure 27.2 Applications of chloroplast transformation technology and the corresponding design of plastid transformation vectors

The inclusion of resistance marker genes has the disadvantage that transformed cells must be traced by stringent methods. Plant regeneration from a cell, embedded in the cellular environment, killed by lethal agents, is a risky technique and not fully understood. In the case of an optical marker like GFP, difficulties arise with the regeneration of a plant from a single GFP-expressing cell. Given the additional problems of genetic contamination, the reporter gene strategy may not be the most ideal solution for selecting transformants. Therefore, alternative selection scenarios that eliminate the use of marker genes should be explored.

(a) For reverse genetics, the selectable marker gene for plastid transformation, *aadA* (a spectinomycin resistance gene) is used to disrupt an endogenous gene of unknown or unclear function. (*ycf*: plastid open reading frame; *ycf*^{mut}: mutated plastid open reading frame)

(b) For gene replacement, an allele of an endogenous gene is linked to the *aadA* marker gene to replace the endogenous allele by homologous recombination. This strategy can also be used to introduce point mutations into plastid genes and open reading frames to study their functions.

(c) For expression of a transgene of interest (GOI), this passenger gene is linked to the *aadA* gene and targeted to a neutral insertion site, typically an intergenic region of the plastid genome. (A, B: flanking plastid DNA regions for integration via homologous recombination; ptDNA: plastid genome; UTR: untranslated region).

Flowers from transplastomic plants were pollinated with pollen from wild-type plants, and the seeds were germinated on MS medium containing spectinomycin. Whereas the wild-type control is clearly spectinomycin-sensitive and all seedlings bleach out (right), F_1 seedlings from the cross of a transplastomic tomato plant with a wild-type plant exhibit uniform resistance to the antibiotic (Figure 27.3).

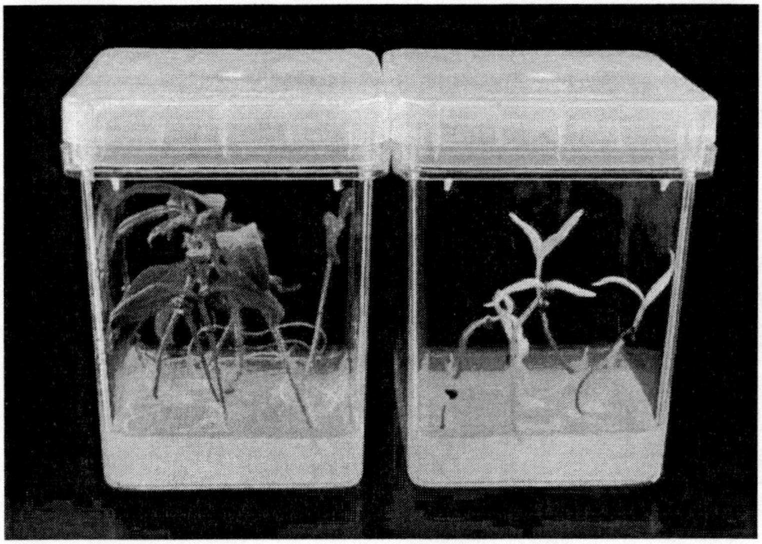

Figure 27.3 Maternal inheritance of the spectinomycin resistance trait in the F_1 progeny of transplastomic tomato plants

Figure 27.4 shows the generation of tomato plants with transgenic plastids.

Figure 27.4a shows primary selection of spectinomycin-resistant tomato calli. A plate with bombarded leaf pieces is shown after three months' incubation on spectinomycin-containing tissue culture medium. The leaf tissue is bleached out as a result of effective inhibition of chloroplast translation. Spectinomycin-resistant calli appear as small yellow or pale green mounts of dividing cells (arrow).

Figure 27.4b shows propagation of spectinomycin-resistant tomato lines. Tissue samples from primary plastid transformants are subjected to additional selection cycles on antibiotic-containing culture medium. On this medium, the tissue grows as undifferentiated callus from which samples are taken at regular intervals for homoplasmy tests (Figure 27.4).

Figure 27.4c shows plant regeneration from homoplasmic transplastomic callus tissue. Shoot regeneration is observed approximately four weeks after transfer of homoplasmic callus material to shoot induction medium.

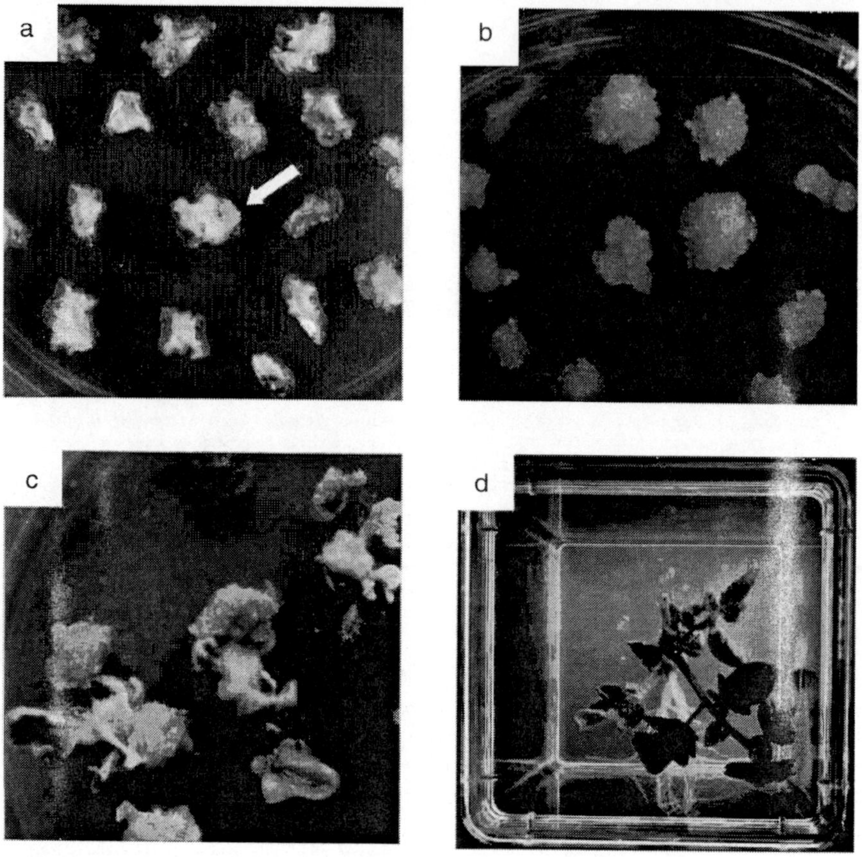

Figure 27.4 Generation of tomato plants with transgenic plastids

Figure 27.4d shows rooting of transplastomic tomato shoots. Shoots induced from homoplasmic calli are transferred to boxes with hormone-free rooting medium. Following successful rooting, plants are transferred to the soil and grown to maturity in the greenhouse.

APPLICATION OF PLASTID ENGINEERING

1. Production of Plastidic Proteins

Current research is exploring the advantages of plastomics for the expression of genes coding for insecticidal proteins or allowing for herbicide resistance. Crops expressing the *Bacillus thuringiensis* (Bt) toxin from the nucleus may only produce suboptimal amounts of toxins giving rise to an enhanced risk of pests developing Bt resistance. In attempts to reduce the development of Bt toxin resistance, the gene (*cry1A*) coding for the Bt toxin, Cry1A(c), was inserted into the plastome. The high amplification rate resulted in an accumulation of the strongly insecticidal Cry1A(c) protoxin.

Recently, overexpression of the *cry2Aa2* Bt gene using chloroplast engineering in tobacco caused 100% mortality of insects that are largely resistant to other Bt proteins. In the transformants, the level of protoxin was 20–30 fold higher than in nuclear transgenic plants. In view of the required containment of the herbicide-resistant genes, a successful strategy was designed to engineer the plastome for glyphosate resistance. Glyphosate is a potent herbicide that is a competitive inhibitor of an essential step in the aromatic amino acid biosynthetic pathway. Integration of the petunia *EPSPS* (5-enol pyruvyl shikimate 3-phosphate synthase) gene into the tobacco plastome resulted in an overproduction of EPSPS leading to glyphosate resistance.

Likewise, the effect of oxyfluorfen, a diphenyl ether herbicide, was counteracted by plastomic insertion of the *Bacillus subtilis* gene encoding protoporphyrinogen oxidase (*protox*). Protoporphyrinogen oxidase catalyses a major step in the production of chlorophyll and heme groups, which is the action site of the diphenyl ether herbicides. Transplastomic lines for the *B. subtilis protox* exhibited a higher degree of oxyfluorfen resistance than the nuclear transgenic lines. The abundant expression capacity of the plastome has the potential for massive production of foreign molecules in various plants.

Transformation of plastids has already been achieved for tobacco, *Arabidopsis* and potato, looks encouraging for rice and will presumably be extended to cover several more species in the near future. It is hoped that plastid-mediated molecular farming will lead to the biofabrication of a range of biopolymers and pharmaceutical proteins. The production of a human somatotropin in a soluble biologically active form and biodegradable protein-based polymers in tobacco are initial steps in this direction.

2. Plastome-encoded Proteins

Many efforts in plastid engineering have so far dealt with the dissection of the plastid genes and their expression. Targets for chloroplast transformation also include the proteins involved in the metabolic pathways of plastids. Amongst these are enzymes engaged in carbon fixation, like Rubisco, and proteins present in the photosynthetic reaction centres. Chloroplast transformation strategies surmounted previous obstacles to mutagenize the gene encoding the large subunit of Rubisco (*rbcL*) in higher plants. Tobacco plants mutated for *rbcL* showed reduced Michaelis constants for CO_2, O_2 and ribulose bisphosphate, and an increased oxygenase activity at limiting O_2 levels.

Additionally, tobacco has also been transformed with the *rbcL* gene from sunflower and that of the cyanobacterium, *Synechococcus*. The latter line produced *rbcL* mRNA but no large subunit protein or enzyme activity. In contrast, the hybrid protein composed of sunflower and tobacco subunits produced a catalytically active hybrid form of the enzyme. In a third line, a chimeric sunflower–tobacco large subunit arising from homologous recombination within the *rbcL* gene had properties similar to the hybrid enzyme.

3. Herbicide Resistance

Glyphosate is a potent, broad-spectrum herbicide that is highly effective against grasses and broad-leaf weeds. Glyphosate works by competitive inhibition of an enzyme in the aromatic amino acid biosynthetic pathway, 5-enol pyruvyl shikimate 3-phosphate synthase (EPSPS). Unfortunately, like most commonly used herbicides, glyphosate does not distinguish crops from weeds, thereby restricting its use. Therefore, engineering of crop plants for resistance to the herbicide is a standard strategy to overcome the lack of herbicide selectivity. However, this approach raises the concern that if the engineered resistance gene escapes via pollen dispersal, it might result in resistant weeds or might cause genetic pollution among other crops.

Engineering foreign genes through chloroplast genomes (which are maternally inherited in most crops) offers a solution to this problem. Although pollen from plants with maternal chloroplast inheritance can contain metabolically active plastids, the plastid DNA is lost during pollen maturation and hence is not transmitted to the next generation. In addition, the target proteins for many herbicides are compartmentalized within the chloroplast. The chloroplast genome can be engineered to confer herbicide resistance by expressing a petunia EPSPS nuclear gene via the chloroplast genome. The resultant transgenic plants are resistant to tenfold higher levels of glyphosate than the lethal dosage, and the transgene is maternally inherited.

Recently, the *Agrobacterium* EPSPS gene (C4) was expressed in tobacco plastids and resulted in 250-fold higher levels of the glyphosate-resistant C4 protein than was achieved via nuclear transformation. Even though C4 expression in plastids was enhanced more than nuclear expression levels, field tolerance to glyphosate remained the same, showing that higher levels of expression do not always proportionately increase herbicide tolerance. In two different studies, transgenic tobacco plants expressing *bar* genes via the chloroplast genome exhibited field-level tolerance to phosphinothricin (PPT). In this case, even plants with the lowest levels of *bar* expression were resistant to the highest levels of PPT tested, and no pollen transmission of the transgene was detected. However, high-level expression of *bar* genes in the chloroplast was not sufficient to allow direct selection of chloroplast transformants on medium containing PPT.

4. Pathogen Resistance

Since plant diseases have impacts on global crop production, it is highly desirable to engineer plants that are resistant to pathogenic bacteria and fungi. Amphipathic peptides such as magainin are known for antimicrobial properties, but their possible uses in agriculture have not been fully explored yet. Recently, a synthetic antimicrobial peptide (MSI-99) was expressed via the chloroplast genome.

MSI-99 is an amphipathic α-helical molecule with affinity for the negatively charged phospholipids found in the outer membrane of bacteria and fungi. Upon contact with these membranes, individual peptides aggregate to form pores, resulting in microbial lysis. Because of the concentration-dependent action of antimicrobial peptides, MSI-99 was expressed via the chloroplast genome to accomplish high-dose release at the point of infection. *In vitro* and *in planta* assays with T_0, T_1 and T_2 plants confirmed that the peptide was expressed at high levels (up to 21.5% total soluble protein) and retained biological activity against *Pseudomonas syringae*, a major plant pathogen. In addition, leaf extracts from transgenic plants inhibited the growth of pre-germinated spores of three fungal species *Aspergillus flavus*, *Fusarium moniliforme* and *Verticillium dahliae* by more than 95% compared to untransformed controls. Significantly, growth and development of the transgenic plants were unaffected by hyperexpression of MSI-99 within chloroplasts. Because the outer membrane is an essential and highly conserved part of all microbial cells, microorganisms are unlikely to develop resistance against these peptides.

Therefore, these results give a new option in the battle against phytopathogens. Pharmaceutical companies are exploring the use of lytic peptides as broad-spectrum topical and systemic antibiotics. Moreover, analogs of magainin are also effective against hematopoietic, melanoma, sarcoma and ovarian teratoma lines. Cystic fibrosis makes patients susceptible to respiratory infections, especially multidrug-resistant *Pseudomonas aeruginosa*. When MSI-99 was tested against *P. aeruginosa in vitro*, cell-free extracts of T_1 chloroplast transgenic plants displayed a 96% level of growth inhibition against *P. aeruginosa*. These studies show that MSI-99 has the potential to become an alternative to current antibiotic treatments for microbial infections.

5. Drought Tolerance

Water stress caused by drought, salinity or freezing is a major limiting factor in plant growth and development. Trehalose is a non-reducing disaccharide of glucose whose synthesis is mediated by the trehalose 6-phosphate (T6P) synthase and trehalose 6-phosphate phosphatase complex in *Saccharomyces cerevisiae*. Because it accumulates under stress conditions such as freezing, heat, salt or drought, there is general consensus that trehalose protects against damage imposed by these stresses. Therefore, engineering high levels of trehalose in plants might confer drought tolerance.

Gene containment in transgenic plants is a serious concern when plants are genetically engineered for drought tolerance because of the possibility of creating drought-tolerant weeds and passing on undesired pleiotropic traits to related crops. To prevent these consequences, it is desirable to engineer crop plants for drought tolerance via the chloroplast genome instead of the nuclear genome.

Recently, the yeast trehalose phosphate synthase (*TPS1*) gene was introduced into the tobacco chloroplast and nuclear genomes to study the resultant phenotypes. Although the chloroplast transgenic and the nuclear transgenic plants expressed significant TPS1 enzyme activity, chloroplast transgenic plants showed up to 25-fold higher accumulation of trehalose than nuclear transgenic plants. Nuclear transgenic plants (T_0) with significant amounts of trehalose accumulation exhibited a stunted phenotype, sterility and other pleiotropic effects, whereas chloroplast transgenic plants (T_1, T_2 and T_3) grew normally and had no visible pleiotropic effects.

Chloroplast transgenic plants also showed a high degree of drought tolerance by growing in 6% polyethylene glycol, whereas respective control plants were bleached. Air-dried chloroplast transgenic seedlings and transgenic plants under extreme drought (not watered for 24 days) successfully rehydrated, whereas similarly treated control plants died. Further, investigations have confirmed that trehalose functions by protecting the integrity of biological membranes rather than regulating water potential. Therefore, this clearly indicates that compartmentalization of trehalose within chloroplasts confers drought tolerance without undesirable phenotypes.

6. Chloroplast Transgenesis without Antibiotic Resistance Markers

One disadvantage of chloroplast genetic engineering technology is the use of the antibiotic resistance gene, *aadA*, as a selectable marker. The *aadA* gene product inactivates its selective agents, spectinomycin and streptomycin, by transferring the adenyl moiety of ATP to the antibiotics. Because these antibiotics are commonly used to control bacterial infection in humans and animals, there is concern that their over-use might lead to the development of resistant bacteria. Therefore, several studies have explored strategies for engineering chloroplasts that are free of antibiotic-resistance markers.

The spinach betaine aldehyde dehydrogenase (*BADH*) gene has been developed as a plant-derived selectable marker to transform chloroplast genomes. The selection process involves conversion of toxic betaine aldehyde (BA) by the chloroplast-localized BADH enzyme to non-toxic glycine betaine, which also serves as an osmoprotectant. Because the BADH enzyme is present only in chloroplasts of a few plant species adapted to dry and saline environments, it is suitable as a selectable marker in many crop plants. The transformation study showed that BA selection was 25-fold more efficient than spectinomycin, exhibiting rapid regeneration of transgenic shoots within two weeks. This is the first report of chloroplast engineering without the use of antibiotic resistance genes. Use of selectable markers that are naturally present in spinach, in addition to gene containment, should ease public concerns over genetically modified crops.

Another approach to develop marker-free transgenic plants is to eliminate the antibiotic resistance gene after transformation. Strategies have been developed to remove genes using endogenous chloroplast recombinases that delete the marker genes via engineered direct repeats. Early experiments with *Chlamydomonas reinhardtii* showed that it is possible to exploit these recombination events to eliminate introduced selectable marker genes. Homologous recombination between two direct repeats, engineered to flank a selectable marker, enabled marker removal under non-selective growth conditions.

A similar approach applied in tobacco was effective in generating *aadA*-free T_0 transplastomic lines while leaving a third unflanked transgene, *bar*, in the genome to confer herbicide resistance. Recently, another strategy to eliminate selectable marker genes has been developed, using the P1 bacteriophage CRE-lox site-specific recombination system. In two separate studies, a marker gene flanked by lox sites was introduced into the tobacco chloroplast genome. Its removal was subsequently induced by expressing the CRE protein from the nucleus. These results show that efficient removal of selectable marker(s) from chloroplast genomes is feasible.

7. Transplastomics for Edible Vaccines

A major cost of biopharmaceutical production lies in purification; for example, in insulin production, chromatography accounts for 30% of the production cost and 70% of the set-up cost. Therefore, oral delivery of properly folded and fully functional biopharmaceuticals should significantly cut down the production cost. Bioencapsulation of pharmaceutical proteins within plant cells offers protection against digestion in the stomach but allows successful delivery to the target tissues.

The β subunits of enterotoxigenic *E. coli* (LTB) and cholera toxin of *Vibrio cholerae* (CTB) are candidate vaccine antigens. Integration of an unmodified CTB gene into the tobacco chloroplast genome results in accumulation of up to 4.1% of total soluble leaf protein as functional CTB oligomers (410-fold higher than the unmodified LTB gene expressed via the nuclear genome). In addition, binding assays confirm that chloroplast-synthesized CTB binds to the intestinal membrane GM1-ganglioside receptor, indicating correct folding and disulphide bond formation of the plant-derived CTB pentamers. Increased production of an efficient transmucosal carrier molecule and delivery system, such as CTB, in transgenic chloroplasts makes plant-based oral pharmaceuticals commercially feasible. Because the quaternary structure of many proteins is essential for their function, this investigation shows the potential for other foreign multimeric proteins to be properly expressed and assembled in transgenic chloroplasts.

IMPORTANCE OF TRANSPLASTOMIC PLANTS

The introduction of genes by engineering the chloroplast has been proposed to remedy the problems associated with transgene dispersal into the wild plant population. As there are no plastids, and hence plastid DNA, in the pollen of most crops (exceptions being alfalfa and possibly rice and pea), any gene introduced into chloroplast-engineered plants is unlikely to be transferred via the pollen to the next generation. Introgression of genes from wild relatives, however, could eventually allow chloroplast transgenes to invade the non-transformed population. The extent to which introgression of genes between wild relatives occurs is at present uncertain, although in general, introgression is unusual. Despite the fact that rates are thought to be low, it has been pointed out by Stewart and Prakash that introgression of the common weed *Raphanus raphanistrum* into *Brassica napus* (oilseed rape) occurred at higher rates than the reciprocal cross of *Brassica napus* pollen into *Raphanus raphanistrum*.

LIMITATIONS OF CHLOROPLAST GENETIC ENGINEERING

One barrier to developing plastid transformation for crop plants, including cereals, has been their regeneration from non-green embryonic cells (containing proplastids) rather than leaf cells (containing chloroplasts). Plastid transformation in rice has been shown using vectors containing a translational fusion of *aadA* with *gfp*. Inspection of streptomycin-resistant rice shoots by confocal microscopy revealed a few proplastids with GFP fluorescence. Although this reveals heteroplasmy, it represents a step in the right direction towards monocot plastid transformation. A major limitation in accomplishing this goal might be the low level of marker gene expression in non-green plastids because of low genome copy number and low rates of protein synthesis. Identification of promoters and UTRs active in non-green tissues helps overcome this limitation.

Another limitation is the lack of information on genome sequences for several important crop species to locate intergenic sequences for integration of transgenes. Because of the conservation of the plastid genome sequence across many plant species, the concept of using targeting sequences from one species to transform the plastid genome of another unknown species was developed. Both potato and tomato were transformed using chloroplast vectors with tobacco targeting sequences, and tobacco was transformed with petunia targeting sequences.

Yet another limitation is the challenge of delivering foreign DNA through the double plastid membrane. Among the methods used to transform plant chloroplasts, only particle bombardment has proven to be efficient in transforming different plant species. PEG treatment of protoplasts has been used to stably transform tobacco chloroplast genomes and although laborious, it is inexpensive when compared to particle bombardment. Microinjection is effective for foreign gene expression in chloroplasts, although application of this technology for stable transformation of chloroplast genomes awaits further investigation. Development of new methods of gene delivery systems into chloroplasts facilitates further progress in this field.

CONCLUSION

Although considerable progress has been made over the past few years, there are still a number of hurdles that have to be overcome before plastome transformation becomes a very attractive device for plant biotechnology. First, the number of plant species to which plastome technology is applicable needs to be increased considerably; so far, only a limited number of culture plants have been transformed successfully by plastome manipulation. Second, the success rate of gene insertion into the plastome has to be increased; this may reduce the need for extensive selection procedures. Third, the screening protocols must be simplified so that they are applicable to a large range of plant species.

Plastome technology may be potentiated by several novel strategies. The stromule-mediated interplastidic migration of macromolecules, which in itself is an intriguing event from a fundamental point of view, may be used as an additional tool for improving the yield of transplastomic products. Also, the transfer of prefabricates through the plastid envelope that are further processed in the cytosol may be a target for further development.

REVIEW QUESTIONS

1. What is plastomic engineering? Write the significance of plastid transformation.
2. Discuss briefly the history of plastid engineering.
3. Write the basic methods of plastid transformation and explain with vector designing.
4. Discuss in detail with examples about the production of transplastomic lines for various traits.
5. Write briefly about the advantages and limitations of plastid engineering.
6. Write short notes on (a) Edible vaccine (b) Plastidic protein classes.

REFERENCES

Boynton, J.E., Gillham, N.W., Harris, E.H., Hosler, J.P., Johnson, A.M., Jones, A.R., Randolph-Anderson, B.L., Robertson, D., Klein, T.M. and Shark, K.B. *et al.* (1988). Chloroplast transformation in Chlamydomonas with high velocity microprojectiles. *Science*. 240: 1534–1538.

Blowers, A.D., Bogorad, L., Shark, K.B. and Sanford, J.C. (1989). Studies on Chlamydomonas chloroplast transformation: foreign DNA can be stably maintained in the chromosome. *Plant Cell*. 1(1):123–132.

Daniell, H., Kumar, S. and Dufourmantel, N. (2005). Breakthrough in chloroplast genetic engineering of agronomically important crops. *Trends in Biotechnology*. 23: 238–245.

Greyich, J.J. and Daniell, H. (2005). Chloroplast genetic engineering: Recent advances and future perspectives. *Critical review of Plant Sciences*. 24: 83–107.

Maliga, P. (2002). Engineering the plastid genome of higher plants. *Current Opinion in Plant Biology*. 5: 164–172.

28

PLANT *IN VITRO* BIOTECHNOLOGY IN AGRICULTURE

INTRODUCTION

The use of the term "biotechnology" has become widespread recently but in its most restricted sense, it refers to the molecular techniques used to modify the genetic composition of a host plant, that is, genetic engineering. Moreover, plant cell and tissue culture comprises a set of *in vitro* techniques, methods and strategies that are part of the group of technologies called plant biotechnology. It involves the production of new plants from pieces of tissue placed on sterile nutrient growth medium. The cell clumps (known as a callus) differentiate eventually and grow into tiny plantlets with roots and stems. When these plantlets are transplanted into soil, they grow into full-sized plants.

Significantly, cell and tissue culture techniques are part of a large group of strategies and technologies, ranging through molecular genetics, recombinant DNA studies, genome characterization, gene-transfer techniques, and aseptic growth of cells, tissues, organs, and *in vitro* regeneration of plants that are considered to be integral part of plant biotechnology. Tissue culture techniques, in combination with molecular techniques, have been successfully used to incorporate specific traits through gene transfer.

Cell culture has been exploited to create genetic variability from which crop plants can be improved in order to improve the state of health of the planted material and to increase the number of desirable germplasms available to the plant breeder. *In vitro* regeneration protocols are available for most crop species, although continued optimization is still required for many crops, especially cereals and woody plants. Most of the time, plantlets regenerated from a single callus are genetically identical, called clones; however, mutations sometimes occur giving rise to a plantlet with different characteristics. Variants that develop by this technique may have useful traits.

In vitro techniques for the culture of protoplasts, anthers, microspores, ovules and embryos have been used to create new genetic variation in the breeding lines, often via haploid production. Cell culture has also produced somaclonal and gametoclonal variants with crop-improvement potential. The culture of single cells and meristems can be effectively used to eradicate pathogens from planting material and thereby dramatically improve the yield of established cultivars. Large-scale micropropagation techniques are providing millions of plants for the commercial ornamental market and the agricultural, clonally propagated crop market.

In agriculture, the applications of various tissue culture approaches to crop improvement, such as, breeding, wide hybridization, haploidy, somaclonal variation and micropropagation have been described in the following sections.

MICROPROPAGATION—A SOURCE OF MASS CLONING

Regeneration of a whole plant through tissue culture is popularly called micropropagation. Within a very short time and space, a large number of plantlets can be produced from callus tissue. Again, it is possible to multiply a large number of callus pieces from the original stock culture during subculture.

A great advantage of micropropagation is that millions of identical plants can be grown from only a small amount of parental stock tissue. Tissue culture is widely used in the commercial production of orchids, house plants, cut flowers, fruit plants and ornamental trees, shrubs, and non-woody plants. As a result, laboratory-scale micropropagation protocols are available for a wide range of species and at present, micropropagation is the widely used sector of plant tissue culture technology. There are hundreds of micropropagation firms, which are working on thousands of different tissue lines (species and varieties) in culture. They are working on tissue culture, *in vitro* multiplication and elimination of pathogens, germplasm storage, genetic modification and plant-breeding programmes for different plant species. A list of *in vitro*-regenerated plants through micropropagation is given in Table 28.1.

There are basically three methods that can be used for micropropagation. They are 1) Enhancing axillary-bud breaking, 2) Production of adventitious buds and 3) Somatic embryogenesis.

Axillary-bud breaking produces the least number of plantlets, as the number of shoots produced is controlled by the number of axillary buds cultured, but remains the most widely used method in commercial micropropagation and produces the most true-to-type plantlets. Adventitious budding has a greater potential for producing plantlets, as bud primordia may be formed on any part of the inoculum. However, in the latter two methods, organized structures arise directly on the explant or indirectly from callus.

Unfortunately, somatic embryogenesis, which has the potential of producing the largest number of plantlets has been proved to be applicable in limited number of plant tissues, and can only presently be induced in a few species. Nevertheless, the production of somatic embryos from cell cultures presents opportunities not available to plantlets regenerated by the organogenesis routes, such as mechanization. One approach envisages the use of bioreactors for large-scale production of somatic embryos and their delivery in the form of seed tapes or artificial seeds. Commercial operation is based on somatic embryogenesis has been started by many organizations. Such embryogenesis is playing an important role in improving herbaceous dicots, herbaceous monocots and woody plants.

Table 28.1 Micropropagated Plants via *in vitro* Culture

Manihot esculenta	*Solanum tuberosum*	*Allium cepa*
Beta vulgaris	Asparagus	Cardamom
Gypsophila paniculata	*Allium sativum*	*Spinacea oleracea*
Amaranthus blitum	*Lycopersicon esculentum*	*Agapanthus sp.*
Cucumis sativus	Pineapple	Turmeric
Solanum melongena	*Petunia sp.*	*Abelmoschus esculentus*
Capsicum annum	*Mangifera indica*	*Annona squamosa*
Ananas comosus	*Achras sapota*	*Musa paradisiaca*
Citrus reticulata	*Citrus sinensis*	*Psidium guajava*
Carica papaya	*Litchi chinensis*	*Punica granatum*
Coffee	*Vitis vinifera*	Rose
Aegle marmelos	*Allium*	*Pyrus malus*
Gerbera	*Prunus persica*	*Citrullus lanatus*
Morus alba	*Raphanus sativus*	*Brassica rapa*
Vanilla	Chrysanthemum	*Daucus carota*
Lily	*Maranta*	

Along with advantages of micropropagation techniques, there are several limiting factors for the application of this technology for *in vitro* multiplication approach, like:

i. The cost of labour needed to transfer tissue repeatedly between vessels during subculture and the need for maintenance of aseptic status of regenerants for longer time can account for up to 70% of the production costs of micropropagation.

ii. Problems of vitrification, acclimatization and contamination can cause great losses in a tissue culture laboratory.

iii. Genetic variations in cultured lines, such as polyploidy, aneuploidy and mutations, have been reported in several systems and resulted in the loss of desirable economic traits in the tissue-cultured products.

EMBRYO CULTURE—AN ALTERNATIVE TO OVERCOME EMBRYO ABORTION

The most common reason for post-zygotic failure of wide hybridization is embryo abortion due to poor endosperm development. Embryo culture has been successful in overcoming this major barrier as well as in solving the problems of low seed set, seed dormancy, slow seed germination, inducing embryo growth in the absence of a symbiotic partner, and the production of monoploids of barley. Moreover, the breeding cycle of *Iris* could be made shorter from 2–3 years to a few months by employing embryo rescue technology.

A similar approach has worked with orchids and roses and is being applied to banana and *Colocasia*. Interspecific and intergeneric hybrids of a number of agriculturally important crops have been successfully produced, including cotton, barley, tomato, rice, and jute, *Hordeum* × *Secale*, *Triticum* × *Secale* and some *Brassicas*. At least six barley cultivars (Mingo, Rodeo, Craig, Winthrop, Lester and TB891-6) have been produced out of material selected from doubled haploids originating through the widely used "bulbosum" method of cross-pollination and embryo rescue.

Moreover, Mingo, in particular, was a breakthrough, as it was the first barley cultivar produced by this technique to be patented, in 1980. Briefly, *Hordeum vulgare* ($2n = 14$) is pollinated with pollen from *H. bulbosum* ($2n = 14$). Normally, in this system, the seeds develop for about 10 days and then abort, but embryo rescue could be possible if the immature embryos are isolated aseptically and are brought to culture under *in vitro* conditions.

Applications of Embryo Culture in Agriculture

i. The degeneration of embryos is observed in seeds obtained from the interspecific crosses such as *Lilium speciosum album* × *L. auratum*. In this case, the microsurgical technique of embryo culture is useful in raising full hybrid plants.

ii. The degeneration of endosperm is observed in the fruits obtained from the plants of the crosses between *Hordeum vulgare* and *H. bulbosum*. As a result, embryos die. Such embryos can be rescued by this technique.

iii. To produce disease resistant varieties of *Lycopersicon esculentum*, there are a series of attempts in crossing this variety with *L. peruvianum*. The seeds formed by the cross show abnormal development and no viability were observed in the seeds.

iv. Hybridization between *Corchorus capsularis* and *C. olitorius* is a major aim of jute breeders but their attempts have failed due to early abscission of pollinated flowers, resulting in a low fruit set. Fruits are associated with premature abortion of embryos. The problem could be overcome by adopting embryo culture technique.

v. Embryo culture can be successfully used to bypass the traditional treatments to overcome seed dormancy and to accelerate germination in certain types of seed. For example, *Lactuca sativa, Nemphia insignis, Phacelia tanacetifolia* and *Citrullus colocynthis.*

vi. The breeding cycle may be shortened by embryo culture method. Rose normally takes a whole year to reach the flowering stage, but it has been possible to produce two generations by embryo culture technique.

vii. The nature of sterility in some economic fruits like banana, kachoo seeds, etc., which never germinate in nature could be overcome by embryo culture.

viii. *Melilotus officinalis* has high coumarin content which is harmful to cattle. By employing embryo culture technique, hybrid plants were obtained by the crossing between *M. officinalis* and *M. alba* (low coumarin), and some success was achieved in producing the harmless variety.

HAPLOIDS PRODUCTION IN AGRICULTURAL CROPS

Haploid plants have the gametophytic (one-half of the normal) number of chromosomes, therefore, haploid plants are of special interest to plant breeders because they allow the expression of simple recessive genetic traits or mutated recessive genes. Also, doubled haploids can be used immediately as homozygous breeding lines. The efficiency in producing homozygous breeding lines via doubled *in vitro* produced haploids represents significant savings in both time and cost compared to other methods. Three possible *in vitro* methods have been used to generate haploids:

i. Gynogenesis Culture of excised ovaries and ovules

ii. Androgenesis Culture of excised anthers and pollens

iii. Bulbosum technique Embryo culture

More than 200 plant species have been used to produce haploid plants by pollen, microspore and anther culture. These include cereals (barley, maize, rice, rye, triticale and wheat), forage crops (alfalfa and clover), fruits (grape and strawberry), medicinal plants (*Digitalis* and *Hyoscyamus*), ornamentals (*Gerbera* and sunflower), oil seeds (canola and rape), trees (apple, litchi, poplar and rubber), plantation crops (cotton, sugar cane and tobacco), and vegetable crops (asparagus, brussels sprouts, cabbage, carrot, pepper, potato, sugar beet, sweet potato, tomato and wing bean).

Haploid wheat cultivars, derived from anther culture, have been released in several countries. It was reported that 5 to 7 years could be saved on producing inbred lines in a Chinese maize-breeding programme by using anther culture-derived haploids. A similar saving has been reported for triticale and the horticultural crop *Freesia*. In asparagus, anther-derived haploids have been used to produce an all-male F_1 hybrid variety. Various stress treatments and the proper stages of responsive pollen grains required to induce androgenesis in agricultural crops are given in Table 28.2.

Table 28.2 Stress treatment and stage that induce androgenesis in agricultural crops

Plant	Stress treatment	Stage of responsive pollen grain
Brassica napus	Heat treatment (32°C for 8 hours), gamma rays, colchicines	Late unicellular to early binucleate pollen grains
Hordeum vulgare	Stress by mannitol, calcium and ABA	
Nicotiana tabacum	Glutamine and sugar starvation (then transfer to high glucose medium)	Young bicellular pollen grains
Triticum aestivum	Cold treatment (3–24 days at 4–10°C)	
Oryza sativa and other millets		

Similarly, ovule culture has been successfully employed in the production of haploid plants in plant species such as *Allium cepa, Helianthus annuus, Solanum tuberosum, Beta vulgaris, Cucumis melo, Brassica oleracea, Petunia axillaris, Melandrium album, Hevea brasiliensis* and *Gerbera jamesonii.* A list of haploid plants regenerated through *in vitro* gynogenesis is mentioned in Table 28.3.

Table 28.3 Haploid Plants raised through *in vitro* Gynogenesis

Plant name	Family
Beta vulgaris	Chenopodiaceae
Gerbera jamesonii and *Helianthus annus*	Asteraceae
Hordeum vulgare	Poaceae
Oryza sativa	Poaceae
Triticum aestivum	Poaceae
Zea mays	Poaceae
Lilium tuberosum	Liliaceae
L. davidii	Liliaceae
Allium cepa	Liliaceae
Nicotiana tabacum	Solanaceae
N. rustica	Solanaceae
Petunia axillaris	Solanaceae
Solanum tuberosum	Solanaceae
Brassica oleracea	Brassicaceae
Cucurbita pepo	Cucurbitaceae

Ovule culture is a boon for the plant breeders in obtaining seedlings from crosses, which are normally unsuccessful because of abortive embryos. It has been observed that in several inter specific crosses, the hybrid embryo of *Abelmoschus* fails to develop beyond the heart or torpedo-shaped embryo. Viable hybrids have been obtained by culture of the fertilized ovule and some of the hybrids which have been produced by ovule culture are as follows:

- *A. esculentus* × *A. ficulneus*
- *A. esculentus* × *A. moschatus*
- *A. tuberculatus* × *A. moschatus*
- *Brassica chinensis* × *B. pekinensis*
- *Lolium perenne* × *Festuca rubra*

Haploid Production through Distant Hybridization

The production of haploid plants is not only achieved by pollen or anther culture, but there are other alternatives of haploid production where selective elimination of chromosome of one species during interspecific crosses of two different species of the same plant and retention of one chromosome of one species is used to obtain haploids in barley and also in wheat. The production of haploids in barley in this manner involves inter-specific hybridization of *Hordeum vulgare* ($2n = 14$) with *H. bulbosum* ($2n = 14$), a cross-pollinated, self incompatible perennial species. About 95% progeny are haploids and remainder being diploid hybrids. A cross between 2× barley and 4 × *H. bulbosum* mainly produces triploid hybrids.

Cytological examination of early embryo and endosperm in VV × BB or BB × VV hybrids revealed gradual elimination of chromosome. While most of the genotypes of barley readily produce haploids and probably chromosome elimination seems to be controlled by genotypes of both parents. Wheat haploids can also be obtained through crosses between wheat and *H. bulbosum* (both 2*x* and 4*x*) and crosses between wheat and maize.

IN VITRO POLLINATION/ FERTILIZATION

This technique has been used to facilitate both interspecific and intergeneric crosses, to overcome physiological-based self-incompatibility and to produce hybrids. A wide range of plant species has been recovered through this technique via pollination of pistils and self- and cross-pollination of ovules. This range includes agricultural crops, such as tobacco, clover, corn, rice, canola, poppy and cotton. The use of delayed pollination, distant hybridization, pollination with abortive or irradiated pollen, and physical and chemical treatment of the host ovary have been used to induce haploidy. Some of the plant species where *in vitro* pollination has been successfully reported:

- *Melandrium album* × *Viscaria vulgaris*
- *M. album* × *Minuartia laricifolia*

- *M. album* × *Dianthus serotinus*
- *N. tabacum* × *Petunia hybrida*

PROTOPLAST FUSION OR SOMATIC HYBRIDIZATION

Protoplast fusion has often been suggested as a means of developing unique hybrid plants which cannot be produced by conventional sexual hybridization. Protoplasts can be produced from many plants, including most crop species. However, any two plant protoplasts can be fused by chemical or physical means but production of unique somatic hybrid plants is limited by the ability to regenerate the fused product and sterility in the interspecific hybrids rather than the production of protoplasts.

Perhaps the best example of the use of protoplasts to improve crop production is that of *Nicotiana*, where the somatic hybrid products of a chemical fusion of protoplasts have been used to modify the alkaloid and disease-resistant traits of commercial tobacco cultivars. Somatic hybrids were produced by fusing protoplasts, using a calcium polyethylene glycol treatment, from a cell suspension of chlorophyll-deficient *N. rustica* with an albino mutant of *N. tabacum*. The wild *N. rustica* parent possessed the desirable traits of high alkaloid levels and resistance to black root rot.

Furthermore, fusion products were selected as bright green cell colonies. The colour appearance is due to the genetic complementation for chlorophyll synthesis in the hybrid cells. Significantly, plants recovered by shoot organogenesis showed a wide range of leaf alkaloid content but had a high level of sterility. However, after three backcross generations to the cultivated *N. tabacum* parent, plant fertility was restored in the hybrid lines, although their alkaloid content and resistance to blue mould and black root rot were highly variable. Interestingly, neither parent was known to possess significant resistance to blue mould.

The fusion of protoplasts from 6-azauracil-resistant cell lines of *Solanum melongena* (aubergine) with protoplasts from the wild species *S. sisymbriifolium* yielded hybrid, purple-pigmented cell colonies that underwent regeneration via organogenesis. As protoplasts from the parental cell suspension cultures could not be regenerated, hybrids could be screened by their 6-azauracil resistance, capacity to synthesize anthocyanins (purple pigment) and ability to undergo shoot organogenesis. The restoration of regeneration ability through complementation has also been observed in *Nicotiana* cell fusion products. The hybrids resulting from this study were found to be resistant to root-knot nematodes and spider mites. However, they were also completely sterile and could not be incorporated into an aubergine-breeding programme.

There are two possible ways of solving this sterility problem, either by "back" fusions of somatic hybrids with the cultivated parents or by initiation of suspension cultures of the hybrid cells so that more of the wild species chromosomes can be eliminated. However, these two methods have so far been unsuccessful with these hybrids. Selection of hybrids and use of protoplast fusion for hybridization in crop plants has been reported in *Brassicas,* citrus, rice, carrot, canola, tomato, and the forage legumes, alfalfa and clover. The production of novel hybrids through protoplast fusion should focus on four areas:

i. Agriculturally important traits.
ii. Achieving combinations that can only be accomplished by protoplast fusion.
iii. Somatic hybrids integrated into a conventional breeding programme.
iv. The extension of protoplast regeneration to a wider range of crop species.

Interestingly, in the case of *Nicotiana,* all of these criteria were met, although this took 12 years.

Distant Somatic Hybrids in Agriculture

In contrast to sexual hybrid cells (zygote), which contain the plasmogens from the female parent only, somatic hybrid cells obtain cytoplasmic components from both the parental species. The plasmogens appear to be distributed randomly during the mitotic cell divisions. As a result, some cells receive chloroplasts from any one parental species and a small proportion of cells also retain the chloroplasts of both species. This is reflected in the plants regenerated from these cells. A somatic hybrid plant may contain chloroplasts from one parental species and mitochondria from the other fusion parent.

i. *Symmetric or near-symmetric hybrids* Somatic hybrid plants retain the full or nearly full somatic chromosome complements of the two parental species; these are called symmetric hybrids. Such hybrids provide unique opportunities for synthesizing novel species, which may be of theoretical and/or practical interest, for example, Pomato or Topato. Some examples of symmetric hybrids that have been produced successfully are listed below:

- *Solanum tuberosum* × *Lycopersicon esculentum*
- *Datura innoxia* × *Atropa belladonna*
- *Arabidopsis thaliana* × *Brassica campestris*
- *Atropa belladonna* × *Nicotiana chinensis*

ii. *Asymmetric hybrids* Many somatic hybrids exhibit the full somatic complement of only one parental species, while all or nearly all chromosomes of the other parental species are lost during the preceding mitotic divisions; such hybrids are referred to as asymmetric hybrids. Some examples of such types of hybrids are as follows:

- *Daucus carota* × *Aegopodium podagraria*
- *Daucus carota* × *Petroselinum hortense*
- *Hyoscyamus muticus* × *Nicotiana tabacum*
- *Datura innoxia* × *Physalis minima*
- *Nicotiana tabacum* × *Daucus carota*

WIDE HYBRIDIZATION

A critical requirement for crop improvement is the introduction of new genetic material into the cultivated lines of interest, whether via single genes, through genetic engineering, or multiple genes, through conventional hybridization or tissue culture techniques. During fertilization in angiosperms, pollen grains must reach the stigma of the host plant, germinate and produce a pollen tube. The pollen tube must penetrate the stigma and style and reach the ovule. The discharge of sperm within the female gametophyte triggers syngamy and the two sperm nuclei must then fuse with their respective partners. The egg nucleus and fusion nucleus then form a developing embryo and the nutritional endosperm, respectively.

This process can be blocked at any number of stages, resulting in a functional barrier to hybridization and the blockage of gene transfer between the two plants. Pre-zygotic barriers to hybridization (those occurring prior to fertilization), such as the failure of pollen to germinate or poor pollen tube growth, may be overcome using *in vitro* fertilization. Post-zygotic barriers (occurring after fertilization), such as lack of endosperm development, may be overcome by embryo, ovule or pod culture. Where fertilization cannot be induced by *in vitro* treatments, protoplast fusion has been successful in producing the desired hybrids.

SOMACLONAL VARIATION

In some callus culture, there is a major tendency of the callus tissue to exhibit a strong tendency toward numerical variation of the chromosomes in the cells after a number of serial subcultures. Somaclonal variation itself does not appear to be a simple phenomenon, and may reflect pre-existing cellular genetic differences or tissue culture-induced variability. The chromosomal instability in the cultured cells plays an important role in polyploidization of cells and genetically variable plants can be raised from such polyploidized cells by subsequent cultures.

Such variation may be generated through several types of nuclear chromosomal re-arrangements and losses, gene amplification or de-amplification, non-reciprocal mitotic recombination events, transposable element activation, apparent point mutations, or re-activation of silent genes in multigene families, as well as alterations in maternally inherited characteristics.

Many of the changes observed in plants regenerated *in vitro* have potential agricultural and horticultural significance. These include alterations in plant pigmentation, seed yield, plant vigour and size, leaf and flower morphology, essential oils, fruit solids, and disease tolerance or resistance. Such variations have been observed in many crops, including wheat, triticale, rice, oats, maize, sugar cane, alfalfa, tobacco, tomato, potato, oilseed rape and celery. The same types of variation obtained from somatic cells and protoplasts can also be obtained from gametic tissue.

One of the major potential benefits of somaclonal variation is the creation of additional genetic variability in co-adapted, agronomically useful cultivars, without the need to resort to hybridization. This method could be valuable if selection is possible *in vitro*, or if rapid plant-screening methods are available. Thus, tissue culture is proving to be a rich and novel source of variability with a great potential in crop improvement without resorting to mutation or hybridization. Such variant plants may show such useful characters as resistance to a particular disease, herbicide resistance, stress tolerance, etc. However, at present few cultivars of any agronomically important crop have been produced through the exploitation of somaclonal variation.

Plants with Somaclonal Variations Produced *In Vitro*

Monocotyledons

- *Allium sativa*
- *Avena sativa*
- *Hordeum* sp.
- *Lolium* hybrids
- *Oryza sativa*
- *Saccharum officinarum*
- *Triticum aestivum*
- *Zea mays*

Dicotyledons

- *Lactuca sativa*
- *Lycopersicon esculentum*
- *Medicago sativa*
- *Solanum tuberosum*

Somaclonal Variation in Agriculture

Somaclonal variation has been suggested as a useful source for breeding programmes in many crop plants.

Rice The somaclonal variation showed significant improvement in the somaclone progeny relative to the parent in terms of seed weight, seed protein percentage, tiller number and time of flowering. The variety tolerant to salinity and aluminium toxicity has also been produced through this process and these have been segregated for characters such as plant height, maturity, heading date and grain yield.

Wheat Heterozygous and homozygous mutants were screened in the primary regenerants of wheat. In wheat, variants have been analysed for traits such as height and alcohol dehydrogenase synthesis, as well as complex gene loci involved in the synthesis of grain amylases.

Potato The usefulness of somaclonal variation in potato improvement programmes was first demonstrated through the recovery of disease resistant plants in potato (resistance against late blight and early blight). The disease resistant character may be transmitted through subsequent tuber generations.

Sugarcane Resistant varieties of sugarcane against eye-spot disease, Fiji disease and downy mildew are somaclones obtained from tissue cultures.

Maize T-toxin resistance, male fertility and mtDNA, etc. were modified in maize by somaclonal variations. Cytoplasmic male sterility and sensitivity to toxin produced by pathogens on maize may be overcome through regeneration of plantlets from the selected cell lines resistant to toxin and infection. Toxin resistant male fertile plants may be regenerated from Cms-T maize culture grown without exposure to toxin. All this conversion is associated with mitochondria, which undergo genetic changes during cell division. In maize, a new fully functional electrophoretic variant at the alcohol dehydrogenase locus has been screened and subsequently characterized as resulting from a single nucleotide substitution.

Tomato The callus tissue of tomato derived from leaf culture was used for somaclonal variation. Variants were screened for a number of characters such as male sterility, jointless pedicel, fruit colour, indeterminate growth, leaf morphology, branching habit, etc.

Brassica sp. The occurrence of somaclonal variation in *Brassica sp.* has been independently reported several times. Variants were found which affected flowering time, growth habit, waxiness, glucosinolates, etc.

Pelargonium sp. A high degree of variability in culture-regenerated plants from 6 cultivars of *Pelargonium zonale* was observed. Changes were found in plant and organ size, leaf and flower morphology, essential oil constituents, etc. One of the variants has been released, as a new cultivar known as "Velvet Rose".

SYNTHETIC SEED TECHNOLOGY IN AGRICULTURE

A synthetic or artificial seed has been defined as a somatic embryo encapsulated inside a coating and is considered to be analogous to a zygotic seed. There are several different types of synthetic seeds like:

i. Somatic embryos encapsulated in a gel.
ii. Dried and coated somatic embryos.
iii. Dried and uncoated somatic embryos.
iv. Somatic embryos suspended in a fluid carrier.
v. Shoot buds encapsulated in a gel.

No large-scale system for producing such seeds has yet been developed, although pilot studies of moderate size, using somatic embryos encapsulated in a gel have been conducted in Japan with F_1 hybrids of celery and lettuce. Several applications for synthetic seeds have been reviewed by several workers. The use of synthetic seeds as an improvement on more traditional micropropagation protocols in vegetatively propagated crops may, in the long term, have a cost saving, as the labour intensive step of transferring plants from *in vitro* to soil/field conditions may be overcome.

Other applications include the maintenance of male sterile lines, the maintenance of parental lines for hybrid crop production, and the preservation and multiplication of elite genotypes of woody plants that have long juvenile developmental phases. However, before the widespread application of this technology, somaclonal variation will have to be minimized, large-scale production of high quality embryos must be perfected in the species of interest, and the protocols will have to be made cost-effective compared with existing seed or micropropagation technologies. The crops in which artificial seeds have been successfully produced for commercial use is given in Table 28.4.

Table 28.4 Crops in which artificial seeds are produced

Alfalfa	*Apium graveolens*	*Zea mays*
Lactuca sativa	*Brassica* sp.	*Velerina* sp.
Spathoglottis plicata	*Azadirachta*	*Cymbidium*
Daucus carota	*Dendrobium*	*Dioscorea floribunda*
Gossypium hirsutum	*Medicago sativa*	*Morus indica*
Picorrhiza kursoa	*Pogostemon patchouli*	*Phalaenopsis*
Rheum emodi	*Santalum album*	

PATHOGEN ERADICATION

In the agricultural world, the production or yield of a crop can fall dramatically as a result of viral infection and render that particular variety no longer saleable or commercially viable. Crop plants, especially vegetatively propagated varieties are generally infected with pathogens. Strawberry plants, for example, are susceptible to over 60 viruses and mycoplasmas and this often necessitates the yearly replacement of mother plants. In many cases, although the presence of viruses or other pathogens may not be obvious, yield or quality may be substantially reduced as a result of the infection. Many valuable contributions have been made by plant tissue culture to problems concerning plant pathology. One outstanding success is virus eradication by apical meristem culture and its application to problems of plant tumours, especially crown gall.

In China, for example, virus-free potatoes, produced by culture *in vitro*, gave higher yields than the normal field plants, with increases up to 150%. As only about 10% of viruses are transmitted through seeds, careful propagation from seed can eliminate most viruses from plant material. Fortunately, the distribution of viruses in a plant is not uniform and the apical meristems either have a very low incidence of virus or are virus-free. The excision and culture of apical meristems (the meristem with one to three of the subjacent leaf primordia), coupled with thermo or chemotherapy, have been successfully employed to produce virus-free and generally pathogen-free material for micropropagation.

Tissue culture techniques could be of value in restoring the original properties of the variety, by removing the infection, and so bringing it back into the commercial market. These virus-tested stocks could provide ideal material for the national and international distribution of plants, either for further propagation or for use as breeding material.

GERMPLASM PRESERVATION IN AGRICULTURE

The need for a programme for the conservation of plant genetic resources arises from the rapid changes occurring in modern agricultural practice. Primitive cultivars and wild relatives of crop plants constitute a pool of genetic diversity which is invaluable for future breeding programmes. These have already led to replacement by new cultivars, which encompass a much narrower range of genetic diversity. As a result, there is a very real danger of future breeding being impeded by the shrinking genetic bases of some crops. Hence, storage of this sort of irreplaceable breeding material or germplasm (gene combinations available for breeding) and establishment of a centralized gene bank are the practical ways to solve these problems.

One way of conserving germplasm, an alternative to seed banks and especially to field collections of clonally propagated crops, is *in vitro* storage under slow-growth conditions (at low temperature and/or with growth-retarding compounds in the medium) or cryopreservation or as desiccated synthetic seed. The plant materials can be stored in liquid nitrogen for the period desired. A list is given in Table 28.5 which indicates that cryopreservation of various agricultural crop tissues in liquid nitrogen at –196°C is possible. This technique could be particularly valuable for storage of any germplasm, which needs to be maintained in a clonal form. The technologies are all directed towards reducing or stopping growth and metabolic activity. Techniques have been developed for a wide range of plants. The most serious limitations are a lack of a common method suitable for all species and genotypes, the high costs and the possibility of somaclonal variation and non-intentional cell-type selection in the stored material.

Cryopreserved Plants

Most of the plant cells, tissues and organs, which are cryopreserved and recovered successfully using this process include the gymnosperms and the angiosperms.

Gymnosperms Specific types of gymnosperms which have been cryopreserved include species of the genera *Abies, Cupressus, Ginkgo, Juniperus, Picea, Pinus, Pseudotsuga, Sequoia, Taxus, Tsuga, Zamia,* etc. Some of the more useful species of *Taxus* include *T. baccata, T. brevifolia, T. canadensis, T. chinensis, T. cuspidata, T. floridana, T. globosa, T. media, T. nucifera and T. wallichiana* are also cryopreserved.

Angiosperms Angiosperms which could be preserved successfully through cryopreservation techniques include monocotyledon plant cells and dicotyledon plant cells.

- Monocotyledon plant cells include a variety of species of the genus *Avena* (oat), *Cocus* (coconut), *Dioscorea* (yam), *Hordeum* (barley), *Musa* (banana), *Oryza* (rice), *Saccharum* (sugar cane), *Sorghum* (sorghum), *Triticum* (wheat) and *Zea* (corn).
- Dicotyledon plants include species of the genus *Achyrocline, Atropa, Brassica* (mustard), *Berberis, Sophora, Lupinus, Capsicum, Catharanthus, Conospermum, Datura, Daucus* (carrot), *Digitalis, Echinacea, Eschscholzia, Glycine* (soybean), *Gossypium* (cotton), *Hyoscyamus, Lycopersicon* (tomato), *Malus* (apple), *Medicago* (alfalfa), *Nicotiana, Panax, Pisum* (pea), *Ruta, Solanum* (potato) and *Trichosanthes.*

Table 28.5 Cryopreservation of some plant cultures in liquid nitrogen at –196°C

Species	Culture system
Acer pseudoplatanus	Cell suspension
Atropa belladonna	Anther
	Pollen embryo
Capsicum annuum	Cell suspension
	Callus
Daucus carota	Cell suspension
	Callus
	Somatic embryo
Nicotiana tabacum	Callus
	Anther
	Shoot tip
	Pollen embryo
Oryza sativa	Anther
	Cell suspension
	Callus
Zea mays	Callus
	Cell suspension
Arachis hypogea	Shoot tip
Cicer arietinum	Shoot tip
Ipomoea batatas	Shoot tip
Malus domestica	Shoot tip
Pisum sativum	Shoot tip
Hordeum vulgare	Zygotic embryo
Lycopersicon esculentum	Zygotic embryo
Triticum aestivum	Zygotic embryo

PLANT BREEDING IN AGRICULTURE

Conventional breeding methods are the most widely used for crop improvement. But in certain situations, these methods have to be supplemented with plant tissue culture techniques, either to increase their efficiency or to achieve an objective not possible through conventional methods. Plant breeding can be conveniently separated into two activities: 1) Manipulating genetic variability and 2) Plant evaluation.

Historically, selection of plants was made by simply harvesting the seeds from those plants that performed best in the field. Controlled pollination of plants led to the realization that specific crosses could result in a new generation that performed better in the field than either of the parents or the progeny of subsequent generations, i.e., the expression of heterosis through hybrid vigour was observed. Because one of the two major activities in plant breeding is manipulating genetic variability, a key prerequisite to successful plant breeding is the availability of genetic diversity. It is in this area, creating genetic diversity and manipulating genetic variability, that biotechnology (including tissue culture techniques) is having its most significant impact.

In spite of the general lack of integration of most plant-biotechnology and plant-breeding programmes, field trials of transgenic plants have recently become much more common. There are therefore reasons to believe that we are on the verge of revolution, in terms of the types and genetic makeup of our crops that has been predicted for more than a decade.

More than 80 different plant species have already been genetically modified, either by vector-dependent (e.g., *Agrobacterium*) or vector-independent (e.g., biolistic, microinjection and liposome) methods. In almost all cases, some type of tissue culture technology has been used to recover the modified cells or tissues. In fact, tissue culture techniques have played a major role in the development of plant genetic engineering. Tissue culture will continue to play a key role in the genetic engineering process for the foreseeable future, especially in efficient gene transfer and transgenic plant recovery.

MUTAGENESIS AND SELECTION OF CELL LINES

Another important approach in crop improvement is the mutation of tissue culture cells to produce a mutant line from which plants can be raised. Production of a mutant line is highly desirable for plant breeding. Callus cells, produced either from a vegetative cell or reproductive tissues, can be subjected to a range of mutagenic chemicals, e.g., N-nitroso-N-methyl urea, or ionizing radiations, such as, gamma rays. The hope is that such treatment will effect permanent changes in the DNA pattern of some cells. Plants could be raised from the treated cultures and any mutant whole plants could be selected from the population either by physical differences or by metabolic/biochemical differences. Biochemical mutants could be selected for disease resistance, resistance to phytotoxin, improvement of nutritional quality, adaptation of plants to stress conditions, such as, saline soils, and to increase the biosynthesis of plant products used for medicinal or industrial purposes. A number of mutagens can be employed to obtain useful mutant cell lines in agricultural crops and these are listed in Table 28.6.

Table 28.6 Useful mutant cell lines obtained in agriculture via *in vitro* techniques

Selective agent	Species	Mutagen	Plant regeneration
Herbicide resistance			
Asulam	*Apium gravaeolens*	None	No
Bentazone	*Nicotiana tabacum*	g-ray	Yes
2,4-D	*Lotus corniculatus*	None	Yes
2,4-D; 2,4,5-T; 2,4-DBa	*Trifolium repens*	None	No
Paraquat	*N. tabacum*	X-ray	Yes
Phenmedifarm	*N. tabacum (haploid)*	γ-ray	Yes
Picloram	*N. tabacum*	None	Yes
Cold tolerance	*Daucus carota* *Nicotiana sylvestris*	EMS None	Yes Yes

Mutagenesis and selection of cell lines in cell suspension culture is another technique of plant improvement. In the field of mutagenesis and selection of cell lines *in vitro* and the exploitation of totipotency, biotechnologists are trying to improve plants, for example, by producing crop species that are more resistant to drought, disease, poor soil conditions, chemical pesticides and herbicides. Scientists can select for specific mutant types by exposing developing callus cells to certain poisonous substances. Those cells that survive have genes that provide resistance to the poison. The resulting plants and their offspring should also show resistance to this chemical.

REVIEW QUESTIONS

1. Write an essay on the role of plant tissue culture in crop improvement.
2. What is the role of micropropagation and embryo culture in agriculture? Give examples.
3. Define the role of plant tissue culture in plant breeding.
4. Give the name of crop plants where somaclonal variation and somatic hybridization techniques are utilized for their improvement.
5. How androgenesis and gynogenesis useful in crop improvement? Give examples.
6. Name the plants and their cultures, which can be cryopreserved.

7. How embryo culture is significant in agriculture?
8. Write the production of haploid in agricultural crops.
9. What is protoplast hybridization? Discuss in detail the application of protoplast hybridization in agriculture.
10. Discuss briefly about the somaclonal variation and its role in agriculture improvement.
11. Write an essay on role of cryopreservation in agriculture.

REFERENCES

Bajaj, Y.P.S. (1990). *In vitro* production of haploids and their use in cell genetics and plant breeding. In: *Biotechnology in Agriculture and Forestry*.Vol. 12. Bajaj, Y.P.S. (ed.). Springer Verlag, Berlin. pp. 3–44.

Bhojwani, S.S. and Razdan, M.K. (1983). Plant Tissue Culture: Theory and Practice. *Developments in Crop Science*.Vol. 5. Elsevier, Amsterdam.

Boxus, P. (1976). Rapid production of virus-free strawberry by *in vitro* culture. *Acta Hortic.* 66: 35–38.

Brown, D.C.W. and Thorpe, T.A. (1995). Crop improvement through tissue culture. *World J. Micr. and Biot.* 11: 409–415.

Cassells, A.C., Minas, G. and Long, R. (1980). Culture of *Pelargonium* hybrids from meristems and explants: Chimeral and beneficially-infected varieties. In: *Tissue Culture Methods for Plant Pathologists.* Ingram, D.S. and Helgeson, J.P. (eds.). Blackwell Scientific Pub., Oxford. pp.125–130.

Chowdhury, M.K.V. and Vasil, I.K. (1993). Molecular analysis of plants regenerated from embryogenic culture of hybrid sugarcane cultivar (*Saccharum* spp.). *Theor. Appl. Genet.* 86: 181–188.

Debergh, P.C. and Zimmerman, R.H. (1991). *Micropropagation*-1. Kluwer Academic Publishers, Dordrecht, The Netherlands.

Harper, G., Osuji, J.O., Heslop-Harrison, J.S. and Hull, R. (1999). Integration of banana streak badanavirus into the *Musa* genome: molecular and cytogenetic evidence. *Virology*. 225: 207–213.

Hendre, R.R., Mascarenhas, A.F., Pathak, M., Nadgir, A.L. and Jagannathan, V. (1975). Growth of virus free sugarcane plants from apical meristem. Indian Phytopathology. 28: 175–178.

Herman, E.B. (ed.). (1995). Regeneration and Micropropagation: Techniques, Systems and Media (1991–1995). *Recent Advances in Plant Tissue Culture*, Vol. 3. Agritech Consultants, Inc., Shrub Oak, UK.

Jain, S.M., Brar, D.S. and Ahloowalia, B.S. (eds.). (2002). *Molecular Techniques in Crop Improvement.* Kluwer Acad. Press, Dordrecht, The Netherlands.

Kodym, A. and Zapata-Arias, F.J. (2001). Low cost alternatives for the micropropagation of banana. *Plant Cell, Tissue and Organ Culture.* 66: 67–71.

Maluszynski, M. and Kasha, K.J. (2002). Mutations, in vitro and Molecular Techniques for Environmentally Sustainable Crop Improvement. Kluwer Acad. Pub., Dordrecht, The Netherlands.

Misawa, M. (1994). Plant tissue culture: an alternative for production of useful metabolite. *FAO Agricultural Services Bulletin.* No. 108. Food and Agriculture Organization of the United Nations.

Murashige, T. (1990). Plant propagation by tissue culture: A practice with unrealized potential. In: Ammirato, P.V., Evans, D.A, Sharp, W.R. and Bajaj, Y.P.S. (eds.). *Handbook of Plant Cell Culture,* Vol. 5: *Ornamental Plants.* McGraw-Hill, New York. pp. 3–9.

Ndowora, T., Dahal, G., Lafleur, D., Harper, G., Hull, R., Olszewski, N. E. and Lockhart, B. (1999). Evidence that badnavirus infection in *Musa* can originate from integrated pararetroviral sequences. *Virology.* 255: 214–220.

Nelke, M., Nowak, J., Wright, J.M. and McLean, N.L. (1993). DNA fingerprinting of red clover (*Trifolium pratense* L.) with Jeffrey's probes: detection of somaclonal variation and other applications. *Plant Cell Rep* 13: 72–78.

Raghavan, V. (1980). Embryo culture. *International Reviews of Cytology Supplement,* lIB. pp. 209–40.

Raghavan, V. (1994). *In vitro* methods for the control of fertilization and embryo development. In: *Plant Cell and Tissue Culture.* Vasil, I.K. and Thorpe, T.A. (eds.). Kluwer Academic Pub., Dordrecht. pp. 173–194.

Sanada, M., Sakamoto, Y., Hayashi, M., Mashiko. T., Okamoto, A. and Ohnishi, N. (1993). Celery and lettuce. In: *Synseeds: Applications of Synthetic Seeds to Crop Improvement.* Redenbaugh, K. (ed.). CRC Press, Boca Raton, FL. pp. 305–327.

Vasil, I.K. and Thorpe, T.A. (1994). *Plant Cell and Tissue Culture.* Kluwer Academic Pub., Dordrecht, The Netherlands.

You, C., Chen, Z. and Ding, Y. (1993). *Biotechnology in Agriculture.* Kluwer Academic Pub., Dordrecht, The Netherlands.

29

PLANT *IN VITRO* BIOTECHNOLOGY IN FORESTRY

INTRODUCTION

In the fact of a growing world population and dwindling non-renewable resources, the demand for wood and wood products is expected to rise over the next several decades. This is especially true for the use of cellulose by the plastics industry and for the production of fuel either by direct burning or by producing charcoal, alcohol, or methane.

The forests account for 29–34% of the land area on earth, on worldwide basis, and for 19.52% (62.20 million ha) of the land area in India, where only 10.88% (35.79 million ha) is covered by closed forests. While on the one hand, forests are a source of goods and services (including firewood, fodder, pulp, and fibre industries), they also maintain climatic stability and conserve water and soil.

Deforestation is the clearance of naturally occurring forests by logging and burning. Deforestation occurs for many reasons—trees or derived charcoal are used as, or sold, for fuel or as a commodity, while cleared land is used as pasture for livestock, plantations of commodities, and settlements. The removal of trees without sufficient reforestation has resulted in damage to habitat, biodiversity loss and aridity. It has adverse impacts on biosequestration of atmospheric carbon dioxide. Deforested regions typically incur significant adverse soil erosion and frequently degrade into wasteland.

Disregard or ignorance of intrinsic value, lack of ascribed value, laxity in forest management and deficient environmental law are again some of the factors that cause deforestation on a large scale. In many countries, deforestation is an ongoing issue that is causing extinction, changes in climatic conditions, desertification, and displacement of indigenous people.

Reforestation is the restocking of existing forests and woodlands, which have been depleted. Reforestation can be used to improve the quality of human life by removing pollution and dust from the air, rebuilding natural habitats and ecosystems, mitigating global warming since forests facilitate biosequestration of atmospheric carbon dioxide, and harvesting of resources, particularly timber. The term "reforestation" is similar to afforestation, the process of restoring and recreating areas of woodlands or forest that may have existed long ago but were deforested or otherwise removed at some point in the past.

The main objectives of forest plantations involve the production of timber and pulp wood and also to increase the forest areas worldwide. Moreover, forest plantations composed of monospecific type of plants are not generally important to preserve and develop nature's biodiversity for proper maintenance of forest ecosystem. However, forest ecosystem can be managed by ways and means to enhance their biodiversity protection functions and they are important providers of ecosystem services such as maintaining nutrient capital, protecting watersheds and soil structure as well as storing carbon. Furthermore, they may also play an important role in alleviating pressure on natural forests for timber and fuel-wood production. There is also considerable potential for increasing forest productivity by planting more and more trees with improved growth rates, form, wood quality, and insect and disease resistance. Thus, forest genetics and tree improvement programmes will further play increasing roles in the forests of the present and future.

VEGETATIVE PROPAGATION

Vegetative propagation offers several advantages for tree improvement. Selected tree genotypes can be moved to seed orchards, where they may be recombined in pedigreed offspring via controlled pollination. Mass propagation by cloning selected superior genotypes for planting stock captures both additive and non-additive gene effects; thus, a significantly higher genetic gain is anticipated than from traditional methods. At least a 10% increase can be expected from planting selected clonal propagules rather than selected half-sib families. A considerable time saving until a seed orchard is in full production can also be expected by using clonal rather than seed propagules for orchard establishment. High frequency cloning may also be useful for forest trees characterized by poor seed set, absence of uniform seed production, and seed prone to genetic damage or loss of viability during storage.

Rooting of cuttings has been the most commonly used method to propagate vegetatively selected material for reforestation purposes. For some genera of hard woods such as *Eucalyptus* and *Quercus* and for most conifers, however, root cuttings from mature trees is difficult, and grafting is usually inappropriate because of labour, cost, or incompatibility problems. For these species and others, cell, tissue or organ culture offers considerable potential for rapid, economical propagation of selected genotypes.

CONVENTIONAL FOREST BREEDING TECHNOLOGY

Conventional breeding technology for forest involves well-known quantitative genetic techniques such as progeny testing and efficient selection for utilizing existing genetic resources and also selection of superior families of clones within particular species. Traditional genetic improvement programmes have been utilized to produce very large genetic gains across the world.

There are a number of constraints in the tree improvement process. The major constraint in the improvement of forest tree is time. Generally, there are four major ways in which time influences the three phases of the tree improvement process: 1) evolutionary time 2) time of harvest 3) time of elapsed before age-stable performance is achieved and 4) time of reproductive maturity. As such evolutionary time cannot be directly influenced by technology; it can influence what technologies are appropriate. The remaining three time factors affect the rate at which progress can be achieved. The existence of long juvenile phases in fruit and forest trees has been one of limiting factors for their genetic improvement, preventing full domestication of economically important woody species.

Tree improvement is a process of managing genetic resources. Through this process, the inherent genetic variability in a particular population is managed by recurrent cycles of selection and breeding. Tree improvement involves three distinct, but related phases: 1) Conservation 2) Selection and breeding, and 3) Propagation. As we progress from one phase to the next, decreasing amount of variations are being managed. Therefore, at any one time, more variation is being conserved or selected than is being deployed through propagation. Conservation is directed towards both wild and managed germplasm resources. Breeding and selection are the central activities of a tree improvement programme and involve formation of mating designs, evaluation of resultant progeny and selection of desirable genotypes for propagation and further breeding for a next generation. Tree improvement programmes are well under way for a wide variety of commercially valuable forest tree species throughout the world.

Under the breeding programs of forest trees and clonal forestry more attention has been paid to some of the major activities where fast growing hard wood forest and forest products are much in demand. Companies world wide are using this to maximize productivity, quality and their profitability status. Today clonal forestry is one of the most exciting and controversial issues faced by the industry. There is no issue whether clonal forestry is efficiently working or not rather the issue is about a proper and responsible assessment. During the last few decades, the world has seen the enormus interest and research efforts in vegetative propagation of commercial timber species. Today forest breeding programmes are in the midst of a technological revolution. Therefore clonal forestry, is a component of the total system of genetic management of forestry, which represents the ultimate achievement in tree breeding.

Traditional forest tree improvement programmes have begun to pay dividends, especially in the pulp and paper industry in some countries. However, these programmes are long term and require large areas of land for testing of selected genotypes and for production of seedlings for reforestation.

Tissue culture techniques offer a new dimension to tree improvement. Although tissue culture technology will not supplant traditional tree breeding efforts, tree improvement researchers may someday be able to use tissue culture techniques to evaluate genotypes in experiments of growth rates, cold hardiness, disease resistance, and tolerance to drought or the chemicals such as soil salts or herbicides, which would have taken much longer and would have required many hectares to complete.

BIOTECHNOLOGY OF TREE IMPROVEMENT

It was predicted that demand for wood would double within the next 50 years. The largest demand should come from tree-poor countries like India. It was postulated that biotechnology offers the potential for revolutionary changes in forest productivity to meet that demand. Forestry would probably realize the greatest improvement of any crop from biotechnology. One of the most important things that biotechnology could do for us in forestry is cut the long periods required for tree improvement using conventional technologies.

The advent of biotechnology does not means that conventional breeding will be swept away. The conventional genetic improvement programmes will still be the methods of choice for improving quantitative and multigenetic traits. Long-term multigeneration breeding programme will still be necessary to obtain improvements in traits such as growth rate and overall environmental adaptation. Biotechnology is a collection of techniques that can be used for enhancing the impact of biological programmes. The major emphasis in forest biotechnology has been on systems of tissue and protoplast culture.

Selection and Clonal Propagation of Trees

It is expected that genetically improved forest trees will not be available for planting for some years even in the advanced countries. It will be necessary meanwhile to clone pre-existing, superior trees. An initial problem encountered involves the identification of those features of a superior adult phenotype that can be attributed to genotype and those that result from its environment. A possible route to cement this problem would be to use molecular probes to screen the genome but such diagnostic tools are not yet available. In the meantime, the most realistic approach will be to select superior phenotypes and assume that, at worst, a tree with a superior phenotype is a genotype that responded well to good management practices, while, at best, in the absence of human inference, the genotype is totally responsible for its enhanced field performance.

A further difficulty is encountered in collection of material from large and old trees for propagation. Since woody material obtained from adult trees is usually difficult or impossible to root, taking the cuttings is from the parent tree would be a wasteful procedure to produce nursery stock plants. A more effective method would be to graft buds onto suitable root-stocks as is traditionally practiced in the propagation of fruit trees and for range of ornamental trees. While clonal root-stock would enhance the uniformity of the propagated stock, such clones would not be available in early experiments with most species and grafting onto a mixed seedling population would be more realistic.

PLANT TISSUE CULTURE IN FORESTRY

Plant tissue culture technology is playing an increasingly important role in basic and applied studies, including tree improvement. The application of tissue culture technology, as a central tool or as an adjunct to other methods, including recombinant DNA techniques, is at the vanguard in plant modification and improvement for agriculture, horticulture and forestry.

Plant tissue culture techniques are of tremendous value to forest tree improvement. The technology is envisaged as playing a complementary role to traditional methods by exploiting spontaneous or induced genetic and epigenetic variability in culture, by use of haploidy and by the use of protoplasts. Haploids and protoplasts will aid in shortening breeding cycles and allow for unconventional crosses, respectively. Clonal propagation is an integral part of any tree improvement programme, and in addition can play an independent role in reforestation, clonal orchard establishment and in energy foresting. The goals, problems and limitations of these applications of tissue culture technology to forest tree improvement are indicated and assessed.

Whole plants have been regenerated from a number of explant sources, including anthers, cotyledons, embryo, hypocotyls, needles, shoot tips, stem pieces, and mega-gametophyte tissue. The capacity for mass propagation from tissue culture is great. For many herbaceous plants, a million-fold increase in production rate over conventional methods is not unrealistic. Analysis suggests that tissue culture will contribute significantly to the improvement of forests through exploitation of existing genotypes and production of new, commercially valuable genotypes. Such changes may significantly influence worldwide management decisions in forestry.

Micropropagation as a Method of Cloning

An alternative approach to cloning is provided by micropropagation. This process is based upon the principle that a bud taken from a given specimen and cultured on a nutrient medium containing growth regulators, usually at least one cytokinin, develops into a shoot and its axillary buds grow out to form branches. Shimazo-Sato and Mori have thoroughly explained its various aspects such as 1) development and potential of axillary meristems 2) biological functions of dormancy and apical dominance, and 3) physiological, genetic and molecular biological approaches that control growth and dormancy in axillary buds of plants. These side branches can be rooted or transferred to fresh multiplication medium in which axillary buds again grow out.

This process can be carried out indefinitely to produce a limitless number of shoots. Following the success with large number of herbaceous species, information is now slowly being gathered on the micropropagation of an increasing range of fruit and timber trees. For forest trees, maximum success has been reported in conifers. Based on published data, explants from mature trees are likely to be more difficult to establish *in vitro* owing to excessive polyphenol oxidation and rapid necrosis of the explants, but in some species, this problem has been overcome. Few laboratories have yet reported on the field performance of micropropagated trees.

Micropropagation as related to forest trees has been reviewed by several workers. Several Indian laboratories have developed protocols for micropropagation of woody/forest plants. The Department of Biotechnology, Govt. of India has supported R&D activities by establishing micropropagation facilities at various geographical regions of the country. Two micropropagation Technology Parks are established at Tata Energy Research Institute, New Delhi and National Chemical Laboratory, Pune. There is need to establish linkages between the institutions (which are developing micropropagation technologies) and user agencies. Plant micropropagation has an outstanding place in biotechnology industry. Department of Biotechnology has published the achievements of R&D activities in tissue culture and micropropagation of plants in India. Plant production through this technique can benefit from the utilization of mycorrhiza, the naturalistic association between plant root and fungi.

Reforestation through Micropropagation

Today, there is a widening gap between the supply and demand of forest-based materials, and the forests are being denuded leading to degeneration of environment in a variety of ways. It has also been shown that the degraded lands can be effectively used and restored by planting forests using trees of wide adaptability and high productivity. In this manner, high yield of biomass and other plant products may be obtained from the degraded lands. For this purpose, it is recommended that cloned propagation of superior genetic stocks rather than seeds of uncertain genetic quality should be used. Even on the degraded land, few trees grow well despite the adverse conditions, and these trees can be used for clonal micropropagation through the technique of tissue culture. It has been estimated that, a multiplication rate of 10,000 to 100,000 per year is technically possible for many species. Towards this objective, superior genotypes of many species, capable of growing well on degraded lands, have been selected for mass multiplication. Clonal multiplication involves the following four steps:

1. Initiation of aseptic culture.
2. Shoot multiplication using shoot apical meristem or buds.
3. Rooting of *in vitro* formed shoots.
4. Acclimatization and transfer of micropropagated plantlets in the field.

It is estimated that in the year 1990, 500 million plants of diverse nature were produced in 60 countries. In New Zealand, large number of micropropagated plantlets of pine (*Pinus radiata*) was used for reforestation, and there were as many as two million plantlets in the field, in 1992. During 1992–2002, increasing number of plants were produced through micropropagation and used for reforestation in different countries. In India also, with the support from Department of Biotechnology (DBT), a number of forest trees were used successfully for micropropagation under two Tissue Culture Pilot Plant (TCPP) facilities that were initiated, one at TERI, New Delhi and the other at NCL, Pune. Nearly 60 lakh plantlets were produced and field planted over an area of 6000 hectares. Some of the trees, which were used in these pilot projects, included the following: *Alnus* sp., bamboo, teak, poplar, *Prosopis, Capparis,* canes (*Calamus*), Neem and *Eucalyptus* sp.

Embryo Culture in Forestry

Embryo culture represents the earliest technique to obtain viable offsprings following interspecific and intergeneric hybridizations where routine fertilization failed to produce a well-defined and full-term embryo. Most extensive use of embryo culture technique has been made for making interspecific and intergeneric crosses within the tribe, *Triticeae* of the family, Poaceae.

The embryos of different developmental stages, formed within the female gametophyte through sexual process, can be isolated aseptically without any damage or dissected embryos from the bulk of maternal tissues of ovule, seed or capsule, are cultured *in vitro* under aseptic and controlled physical conditions in glass vials/tubes containing nutrient medium to grow directly into plantlets.

For the first time, successful embryo culture to obtain an interspecific cross between *Linum perenne* and *L. austriacum* was demonstrated by Laibach in 1929. Since then, several refinements have been made in embryo culture techniques which have been a popular approach for raising hybrids from a number of incompatible crosses. Embryo culture allows one to investigate the factors that influence the embryonic growth under controlled conditions. Zygotic embryo can be used in artificial culture to investigate the factors that influence the embryonic growth.

The young or immature embryos are removed from developing seeds and inoculated on a suitable medium to obtain plantlets. The young embryos in culture generally do not complete their natural course of development and germinate prematurely. Sometimes, embryos from mature seeds may also be used for embryo culture. So far, it has not been possible to culture embryos before a certain stage of development (e.g., before the globular stage, particularly in case of cereals and millets). However, in some plant species, even very young embryos of few cells have been successfully cultured. Young embryos need to be isolated with their intact suspensor as they provide gibberellins to the developing embryo.

Embryo culture is generally applied for obtaining rare hybrids. Some *Pinus* sp. such as *P. lambertiana* is very susceptible to blister rust disease whereas *P. armandi* and *P. koraiensis* are resistant to such disease. An attempt has been made to cross *P. lambertiana* × *P. armandi* and *P. lambertiana* × *P. koraiensis* to bring the disease-resistant character in the hybrid. In both cases, fertilization occurs, but seeds are not germinable.

In such cases, by successful rearing of excised premature hybrid embryos through tissue culture, it is possible to grow a hybrid disease resistant *Pinus* plant.

Biotechnology of Somatic Embryogenesis

Since 1985, researchers have made a great deal of progress in the development of somatic embryogenesis for conifer reforestation. It has been predicted that the biotechnology of somatic embryogenesis will be the technology of 21st century in forestry. Bioprocessing for tree production in forest industry was reviewed by Timmis of Weyerhaeuser Company, USA. In India, research on induction of somatic embryogenesis has been done by several laboratories including, Bhabha Atomic Research Centre (BARC), Mumbai; National Botanical Research Institute (NBRI), Lucknow; Banaras Hindu University (BHU), Varanasi and Biotechnology Centre of J.N.V. University, Jodhpur.

Successful protocols for axillary budding (particularly for hardwoods) and adventitious budding (particularly for conifers) exist for a large number of forest tree species, and the number of species for which somatic embryogenesis has been reported is increasing, particularly for conifers. Compared to cuttings, the higher multiplication rates available through micropropagation offer advantages with respect to the capture of genetic gains through clonal forestry. One major factor impeding early application in many industrial plantation programmes is that breeding is not sufficiently advanced for clonal forestry to be contemplated.

Protoplast Fusion Technology in Forestry

This is a field with a long history of research. There have been several successes, notably with species of the Brassicaceae and Solanaceae. Severe limitations are imposed, however, by the taxonomic relationship of parents, and the requirement for regeneration from protoplasts. Expectations of success with a new pair of species is expected to be low. This is a field with little apparent applicability to forest tree species, particularly those (e.g., some tropical hardwoods and non-industrials) for which investigations of the possibilities for traditional hybridization remain minimal. In the long term, the objectives of protoplast fusion may be better served by manipulations at the DNA level.

Anther Culture of Forest Trees

Although anther culture has been used for the rapid production of homozygous lines in the breeding of some self-pollinated cereals and vegetable crops, few reports exist of regeneration from gametophyte cultures of forest tree species. Research with this objective would be likely to be long term with a relatively high risk of an unsuccessful result. Induction of haploid plants has no immediate application in forest tree improvement programmes. Such plants may be of some use in basic genetic studies, e.g., for studies of heterosis in forest tree species. As a long-term strategic research objective for industrial species, induction of haploid plants should be of low priority until such time as methods for early selection and the promotion of early flowering are available. The breeding of non-industrial species is unlikely to warrant this level of sophistication.

Somaclonal Variation: A Tool of Forest Improvement

Variation induced during cell or callus cultures has been reported for many species. For some trees, variants have been produced showing economically useful levels of traits such as resistance to disease and increased levels of salt. The phenomenon is still not clearly understood, and persistence of modifications through subsequent sexual generations has been demonstrated in some cases but not in others. This is not a research area where favourable results could be predicted with any confidence, and use of the approach is furthermore dependent on being able to regenerate whole plants from the cells or callus. Research with new species will be more soundly based when the phenomena are better understood in the model species already under study.

Assuming stability of the characteristics, the approach is of most value where selection can be done at the cell level (e.g., by exposure to a phytotoxin or to high mineral levels), thus enabling screening of large numbers of genotypes, and where the level of the trait sought lies outside the range available naturally in the species. Cold tolerance in the *Eucalyptus* sp. is an example which may meet these criteria. No immediate applicability is evident to the little known tropical hardwoods or to non-industrial species, for which genetic variation naturally available is generally poorly defined.

TREES OF ARID AND SEMI-ARID REGIONS

Several tree species contribute to production of biomass in arid and semi-arid regions in India. These include *Anogeissus* sp., *Acacia* sp., *Azadirachta indica*, *Aegle marmelos*, *Capparis* sp., *Boswellia serrata*, *Calligonum polygonoides*, *Balanites aegyptiaca*, *Dalbergia* sp., *Maytenus emarginata*, *Prosopis* sp., *Slavadora* sp., *Sterculia urens*, *Tecomella undulate*, *Terminalia* sp., *Tecomella undulate*, *Zygypus* sp., *Wrightia* sp., etc. Considerable work has been done on tissue culture of some of the desert tree species. However, there is meagre knowledge about several important aspects of biology of even useful trees. It is suggested that research on tree genomics should be initiated in the country.

There is a need for taking up projects on expressed sequence tags (ESTs) and microsatellites as molecular markers for *Anogeissus* sp., *Capparis* sp., *Prosopis cineraria*, *Tecomella undulate* and *Terminalia* sp. These species bear important traits and genes that can be characterized and identified for future prospecting.

CRYOPRESERVATION AND *IN VITRO* STORAGE OF FOREST TREES

Although increasingly used process for the storage of germplasm threatened of agricultural species, *in vitro* storage and cryopreservation have little to offer for this purpose in forest tree species. Gene pools of most of the established industrial species are reasonably well conserved *in situ* and *ex situ* stands and preserved in seed stores. Undoubtedly, an urgent gene conservation problem exists for many tree species, particularly among the tropical hardwoods and non-industrial species. Distributions of these species are poorly known, as also their biological characteristics.

Although seeds of some species are recalcitrant, many are orthodox. Major impediments to the conservation of forest tree germplasm are the availability of resources sufficient for only a very small fraction of the survey and collection work which would be required before any germplasm could be stored, and the unreliability of many existing seed storage facilities. Replacement of existing facilities with more sophisticated technology is not likely to make a positive impact on the gene conservation problem with tropical tree species. Even for the recalcitrant species, activities should favour the establishment of *ex situ* plantings which will facilitate evaluation of the material. In the long term, cryopreservation and *in vitro* storage may have some application as a back-up conservation strategy for populations of well surveyed species which are recalcitrant. The poplars and walnut currently fall into this category.

Cryopreservation warrants much more attention as a means of maintaining juvenility and capturing genetic gains offered by clonal forestry with industrial species. The technology is thus applicable mainly to programmes where good breeding programmes are in place, clonal forestry is a realistic goal, and "rejuvenation" is difficult, in particular, for the conifers.

With the increasing emphasis on international breeding of both industrial and non-industrial forest tree species in tropical countries, international movement of material will be important. In the short term, such movement is likely to be mainly in the form of seed for species and provenance testing. Movement of vegetative material, however, is likely to become more important as breeding programmes advance, and *in vitro* approaches could find application in the long term.

GENETIC TRANSFORMATION IN FOREST TREES

It was the last decade of 20th century that research had been conducted on the characterization of DNA from trees and isolation and characterization of specific genes. Serious experiments in genetic engineering with tree species were attempted in the developed countries. Genetic transformation has been reported in a number of tree species including hybrid poplar, loblolly pine, aspen, *Citrus* sp., apple, etc. Rugh *et al.* developed transgenic yellow poplar for mercury phytoremediation. Phytoremediation of trichloroethylene with genetically engineered poplar has also been developed. Transgenic aspen trees are developed in which expression of lignin biosynthesis pathway gene *PtCL* 1, encoding coumarate: coenzyme A ligase (4CL) has been downregulated by anti-sense inhibition. The transgenic *Citrus* produced fertile flowers and fruits as early as the first year.

It is important that genetically engineered genotypes be of high quality with respect to other traits as well. The clonal test is the most logical basis for integration of genetic engineering into traditional tree improvement programmes. For these reasons, genetic engineering is most appropriately conducted with species where breeding programmes are advanced and clonal forestry can be realistically contemplated. Research on this subject should not assume a high priority with species for which natural variation available within the taxon remains poorly investigated.

GENETIC MAPPING OF WOODY PLANTS

There have been major changes during the last 20 years in plant biotechnology. The genome sequencing of *Arabidopsis* has been completed. This is rightly heralded as a landmark event, as plant biologists enter a new era in which genomics promises to address fundamental questions in botany. In the post-genomic era, rice and *Arabidopsis* are often presented as examples of a typical monocot and a typical dicot, respectively. It is important to realize that the intensive studies of these model systems also have their limitations. Developmental regulation of trees, in general and leguminous and gymnospermous trees, in particular, offers different areas of research.

Genomic approaches and tools of molecular biology are revolutionizing the study of plants to discover the mechanisms, development and control of developmental processes. Among the tree legumes, the black locust (*Robinia pseudoacacia*) has been investigated by the scientists of developed countries. These genomic studies may be evaluated but in order to have better perspectives, efforts should be made towards genomics or trees, which have wide ranges of environmental adaptations and have more economic and ecological values.

Molecular Markers for Forest Trees

Techniques for the analysis of isozymes, RFLPs and RAPDs are now well established, and the adaptation of procedures to suit new species could be expected to be a relatively short-term undertaking. Markers have important immediate applications in supportive research for advanced breeding programmes with industrial species, mainly in relation to quality control, e.g., checking of clonal identification, orchard contamination and within-orchard mating patterns by "fingerprinting". Isozymes will be satisfactory for many, but not all of these purposes.

Markers also have important immediate application in supportive research for tropical hardwoods and non-industrial species, in particular for essential taxonomic studies and investigations of mating systems. Both isozymes and DNA markers will be useful for these purposes. Markers will also be useful for the quantification of genetic variation to aid in sampling strategies for gene conservation and breeding population collections, although, due to modest correlations with patterns of variation for adaptive traits, they must be used conservatively as complements to traditional methods.

Realistically, application of marker-assisted selection in the short- or medium-term is likely to be very limited. Cheaper markers would be required and, even when these were available, the technology would apply mainly to advanced and sophisticated breeding programmes—those in which creation and maintenance of the appropriate population structures could be afforded, and where clonal forestry is achievable. A small number of programmes fall into this category and, for these, some effort aimed at developing marker-assisted selection can be justified. For most species, current resources would be far better directed towards moving breeding programmes to this stage of advancement, rather than to the development of marker-assisted selection. Marker-assisted selection is unlikely to have much application to non-industrial species, although, by virtue of a very short generation time, taxa such as *Gliricidia* may be useful model species for experimentation.

The major current value of markers lies in long-term strategic research, in the great contributions which marker studies are making to advances in the understanding of basic genetic mechanisms and genome organization at the molecular level. For forest tree species, the study of quantitative traits will be an important emphasis of this work in the coming years. This work will be most efficiently concentrated on a few model species, such as *Pinus taeda* and in particular hybrids such as *P. elliottii* × *P. caribaea*.

REVIEW QUESTIONS

1. Write an essay on role of plant tissue culture in forestry.
2. How micropropagation technology is applicable for forestry improvement?
3. What is somatic embryogenesis? Write its significance in forestry.
4. Discuss about the role of somaclonal variation in forestry.
5. Write notes on
   i. Molecular markers in forestry
   ii. Vegetative propagation
   iii. Conventional forest breeding
   iv. Genetic mapping of forest trees.

REFERENCES

Debergh, P.C. and Zimmerman, R.H. (1990). *Micropropagation Technology and Application.* Kluwer Academic Publications, Dordrecht.

Evans, D.A., Sharp, W.R and Ammirato, P.V. (1986). *Handbook of Plant Cell Culture.* Vol. 4. *Techniques and Applications.* Macmillan Publishing Company, New York.

Hammond, J., McGarvey, P and Yusibov, V. (2000). *Plant Biotechnology.* Springer-Verlag, Berlin.

Harry, I. S. and Thorpe, T. A. (1994) *In vitro* methods for forest trees. In: *Plant Cell and Tissue Culture* Vasil, I. K. and Thorpe, T. A. (eds.). Kluwer Academic Publishers, Dordrecht. pp. 539–560.

Hayward, M.D., Bosemark, N.O. and Romagosa, I. (1993). *Plant Breeding.* Chapman and Hall, New York.

Jain, S.M. and Ishii, K. (2003). *Micropropagation of woody plants and fruits.* Kluwer Academic Publications, Dordrecht.

Kodym, A., Hollenthoner S and Zapata-Arias F.J. (2001). Cost reduction in the micropropagation of banana by using tubular skylights as source for natural lighting. *In vitro Cell. Dev. Biol. Plant.* 37: 237–242.

Misawa, M. (1994). Plant tissue culture: an alternative for production of useful metabolite. *FAO Agricultural Services Bulletin.* No. 108. *Food and Agriculture Organization of the United Nations.*

Thorpe, T.A., Harry, I.S. and Kumar P.P. (1991). Application of micropropagation to forestry. In: *Micropropagation.* Debergh, P.C. and Zimmerman, R.H. (eds.). Kluwer Academic Publishers. Dordrecht. pp. 311–336.

Zimmerman, R.H. (1986). Regeneration in woody ornamentals and fruit trees. In: *Cell Culture and Somatic Cell Genetics of Plants.* Vol. 3. Vasil, I.K. (ed.). Academic Publishers, New York. pp. 243–258.

30

PLANT *IN VITRO* BIOTECHNOLOGY IN INDUSTRY

INTRODUCTION

Human beings depend on plants for many compounds in addition to food, such as medicines, pigments, vitamins, hormones, flavouring agents, latex and tannins. If most plant somatic cells are totipotent, it should be possible to take a culture of cells from a plant that naturally produces a certain biochemical and cause the culture to produce that chemical under *in vitro* conditions. The main difficulty is that we do not yet understand the regulation mechanisms that control the production of most biochemical substances and so cannot manipulate them. Even so, a surprising number of cell cultures have been found that do produce specialized biochemicals found in the intact parent, including alkaloids such as nicotine, atropine, ephedrine, caffeine, and codeine, and their precursors and derivatives. Production of cardiac glycosides and other steroids, benzoquinones, latex, phenolics, anthocyanins, organic acids, antitumour agents, antimicrobials and various flavours and odours has also been reported.

The first patent for producing biochemicals commercially by large-scale plant cell culture was issued in 1956. The general approach since then has been to select high product-yielding cell lines, preferably as suspension culture, and to enhance their efficiency, either by feeding them inexpensive product precursors or by manipulating their biosynthetic control mechanisms. Here, the expertise of the microbial product industry is invaluable. Large-volume automated culture vessels, called fermentors, have been used successfully to mass produce cultured plant cells. This technology will soon be producing selected pharmaceuticals and other high-cost biochemicals commercially.

The first success was demonstrated by the Mitsui Petrochemical Industries (Tokyo) in their ability to produce shikonin commercially in 750-litre fermentors from *Lithospermum* plant cells. Shikonin is a dye and an antibacterial product, valued at approximately \$4000/kg. The research and development effort required is well worth the investment to achieve the *in vitro* production of not only specialty biochemicals, but potentially, food, spices and industrial commodities. Such efforts are needed not only for the production and biotransformation of biochemicals but also for the study of photoautotrophic aseptic/non-aseptic cultures, florogenesis, de-repression, that is, the activation of dormant genes.

PHARMACEUTICAL SUBSTANCES OF PLANT ORIGIN

The vast bulk of early medicinal substances were plant-derived. An estimated 3 billion people worldwide continue to use traditional plant medicines as their primary form of health care. At least 40% of all prescription drugs sold in India contain active substances which were originally isolated from plants (or are modified forms of chemicals originally isolated from plants). Many of these were discovered by a targeted knowledge-based ethnobotanical approach. Researchers simply recorded plant-based medical cures for specific diseases and then analysed the plants for their active ingredients. Plants produce a wide array of bioactive molecules via secondary metabolic pathways. Most of these probably evolved as chemical defense agents against infections or predators.

While some medicinal substances continue to be directly extracted from plant material, in many instances plant-derived drugs can now be manufactured, at least in part, by direct chemical synthesis. In addition, chemical modifications of many of these plant "lead" drugs have yielded a range of additional therapeutic substances. The bulk of plant-derived medicines can be categorized into a number of chemical families, including alkaloids, flavonoids, terpenes and terpenoids, steroids (e.g., cardiac glycosides), as well as coumarins, quinines, salicylates and xanthines. The details of these chemicals and their production at industry level by the use of bioreactors are discussed in this chapter.

Large-scale Culture of Ginseng Cells

Ginseng roots have been widely used as a tonic and precious medicine since ancient times, particularly in oriental countries including Korea and China. It has been known that the root contains various saponins and sapogenins. Among them, ginsenoside-Rb has a sedative activity, while Rg has a stimulative activity.

Although there are several species of ginseng, the commercially important species is *P. ginseng*. The cultivation of ginseng in the field requires four to seven years, and it is impossible to plant consecutively for 20 to 50 years but the demand for the plant has increased dramatically in the world, and its price has soared. These are reasons why many researchers have tried to produce ginseng cells through plant tissue cell cultures.

The Japanese company, Nitto Denko Corporation, constructed a 20 litre fermentor to scale-up cultivation of ginseng cells and optimized various environmental conditions using 30 litre jar fermentors for the cell line having partly differentiated tissues. Glucose in the medium promoted cell growth in the initial stages of the fermentation and sucrose fed during the growth cycle stimulated the production of

Table 30.1 Continued

Drug	Chemical type	Indication	Plant producer
Quinine	Alkaloid	Malaria	*Cinchona pubescens*
Reserpine	Alkaloid	Antihypertensive (reduces BP)	*Rauvolfia serpentina*
Scopalamine	Alkaloid	Motion sickness	*Datura stramonium*
Taxol	Terpenoid	Ovarian, breast cancer	*Taxus brevifolia*
Theophylline	Xanthine	Anti-asthmatic, diuretic	*Camellia sinensis*
Vinblastine	Alkaloid	Hodgkin's disease	*Catharanthus roseus*
Vincristine	Alkaloid	Leukaemia	*Catharanthus roseus*

Tropane Alkaloids

Scopolamine and hyoscyamine alkaloids occur in leaves of (Solanaceae) plants including *D. myoporoides* and *D. leichhardtii*. *Scopolia, Atropa, Hyoscyamus* and *Datura* also contain tropane alkaloids. The concentrations of scopolamine and hyoscyamine in cultured cells are generally very low in spite of many efforts to increase the yield using various approaches. Therefore, the plant cell culture has not yet been employed to manufacture these alkaloids. Tabata *et al.* added tropic acid into *Scopolia japonica* suspension cultures as a precursor and increased the level of alkaloids up to 15 times. Mitsuno *et al.* selected a high tropane alkaloid producing strain of *Hyoscyamus niger,* which produced about 7 times more hyoscyamine, than that of the parent strain. According to their results, there was no direct correlation between high-producing ability and variation of chromosomal numbers.

Morphinan Alkaloids

Codeine is an analgesic and cough-suppressing drug and *Papaver somniferum* is a traditional commercial source of codeine and morphine which can be converted to codeine. The efficient production of thebaine and codeine using cell culture systems by *de novo* synthesis was not successful; therefore, studies were conducted for the biotransformation of codeinone to codeine using immobilized cells of *P. somniferum*. The conversion yield was 70.4% and about 88% of codeine converted was excreted into the medium. The level of alkaloid increased 26 times in the presence of the elicitor (mycelium of *Botrytis* sp.), 29% of the dry cell weight, compared to the medium without it.

Vinca Alkaloids Production

Vinblastine and vincristine are currently produced commercially by extraction from *Catharanthus roseus* (Apocynaceae) plants, but the process is not efficient because of very low concentrations of the alkaloids in the plant. It was reported that the concentration of both vinblastine and vincristine was only 0.0005% as a dry weight basis.

In order to produce these useful anticancer drugs much more efficiently, many scientists have tried to apply plant tissue culture technology. Misawa and his colleagues established an economically feasible process consisting of production of catharanthine by plant cell fermentation and a simple chemical or an enzymatic coupling. The vinblastine molecule is derived from two monomeric alkaloids, catharanthine and vindoline. The concentration of vindoline in the intact *C. roseus* plant is approximately 0.2% as a dry weight basis, which is much a higher level than catharanthine, and moreover vindoline is less expensive compared to catharanthine and vinblastine.

The MS medium was the most favourable for catharanthine production but the optimal levels of phytohormones for the growth and the production were varied in different cell lines. For example, one line required no phytohormones but another line required 0.1 mg/l NAA and 0.1 mg/l kinetin. Addition of various chemically defined compounds to the medium as "inducers" was found to stimulate the production efficiently. Among them, the effects of vanadyl sulphate, abscisic acid and NaCl on the production of catharanthine were significant. When abscisic acid was added to the culture as an elicitor on the 7th day of cultivation, the final titre of catharanthine was raised to 85 mg/l in a 30-litre fermentor.

In the second stage, the researchers tried to couple enzymatically or chemically catharanthine produced by the cell culture process with commercially available vindoline. As an enzyme source for the coupling, a crude preparation obtained by 70% ammonium sulphate precipitation from the cultured cells of *C. roseus* was used. The reaction mixture containing both the monomeric alkaloids, Tris buffer (pH 7.0) and the enzyme preparation was incubated at 30°C for 3 hours. It was determined that the enzyme reaction gave various dimeric alkaloids including vinamidine, 3-(R)-hydroxyvinamidine and 3'4'-anhydrovinblastine. Although neither vinblastine nor vincristine was detected in the mixture, it was recognized that a substantial amount of anhydrovinblastine was formed as a major coupling product when an excess amount of sodium borohydride was added to the mixture after incubation.

Formation of vinblastine from vincristine is detected using a crude enzyme preparation obtained from suspension cultured cells of *C. roseus*. The highest yield of conversion obtained was 13% from 0.13 mg anhydrovinblastine in 1 ml of the reaction mixture after 3 hours of incubation at 30°C, at pH 7.0. Later it was found that ferric ion catalysed the coupling reaction significantly in the absence of the enzyme. It is of interest that the products of the chemical coupling were not only anhydrovinblastine but also vinblastine. The yields of both alkaloids were 52.8% and 12.3%, respectively after 3 hours of incubation at 30°C, at pH 7.0. Circular dichroism confirmed that α-coupling exists between the 2 monomeric units of both vinblastine and vincristine produced either enzymatically or chemically.

The chemical-coupling reaction is a novel and an efficient process to produce vinblastine, and is applied commercially. Mitsui Petrochemical increased the yield of catharanthine up to 150 mg/l in the

MS medium supplemented with 1 mg/l NAA and 0.1 mg/l kinetin using the best producing cell line. The stimulating activity of NaCl and KCl on alkaloid production was also confirmed. Furthermore, the scientists of Mitsui employed high-cell density cultures and reported yields of catharanthine up to 230 mg/l/week. The addition of ferric chloride, oxalate, maleate and sodium borohydrate stimulated the yield of vinblastine from anhydrovinblastine up to 50%.

Taxol

Taxol is obtained by extraction from the bark of wild-grown *T. brevifolia* trees. The demand for taxol is undoubtedly increasing since it is applied for cancer treatment, but its supply is limited. Other related plants such as *T. canadensis* and *T. cuspidata* also contain taxol and other related compounds. Generally, the concentration of taxol in *Taxus* plants is very low. Therefore, harvesting *Taxus* trees for production of taxol commercially causes a serious problem from the environmental point of view.

Plant tissue and cell cultures as well as chemical transformation processes are the alternative ways to extract baccatin from needles of the plant. A cell line of *T. brevifolia* in suspension produced 3.9 mg/l taxol in the medium after 26 days of culture. The level of taxol was determined by reverse-phase HPLC. It is of interest that all taxol produced was secreted into the medium, which was very unusual for plant cell cultures. Researchers also established callus cultures and suspension cultures from *T. brevifolia* cv *Repandens, T. cuspidata, T. media*. Though some cell lines grew fast and their doubling times were 9–14 days, the levels of taxol were too low for commercial production. Induction of the hairy roots of *Taxus* plants has been tried.

Japanese researchers described that they detected and identified taxol in the callus of *T. cuspidata* and its level was 0.02 + 0.005% in dry weight after 2 months in culture. Suspension cultures of *T. cuspidata* were also established from the callus cultures and subsequently immobilized onto glass fibre mats. The cells were maintained as immobilized cultures for 6 months. The level of taxol in the immobilized cells was 0.012 + 0.007% of the extracted dry weight.

Taxol is one of the most suitable and desirable plant products to plant cell culture research because the shortage of its supply and its high value. Interestingly, a fungus, *Taxomyces andreanae* (isolated growing on *T. brevifolia*) also produces taxol, although at very low levels. Genetic manipulation of this fungus may, however, yet yield mutants capable of synthesizing taxol in quantities rendering production by this means economically viable.

Cardenolides

The major cardiac glycosides that have found medical use have been isolated from species of *Digitalis*. "Digitalis" in pharmaceutical circles mean "a crude extract of dried *Digitalis* leaves. This contains two glycoside components digoxin and digitoxin, which increase heart muscle contraction. For commercial production, *Digitalis* plants are being cultivated in the fields of several countries. The studies of plant tissue and cell cultures for production of cardiac glycosides were begun more than 30 years ago and the report presented by Hildebrandt and Riker in 1959 was probably the first one in this field. Furthermore, Staba investigated the nutritional requirements of tissue cultures of *Digitalis lanata* and *D. purpurea* in 1962.

The production of cardiac glycosides in *Digitalis* tissue cultures, was generally very low, and during the successive transfers of the cultured cells, the amount of cardenolides often decreased and disappeared completely. But, many researchers indicated that morphological differentiation caused an increase in productivity. For example, organ cultures of *D. lanata* leaves and roots produced cardenolides and during cultivation, the level of digoxin in the tissues rose with increasing age. The renewed organ differentiation from callus tissues of *D. purpurea* led to a new formation of cardenolides.

In contrast to the *de novo* synthesis, the biotransformation process with *Digitalis* plant cells seems to be more promising from a commercial point of view. It was found that *D. lanata* and *D. purpurea* callus cultures rapidly transformed progesterone to pregnane. Leaf and root cultures of *D. lanata* and shoot forming callus tissues of *D. purpurea* accumulated an increased level of digoxin and/or digitoxin when progesterone was added to the cultures. Stohs and Staba studies on the biotransformation of cardenolides by *Digitalis* cells and recognized that glycosylation reaction occurred.

Among many studies, the biotransformation from digitoxin to digoxin using *Digitalis lanata* cells is the most interesting approach in terms of commercial application since digoxin has a higher demand as a drug for heart diseases than digitoxin. It is advantageous that *Digitalis* leaves contain a larger amount of digitoxin which can be used as a substrate. It is a hydroxylation reaction at the 12 β-position of digitoxin, and β-methyldigitoxin was the most suitable substrate in this biotransformation as methyldigoxin is the major product. To reduce the production cost, the use of immobilized *D. lanata* cells and semi-continuous culture could be an alternative approach.

ORGAN CULTURES AND PHARMACEUTICALS

Since production of secondary metabolites is generally higher in differentiated tissues, attempts are made to cultivate shoot cultures and root cultures for the production of medicinally important compounds since these organ cultures are relatively more stable. There are a number of medicinal plants whose shoot cultures have been studied for metabolites (Table 30.2). Similarly, root cultures are valuable sources of medicinal compounds; root systems of higher plants generally exhibit slower growth and are difficult to harvest.

According to Wienmann, a close correlation existed between the expression of secondary metabolism and morphological and cytological differentiation, but he concluded that it is not yet clear to what extent secondary metabolism depends on the development of specific structures; it is unknown whether these two processes are genetically and/or physiologically linked.

Table 30.2 Shoot cultures of medicinal plants

Plant species	Product
A. annua	Artemesinin
A. belladonna	Atropine
Begonia sp.	–
Catharanthus roseus	Vindoline
Cinchona spp.	Vinblastine
D. purpurea	Cardenolides
Pelargonium tomentosum	Essential oils
Withania somniferum	Withanolides

The development of a certain level of differentiation is considered to be important in the successful production of phytochemicals by cell cultures. There are many examples demonstrating a relationship between differentiation and secondary metabolite accumulation. Hiraoka and Tabata successively transferred the callus of *Datura meteloides* from a medium with auxins to one without auxin and then cultivated continuously. As seen in Table 30.3 shoots, stems and roots were differentiated by turns and tropane alkaloid levels increased.

Table 30.3 Relationship between productivity of Alkaloids and differentiations

Plants	Alkaloid concentration (% dry wt.)
Callus	1×10^{-2}
Shoots-forming callus	1.5×10^{-2}
Growing shoots	1×10^{-2}
Root-forming shoots	3×10^{-2}
Leaf of young plant	1×10^{-1}
Leaf of matured plant	1×10^{-1}

Source Hiraoka, N. and Tabata, M. Phytochem. 13 1671 (1974).

There is a correlation between the stages of morphological differentiation and producing ability of the alkaloids using *P. somniferum*. A green callus, which differentiates epidermis or vascular bundles produced the alkaloids. A limited degree of tissue differentiation occurred and the cell contained codeine as a main alkaloid while the level of morphine increased as differentiation progressed.

The axenic callus and shoot cultures of *Pyrethrum cinerarifolium* had an ability to produce pyrethrin. It was also concluded that differentiated culture tended to produce more pyrethrins than callus cultures. Using an established shoot culture derived from the disc floret of the plant, 341.8 mg of pyrethrins per 100 gram wt. were obtained.

An interesting relationship was shown between differentiation and anthocyanin production using *Daucus carota* cells. The cells were fractionated by Ficoll density gradient centrifugation. In the density fraction (>14% of Ficoll), somatic embryos were formed in a medium containing 10^{-7} M zeatin but anthocyanin was scarcely produced. On the other hand, the cells in the lower density fraction (>12% of Ficoll) synthesized anthocyanin in the same medium but formed few embryos. About 40 to 50% of the total cells in the higher cell fraction synthesized anthocyanin to the maximum.

The use of organs as opposed to cells or cell aggregates might necessitate an adaptation of cultured scale-up technologies, but in general, these are not considered to be insurmountable. Indeed, root cultures of *Panax ginseng* have successfully been grown up to a volume of 20 litres and high levels of ginsenosides were obtained. A semi-continuous production system using differentiated cultures has been used, in which a series of vats with a 10% aliquot being transferred to the next while still in the dividing stage, with the remainder being left to grow and accumulate product.

HAIRY ROOT CULTURES AND PHARMACEUTICALS

The ability of *Agrobacterium rhizogenes* to induce hairy roots in a range of host plants has lead to studies on it, as a source of root-derived pharmaceuticals. Hairy roots are induced by transfer of T-DNA from the plasmid of *A. rhizogenes* to host tissue, resulting in root formation. These roots can be cultured in hormone-free medium and there are several examples of enhanced accumulation of secondary products, relative to non-transformed tissue. A number of examples are given in Table 30.4.

Table 30.4 Hairy root cultures producing pharmaceutical products of interest

Plant species	Product
Bidens sp.	Polyacetylenes
Datura sp.	Tropane
Cassia sp.	Anthroquinones
Glycyrrhiza uralensis	Glycyrrhizin
Panax ginseng	Saponin

Hairy root clones of *Hyoscyamus* plants grew faster than ordinary root cultures and produced the similar level of tropane alkaloids to that accumulated in the intact plants. About 17 days old, cultured hairy roots of *Beta vulgaris* had twice as much betacyanin and three times as much betaxanthin as seedling roots. On an mg/gm dry wt. basis, the concentrations of these betalains were equal to or greater than those reported for storage roots.

The hairy root cultures of *Lupinus polyphyllus* and *L. hartweigii* produced higher amounts of isoflavine glucosides than those of hormone-dependent cells. The hairy root cultures of *Peganum harmala* synthesized higher levels of β-carboline glucoid, ruine and serotonin. *Valeriana officinalis* var. *sambucifolia* hairy roots produced 44.3 mg/gm cells dry wt. of valepotriates having spasmolytic and sedative activities. The productivity was 0.9 mg/gm/day and the concentration was approximately 4 times higher than that of normal roots.

The levels of nicotine in *Nicotiana rustica* hairy roots and of betanine in *Beta vulgaris* were shown to be comparable or slightly higher than their intact plants. Hairy root cultures of *Datura stramonium* and related *Datura* species were reported by Rhodes *et al.* to produce hyoscyamine as the major tropane alkaloid and small amounts of other tropanes including atropine and hyoscine (scopolamine).

To cultivate hairy roots in large-scale fermentors, Rhodes *et al.* designed and patented a fermentor comprising a vessel defining a fermentation chamber and a space-filling wire lattice of inoculation points designed to be located within the chamber. At the start of the fermentation process, the fermentation vessel is inoculated with suitable small lengths of hairy roots which then grow to substantially fill the vessel. It has been found to be highly advantageous if the lengths of roots which form the inoculants are distributed throughout the vessel. To achieve this, a space-filling lattice of inoculation points is provided in the fermentation vessel. The inoculating lengths are suspended in a suitable medium which is then passed into the vessel. The inoculants lengths lodge at the inoculation points and thereafter grow in a conventional fashion.

The biotechnological application of hairy root cultures is promising for a number of reasons: 1) stable, high level production; 2) fast auxin-independent growth and 3) the suitability for adaptation to fermentor systems. It is therefore one of the most feasible techniques from an industrial point of view.

AUTOMATED MICROPROPAGATION AT INDUSTRIAL SCALE

In order to reduce the cost of plant tissue culture either for micropropagation or for production of bioactive compounds, automatic plants having bioreactors and bioprocessors are being developed. The production rate can be enhanced tens of thousands of times, so that elite plants are multiplied by plant biotech companies at a commercial scale. This has become necessary, particularly in the developed countries, where the major part of the operational cost is due to labour. In India, at BARC, airlift bioreactors of 20 litre capacity have been fabricated and used for the production of several bioactive compounds and true-to-type plant material.

In vitro propagation is based on enhanced axillary bud proliferation and on the ability of differentiated, often mature plant cells, to redifferentiate, and develop new meristematic centres that are capable of regenerating fully normal plants. Regeneration is potentiated through two morphogenic pathways:

1. *Organogenesis* The formation of unipolar organs in this pathway.

2. *Somatic embryogenesis* The production of bipolar structures, somatic embryos with a root and a shoot meristem.

Micropropagation (*in vitro* propagation of axillary and/or adventitious buds as well as somatic embryos) is presently used as an advanced biotechnological system for the production of identical pathogen-free plants for agriculture and forestry. The technique, however, is still costly due to intensive hand manipulation of the various culture phases and is not used commercially for all plant species. In addition, in some plants, the initial stage of establishment and response is slow and the survival of the plants in the final stage *ex vitro* is often poor, which further reduces the micropropagation production potential.

Efficient commercial micropropagation depends on rapid and extensive proliferation along with the use of large-scale cultures for the multiplication phase. Furthermore, normal plant development during the acclimatization and hardening stage is mandatory to ensure a high percentage of survival after transplanting to the greenhouse. Mechanization and automation of the micropropagation process can greatly contribute to overcoming the limitation imposed by existing conventional labour-intensive methods. Considerable attention has been directed towards automation of the repeated cutting, separation, subculture, and transfer of buds, shoots, or plantlets during the multiplication and transplanting.

Progress in tissue culture automation will depend on the use of liquid cultures in bioreactors, which will allow fast proliferation, mechanized cutting, separation, and automated dispensing. These techniques were reported for some plants and were shown to reduce hand manipulation and thus reduce *in vitro* plant production costs. One major problem addressed was microbial contamination as affected by both the introduced plant material and the operation procedures of large-scale bioreactors. Liquid media have been used for plant cells, somatic embryos and organ cultures, in both agitated flasks and various types of bioreactors.

Although the use of bioreactors has been directed mainly for cell suspension cultures and secondary metabolites production, research directed at improving bioreactors for somatic embryogenesis has been reported for several plant species. Bioreactors were also used for the cultivation of hairy roots mainly as a system for secondary metabolite production. The use of a simplified acoustic window mist bioreactor for transformed roots and for carnation shoots was reported to improve biomass growth.

Information on the use of bioreactors as a system for plant propagation (Table 30.5) through the organogenic pathway, although limited to a small number of plant species, is presently being applied to several ornamental and some vegetable and fruit crop plants.

Table 30.5 Plants propagated in bioreactor cultures

Species	Response
Amaryllis hippeastrum	Buds, plants, bulblets
Ananas comosus	Shoot clusters
Apium graveolens	Somatic embryos
Araceae species	Shoots, plants
Brodiaea complex	Bud clusters, corms
Coffea arabica	Shoot clusters, plants
Cyclamen persicum	Callus, somatic embryos
Daucus carota	Callus, somatic embryos
Dianthus caryophyllus	Shoots, plants
Eschcholtzia californica	Somatic embryos
Euphorbia pulcherrima	Somatic embryos
Fragaria ananasa	Shoots, plants
Gladiolus grandiflorum	Bud clusters, plants, corm
Hevea brasiliensis	Buds, plants
Hyacinthus orientalis	Bulblets, plants
Lilium spp.	Plants, bulblets
Medicago sativa	Callus, somatic embryos
Musa spp.	Buds, plants, Bud clusters
Nephrolepis exaltata	Buds, plants
Nerine sarniensis	Proembryogenic clusters, somatic embryos
Ornithogalum dubium	Shoot clusters, plants, bulblets
Populus tremula	Bud clusters, shoots
Picea glauca	Somatic embryos
Picea glaucaengelmannii	Somatic embryos
Picea marianna	Somatic embryos
Solanum tuberosum	Plants, tubers, bud clusters

However, liquid cultures confer several problems associated with abnormal plant development, a phenomenon described as hyperhydricity, previously termed "vitrification", which causes poor plant development *in vitro* and later in *ex vitro*. Attempts were carried out to overcome hyperhydricity in liquid cultures by the use of growth retardants, inhibitors of gibberellin biosynthesis, which decreased hyperhydricity, reduced shoot elongation, and induced bud or meristem cluster formation. The clusters were shown to be an alternative propagation system for bioreactor cultures, providing a biomass with limited leaf elongation.

PLANT DEVELOPMENTAL PATHWAYS IN BIOREACTORS

Somatic Embryogenesis Pathways

The use of liquid cultures for the cultivation of somatic cells in recapitulating embryogeny was reported by Steward *et al.* and Reinert as early as 1958 in carrot. The underlying concept for cell totipotency and the ability to renew growth and express morphogenesis was based on the assumption that isolation and bathing of the cells in a specific medium will simulate the zygotic embryo conditions in the ovary. The bathing liquid medium that contains various nutrients and growth regulators was assumed to be a source for similar stimuli that were present in the zygotic embryos' immediate environment. In addition, the segregation of embryogenic from non-embryogenic cells amplified the expression of cell totipotency. Somatic embryogenesis was achieved by the initial use of an induction medium containing an auxin, usually 2,4-dichlorophenoxyacetic acid (2,4-D), and coconut water; the latter was mainly a source of cytokinins, inositol and reduced nitrogen compounds. Once the pro-embryonic clusters formed, the removal or lowering of the auxin concentration initiated a sequence of events similar to zygotic embryogeny.

Somatic embryogenesis was observed at first, mainly in members of the Umbelliferae, cultured in liquid media and has since been reported in gymnosperms and angiosperms, including several ornamentals, and vegetable and field crops, as well as perennial woody plants. The embryogenic pathway, as opposed to the organogenic pathway, can be a more efficient and productive system for large-scale clonal propagation. Somatic embryogenesis may result in less variation through chimerism. However, somaclonal variation will be less likely to take place in direct regeneration, but might be more pronounced in somatic embryos developing from continuously cultured callus tissue. Since the embryo contains both a root and an apical shoot meristem, the rooting stage required in conventional *in vitro* bud or shoot propagation technology is obviated. Somatic embryos are small and can be adequately handled in scaled-up procedures. They are amenable to sorting and separation by image analysis, dispensing by automated systems, and can be encapsulated and either stored or planted directly, with the aid of mechanized systems.

Organogenic Pathways

Presently most plants are being commercially propagated *in vitro* through organogenic pathways of regeneration on agar-gelled solid nutrient media, however, this mode of propagation is highly tedious, costly and time-consuming. Therefore, alternative approaches are required to minimize the difficulties in organogenic regeneration on agar-supplemented media. The process has to be initiated using liquid cultures in bioreactors to transform it to automation. Unfortunately the knowledge on the use of bioreactors for unipolar structures such as protocorms, buds, or shoots is still limited, mainly because of the problems of hyperhydricity of the leaves and shoots. One of the first attempts in using liquid cultures for micropropagation of buds was made on orchids that formed protocorms with minimal shoot elongation during the liquid culture stage.

These protocorms further could be separated and induced to form new plants after subculture to agar-gelled medium. Several other ornamental plants also have been propagated subsequently through the organogenic pathway in liquid shake cultures and bioreactors. By controlling shoot growth and providing suitable culture conditions that reduced abnormal leaf growth and also enhanced the formation of bud or meristematic clusters, as similar to potato, gladiolus and *Ornithogalum dubium* where, a high proliferation rate was achieved without the phenomenon of hyperhydricity.

PLANT CELL AND TISSUE GROWTH IN BIOREACTORS

The advantages provided by aerated liquid cultures in bioreactors include better contact between the plant biomass and the medium, no restrictions of gas exchange, control of composition of both the medium and gaseous atmosphere, and ability to manipulate the plant biomass in relation to the medium volume. The need for efficient circulation and mixing of the plant biomass, especially for cluster and embryogenic tissue, is essential to prevent sedimentation and allow optimal growth.

Various types of bioreactors with mechanical or gas-sparged mixing were used for plant cell cultures, to provide stirring, circulation and aeration. Mechanically stirred bioreactors depend on impellers, including a helical ribbon impeller, magnetic stirrers or vibrating perforated plates. Aeration, mixing and circulation in bubble column or airlift bioreactors is provided by air entering the vessel from a side or bottom opening through a sparger; as the air bubbles rise, they lift the plant biomass and provide the oxygen gas required. Mechanically stirred bioreactors are used for effective mixing, aeration, dispersion of air bubbles, and prevention of large cell aggregates formation in cell suspension cultures.

In recent years the use of airlift and bubble column bioreactors has become essential in order to initiate the cultivation of plant cells, embryos or organs at large scale level. Simultaneously, to some extent mechanically stirred tank bioreactors also could be effective but relatively at lesser level because of higher shearing properties in comparison to bubble column or airlift bioreactors. Plant cells are relatively large in size, have sensitive cell walls which are highly susceptible to shearing forces, and can be easily damaged. Furthermore, since plant cells, unlike microbial cultures, do not require high O_2, therefore, the use of bubble column or airlift bioreactors was found to be adequate and advantageous for plant tissue cultured in liquid media. In addition, it is shown that mixing by gas sparging in bubble column or airlift bioreactors lacking impellers or blades is far less damaging for clusters than mechanical stirring, since they have shown to have a lower shearing stress.

The main advantage of airlift bioreactors is their relatively simple construction and the lack of regions of high shear. There is more advantage in terms of reasonably high mass and heat transfer, along with relatively high yields at low input rates. A bubble-free oxygen supply bioreactor with silicone tubing was found to be suitable for embryogenic cell suspensions and provided foam-free cultures. For hairy root culture, an acoustic mist bioreactor was found to be the most suitable bireactor to increase root biomass significantly.

Significantly, the bioreactor configuration and volume should be determined in accordance with their mixing and aeration requirements. These approaches could be effective in achieving efficiently large-scale cultures of both somatic embryos and organogenic plant tissues. Moreover, proper mixing and aeration is required for the specific plant or tissue to be propagated as well as to minimize the intensity of the shear stress.

However, microbial contamination is one of the major problems that have been observed during large-scale liquid cultures. Fungi, bacteria, yeast, and insects are a source of serious contamination, causing heavy losses of plant material in general laboratories and commercial industries also. In liquid cultures, the losses are even greater, as the source of contamination due to manipulation of the bioreactor apparatus is dependent on the various stages of preparing and maintaining the equipment. However, in several laboratories, the operation area is being kept being sterile by a positive pressure airflow, which decreases contamination risks. In addition, constant screening of the plant tissue for contaminants and continuous indexing is another precautionary measure to improve the required results.

REVIEW QUESTIONS

1. Write an essay on role of plant tissue culture in pharmaceutical industry.
2. What are the major alkaloids which can be produced through *in vitro* technology?
3. Write briefly about the some pharmaceutical substances which are of plant origin.
4. What is organ culture and how it can be applicable in pharmaceutical industry?
5. Discuss about the role of automated micropropagation techniques in industry.
6. What are the bioreactors? Write about the major plant developmental pathways in bioreactor.

REFERENCES

Balandrin, M.F. and Klocke, J. A. (1988). Medicinal, aromatic and industrial materials from plants, In: *Biotechnology in Agriculture and Forestry, Medicinal and Aromatic Plants I* (ed.). Bajaj, Y.P.S., Springer-Verlag, Berlin. pp. 3–36.

Chen, C.C., Chen, S.J., Sagare, A.P. and Tsay, H.S. (2001). Adventitious shoot regeneration from stem internode explants of *Adenophora triphylla* (Thunb.) A.DC. (Campanulaceae) an important medicinal herb. *Botanical Bulletin of Academia Sinica.* 42:1–7.

Cox, P. A. and Balick, M. J. (1994). The ethnobotanical approach to drug discovery. *Sci. Am.* June, 6CL-65.

Hagendoorn, M. J.M., Jamar, D. C. L., Meykamp, B. and Van der Plas, L. H. W. (1996). Cell division versus secondary metabolite production in *Morinda citrifolia* cell suspensions. *J. Plant Physlol.* 150: 325–330.

Misawa, M. (1994). Plant tissue culture: an alternative for production of useful metabolite. *FAO Agricultural Services Bulletin.* No. 108. *Food and Agriculture Organization of the United Nations.*

Moms, P. (1985). Regulation of product synthesis in cell cultures of *Catharanthus roseus* II Comparison of production media. *Planta Medica.* 52, 121–126.

Morton, J.F. (1992). The ocean-going Noni or Indian Mulberry (*Morinda citrifolia,* Rubiaceae) and some of its "colorful" relatives. *Econ Bot.* 46: 241–256.

P.R.H., Van der Hijden, R, and Verpoorte, R. (1995). Cell and tissue culture of *Catharanthus roseus* a literature survey II updating from 1988 to 1993. *Plant Cell Tiss. Organ Cult.* 42: 1–25.

Tom, R., Jardin, B., Chavarie, C and Archambault, J. (1991). Effect of culture process on alkaloid production by *Catharanthus roseus. J Biotechnol.* 21: 1–20.

Van der Hijden, R., Verpoorte, R. and Ten Hoopen, H.J.G. (1989). Cell and tissue cultures of *Catharanthus roseus* (L) G.Don. A Literature survey. *Plant Cell Tiss. Organ Cult.* l: 231–280.

Verpoorte, R., Van der Heijden, R. and Schrtpesma, J. (1993). Plant cell technology for the production of alkaloids: present status and prospects. *J Nat. Prod.* 56: 186–207.

Zhong, J.J. (1995). Recent advances in cell cultures of *Taxus* sp. for production of the natural anticancer drug taxol. *Plant Tissue Biotechnol.* 1, 75–80.

GLOSSARY

Abiotic Absence of living organisms (non-living).

Acetosyringone A phenolic compound produced by wounded tissues of dicot plant species that activates the *vir* genes of *Agrobacterium.*

Abiotic Stresses The stress caused (e.g., to living organisms) by non-living or environmental factors such as cold, drought, flooding, salinity, ozone and toxic metals (e.g., aluminium, Hg, etc.).

Abscisic acid A phytohormone (plant hormone) utilized to control the size of stomatal pores.

Acclimatization The biological process whereby an organism adapts to a new environment.

Adventitious Developing from unusual points of origin, such as shoot or root tissues, from callus or embryos and from sources other than zygotes.

Adventive embryony Embryo formation and development resulting from asexual cells as that occurring in certain members of the Rutaceae family.

Agar A complex mixture of polysaccharides obtained from marine red algae. It is also called agar-agar. Agar is used as a solid substrate for the laboratory culture of cells and tissues and as an emulsion stabilizer in foods and as a sizing agent in fabrics.

Agarose A highly purified form of agar used as a stationary phase (substrate) in some tissue culture, chromatography and electrophoresis experiments.

Aging The process, affecting organisms and most cells, whereby each cell division (mitosis) brings that cell (or organism composed of such cells) closer to its final cell division (i.e., cell death).

Agrobacterium tumefaciens A naturally occurring soil-surface gram-negative bacterium that is capable of inserting its DNA (genetic information) into plant cell.

Agroinfection Introduction of a viral genome into plant cells using *Agrobacterium.*

Alkaloids A class of toxic compounds that are naturally produced by some organisms (e.g., ants, certain plants such as lupines and certain fungi such as ergot).

Androgenesis Development of plant *in vitro* from male gametophyte.

Antisense RNA RNA produced by transcription of sense strand; complementary to the mRNA.

Antisense RNA technology The integration of an antisense construct of the target gene into the genome to suppress an endogenous gene.

Apical meristem Small group of meristematic cells that are located at the tips of the shoot and root axis and are progenitors of these structures.

Artificial seed A gel bead containing somatic embryo or shoot bud, necessary nutrients, etc.

Artificial seed, desiccated Somatic embryos hardened and encapsulated in a suitable coating material and desiccated.

Artificial seed, hydrated Somatic embryos encapsulated in hydrated gels, which remain hydrated.

Aseptic culture The purposeful culture of different types of cells and tissues of higher plants without any contamination of bacteria, fungi, fungal spores, etc.

Aseptic technique Procedures used to prevent the introduction of fungi, bacteria, viruses, mycoplasma or other microorganisms into cultures.

Autoclave A machine capable of sterilizing wet or dry items with steam under pressure. Pressure cookers are a type of autoclaves.

Autoradiography Imaging a radioactive component onto an X-ray film; used to locate the presence of the radioactive component within a tissue, cell or organelle.

Auxin A group of plant growth regulators (Greek, *auxein* means "to increase") that promotes callus growth, cell division, cell enlargement, adventitious buds and lateral rooting. Endogenous auxins are auxins that occur naturally. Indole 3-acetic acid (IAA) is a naturally occurring auxin. Exogenous auxins are auxins that are manmade or synthetic. Examples of exogenous auxins are 2,4-dichlorophenoxyacetic acid (2,4-D), indole 3-butyric acid (IBA), α-naphthaleneacetic acid (NAA) and 4-chlorophenoxyacetic acid (CPA).

Batch culture A type of cell suspension culture in which cells or small cell aggregates grow in a definite volume of nutrient medium.

Ballistic method *See*, Particle Gun Method.

Biolistic method *See*, Particle Gun Method.

Biotechnology The various ways or means of manipulating life forms (organisms) to provide desirable products for man's use.

Blotting Transfer of nucleic acid (DNA/RNA) fragments or proteins from gel on to a suitable solid support.

Callus An unorganized or proliferated mass of undifferentiated plant cells, as a wound response in culture.

c-DNA Copy or complementary DNA produced by using mRNA as a template; reaction catalysed by reverse transcriptase or RNA-dependent DNA polymerase.

c-DNA library A population of bacterial transformants or phase lysates in which each mRNA isolated from an organism is represented as its cDNA insertion in a vector.

Cell culture The *in vitro* (outside of body, in a test tube) propagation of cells, isolated from living organisms.

Cell density Number of cells per unit volume; usually, expressed as cells/ml.

Cell differentiation The process by which descendants of a common parental cell achieve and maintain specialization of structure and function.

Cell fusion The combining of cell contents of two or more cells to form a single cell.

Cellulase The enzyme that digests cellulose to simple sugars such as glucose.

Chemically defined medium A nutritive solution for cell culture, in which each component is specifiable and ideally of known chemical structure.

Chemostate It is a type of cell suspension culture in which growth rate and cell density remain constant by a fixed rate of input of growth-limiting nutrient.

Chimera Organisms or tissue that is not genetically homogeneous.

Clone (an organism) A group of individual organisms (or cells) produced from one individual cell through asexual processes that do not involve the interchange or combination of genetic material. As a result, members of a clone have identical genetic compositions.

Clonal propagation *In vitro* propagation of plants that are considered to be genetically uniform and originated from a single individual or explant.

Contaminant By definition, any unwanted or undesired organism, compound or molecule present in a controlled environment.

Contamination Being infested with unwanted microorganisms such as bacteria or fungi.

Cryobiology The science of preserving cells, tissues, etc. in frozen state in liquid nitrogen (at –196° C).

Cryopreservation Storage of cells, tissues or organs in liquid nitrogen at –196°C.

Cryoprotectant A compound used to protect cells from cryoinjury.

Culture Growth and development of a cell, tissue or organ under laboratory conditions (e.g., a plant growing *in vitro*).

Culture medium Any nutrient system for the artificial cultivation of bacteria or other cells. The medium usually consists of a complex mixture of organic and inorganic materials.

Cybrid When two or more protoplasts are subjected to fuse, their cytoplasm always fuse but their nuclei may not fuse. If, in a binucleate heterokaryon, one of the nuclei disappears then it is called a cybrid or a cytoplasmic hybrid as the cytoplasms remain in a fused state.

Cytokinin A group of plant growth regulators that regulate growth and morphogenesis and stimulate cell division. Endogenous cytokinins occuring naturally, include zeatin and 6-γ, γ-dimethylallylaminopurine (2iP). Exogenous cytokinins are man-made or synthetic; include 6-furfurylaminopurine (kinetin) and 6-benzylaminopurine (BAP).

Cytodifferentiation Development of different cell types and origin of modifications in cellular structures.

Cytoplast A protoplast devoid of nucleus; prepared by centrifugation or other approaches.

Dedifferentiation Change of differentiated plant cells into meristematic ones.

Differentiation Process of change in cell, tissue or organ, resulting in the variety of structure and function found in the adult or other phases in the life history.

Dihaploid Haploid from a tetraploid species.

Direct gene transfer Stable transformation of cells without mediation of a biological agent like *Agrobacterium*; e.g., electroporation, particle gun, etc.

Disarming In Ti and Ri plasmid vectors, it refers to deletion of genes from T-DNA concerned with cytokinin and auxin biosynthesis.

DNA cloning Introduction of a recombinant DNA into a suitable host for its multiplication, expression or integration into the host chromosomes.

Doubled haploid Individual obtained by chromosome doubling of a haploid.

Electrofusion Protoplast fusion due to a brief high-voltage electrical pulse.

Electroporation A process utilized to introduce a foreign gene into the genome of an organism by the use of electricity.

Embryo culture *In vitro* culture of isolated zygotic immature or mature embryos.

Embryo-nurse endosperm technique In plant embryo culture; young embryos transplanted into developing endosperms cultured *in vitro*.

Embryo pollen An embryo developing from a pollen grain.

Embryo rescue Refers to the tissue culture techniques/technologies utilized to enable the fertilized embryo resulting from a "wide cross" (between two non-sexually compatible plant species) to grow and mature into a seed producing plant.

Embryogenesis, somatic The process of development of an embryo from a somatic cell.

Embryoid A structure comparable to the zygotic embryo which is produced in tissue culture by dividing somatic cells.

Endogenous gene A gene naturally present in an organism.

Endonuclease An enzyme that produces internal cuts in a DNA molecule.

Endosperm The interior portion of a plant seed, beneath the outer hull, which is the source of nutrition (the portion that people tend to eat, in food crops). In grains (e.g., rice or corn/maize), the endosperm consists primarily of starch.

Epigenetic changes Variations quite stable during cell culture, but not expressed in plants regenerated from them or in the sexual progeny of such plants.

Explant The cell or tissue isolated from the intact plant and that is used to initiate the cultures.

Fermentation A term first used with regard to the foaming that occurs during the manufacture of wine and beer.

Fluorescence The reaction of certain molecules upon absorption of specific amount/wavelength of light, in which those molecules emit (re-radiate) light energy possessing a longer wavelength than the original light absorbed.

Freezing (Cryopreservation) Cells/ tissues cooled slowly or rapidly and finally immersed in liquid nitrogen; stored at –196°C in a liquid nitrogen refrigerator.

Fusogen In protoplast fusion, it refers to an agent used to induce fusion between protoplasts, e.g., PEG.

Gametoclonal variation Genetic variability, present among cells/plants derived from pollen.

Gene A segment of DNA which governs some trait usually by producing a polypeptide or only RNA (tRNA, rRNA).

Genetic engineering The selective, deliberate alteration or manipulation of genes (genetic material) of an organism.

Gene silencing Suppression of transgene expression in transgenic plants usually due to transgene methylation.

Gene tagging Use of *Agrobacterium* T-DNA, transposons, etc. for identification and/or isolation of a gene by first integrating it into the gene concerned.

Genome Complete haploid set of genes or chromosomes of a cell or organism.

Genomic library A collection of recombinant plasmid clones or phase lysates; the sum total of DNA inserts ideally represents the entire genome of the concerned organism.

Genomics Determination of the structure (including DNA sequence) and function of the genome of an organism.

Germplasm The total genetic variability to an organism, represented by the total available pool of germ cells or seed.

Gibberellins A plant growth regulator that influences cell enlargement. Endogenous forms of gibberellin include gibberellic acid (GA_3).

Gynogenesis The process of haploid production from cultured unfertilized ovaries.

Habituation The ability of cultured tissue or organ to grow on nutrient medium without addition of any plant growth hormone.

Haploid A cell with one set of chromosomes; half as many chromosomes as normal somatic body cell contains.

Hairy root Roots densely covered with hairs, induced by *Agrobacterium* rhizogenes infection.

Helper plasmid A plasmid used for providing the functions missing from another plasmid.

Helper Ti Plasmid A Ti plasmid containing functional *vir* region helps in transfer of T-DNA (containing DNA insert) from mini-Ti which lacks the vir region.

Heterokaryon A cell obtained by fusion of two cells/ protoplasts of two different species/strains; the nuclei are not yet fused.

Homokaryon A cell obtained by fusion of two or more cells/protoplasts of the same species/genotype; nuclei are still not fused.

Hybridization The mating of two plants from different species or genetically very different members of the same species to yield hybrids (first filial hybrids) possessing some of the characteristics of each parent.

Hybridization, nucleic acid Complementary base-pairing between single strands of DNA/RNA.

Hybrid cell/ protoplast A cell or protoplast derived by fusion of two different cells; the nuclei are also fused to yield only a single nucleus.

Hybrid rescue Embryo culture to save hybrid embryos that would otherwise die due to endosperm degeneration.

Hybrid seed Seed produced by crossing distinct lines or strains.

Inoculum A portion of aliquot of suspension culture used for its subculture.

Inoculation Introduction of appropriate number of cells, etc. into culture medium to obtain their growth and multiplication.

In situ In the natural or original position.

In vitro Latin phrase meaning "to be grown in glass." Propagation of plants in a controlled and artificial environment using plastic or glass culture vessels, aseptic techniques and a defined growing medium.

***In vitro* fertilization** In plants, isolated sperms and egg cells are fused *in vitro* and the zygote cultured *in vitro* to obtain embryos.

In vivo Latin phrase meaning "to be grown naturally." Cell and tissue when they remain integrated into the whole plant.

Macronutrients Elements required in >0.5 mmol l^{-1} concentration and include N, P, K, Ca, S and Mg.

Marker gene A gene, whose transmission /expression can be easily and efficiently monitored.

Marker, scorable A gene, which permits easy and rapid identification of the organism/cell in which it is present and expressed.

Marker, selectable A gene, which permits easy and rapid selection of the organism/cell in which it is present and expressed.

Medium A nutritive solution, solid or liquid, to culture the cells or tissues (Plural, Media).

Medium, agar Medium gelled with agar (6–8 g/l).

Meristemoid A localized group of meristematic cells that appear within the callus tissue during culture and such cell's may give rise to roots or shoots or both.

Messenger RNA (mRNA) The RNA used for polypeptide synthesis by an assembly of ribosomes, tRNA and various factors concerned with translation.

Micronutrients Elements required by plants in <0.5 mmol l^{-1} concentration and include Fe, Zn, Mn, Cu, B and Mo.

Microprojectile In case of particle gun, it refers to 1–2 µm diameter tungsten/gold particles used to deliver DNA.

Micropropagation *In vitro* clonal propagation of plants from shoot tips or nodal explants, usually with an accelerated proliferation of shoots (mass-produce) during subcultures.

Mutagenesis Induction of mutations by applying chemical/physical mutagens.

Mutation A sudden, heritable change in a trait of an organism.

Mutation, induced Mutation produced by some chemical or physical treatment by man.

Mutation, spontaneous Mutation occurring in nature without any specific treatment.

Opines Unique nitrogenous compounds produced by plant cells transformed by *Agrobacterium*; not produced or used by plant cells themselves; e.g. octopine, nopaline, agropine, mannopine, etc.

Organ culture In plants, it refers to organs like shoot tip being cultured and generally whole plants are obtained.

Organic additives In plant tissue culture, it refers to complex organic materials that are added to the medium, e.g., coconut milk, yeast extract, casein hydrolysate, etc., used only when basic synthetic media prove to be inefficient.

Organogenesis The development of primordia or organs from undifferentiated cell masses in tissue culture.

Ovary culture Culture of unfertilized ovaries to obtain haploid plants or for *in vitro* pollination.

Particle gun A device to accelerate 1–2 µm tungsten or gold particles coated with DNA; the particles enter cells/nuclei and deliver the DNA.

Particle gun method Use of a particle gun for genetic transformation experiment.

Permanent transformation Integration of a gene present in a vector into a chromosome of the host cell.

Pathogen A disease-causing organism.

Pathogenic Capable of causing a disease.

Plant hormone An organic compound synthesized in minute quantities by plants. It influences and regulates plant physiological processes, also called a phytohormone.

Plant tissue culture The growth or maintenance of plant cells, tissues, organs or whole plants *in vitro*.

Plant vitrification solution (PVS) A highly concentrated solution of cryoprotectants used for partially desiccating the tissues/cells before freezing them in liquid nitrogen.

Plasmagene A gene located in the cytoplasm (in mtDNA or cpDNA).

Plasmid Circular DNA other than bacterial chromosome capable of independent replication and transmission.

Pollen dimorphism Occurrence of larger and smaller (s-grains) pollen grains in the same anther.

Pollination Transfer of mature functional pollen onto mature stigma.

Pollination, *in vitro* Applying pollen to the stigma or ovules of cultured ovary/ovules.

Polymerase chain reaction Amplification of a DNA segment by using a thermostable DNA polymerase enzyme and two primers specific to the 3'-ends of this segment and thermal cycler.

Polyploidy More than two copies of a single genome or two copies each of two or more genomes present.

Preculture Culture of explants under specified conditions for a few days before they are used for some specific purpose, e.g., for cryopreservation, etc.

Pretreatments Exposure of cells/tissues to specific factors before they are used for culture.

Primary culture A culture started from explants taken directly from the plant body.

Proembryogenic mass A group of meristematic cells that would give rise to a somatic embryo under suitable conditions.

Promoter A DNA sequence to which RNA polymerase first binds during transcription.

Proteome The protein content of a cell, which reflects the biochemical capability of the cell.

Protoplast A structure consisting of the cell membrane and all of the intracellular components, but devoid of a cell wall.

Protoplast fusion Joining of two protoplasts in laboratory to achieve somatic hybrid cell.

Pure culture *In vitro* growth of only one type of cells.

Recombinant DNA Ordinarily, a vector containing a foreign DNA insert.

Recombinant DNA technology Construction of recombinant DNA and often cloning of this DNA.

Reculture In cryopreservation, culture of cells/tissues after thawing.

Regeneration In plant cultures, a morphogenetic response to a stimulus that results in the formation of organs, embryos or whole plants.

Regeneration, shoot Development of shoot buds from cultured plant cells.

Regulator gene A gene that encodes a protein, which binds to the operator region of an operon.

Reporter gene *See*, Marker Gene.

Restriction endonuclease An endonuclease, which cleaves DNA molecules either within or close to specific sequences called recognition sequences.

Rhizogenesis Root regeneration from cultured plant cell/shoots, etc.

Secondary metabolite Metabolite that is not required by the producing organism for its life-support system.

Selection, cell Use of a selection pressure, which ultimately kills wild type cells and allows only the variant cells to survive and multiply.

Selection, direct Cells resistant to the selection pressure survive and multiply; wild type cells are killed.

Selection, stepwise The level of selection pressure is increased in a stepwise fashion to a cytotoxic level; often leads to amplification of the concerned gene.

Sense RNA RNA transcribed from the antisense strand of gene; the mRNA and its primary transcript.

Sense strand The nontranscribed strand of a gene; its sequence is identical to that of mRNA.

s-Grains Smaller pollen grains that produce embryos/callus during culture and are smaller in size, poorer in starch accumulation and stain faintly with acetocarmine.

Shoot apical meristem Undifferentiated tissue, located within the shoot tip, generally appearing as a shiny dome-like structure distal to the youngest leaf primordium and measuring less than 0.1 mm in length when excised.

Shoot-tip Terminal portion of shoot; contains shoot apical meristem and leaf primordia.

Single-cell protein (SCP) Protein derived from single-celled organisms with high protein content (e.g., yeast).

Somaclonal variation Phenotypic variation, either genetic or epigenetic in origin, displayed among cultured cells and tissue.

Somaclones Plants derived from any form of cell culture involving the use of somatic plant cells.

Somatic cells In eukaryotes, all body cells except the gametes are somatic cells.

Somatic embryogenesis Development of embryos from somatic cells in culture whose structure is similar to zygotic embryos found in seeds and to analogous embryonic organs such as cotyledons or cotyledonary leaves.

Somatic embryo, primary Somatic embryos regenerating from callus or explant.

Somatic embryo, secondary Somatic embryos regenerating from the tissues of preformed somatic embryos.

Somatic hybrid, asymmetric Somatic hybrid plants having the complete genome of one species but only one to few chromosomes of the other fusion partner.

Somatic hybrid, symmetric Somatic hybrid plants possessing the complete or nearly complete genomes of the two fusion partners.

Somatic hybridization Production of hybrid cells/individuals by fusion between somatic cells or protoplasts from somatic cells of two species/strains.

Southern hybridization Technique due to E.M.Southern; restriction enzyme-digested DNA subjected to gel electrophoresis, denatured, blotted onto a nitrocellulose/nylon membrane, fixed, hybridized with a single- stranded DNA/RNA probe; and hybridization detected by autoradiography.

Somatic variants Regenerated plants (e.g. clones) derived from cells that originally came from the same plant, but are not genetically identical. Such plants (clones) are called "sports" or somatic variants because they vary (genetically) from the parent plant.

Sterilization Inactivation or removal of all living organisms from a substance or surface.

Sterile techniques The practice of working with cultures in an environment, free from microorganisms.

Subculture This is the process of transfer of grown cultures from old culture medium into fresh culture medium.

Surface sterilization Inactivation of microbes present on the surface of explants.

Suspension culture Culture of cells and cell aggregates suspended in liquid medium.

Target gene A gene the expression of which is targeted to be manipulated.

Targeted gene transfer Integration of a transgene in the host genome at the precise site of its allele (or at some specified location).

T-DNA-transferred DNA The segments (about 23 kb) of Ti and Ri plasmids of *Agrobacterium* transferred into plant genome, flanked by the 24 bp right and left borders.

Tetrazolium test Cells treated with 1–2% solution of 2,3,5-triphenyl tetrazolium chloride; live cells reduce TTC to formazon and become stained red.

Thawing In cryopreservation of plant cells, this refers to frozen material being plunged in a water bath at 37–40°C for 90 seconds and then kept in ice bath till reculture.

Tissue culture Culture of whole organs, tissue fragments or dispersed cells on a suitable nutrient medium.

Tissue-specific genes Genes that express only in specific tissues.

Totipotency The ability to grow/differentiate into all the types of cells/tissues constituting an (adult) organism's body.

Transcription Production of RNA copy of a DNA strand; catalysed by RNA polymerase enzyme.

Transgene A gene transferred into an organism using recombinant DNA technology.

Transgenic An animal or plant containing a gene transferred by recombinant DNA technology.

Transgenic plant A plant containing a transgene.

Translation Production of a polypeptide based on the base sequence of mRNA, carried out together by ribosomes, tRNA and several other factors.

Trehalose A disaccharide (simple sugar) that is naturally synthesized (manufactured) by many plants and animals in response to the stress of freezing, heating or drying.

Triploid Refers to organisms that possess three sets of chromosomes, instead of the normal two sets.

Turbidostat A type of suspension culture; when culture reaches a predetermined cell density, a volume of culture is replaced by fresh medium and works well at growth rates close to the maximum.

Two-phase culture system In biochemical production from plant tissue cultures; some compounds (constituting the second phase) are added to the culture medium (constituting the first phase), which absorb, dissolve or entrap the biochemical released in the medium.

Unstable variants Variants that were resistant to a stress and that become susceptible after a period of growth in the absence of selection pressure. This may be due to gene amplification or altered gene expression.

Vaccine, edible The product of such transgenic plants that produce an antigen of the pathogen. The produce is fed raw to animals (or humans) to induce an immune response against a pathogen.

Vector A DNA molecule capable of independent existence and replication and into which DNA to be cloned is inserted for replication and / or expression.

Vector, binary Ti (or Ri) plasmid vectors. A pair of plasmids must be used together for plant cell transformation—one plasmid (mini- Ti) is an *E. coli* plasmid having disarmed T-DNA, the other plasmid is a helper Ti plasmid (having vir region but lacking T-DNA).

Vector, cloning All vectors used in recombinant DNA technology; used to propagate DNA inserts in a suitable host.

Vector, cointegrate A Ti vector produced by integrating by homologous recombination an intermediate vector into the disarmed T-DNA of a modified Ti plasmid.

Vitrification In tissue culture, it refers to the brittle, glassy and water-soaked appearance of *in vitro*-raised shoots; in cryopreservation, it refers to the phenomenon of freezing of all cytoplasm into an amorphous glass-like structure, usually, after pretreatment with a plant vitrification solution.

Wild type The form of a character normally found in a species.

Wound response Production of phenolic compounds, especially, acetosyringone and *para*-hydroxyacetosyringone, by wounded tissues of most dicot tissues.

Zygote A fertilized egg formed as a result of the union of the male (sperm) and female (egg) sex cells.

INDEX

E

F

U

V

W

X

Z

PLATE 1

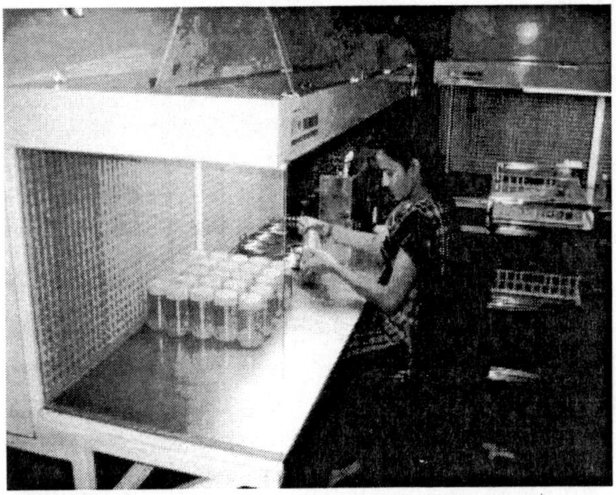

Plate 1.1 Laminar Air Flow cabinet for culture work

Plate 1.2 Media preparation room

Plate 1.3 A constructed greenhouse

Plate 1.4 A net house for acclimatization of regenerated plants

Plate 1.5 Fan section of a greenhouse

Plate 1.6 Arrangement of cultures in an incubation chamber

Plate 1.7 Regeneration of multiple shoots at multiplication stage

Plate 2

Plate 2.1 Regeneration from nodal shoot segments

Plate 2.2 Regeneration of roots from *in vitro* formed young shoots on rooting media

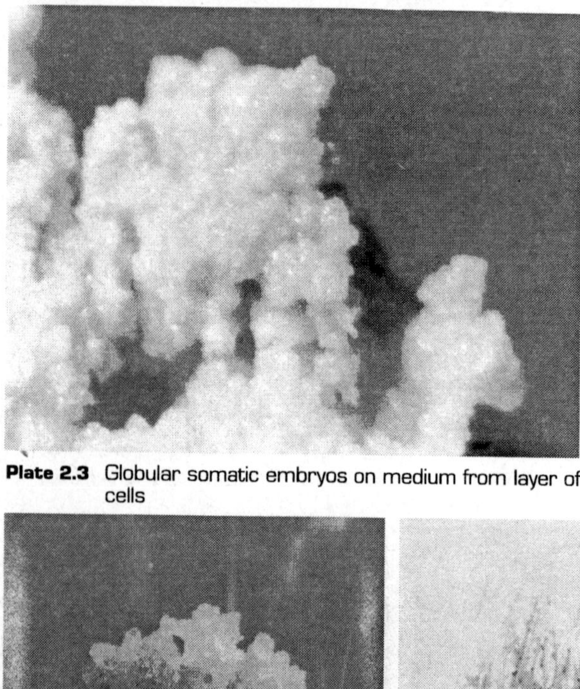

Plate 2.3 Globular somatic embryos on medium from layer of cells

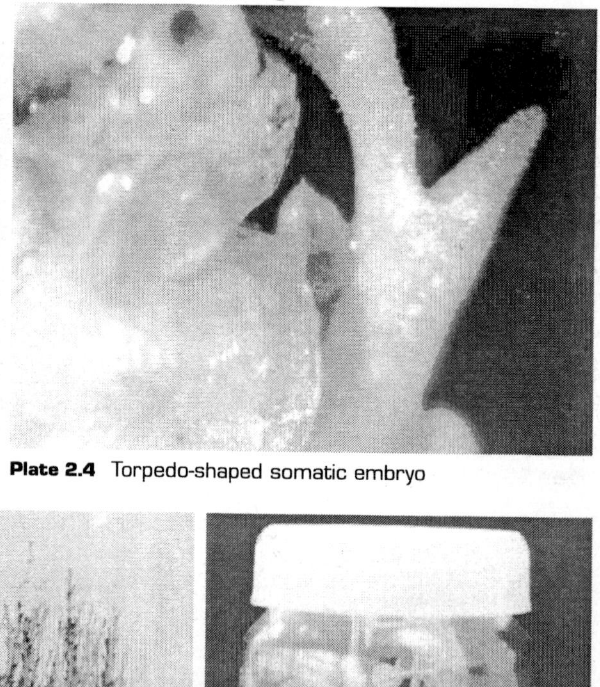

Plate 2.4 Torpedo-shaped somatic embryo

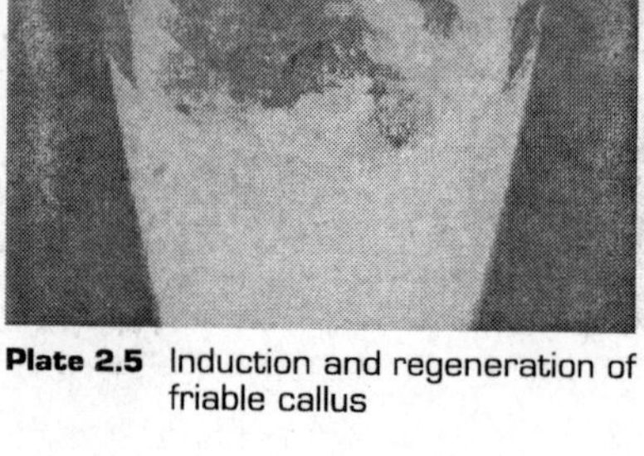

Plate 2.5 Induction and regeneration of friable callus

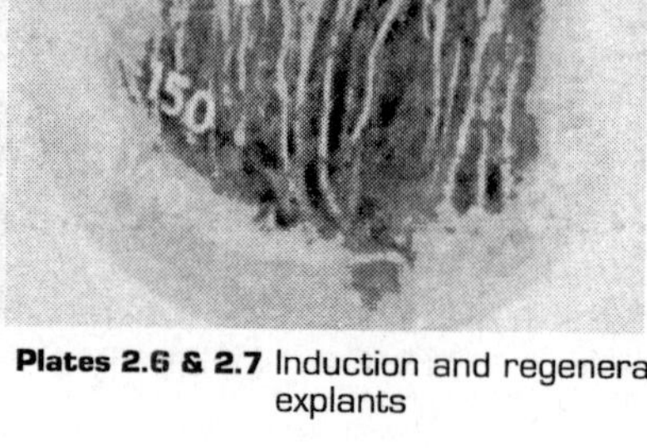

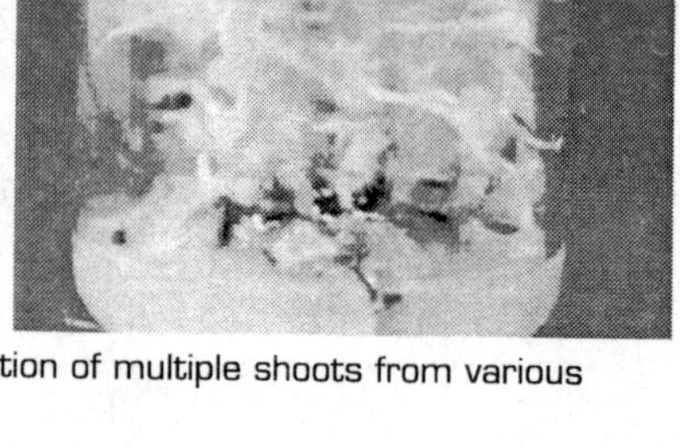

Plates 2.6 & 2.7 Induction and regeneration of multiple shoots from various explants

PLATE 3

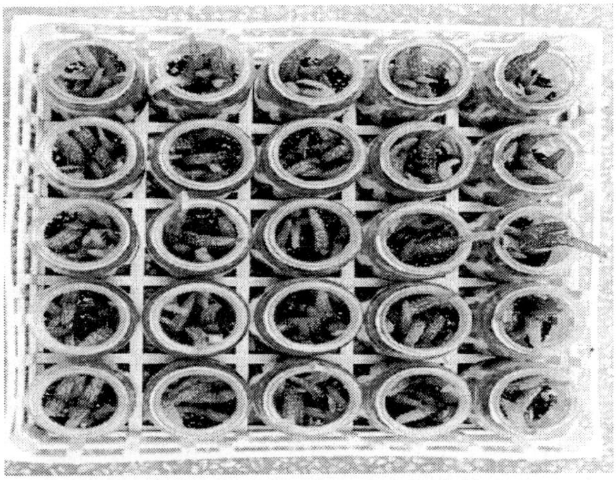

Plate 3.1 Hardening of *in vitro* regenerated *Aloe vera* plantlets in greenhouse

Plate 3.2 *In vitro* regenerated plantlets under hardening conditions

Plate 3.3 Hardened plantlets ready for field transfer

Plate 3.4 Friable callus with greenish structures appearing on the surface in a dicot plant

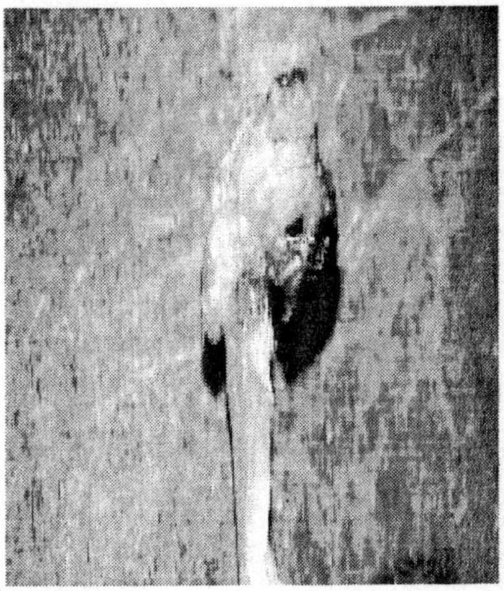

Plate 3.5 A callus formation at one end of leaf-base segment

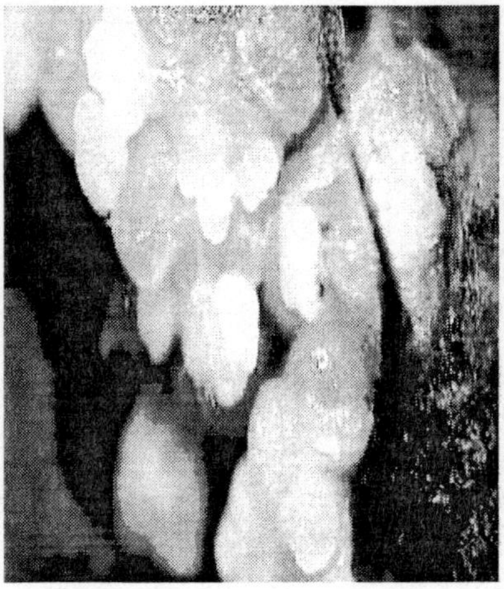

Plate 3.6 Formation of numerous somatic embryos in a millet crop

Plate 3.7 *In vitro* regenerated complete plantlets with root and shoot

PLATE 4

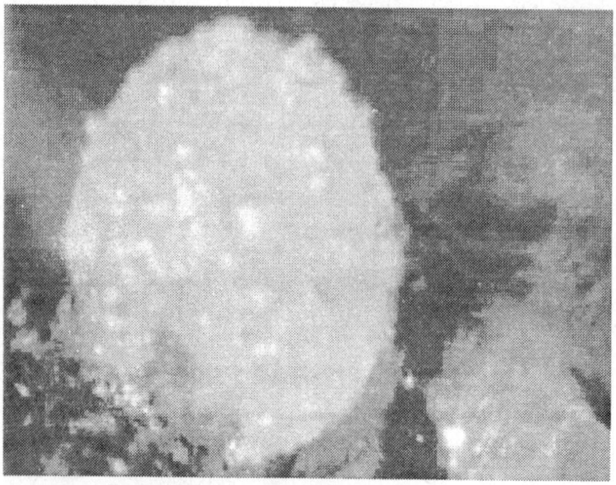

Plate 4.1 Proembryonic mass of cells from explant

Plate 4.2 Differentiation of globular-stage and heart-shaped somatic embryos

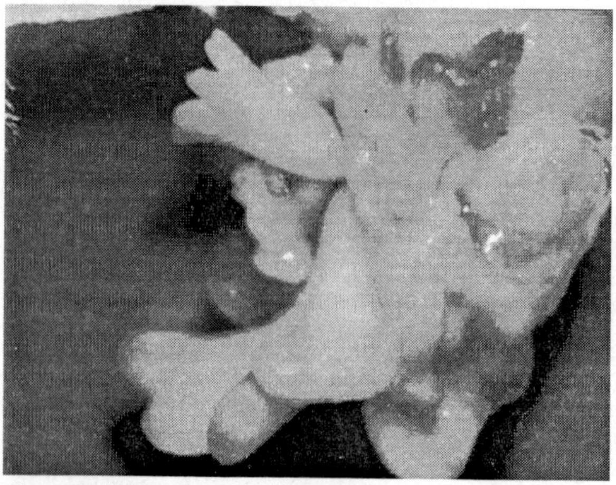

Plate 4.3 Various stages of somatic embryogenesis in a dicot plant

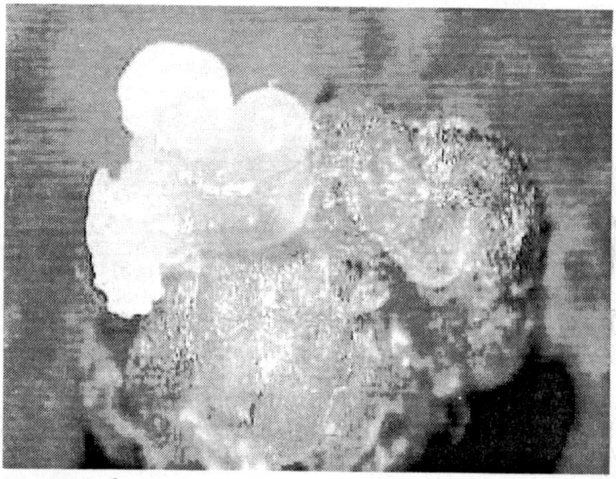

Plate 4.4 A compact callus with somatic embryos in a monocot crop

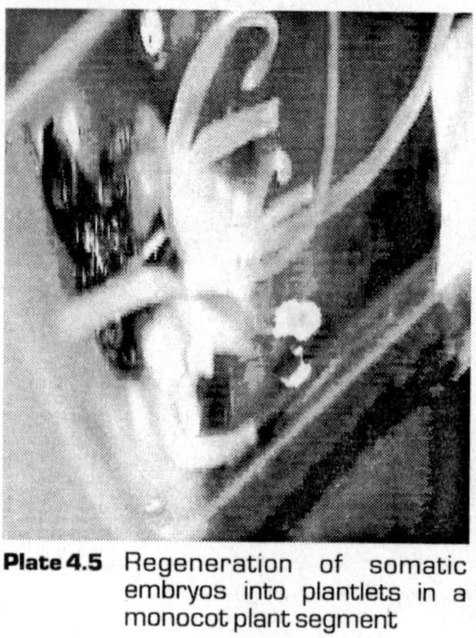

Plate 4.5 Regeneration of somatic embryos into plantlets in a monocot plant segment

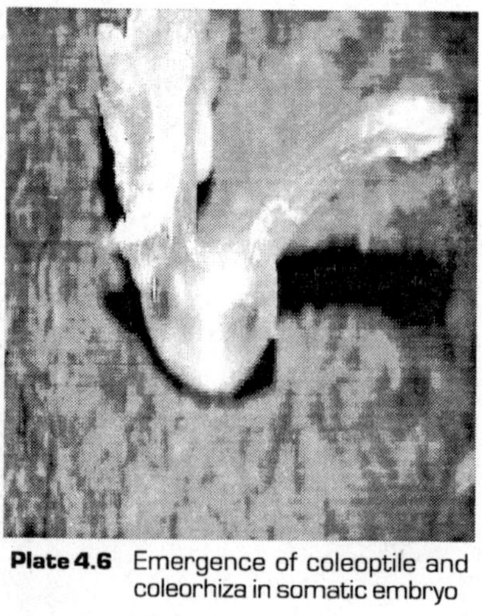

Plate 4.6 Emergence of coleoptile and coleorhiza in somatic embryo

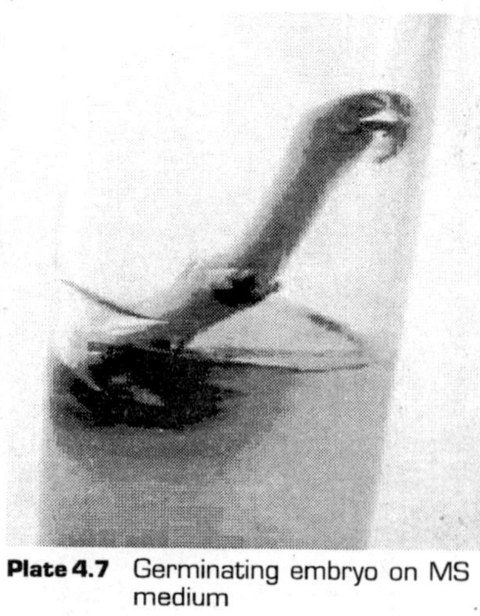

Plate 4.7 Germinating embryo on MS medium